Palaeoecology

Palaeoecology

Ecosystems, environments and evolution

Patrick J. Brenchley
Department of Earth Sciences
University of Liverpool
UK

and

David A.T. Harper
University College Galway
Ireland

CHAPMAN & HALL
London · Weinheim · New York · Tokyo · Melbourne · Madras

Published by Chapman & Hall, an imprint of Thomson Science, 2–6 Boundary Row, London SE1 8HN, UK

Thomson Science, 2–6 Boundary Row, London SE1 8HN, UK

Thomson Science, 115 Fifth Avenue, New York, NY 10003, USA

Thomson Science, Suite 750, 400 Market Street, Philadelphia, PA 19106, USA

Thomson Science, Pappelallee 3, 69469 Weinheim, Germany

First edition 1998

Thomson Science is a division of International Thomson Publishing I(T)P®

Typeset in Times Ten 10/12 pt
by Florencetype Ltd, Stoodleigh, Devon

Printed in Great Britain by The Alden Press, Oxford

ISBN 0 412 43450 4

A catalogue record for this book is available from the British Library

To our families, who have in their various ways enriched the habitats in which this book was written.

Contents

Preface

Palaeoecology uses the fossil record to reconstruct the life habits of past organisms, their association in communities and their relationship to the environments in which they lived. Traditionally, studies have concentrated on case histories, using an array of fossil assemblages to reconstruct past ecosystems. However, in recent years the focus of palaeoecology has been broadening as it has become more fully appreciated that the biosphere is intimately linked to the history of the planet as a whole. Progressively, there have been more studies of the evolution of the ecological world. now comprising a sub-discipline of evolutionary palaeoecology. Other studies have focussed on the interplay between the physical and biological world to show that, as in a Gaian system, influences may work both ways; climate may influence the prevailing ecosystems, but they in their turn may modify the climate systems. These are the big themes with which we introduce the book. In the following sections we have covered the traditional fields of palaeoecology, but have also explored the ways in which palaeoecology can contribute to a broader understanding of the evolution of the Earth. We have described how basic ecological information comes from applying knowledge of living organisms to the study of past ecosystems and to the reconstruction of the past ecological world, at population, community and biogeographic levels. We have also shown how preservation affects the record on which such reconstructions are based. Such palaeoecology knowledge is the basis for a better understanding of past environments, but it is also the basis for a broader appreciation of the role of ecological systems in the evolution of our planet. In the later parts of the book the nature of ecological evolution is explored in both marine and terrestrial environments with emphasis on changes in diversity, periods of radiation and times of extinction and how these might relate to climate change, perturbations of the ocean, bolide impact, but also to the momentum of organic systems themselves.

We are grateful to numerous colleagues and friends who have contributed to this book with advice, discussion and support; in particular Jim Marshall, Charlie Underwood, Euan Clarkson and Mike Benton, who were generous with both their time and help. Special thanks go to Mike Benton, Euan Clarkson, Andy Jeram, Paddy Orr, Alan Owen, Chris Paul, Ron Pickerill, John Hudson, Derek Briggs, Andrew Smith, Paul Wignall, Jim Marshall and Charlie Underwood who read chapters and offered good advice.

We would like to thank Dianne Edwards, Ed Jarzembowski, Andy Jeram, Martin Lockley, Ron Pickerill, Euan Clarkson, John Hudson and Richard Fortey for providing photographs for inclusion in the book.

The book arose from initial discussions with Simon Tull and we were encouraged to continue writing by Ruth Cripwell and Ian Francis who has seen the book through to completion. To these we offer our thanks.

Lastly, but very importantly we would like to acknowledge the work of Kay Lancaster who has prepared all the figures for publication and so has given her special stamp to the finished work. We have greatly appreciated her skill, advice and perseverance.

Figure and table acknowledgements

CHAPTER 1

Figure 1.1 Replotted from Phillips, J. 1860. *Life on the Earth: Its origin and succession.* Cambridge and London; Figure 1.2 Redrawn from various sources; Table 1.1 Assembled from various sources; Table 1.2 From Schäfer, W. 1972. *Ecology and paleoecology of marine environments.* University of Chicago Press, Chicago. 568 pp.; Figure 1.3 Based on McKerrow, W.S. 1978. *Ecology of fossils.* Duckworth Press, London. 383 pp.; Figure 1.4 A Based on Kershaw, S. 1990. Evolution of the Earth's atmosphere and its geological impact. *Geology Today*, March/April 1990, 55–60. B. Based on Watson, J. 1983. *Geology and Man. An introduction to applied earth science.* George Allen and Unwin, London. 150 pp.; Figure 1.5 Replotted from Van Andel, T.H. 1994. *New views on an old planet.* 2nd edition. Cambridge University Press, Cambridge. 439 pp.; Figure 1.6 Replotted from a, Fischer, A.G. 1984. The two Phanerozoic supercycles. *In* Berggren, W.A. and van Couvering, J.A. (eds). *Catastrophes and Earth history.* Princeton University Press, Princeton, 129–150 and b, Frakes, L.A., Frances, J.E. and Syktus, J. 1992. *Climate modes of the Phanerozoic.* Cambridge University Press, New York; Figure 1.7 Modified from Fischer, A.G. 1984. The two Phanerozoic supercycles. *In* Berggren, W.A. and van Couvering, J.A. (eds), *Catastrophes and Earth history.* Princeton University Press, Princeton, 129–150; Figure 1.8 Replotted from various sources; Figure 1.9 Modified from Sandberg, P.A. 1983. An oscillating trend in Phanerozoic non-skeletal carbonate. *Nature* 305, 19–22.; Figure 1.10 Redrawn from Kershaw, S. 1990. Evolution of the Earth's atmosphere and its geological impact. *Geology Today*, March/April 1990, 55–60; Figures 1.11–1.13 Replotted from Lovelock, J.E. 1989. Geophysiology. *Transactions of the Royal Society of Edinburgh: Earth Sciences* 80, 169–175

CHAPTER 2

Figure 2.3 Modified from Barnes, R.S.K. and Mann, K.H. 1991. *Fundamentals of aquatic ecology.* Blackwells Scientific Publications, Oxford, 270 pp.; Figure 2.4 Modified from Open University Course

Team 1989, *Seawater: its composition,properties and behaviour,* Pergamon Press and Open University, 165pp.; Figure 2.5 Modified from Lees, A. and Miller, J. 1985. Facies variations in Waulsortian buildups, Part 2: Mid-Dinantian buildups from Europe and North America. *Geological Journal* 20, 159–180.; Figure 2.6 Modified from Lees, A. and Miller, J. 1985. Facies variations in Waulsortian buildups, Part 2: Mid-Dinantian buildups from Europe and North America. *Geological Journal* 20,159–180.; Figure 2.7 From Barnes, R.S.K. and Mann, K.H. 1991. *Fundamentals of aquatic ecology.* Blackwells Scientific Publications, Oxford, 270pp.; Figure 2.10 Modified from Rhoads, D.C. and Morse, I.W. 1971. Evolutionary and ecological significance of oxygen deficient marine basins: *Lethaia* 4, 413–428.; Figure 2.11 From Savrda, C.E. and Bottjer. D.J. 1991. Oxygen-related biofacies in marine strata: an overview and update. *In:* Tyson, R.V. and Pearson, T.H. (eds), *Modern and Ancient Continental Shelf Anoxia.* Geological Society Special Publications. 58, Geological Society, Bath, 201–219.; Figure 2.12(a) and (b) Modified from Kauffman, E.G. 1981. Ecological Reappraisal of the German Posidonienschiefer. *In:* Gray. J.B., A.J. Berry,W.B.N (eds), *Communities of the Past.* Hutchinson Ross Publishing Company, Stroudsburg, 311–382.; Figure 2.12(c) From Seilacher, A.1960. Epizoans as a key to ammonoid ecology. *Journal of Paleontology* 34, 189–193.; Figure 2.13 Modified from Berger, W.H. and Vincent, E, 1986. Deep-sea carbonates: Reading the carbon-isotope signal. *Geologische Rundschau* 75, 249–269.; Figure 2.16 Modified from CLIMAP, 1984. The last interglacial: *Quaternary Research* 21, 123–224.; Figure 2.17 From CLIMAP, 1981. Seasonal reconstructions of the Earth's surface at the last glacial maximim: *Geological Society of America Map and Chart Series* MC-36, 1–18.; Figure 2.18 Modified from Anderson, T.F. and Arthur, M.A. 1983. Stable isotopes of oxygen and carbon and their application to sedimentologic and paleoenvironmental problems. *In:* Arthur, M.A., Anderson, T.F., Kaplan, I.R., Veizer, J. and Land, L.S. (eds), *Stable Isotopes in Sedimentary Geology.* S.E.P.M. Short Course, 10, S.E.P.M., Tulsa, 1.1–1,151.; Figure 2.19 Modified from Siegenthaler, U. 1979. Stable hydrogen and oxygen isotopes in the water cycle. *In:* Jager, E. and Hunziger, J.C. (eds), *Lectures in Isotope Geology.* Springer-Verlag, Berlin, 264–270.; Figure 2.20 Modified from Hudson, J.D. 1990. Salinity from faunal analysis and geochemistry. *In:* Briggs, D.E.G. and Crowther, P.R. (eds), *Palaeobiology: a synthesis.* Blackwell Scientific Publications, Oxford, 406–408.; Figure 2.21 Modified from Robertson, J.D. 1989. Physiological constraints upon marine organisms. *Transactions of the Royal Society of Edinburgh: Earth Sciences* 80, 225–234.; Figure 2.22 From Hudson, J.D., Clements, R.G., Riding, J.B., Wakefield, M.I. and Walton, W. 1995. Jurassic palaeosalinities and brackish-water communities – A case study. *Palaios* 10, 392–407.; Figure 2.24 From Brett, C.E. and Liddell, W.D. 1978. Preservation and paleoecology of a Middle Ordovician hardground community. *Paleobiology* 4, 329–348.; Figure 2.25 From Wilson, J.B. 1986. Faunas of tidal current and wave-dominated continental shelves and their use in the recognition of storm deposits. *In:* Knight, R.J. and McLean, J.R. (eds), *Shelf Sands and Sandstones.* Memoirs of the Canadian Society of Petroleum Geologists, 11, Canadian Society of Petroleum Geologists, Calgary, 313–326. after Werner, Arntz, and Tauchgruppe Kiel,1974); Figure 2.26. Modified from Brenchley, P.J. and Pickerill, R.K. 1993, Animal-sediment relationships in the Ordovician and Silurian of the Welsh Basin. *Proceedings of the Geologists' Association* 104, 81–93.

CHAPTER 3

Figure 3.5 Modified from Brett, C.E. and Baird, G.C. 1986. Comparative taphonomy: A key to palaeoenvironmental interpretation based on fossil preservation. *Palaios* 1, 207–227.; Figures 3.7 and 3.8 Modified from Seilacher, A., Reif, W.-E. and Westphal, F. 1985. Sedimentation, ecological and temporal patterns of fossil Lagerstätten *Extraordinary Fossil Biotas: Their Ecological and Evolutionary significance. Philosophical Transactions of the Royal Society, London* B. 311, 5–26.; Figure 3.9. From Kidwell, S.M. 1991. The stratigraphy of shell concentrations. *In:* Alison, P.A. and Briggs, D.E.G. (eds), *Taphonomy: releasing the data locked in the fossil record.* Plenum Press, New York,115–209.; Figure 3.11 From Kidwell, S.M. and Holland, S.M. 1991. Field description of coarse bioclastic fabrics. *Palaios* 6, 426–434.; Figure 3.12 From Kidwell, S.M., Fursich, F.T. and Aigner, T. 1986. Conceptual framework for the analysis amd classification of fossil concentrations. *Palaios* 1, 228–238.; Figure 3.13. From Speyer, S.E. and Brett, C.E. 1988. Taphofacies models for epeiric sea environments: Middle palaeozoic examples. *Palaeogeography, Palaeoclimatology, Palaeoecology* 63, 225–262.; Figure 3.14 From Martill, D.M. 1985. The preservation of marine vertebrates in the Lower Oxford Clay (Jurassic) of central England.*In:* Whittington, H.B. and Conway Morris, S. (eds), *Extraordinary fossil biotas: their ecological and evolutionary significance.* Philosophical Transactions of the Royal Society of London, B. 311, The Royal Society, London,155–165.; Figure 3.15 Modified from Simms, M.J. 1994. Emplacement and preservation of vertebrates in caves and fissures. *Zoological Journal Of the Linnean Society* 112, 261–283.; Figure 3.21 Modified from Rex, G.M. 1986. Further experimental investigations on the formation of plant compression fossils. *Lethaia* 19,143–159.

CHAPTER 4

Figure 4.1 Redrawn from various sources. Figure 4.2 After Seilacher, A. 1992. Self-organization of morphologies. *In* Álvarez, F. and Conway Morris, S. (eds) *Palaeobiology: preparing for the Twenty-First Century.* Centro de Reunions Internacionales Sobre Biología 3, 91–100. Figure

4.3 After Clarkson, E.N.K., Harper, D.A.T. and Peel, J.S. 1995. The taxonomy and palaeoecology of the mollusc *Pterotheca* from the Ordovician and Silurian of Scotland. *Lethaia* 28, 101–114. Figure 4.4 Redrawn from Benton, M.J. 1988. Recent trends in vertebrate palaeontology. *In* Harper, D.A.T. (ed.), *William King D.Sc. a palaeontological tribute*, 62–79. Galway University Press, Galway. Figure 4.5 After Rudwick, M.J.S. 1964. The function of the zig-zag deflection in brachiopods. *Palaeontology* 7, 135–171. Figure 4.6 After Rudwick, M.J.S. 1961. The feeding mechanism of the Permian brachiopod *Prorichthofenia. Palaeontology* 3, 450–457. Figure 4.7 After Raup, D.M. 1966. Geometric analysis of shell coiling: general problems. *Journal of Paleontology* 40, 1178–1190. Figure 4.8 Redrawn from Swan, A.R.H. 1990. A computer simulation of evolution by natural selection *Journal of the Geological Society of London* 147, 223–228. Figure 4.9 Modified and redrawn from Skelton, P.W. 1985. Preadaptation and evolutionary innovation in rudist bivalves. *Special Papers in Palaeontology* 33, 159–173. Figure 4.10 Redrawn from Benton, M.J. 1990. Evolution of large size. *In:* Briggs, D.E.G. and Crowther, P.R. (eds) *Palaeobiology: a synthesis*. Blackwell Scientific Publications, Oxford, 147–152. Figure 4.11 Modified and redrawn from McNamara, K.J. 1990. Heterochrony. *In:* Briggs, D.E.G. and Crowther, P.R. (eds) *Palaeobiology: a synthesis*. Blackwell Scientific Publications, Oxford, 111–119. Figure 4.12 Based on various sources. Figure 4.13 After Seilacher, A. 1992. Vendobionta: strangest organisms on Earth and evolution of trace fossils. *In:* Álvarez, F. and Conway Morris, S. (eds) *Palaeobiology: preparing for the Twenty-First Century*. Centro de Reunions Internacionales Sobre Biología 3, 59–73. Figure 4.14 Redrawn from Clarkson, E.N.K. 1993. *Invertebrate palaeontology and evolution*. Chapman & Hall, London. 434 pp. Figure 4.15 Modified and redrawn from Wood, R., Zhuravlev, A.Yu. and Anaaz, C.T. 1993 The ecology of Lower Cambrian buildups from Zuune Arts, Mongolia: implications for early metazoan reef evolution. *Sedimentology* 40, 829–858. Figure 4.16 Redrawn from Clarkson, E.N.K. and Taylor, C. 1995. The lost world of the olenid trilobites. *Geology Today* 11, 147–154. Figure 4.17 Modified and redrawn from Neuman, B. 1988. Some aspects of the life strategies of early Palaeozoic rugose corals. *Lethaia* 21, 97–114. Figure 4.18 Redrawn from Harper, D.A.T. and Moran, R. 1997. Brachiopod life styles. *Geology Today* (in press). Figure 4.19 A. Modified and redrawn form Donovan, S.K. 1980. Potential applications of crinoid columnals in palaeontology. *The Amateur Geologist* 10, 20–31; B. After various sources; C. Redrawn from Harper, D.A.T., Owen, A.W. and Doyle, E.N. 1986. Life in the late Ordovician seas. In: Harper, D.A.T. and Owen, A.W. (eds), *Upper Ordovician Fossils*, 8–17. Palaeontological Association, London; D. After Brett, C.E. 1984. *Special Papers in Palaeontology* 32, 44, 301–344. Figure 4.20 Modified and redrawn from Cowen, R. and Rider, J. 1972. Functional analysis of fenestellid bryozoan colonies. *Lethaia* 5, 147–164. Figure 4.21 Redrawn from Clarkson, E.N.K. 1993.

Invertebrate palaeontology and evolution. Chapman & Hall, London. 434 pp. Figure 4.22 Modified and redrawn from Fortey, R.A. and Owens, R.M. 1990. Trilobites. *In:* McNamara, K.J. (ed.), *Evolutionary Trends*. Belhaven Press, London, 121–142. Figure 4.23 After Bates, D.E.B. and Kirk, N.H. 1984. Autecology of Silurian graptolites. *Special Papers in Palaeontology* 32, 121–139. Figure 4.24 Modified and redrawn from Stanley, S.M. 1970. Relation of shell form to life habits of the Bivalvia. *Memoir of the Geological Society of America* 125, 1–296. Figure 4.25 Redrawn from Savazzi, E. 1994. Functional morphology of boring and burrowing invertebrates. In Donovan, S.K. (ed.) *The Palaeobiology of Trace Fossils*. John Wiley and Sons, Chichester, 43–82. Figure 4.26 After Kier, P.M. 1972. *Smithsonian Contributions to Paleobiology* 13, XX–XX.

CHAPTER 5

Figure 5.1 Redrawn from Leakey, M.D. and Hay, L. 1979. Pliocene footprints in the Laetoli Beds at Laetoli, Northern Tanzania. *Nature* 278, 317–323. Figure 5.2 Photograph courtesy of R.K. Pickerill. Figure 5.3 Redrawn from Bromley, R.G. 1996. *Trace fossils biology, taphonomy and applications*. Chapman & Hall, London. 361 pp. Figure 5.4 Redrawn from Ekdale, A.A., Bromley, R.G. and Pemberton, S.G. 1984. *Ichnology: the use of trace fossils in sedimentology and stratigraphy*. Society of Economic Paleontologists and Mineralogists, Short Course 15, 317 pp. Tulsa, Oklahoma. Figure 5.5 Redrawn from Ekdale, A.A., Bromley, R.G. and Pemberton, S.G. 1984. *Ichnology: the use of trace fossils in sedimentology and stratigraphy*. Society of Economic Paleontologists and Mineralogists, Short Course 15, 317 pp. Tulsa, Oklahoma. Figure 5.6 Redrawn from Ekdale, A.A., Bromley, R.G. and Pemberton, S.G. 1984. *Ichnology: the use of trace fossils in sedimentology and stratigraphy*. Society of Economic Paleontologists and Mineralogists, Short Course 15, 317 pp. Tulsa, Oklahoma. Figure 5.7, 5.8, 5.9, 5.10 Redrawn from Frey, R.W. and Pemberton, S.G. 1984. Trace fossil facies models. *In:* Walker, R.G. (ed.) *Facies Models*. 2nd edition, 189–207. Geoscience Canada Reprint. Figure 5.11 Photographs courtesy of R.K. Pickerill. Figure 5.12 Replotted from Hofmann, H.J. 1990. Computer simulation of trace fossils with random patterns, and the use of goniograms. *Ichnos* 1, 15–22. Figure 5.13 Redrawn from Bromley, R.G. 1996. Bromley, R.G. 1996. *Trace fossils biology, taphonomy and applications*. Chapman & Hall, London. 361 pp. Figure 5.14 Bottjer, D.J. and Droser, M.L. 1992. Palaeoenvironmental patterns of biogenic sedimentary structures. In Maples, C.G. and West, R.G. (eds), *Trace Fossils*. Short Courses in Paleontology 5, 130–144. Paleontological Society. Figure 5.15 Bottjer, D.J. and Droser, M.L. 1992. Palaeoenvironmental patterns of biogenic sedimentary structures. In Maples, C.G. and West, R.G. (eds), *Trace Fossils*. Short Courses in Paleontology 5, 130–144. Paleontological Society. Figure

5.16 Orr, P.J. 1994. Trace fossil tiering within event beds and preservation of frozen profiles: an example from the Lower Carboniferous of Menorca. *Palaios* 9, 202–210. Figure 5.17 Redrawn from Bromley, R.G. 1990. *Trace fossils biology and taphonomy*. Special Topics in Palaeontology, 280 pp. Unwin Hyman, London. Figure 5.18 Modified and redrawn from Lockley, M.G. 1991. *Tracking dinosaurs*. Cambridge University Press. Figure 5.19 Modified and redrawn from Lockley, M.G. 1991. *Tracking dinosaurs*. Cambridge University Press. Figure 5.20 Replotted from Lockley, M.G. 1991. *Tracking dinosaurs*. Cambridge University Press. Figure 5.21 Redrawn from Schult, M.F. and Farlow, J.O. 1992. Vertebrate trace fossils. In Maples, C.G. and West, R.G. (eds), *Trace Fossils*. Short Courses in Paleontology 5, 34–63. Paleontological Society. Figure 5.22 Redrawn from Scott, A.C. 1992. Trace fossils of plant-arthropod interactions. In Maples, C.G. and West, R.G. (eds), *Trace Fossils*. Short Courses in Paleontology 5, 197–223. Paleontological Society. Figure 5.23 Redrawn from Scott, A.C. 1992 Trace fossils of plant-arthropod interactions. In Maples, C.G. and West, R.G. (eds), *Trace Fossils*. Short Courses in Paleontology 5, 197–223. Paleontological Society. Figure 5.24 Redrawn from Bottjer, D.J. and Droser, M.L. 1994. The history of Phanerozoic bioturbation. *In* Donovan, S.K. (ed.), *The palaeobiology of trace fossils*. John Wiley & Sons, Chichester, 155–176.

CHAPTER 6

Figure 6.1 Modifed from Pickerill, R.K. and Brenchley, P.J. 1991. Benthic macrofossils as paleoenvironmental indicators in marine siliciclastic facies. *Geoscience Canada* 18, 119–138. Figure 6.2 Based on Ziegler, A.M. 1965 Silurian marine communities and their significance. *Nature* 207, 270–272. Figures 6.3, 6.4, 6.5, and 6.6 From McKerrow, W.S. 1978. *The Ecology of Fossils*. Duckworth, London, 383 pp. Figure 6.7 Oschmann, W. 1988 Upper Kimmeridgian and Portlandian marine macrobenthic associations from southern England and northern France. *Facies* 18, 49–82. Figure 6.8 From Bromley, R.G. 1996. *Trace Fossils: Biology, Taphonomy and Applications*., 2nd edn. Chapman & Hall, London, Fig. 12.4, based on Pollard (unpublished) in Goldring, R. 1991. *Fossils in the Field*. Longman, Harlow, 218pp. Figure 6.9 From Brenchley, P.J. and Pickerill, R.K. 1993. Animal-sediment relationships in the Ordovician and Silurian of the Welsh Basin. *Proceedings of the Geologists' Association* 104, 81–93. Figure 6.10 Modifed from Fürsich, F.T. and Oschmann, W. 1993. Shell beds as tools in basin analyses: the Jurassic of Kachchh, western India. *Journal of the Geological Society*, London 150, 169–185. Figure 6.11 Modified from Pickerill, R.K. and Brenchley, P.J. 1991 Benthic macrofossils as paleoenvironmental indicators in marine siliciclastic facies: *Geoscience Canada* 18, 119–138. Figure 6.12 From Brett, C.E., Boucot, A.J. and Jones, B. 1993. Absolute depth of Silurian benthic assemblages. *Lethaia* 26,

23–40. Figure 6.13 From Underwood, C.J. 1994. Faunal transport within event horizons in the British Upper Silurian. *Geological Magazine* 131, 485–498. Figure 6.14 Modified from Brasier, M.D. 1980. *Microfossils*. Chapman & Hall, London. 193 pp. Figure 6.15 Modified from Corliss, B.H. and Chen, C. 1988. Morphotype patterns of Norwegian Sea deep-see benthic foraminifera and ecological implications. *Geology* 16, 716–719. Figure 6.16 From Tucker, E.M. and Wright, V.P. 1990. *Carbonate Sedimentology*. Blackwells Scientific Publications, Oxford, 482 pp. Figure 6.18 From Tucker, E.M. and Wright, V.P. 1990. *Carbonate Sedimentology*. Blackwell Scientific Publications, Oxford, 482 pp. Figure 6.19 From James, N.P. 1984. Reefs. *In:* Walker, R.G. (ed.), *Facies Models*. Geological Association of Canada, Toronto, 229–244. Figure 6.20 Wignall, P.B. 1993. Distinguishing between oxygen and substrate control in fossil benthic assemblages: *The Journal of the Geological Society* 150, 193–196. Figure 6.21 Based on Demaison, G.J. and Moore, G.T. 1980. Anoxic environments and oil source bed genesis: *American Association of Petroleum Geologists Bulletin* 64, 1179–1209. Figure 6.22 From Wignall, P.B. 1994. *Black Shales*. Clarendon Press, Oxford, 127pp. Figure 6.23 From Fürsich, F.T. 1994. Palaeoecology and evolution of Mesozoic salinity-controlled benthic macroinvertebrate assemblages. *Lethaia* 26, 327–346. Figure 6.24 From Calver, M.A. 1968. Distribution of Westphalian marine faunas in Northern England and adjoining areas. *Proceedings of the Yorkshire Geological Society* 37, 1–72. Figure 6.25 Wignall, P.B. 1993. Distinguishing between oxygen and substrate control in fossil benthic assemblages, *Journal of the Geological Society*, London, 150, 193–196. Figure 6.27. From Kidwell, S.M. 1985. Palaeobiological and sedimentological implications of fossil concentrations. *Nature* 318, 457–460.

CHAPTER 7

Figure 7.2 Adapted from Kurtén, B. 1954. Population dynamics – a new method in paleontology. *Journal of Paleontology* 28, 286–292. Figure 7.3 Original based on various sources. Figure 7.4 from Ryan, P.D., Harper, D.A.T. and Whalley, J.S. 1994. PALSTAT. Chapman & Hall, London, 74 pp. Figure 7.5 From James, N.P. 1984. Reefs. *In:* Walker, R.G. (ed.), *Facies Models*. Geological Association of Canada, 229–244. Figure 7.6 From Thorson, G. 1957. Bottom Communities (sublittoral or shallow shelf). *In:* Hedgpeth, J. (ed.), *Treatise on Marine Ecology and Paleoecology*. Memoir of the Geological Society of America 67, 461–534. Figure 7.7 From Thorson, G. 1957. Bottom Communities (Sublittoral or shallow shelf). *In:* Hedgpeth, J. (ed.), *Treatise on Marine Ecology and Palaeoecology*. Memoir of the Geological Society of America 67, 461–534. Figure 7.8 Modified from Pickerill, R.K. and Brenchley, P.J. 1975. The application of the community concept in palaeontology: *Maritime Sediments* 11. Figure 7.9 From Lockley, M.G. 1980. The Caradoc faunal associations of the area between Bala and Dinas Mawddwy, North Wales. *Bulletin of the British Museum (Natural History), Geology Series* 33, 165–235. Figure 7.10 Modified from Cocks, L.R.M. and McKerrow 1984. Review of the distribution of the commoner animals in the Lower Silurian marine benthic communities. *Palaeontology* 27, 663–670. Figure 7.11 Modified and redrawn from Ryan, P.D., Harper, D.A.T. and Whalley, J.S. 1994. PALSTAT. Chapman and Hall, London, 74 pp. Figure 7.12 Modified from Springer, D.A. and Bambach, R.K. 1985. Gradient versus cluster analysis of fossil assemblages – a comparison from the Ordovician of southwestern Virginia. *Lethaia* 18, 181–198. Figure 7.13 Modified and redrawn from Brower, J.C. and Kile, K.M. 1988. Seriation of an original data matrix as applied to palaeoecology. *Lethaia* 21, 79–93. Figure 7.14 Redrawn from Patzkowsky, M.E. 1995. Gradient analysis of Middle Ordovician brachiopod biofacies: biostratigraphic, biogeographic, and macroevolutionary implications. *Palaios* 10, 154–179. Figure 7.15 Various sources. Figure 7.16 From Copper, P. 1988b. Paleoecology: paleosystems, paleocommunities. *Geoscience Canada* 15, 199–208. Figure 7.17 From Copper, P. 1988b. Paleoecology: paleosystems, paleocommunities. *Geoscience Canada* 15, 199–208. Figure 7.18 From Copper, P. 1988b. Paleoecology: paleosystems paleocommunities. *Geoscience Canada* 15, 199–208. Figure 7.19 From Copper, P. 1988b. Paleoecology: paleosystems, paleocommunities. *Geoscience Canada* 15, 199–208. Figure 7.20 From Martill, D.M., Taylor, M.A. and Duff, K.L. 1994. The trophic structure of the biota of the Peterborough Member, Oxford Clay Formation (Jurassic), UK. *Journal of the Geological Society of London* 151, 173–194. Figure 7.21 Modified from Martill, D.M., Taylor, M.A. and Duff, K.L. 1994. The trophic structure of the biota of the Peterborough Member, Oxford Clay Formation (Jurassic), UK. *Journal of the Geological Society of London* 151, 173–194. Figure 7.22 Modified from Brett, C.E. 1990. Predation, Marine. *In:* Briggs, D.E.G. and Crowther, P.R. (eds) *Palaeobiology: a synthesis*. Blackwell Scientific Publications, Oxford, 368–372. Figure 7.23 From Ager, D.V. 1963. *Principles of Paleoecology*. McGraw-Hill, New York, 319 pp. Figure 7.24 From Aberhan, M. 1994. Guild-structure and evolution of Mesozoic benthic shelf communities. *Palaios* 9, 516–545. Figure 7.25 From Copper, P. 1988, Paleoecology: paleosystems, paleocommunities. *Geoscience Canada* 15, 199–208. Figure 7.26 Modified from Snow, D.W. 1981. Coevolution of birds and plants. *In:* Forey, P.L. (eds), *The Evolving Biosphere*. British Museum of Natural History and Cambridge University Press, Cambridge, Figure 15.2. Figure 7.27 From Alberstadt, L.P., Walker, K.R. and Zurawski, R.P. 1974c, Path reefs in the Carters Limestone (Middle Ordovician) in Tennessee, and vertical zonation in Ordovician reefs. *Geological Society of America Bulletin* 85, 1171–1182. Figure 7.28 Modified from Copper, P. 1988. Ecological succession in Phanerozoic reef ecosystems: Is it real? *Palaios* 3, 136–152. Figure 7.29 Modified from Copper, P. 1988. Ecological succession in Phanerozoic reef ecosystems: Is it real? *Palaios* 3, 136–152. Figure 7.30 Based on data in Sanders, H.L. and

Hessler, R.L. 1969. Ecology of the deep-sea benthos. *Science* 28, 1419–1424. Figure 7.31 Modified from Rex, M.A. 1981. Community structure in deep-sea benthos. *Annual Review of Ecological Systematics* 12, 332-353. Figure 7.32 From Pickerill, R.K. and Brenchley, P.J. 1991. Benthic macrofossils as paleoenvironmental indicators in marine siliciclastic facies. *Geoscience Canada* 18, 119–138. Figure 7.33 Redrawn from Tipper, J.C. 1979. Rarefaction and rarefiction – the use and abuse of a method in palacoecology. *Paleobiology* 5, 423–434. Figure 7.34 Modified from Walker, K.R. and Laporte, L.F. 1970. Congruent fossil communities from Ordovician and Devonian carbonates of New York. *Journal of Paleontology* 44, 928–944. Figure 7.35 Modified from Boucot, A.J. 1975. *Evolution and Extinction Rate Controls*, Elsevier, Amsterdam, 427 pp.

CHAPTER 8

Figure 8.1 Adapted from Smith P.L. 1990. Paleobiogeography and plate tectonics. *Geoscience Canada* 15, 261–279. Figure 8.2 Adapted from various sources. Figure 8.3 From Hansen, T.A. 1980. Influence of larval dispersal and geographic distribution on species longevity in neogastropods. *Paleobiology* 6, 193–207. Figure 8.4 Adapted from Grande, L. 1990. Vicariance biogeography. *In:* Briggs, D.E.G. and Crowther, P.R. (eds) *Palaeobiology: a synthesis*. Blackwell Scientific Publications, Oxford, 401–403. Figure 8.5 Modified and redrawn from Smith, P.L. and Tipper, H.W. 1986. Plate tectonics and paleobiogeography: Early Jurassic (Pliensbachian) endemism and diversity. *Palaios* 1, 399–412. Table 8.1 Data from Smith, P.L. and Tipper, H.W. 1986. Plate tectonics and paleobiogeography: Early Jurassic (Pliensbachian) endemism and diversity. *Palaios* 1, 399–412. Figure 8.6 Original ordinations based on data from Smith, P.L. and Tipper, H.W. 1986. Plate tectonics and paleobiogeography: Early Jurassic (Pliensbachian) endemism and diversity. *Palaios* 1, 399–412. Figure 8.7 Original dendrograms based on data from Smith, P.L. and Tipper, H.W. 1986. Plate tectonics and paleobiogeography: Early Jurassic (Pliensbachian) endemism and diversity. *Palaios* 1, 399–412. Figure 8.8 Original cladograms based on data from Smith, P.L. and Tipper, H.W. 1986. Plate tectonics and paleobiogeography: Early Jurassic (Pliensbachian) endemism and diversity. *Palaois* 1, 399–412. Figure 8.9 Based on various sources. Figure 8.10 Redrawn from Fritel, P.H. 1903. *Histoire naturelle de la France. Part 24. Paléobotanique (Plantes fossiles)*. Deyrolle, Paris. 325 pp. Figure 8.11a After Smith, P.L. 1990. Paleobiogeography and plate tectonics. *Geoscience Canada* 15, 261–279; b. After Hallam, A. 1986. Evidence of displaced terranes from Permian to Jurassic faunas around the Pacific margins. *Journal of the Geological Society of London* 143, 209–216. Figure 8.12 Modified and redrawn from Valentine, J.W. 1973. *Evolutionary paleoecology of the marine biosphere*. Prentice-Hall, Englewood Cliffs, N.J. 511 pp.

Figure 8.13 Redrawn from Smith, P.L. 1990. Paleobiogeography and plate tectonics. *Geoscience Canada* 15, 261–279. Figure 8.14 Redrawn from Benton, M.J. 1990. Vertebrate palaeontology. Chapman & Hall, London. 377 pp. Figure 8.15 Modified from Williams, H. 1984. Miogeoclines and suspect terranes of the Caledonian–Appalachian origin: tectonic patterns in the North Atlantic region. *Canadian Journal of Earth Sciences* 21, 887–901. Figure 8.16 Replotted from MacArthur, R.H. and Wilson E.O. 1963. An equilibrium theory of island biogeography. *Evolution* 17, 373–387. Figure 8.17 After Jablonski, D. 1986. Background mass extinctions: the alternation of macroevolutionary regimes. *Science* 231, 129–133.

CHAPTER 9

Figure 9.2 Modified from Tucker, M.E. 1992 The Precambrian-Cambrian boundary: seawater chemistry, ocean circulation and nutrient supply in metazoan evolution, extinction and biomineralisation. *The Journal of the Geological Society* 149, 655–668. Figure 9.3 From Sepkoski, J.J.Jr. 1990. Evolutionary Faunas. *In:* Briggs, D.E.G. and Crowther, P.R. (eds), *Palaeobiology: a Synthesis*. Blackwell Scientific Publications, Oxford, 37–41. Figure 9.4 From Sepkoski, J.J.Jr. 1990. Evolutionary Faunas. *In:* Briggs, D.E.G. and Crowther, P.R. (eds), *Palaeobiology: a Synthesis*. Blackwell Scientific Publications, Oxford, 37–41. Figure 9.5 Modified from Sepkoski, J.J.Jr. 1984. A kinetic model of Phanerozoic kinetic diversity. III Post Paleozoic families and mass extinctions. *Paleobiology*, 10, 246–267. Figure 9.6 From Smith, A.B. 1994. *Systematics and the fossil record*. Blackwell Scientific Publications, Oxford, 223 pp. Figures 9.7, 9.8 and 9.9 Modified from Signor, P.W. 1985. Real and apparent trends in species richness through time. *In:* Valentine, J.W. (eds), *Phanerozoic diversity patterns*. Princeton University Press, Princeton, N.J., 129–150. Figure 9.10 From Bambach, R.K. 1977. Species richness in marine benthic habitats through the Phanerozoic. *Paleobiology* 3, 152–167. Figure 9.11 Modified from Raup, D. and Sepkoski, J.J.Jr. 1982. Mass extinctions in the marine fossil record. *Science* 231, 833–836. Figure 9.13 Modified from Raup, D.M. 1989. The case for extra-terrestrial causes of extinction. *Philosophical Transactions of the Royal Society of London* B 325, 421–435. Figure 9.14 From Rosenzweig, M.L. 1995. *Species diversity in space and time*. Cambridge University Press, Cambridge, 463 pp. Modified from Fox, B.J. 1983. Mammal species diversity in Australian heathlands: the importance of pyric succession and habitat diversity. *In:* Kruger, F.J., Mitchell, D.T. and Jarvis, J.U.M. (eds), *Mediteranean-type ecosystems: the importance of nutrients*. Springer-Verlag, Berlin, 473–489. Figure 9.16 from Joachimski, M.M. and Buggisch, W. 1993. Anoxic events in the late Frasnian – Causes of the Frasnian-Famennian faunal crisis.: *Geology* 21, 675–678. Figure 9.17 Modified from Erwin, D.H. 1994. The Permo-Triassic extinction. *Nature*, 367, 231–236. Figure

9.18 Modified from Benton, M.J. 1986, More than one event in the Late Triassic mass extinction. *Nature* 321, 857–861. Figure 9.19 Modified from Jarvis, I., Carson, G.A., Cooper, M.K.E., Hart, M.B., Leary, P.N., Tocher, P.A., Horne, D. and Rosenfeld, A. 1988. Microfossil assemblages and the Cenomanian-Turnonian (late Cretaceous) oceanic oxygen event: *Cretaceous Research* 9, 3–103. Figure 9.20 Modified from Keller, G. and Perch-Nielsen, K.v.S. 1995. Cretaceous-Tertiary (K/T) mass extinction: Effect of global change on calcareous microplankton. *In:* Stanley, S.M., Kennett, J.P. and Knoll, A.H. (eds), *Effects of past global change on life*. National Academy Press, Washington, 72–93. Figure 9.21 Modified from Sheehan, P.M. 1992. Patterns of synecology during the Phanerozoic. *In:* Dudley, E.C. (ed.), *The Unity of Evolutionary Biology*. Discorides Press, Portland, 103–118.

CHAPTER 10

Figure 10.1 Replotted from Edwards, D. and Burgess, N.D. 1990. Plants. *In:* Briggs, D.E.G. and Crowther, P.R. (eds), *Palaeobiology – a synthesis*. Blackwell Scientific Publications, Oxford, 60–64. Figure 10.2 Photograph courtesy of Dianne Edwards). Figure 10.3 Replotted from Selden, P.A. and Edwards, D. 1989. Colonisation of the land. *In:* Allen, K.C. and Briggs, D.E.G. (eds), *Evolution and the fossil record*. Belhaven Press, London, 122–152. Figure 10.4 Photograph courtesy of A.J. Jeram (© A.J. Jeram). Figure 10.5 Redrawn from Milner, A.C. 1990. *In:* Briggs, D.E.G. and Crowther, P.R. (eds), *Palaeobiology – a synthesis*. Blackwell Scientific Publications, Oxford, 68–72. Figure 10.6 D.A.T. Harper, original photograph. Figure 10.7 Photograph courtesy of E. Jarzembowski. Figure 10.8 Redrawn from Benton, M.J. 1990. Vertebrate palaeontology. Chapman & Hall, London. 377 pp. Figure 10.9 Replotted from Niklas, K.J., Tiffney, B.H. and Knoll, A.H. 1983. Patterns in vascular land plant diversification. *Nature* 303, 614–616. Table 10.1 Replotted from Gray, J. 1993. Major Palaeozoic land plant evolutionary bio-events. *Palaeogeography, Palaeoclimatology, Palaeoecology* 104, 153–169. Figure 10.10 Replotted from Benton, M.J. 1985. Mass extinction among non-marine tetrapods. *Nature* 316, 811–814. Figure 10.11 Modified from Bakker, R. 1986. *The Dinosaur heresies*. Penguin Books, London. 481 pp. Figure 10.12 Redrawn from Trewin, N.H. 1993. Depositional environment and preservation of biota in the Lower Devonian hot-springs of Rhynie, Aberdeenshire, Scotland. *Transactions of the Royal Society of Edinburgh: Earth Sciences* 84, 433–442. Figure 10.13 Redrawn from Clarkson, E.N.K., Milner, A.R. and Coates, M.I. 1993. Palaeoecology of the Viséan of East Kirkton, West Lothian, Scotland. *Transactions of the Royal Society of Edinburgh* 84, 417–425. Figure 10.14 Redrawn from Benton, M.J. 1990. *Vertebrate palaeontology*. Chapman & Hall, London. 377 pp. Figure 10.15a, b Redrawn and replotted from Benton, M.J., Warrington, G.,

Newell, A.J. and Spencer, P.S. 1994. A review of British Middle Triassic terapod assemblages. In Fraser, N.C. and Sues, H.-D. (eds) *In the shadow of the dinosaurs*. Cambridge University Press, Cambridge, 131–160. Figure 10.16a, b. Redrawn and replotted from Metcalfe, S.J., Vaughan, R.F., Benton, M.J., Cole, J., Simms, M.J. and Dartnall, D.L. 1992. A new Bathonian (Middle Jurassic) microvertebrate site, within the Chipping Norton Limestone Formation at Hornsleasow Quarry, Gloucestershire. *Proceedings of the Geologists' Association* 103, 321–342. Figure 10.17 Photograph courtesy of M.G. Lockley. Figure 10.18 Redrawn form McKerrow, W.S. 1978. *Ecology of Fossils*. Duckworth Press, London. 383 pp. Figure 10.19 Replotted from Sloan, R.E., Rigby, J.K., Van Valen, L.M., Gabriel, D. 1986. Gradual dinosaur extinction and simultaneous ungulate radiation in the Hell Creek Formation. *Science* 232, 629–633. Figure 10.20 Redrawn from Nisbet, E.G. 1991. *Living Earth*. HarperCollins Academic, London. 237 pp. Figure 10.21 Redrawn from Benton, M.J. 1990. *Vertebrate palaeontology*. Chapman & Hall, London. 377 pp., Figure 10.22. Plotted from various sources.

Investigating the history of the biosphere

Life has been evolving on Earth for nearly 4 billion years. Estimates of current biodiversity suggest there are now anywhere between 3 and 10 million living species (Wilson, 1992), yet probably due to the vagaries of preservation and incomplete labours of palaeontologists, less than 200,000 fossil species have ever been described. Nevertheless, throughout geological time life has evolved in terms of both complexity and diversity. Not only have individual taxa evolved but also the biosphere, as a whole, has changed since its origin over 3800 my ago. The complex interactions between past organisms and their environments have formed an important aspect of the planet's evolution. Biotic systems have evolved with time and must be viewed as an integral and interacting part of the entire planet within a deep-time framework.

1.1 INTRODUCTION

Palaeoecology studies the relationships of fossil organisms to past physical and biological environments. The subject is multidisciplinary, based first on taxonomic palaeontology but also involving geochemistry, palaeobiology, palaeoclimatology, palaeoceanography and of course geology.

Although Neanderthal man collected and displayed fossils in caves, Greeks such as Herodotus and Xenophanes were among the first to document their significance. Nevertheless apart from studies by Leonardo da Vinci (1452–1519), Nicolaus Steno (1638–1686) and Robert Hooke (1625–1703) palaeontology was not formally established until the late 1700s and the early 1800s. During the 19th century the exciting discoveries of dinosaur bones, marine reptiles such as the ichthyosaurs and plesiosaurs, the first bird *Archaeopteryx* and Neanderthal man, together with a wealth of detailed monographs, firmly established palaeontology as an investigative science; whereas the development of the concepts of evolution and adaptation by Charles Darwin and extinction by George Cuvier promoted a more theoretical base to the science. Additionally polymaths like Alexander von Humboldt provided a basis for biogeography during investigations of the distribution of vegetation.

Palaeontologists have been describing new taxa and attempting to reconstruct the anatomy of fossil animals and plants for over 200 years; but palaeoecology, the study of the relationships of organisms to each other and the environment, is a relatively new discipline, deriving real momentum as recently as the 1950s. Nevertheless palaeoecological studies are not a recent innovation. Long before Ernest Haeckel proposed the term **ecology** for, literally, the household or economy of nature, in the late 1860s, the foundations of modern ecology had been established by the work of many biologists and geologists including Darwin, Barrett, Carpenter, Forbes, Wyville Thomson and von Humboldt. Living marine communities were mapped against environmental gradients by Edward Forbes and his colleagues around the Scottish coasts and the Aegean and by Wyville Thomson and William Carpenter in the Atlantic and Mediterranean, whereas Charles Darwin (1859) in his *Origin of Species* emphasized the relationship between adaptation in organisms and their environment through natural selection. Darwin's treatise was illustrated with palaeontological examples, whereas Edward Forbes and Joseph Prestwich had extended knowledge of modern marine faunas to the fossil assemblages and sediments, respectively, in the Cretaceous and Eocene strata of the Isle of Wight (Lyell, 1850). The versatile Alexander von Humboldt addressed the origin and geographic distribution of living and fossil floras based on his travels throughout Europe and in the tropical rainforests; while in an effort to understand Cenozoic molluscs and shell beds in the Caribbean region, Lucas Barrett investigated, during the early 1860s, living coral and mollusc communities off Jamaica, with tragic consequences.

Painted landscapes and seascapes, of course, provide an animated synthesis of both adaptational and community data. *Duria Antiquior*, or Ancient Dorset, was an accurate attempt by Sir Henry De la Beche, in the 1840s, to reconstruct an early Jurassic seascape. Although representations of the 'Carboniferous and Transition' era together with one of the 'Fresh-water' period were planned, they may not have been completed. A more vivid and actual diorama was provided by the 'Geological Illustrations' constructed by Waterhouse Hawkins and others at Crystal Palace during 1854. The life-size models of the marine and terrestrial reptiles of the Secondary (Mesozoic) Era and mammals of the Tertiary (Cenozoic) attempted realistic and spectacular reconstructions of ancient life (Owen, 1854). These studies, however, also helped establish an important trend; the graphical illustration of both marine and terrestrial palaeocommunities remains the summit of many palaeoecological investigations (McKerrow, 1978).

It was not, however, until the early 20th century that more focussed and detailed ecological studies on fossil material were published. Studies in the 1920s by, for example, Sardeson (1929) and in the 1930s by Lamont (1934) related the shell shapes of brachiopods to various life modes in different environments. Elias (1937) presented one of the first integrated discussions of palaeocommunities; his analysis of the Carboniferous faunas of the Big Blue Group in Kansas related the distribution of marine faunas to changing water depth and other environmental factors. These papers formed the basis for a number of further studies during the 1940s and 1950s. By the late 1950s and early 1960s palaeoecology was firmly established as a new and exciting branch of palaeontology (Ager, 1963; Hecker, 1965). Nevertheless the German school of **Actuopaläontologie** continued to develop the more biological aspects of palaeoecology (Richter, 1928) emphasizing a uniformitarianism approach to fossil organisms and communities through the study of the living analogues of fossil representatives (Schäfer, 1965, 1972) .

During the 1960s **community palaeoecology** advanced with studies of depth-related Silurian brachiopods (Ziegler, 1965) and substrate-related Mesozoic brachiopods (Ager, 1965) appearing simultaneously. More recent studies have developed community ecology with the statistical analyses of the composition and distribution of animal and plant associations together with investigations of **biomass** and **trophic structures** of ancient communities (Boucot, 1981). The ecological studies of individual taxa have been aided by more rigorous investigative techniques together with computer and physical modelling. But like palaeontology, palaeoecology too needs a time frame.

The history of life cannot be explained without reference to the theory of evolution. Similarly the development of adaptation, behavioural patterns and biological relationships together with long-term changes in community patterns must also be modelled by evolutionary paradigms (Boucot, 1983). Long-term and large-scale biological changes have been apparent since the late 1800s. John Phillips (1800–1874) had already defined three major eras based on the fossil content and diversity of most of the geological systems (Fig. 1.1): Palæozoic (old life), Mesozoic (middle life) and Cænozoic (new life). Much more recently John Sepkoski's three **evolutionary faunas** (see Chapter 9), the Cambrian, Palaeozoic and Modern faunas, amplified the same concept although Sepkoski's faunas were defined differently, relying on the dominance of particular groups of organisms for definition of each of his three units.

The majority of palaeoecological studies, to date, have employed a reductionist strategy, focussed on the detailed analysis of relatively small biotic units. Importantly, palaeoecology may have a major input into planetary-scale phenomena. Changes through time in the atmosphere, hydrosphere and the lithosphere are intimately linked and related to the evolution of the biosphere. The development of palaeoecology has helped to clarify the patterns of evolution and extinction against a background of environmental change, including for example the controls on resilient and vulnerable organisms and communities. Large subject areas such as evolutionary biology and the study of human-induced biotic change may in the future rely heavily on studies of **evolutionary palaeoecology**. Suggestions that environmental change may partly drive evolution (Vrba, 1983) and moreover that selection may take place above the species level (Stanley, 1975) complement

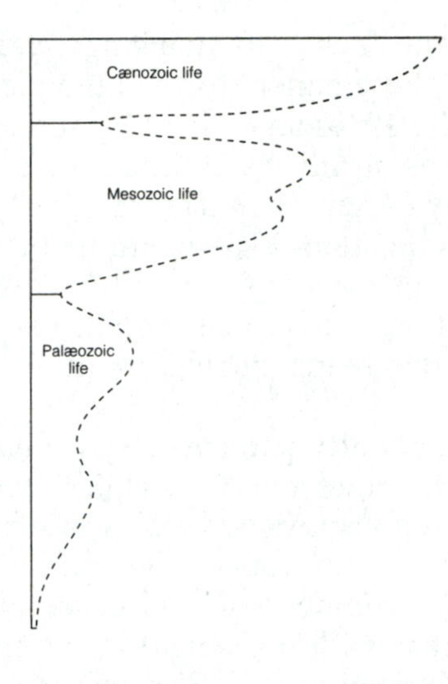

Figure 1.1 John Phillips' divisions of the Phanerozoic based on evolving life in groups of geological systems. This approach contrasts with differently defined evolutionary faunas of Sepkoski (see Chapter 9), based on the relative abundance of groups or organisms through time. (Replotted from Phillips, 1860.)

the importance of extinction events (Raup and Sepkoski, 1984) in shaping biotic change. In addition many biological processes may be **scale independent** (Aronson, 1994); changes at the population and community level may be highly significant in the generation of large-scale biotic change. A more holistic approach involving a view of entire ecosystems against a background of chemical and physical change in the planet's other systems is long overdue. Investigation of the changing dynamics of ecosystems through time provides a logical and timely expansion of palaeoecology.

1.2 DEFINITIONS AND PRINCIPLES

Palaeoecology attempts to discover and elucidate the **life modes** and **relationships** of fossil organisms to each other and their surrounding environments. Two main branches of the science have developed: first, **palaeoautecology** addresses the functions and life styles of individual organisms and secondly **palaeosynecology** is concerned with the operation of associations or communities of organisms. For example autecology tackles the function and operation of an individual coral, whereas synecology is concerned with the workings of an entire coral reef. Evidence of the activity and behaviour of organisms preserved as trace fossils, the study of **ichnology**, is another strand of palaeoecology as is the investigation of the geographical distribution of animals and plants through the study of **palaeobiogeography**.

Palaeoecological studies have a number of limitations. Animals and plants are rarely preserved in life position. Moreover, once dead, the soft tissues of skeletal organisms together with the soft-bodied components of communities rapidly decay or may be removed by scavengers. Clearly some environments are more likely to be preserved than others and the geological record is selective, with only a relatively small and biased sample of the planet's biological history available for study. Deductions regarding the life modes of extinct organisms are difficult, while the strict application of uniformitarianism is often dangerous against a background of changing life styles and habitats through time.

Nevertheless, some basic principles can be applied to most palaeoecological studies. All organisms are adapted and restricted to a particular environment; the environment, however, may range across an entire continent or be limited to part of the intestine of a mouse. Animals and plants are usually adapted to a specific life style and virtually all organisms have a direct or indirect dependence on other organisms. But in marked contrast to ecological studies the functions and relationships of fossil organisms cannot be observed directly. Palaeoecology has one major advantage over ecological studies of living biotas – individuals and communities can be described and analysed through time. Description and analysis of evolutionary palaeoecology, changes in the planet's biota and its relationship to environment through geological time, is a principal focus of this book.

There are a number of schemes for classifying organisms in an ecological context. Fossil organisms can be associated with a particular environment or a range of environments, whereas the life mode or strategy of an organism can help define ecogroups. Feeding strategies or the position of an organism in the food chain may usefully describe a range of trophic categories and the entire trophic structure of communities as a whole.

1.3 MARINE AND TERRESTRIAL ENVIRONMENTS

The majority of fossils and fossil communities are preserved in **marine environments** associated with a range of habitats in estuaries and tidal marshes, rocky and sandy shores, to littoral and open ocean (Fig. 1.2).

Terrestrial environments are mainly governed by elevation, humidity, latitude and temperature, although organisms have inhabited a wide range of continental environments ranging from the Arctic tundras to the lush forests of the tropics (Table 1.1). Terrestrial habitats include lakes, ponds, streams and wetlands together with a wide range of settings associated with grasslands, forests, shrub land and deserts. Their development varies through the tropics and subtropical and temperate latitudes to the poles.

Some of the most abundant and diverse communities inhabit the littoral zone where rocky shores hold some of the most varied and extensively studied of all faunas. For example nearly 2000 individual

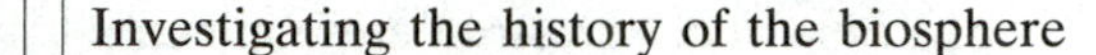

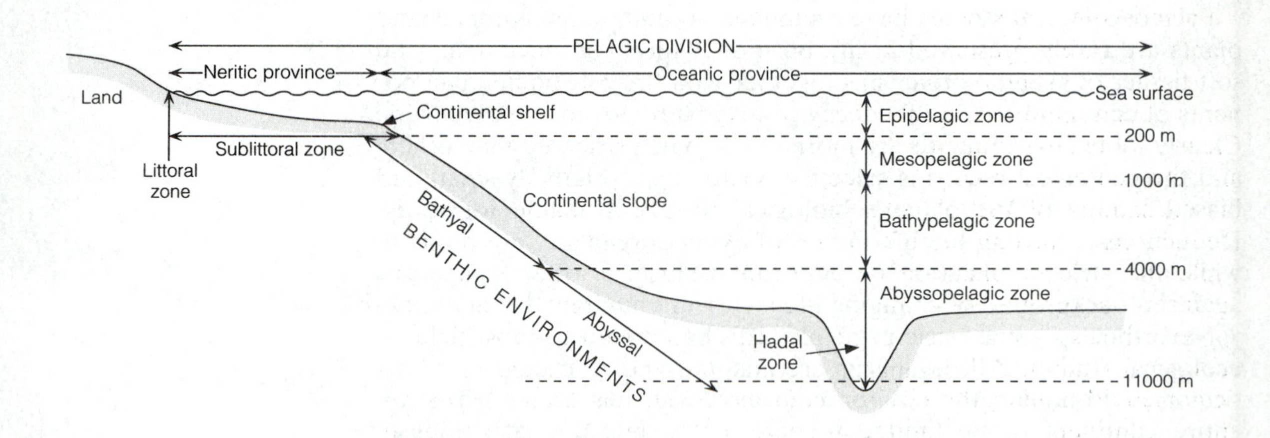

Figure 1.2 Spectrum of marine environments from marine marginal conditions to the abyssal depths; both benthic and pelagic zones are indicated. (Redrawn from various sources.)

organisms have been recorded from a 25 cm^2 sampling grid or quadrat on an exposed wave-battered platform around the Scottish island of Oronsay. Unfortunately few rocky coasts have been recorded from the geological record, although Johnson (1988) has reviewed this environment through geological time, commenting on its neglect by palaeontologists.

Environments have been classified according to **taphonomic conditions** (Table 1.2). The degree of bioturbation, relative autochthoneity to allochthoneity of the biotas together with the homogeneity or

Table 1.1 Classification of terrestrial environments

Continental
- Inland basins (which also include the other categories)
- Soils (in special circumstances)
- Aeolian deposits (such as loess and sand-dunes)
- Volcanic ashes and similar deposits
- Glacial deposits

Fluvial
- Piedmont
- River deposits (including levees and flood plain sediments)
- Deltas
- Estuaries

Paludal
- Swamp deposits of various kinds, lake, backwater, paralic

Lacustrine
- Fresh-water (various categories)
- Saline
- Lagoons

Spelaean
- Cave deposits
- Fissure fillings

Table 1.2 Schäfer's taphonomic classification of environments, based on studies in the North Sea area

1. Vital nonstrate (astrate)	Indigenous benthos No bedding or current structures
2. Vital heterostrate (lipostrate)	Indigenous benthos, often reworked Burrows and trails Some accumulated pelagics Mechanical reworking
3. Lethal heterostrate (lipostrate)	No indigenous benthos No burrows or trails Accumulated transported benthics and pelagics Mechanical reworking of sediment Current structures Sorting and separation of valves
4. Vital isostrate (pantostrate)	Indigenous benthos Burrows and trails Accumulated pelagics Continuous sedimentation; some bio-reworking; no current reworking
5. Lethal isostrate (pantostrate)	No indigenous benthos No burrows or trails Accumulated pelagics Continuous sedimentation; well bedded Benthos transported No current disturbance

heterogeneity of the surrounding sediments are all useful environmental indicators.

1.4 LIFE MODES AND TROPHIC STRATEGIES

Animals and plants can also be identified on the basis of their life modes and strategies together with their mode of feeding and their diet. Organisms can be characterized by their life style and their occupation. For example, animals in benthic communities may be fixed or active, they may be infaunal or epifaunal, they can be suspension feeders, detritus feeders or carnivores. As benthic communities developed and expanded, tiering strategies evolved both above and below the sediment–water interface, while a spectrum of crawling, nektobenthic organisms patrolled the seafloor itself. The water column was populated by a range of nektonic organisms capped by surface waters punctuated by plankton and epiplankton. Benthos contains a variety of suspension- and detritus-feeding organisms together with predators, scavengers and some herbivores and omnivores. Nevertheless organisms do change their life styles and diet during their ontogeny and, in the longer term, during phylogeny.

The offshore muddy sand community of the present-day Irish Sea is dominated by detritus feeders, many infaunal with some epifaunal

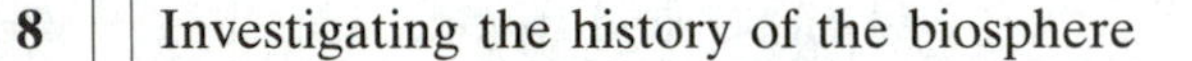

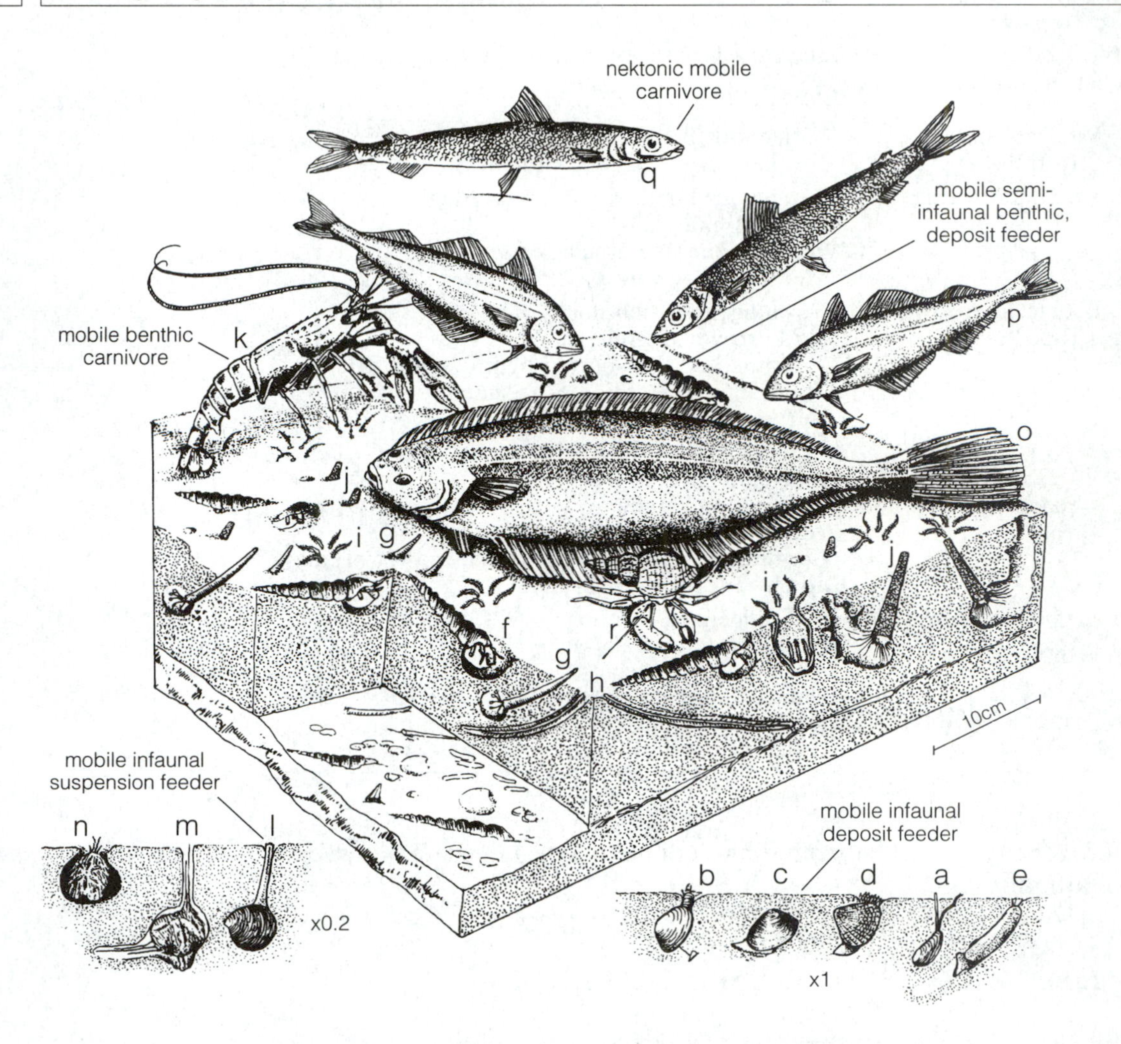

Figure 1.3 Life modes of marine organisms. Reconstruction of a living offshore muddy sand community in the Irish Sea with a range of bivalves (a, b, c, d, e, l), gastropods (f), scaphopods (g), annelids (h, j), asterozoans (i), crustaceans (k, r), echinoids (m, n) and fishes (o, p, q); with insets illustrating large and small burrowers. This Modern biota is dominated by deposit-feeders. (Based on McKerrow, 1978.)

deposit feeders, mobile benthic and nektonic carnivores (Fig. 1.3); relatively few are suspension feeders.

1.5 GEOLOGICAL FRAMEWORK

Palaeoecology and palaeoecological change must be viewed within a geological framework; this structure is provided by stratigraphy (Doyle *et al.*, 1994). The fundamental discovery of 'superposition', by Nicolaus Steno, was enhanced by the ability of Giovanni Arduino to organize rocks into Primary, Secondary and Tertiary systems. The concept of deep time demonstrated by James Hutton hinted at the immense length of geological time, whereas William Smith established the use of fossils in correlation, laying the foundations for biostratigraphy. Through the subsequent 1800s all the geological systems had been defined, providing suitable reference sections for the geological periods. These and smaller divisions now form the basis of chronostratigraphy, a global standard for geological time, although each time unit is subject to precise definition and refinement (Holland, 1986).

Spatial control is provided by plate tectonic models, providing mechanisms to move crustal plates. These plates split apart to form new oceanic floors, collide to either trigger the obduction and subduction of oceanic crust or in the case of continent–continent collision form huge mountain ranges; plates may merely move past each other along strike slip fault zones. Plate tectonic processes have been fundamental in driving both climatic and environmental change (Windley, 1995).

1.6 GLOBAL CHANGE

There have been striking changes to both the physical and biological nature of the Earth, from the mainly sterile conditions of the early-mid Precambrian, lit by a dim red sun with orange skies and brown oceans, to the bright, colourful landscapes and seascapes of the planet today (Fig. 1.4). All the main systems of the planet, the lithosphere, the atmosphere, the hydrosphere and the biosphere have changed and each one has interacted with another.

1.6.1 The Atmosphere

The atmosphere largely owes its nature to the gases derived from the lithosphere and those that result from the activity of plants. The primary atmosphere was probably dominated by CO_2 and water vapour together with smaller proportions of H_2S, CO, CH_4, NH_3, HF, HCL and Ar. This atmosphere was derived from volcanic exhalations during the Archaean and apparently lacked significant amounts of free molecular oxygen and so was a reducing atmosphere. Two main processes appear to have initially generated oxygen, which distinguishes the Earth's atmosphere from that of other planets: first, water can be broken down into hydrogen and oxygen by ultra-violet radiation, and secondly, photosynthesis splits water molecules into O_2 and H_2 while CO_2 and H_2 combine to form organic material. The subsequent history of the atmosphere can be viewed as a progressive enrichment in O_2 through the Precambrian until it approached a level similar to that of the present day (Fig. 1.5) and smaller variations about this level which can be recognized at least through the Phanerozoic. Free oxygen in the atmosphere had risen substantially by about 2.4–1.9 Ga,

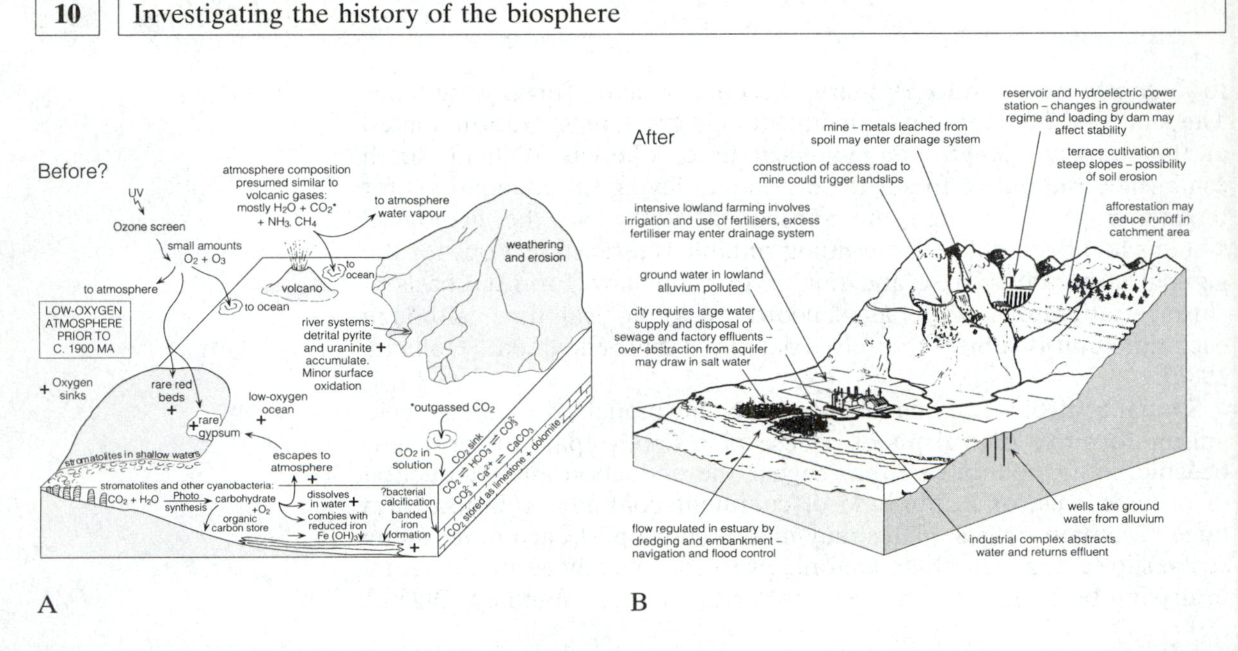

Figure 1.4 A. A barren mid Precambrian landscape with intense volcanic and chemical activity, low oxygen levels and limited biological activity signalled by stromatolites. B. A modern landscape, some 2.5 Gy later, dominated by the affects of human colonists and evolving technology. (Based on Kershaw, 1990 and Watson, 1983.)

suggesting oxygenic photosynthesis was active in the late Archaean. Banded ironstone formations (BIF), red beds and evaporites all provide geological evidence for increasing levels of oxygen in the Precambrian. Most animals take up oxygen by diffusion so that it was not until the Earth's atmosphere attained an oxygen level of about 5–10% of present day values that larger and more complex multicellular organisms could evolve (Fig. 1.5). Later, there was also a significant increase in oxygen in the late Palaeozoic (Fig. 1.6) which is thought to be related to the colonization of terrestrial environments by land plants. There has been speculation that this increased level of oxygen might have promoted a synchronous radiation in terrestrial and possibly marine faunas and floras. If so there are clearly important feed-back relationships between the atmosphere and the biosphere.

The proportion of CO_2 in the atmosphere has also varied significantly throughout geological time. The initial generation of biological O_2 during the Archaean may have oxidized reducing gases such as CO and CH_4, raising levels of CO_2 in the atmosphere and the oceans. High concentrations of CO_2 may have promoted a greenhouse effect inhibiting early Precambrian refrigeration resulting from the low solar luminosity of the primitive sun. The Phanerozoic variations in CO_2 (Fig. 1.6) result from the interaction of the lithosphere, the biosphere and the hydrosphere. Large amounts of CO_2 are generated by degassing of lavas at the mid-ocean ridges, and by the metamorphism of limestones and sediments rich in organic carbon. CO_2 is lost from

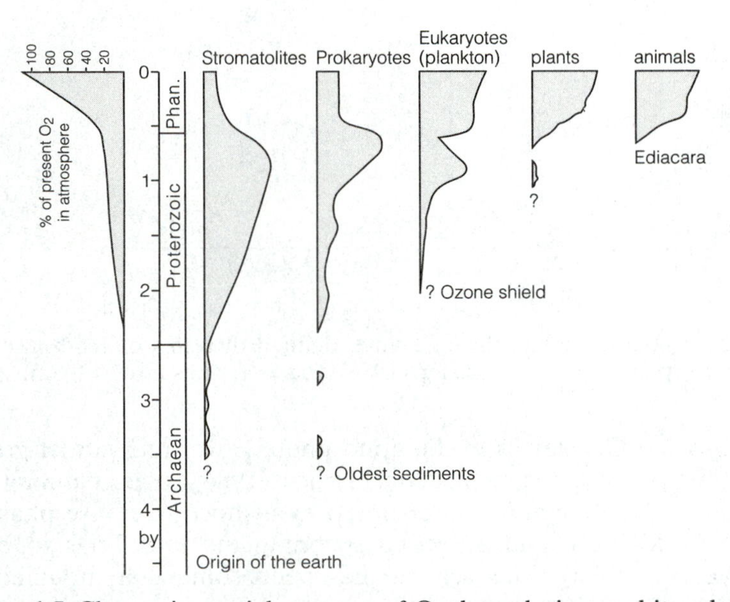

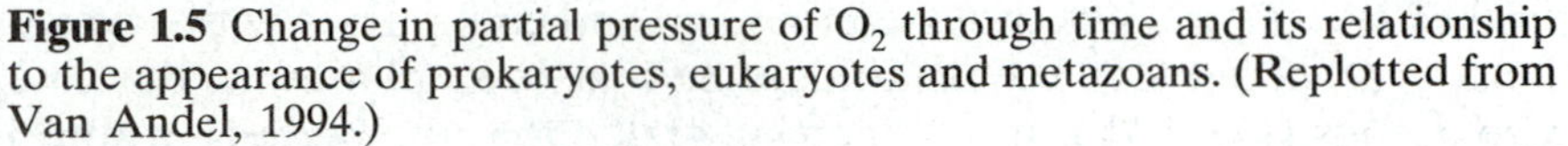
Figure 1.5 Change in partial pressure of O_2 through time and its relationship to the appearance of prokaryotes, eukaryotes and metazoans. (Replotted from Van Andel, 1994.)

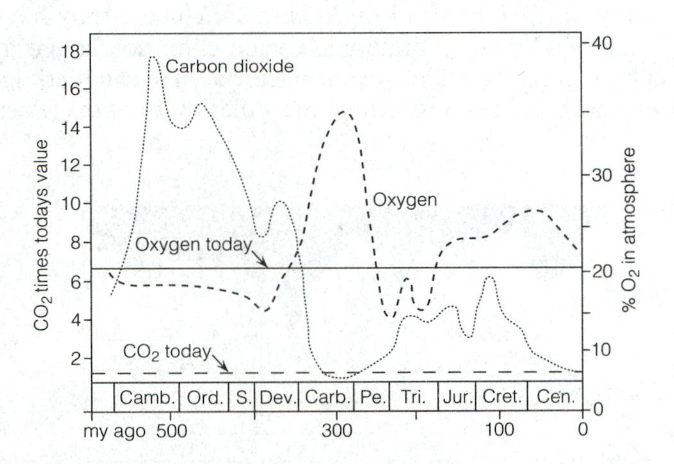

Figure 1.6 Interpreted changes in O_2 and CO_2 levels through the Phanerozoic relative to current gaseous levels. (Replotted from Van Andel, 1994.)

the atmosphere during the weathering of silicates, the formation of limestones and when organic-rich muds are buried. The balance between CO_2 production and loss determines the CO_2 levels in the atmosphere and whether a greenhouse or ice-house climate is likely to prevail. There appear to have been major switches from greenhouse to colder conditions through the Phanerozoic which broadly correlate

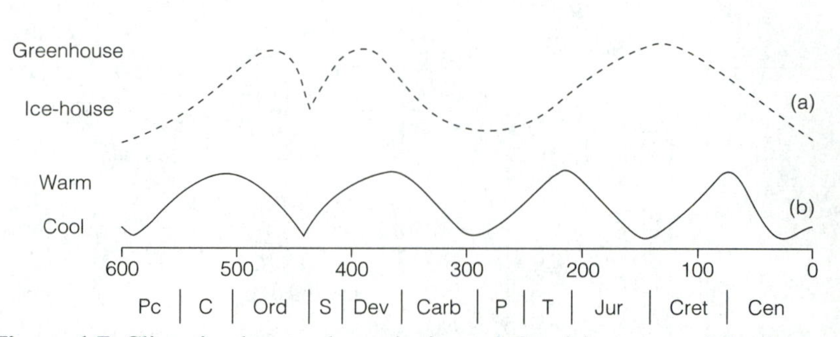

Figure 1.7 Climatic change through time, defined in terms of Ice-house and Greenhouse phases from a, Fischer (1984) and b, Frakes *et al.* (1992).

with estimated CO_2 levels in the atmosphere. Two intervals of greenhouse climate were recognized by Fischer (1984), one spanning the Cambrian to Devonian, but interrupted by a short ice-house phase in the late Ordovician, and a second spanning the late Trias to early Cenozoic (Fig. 1.7a). This scheme has been somewhat modified by Frakes *et al.* (1992) who referred to warm and cool phases with a wavelength of about 150 my and identified a cool phase in the late Cretaceous (Fig. 1.7b). It has been suggested that the periods of warm climate coincide with times of increased volcanicity, high CO_2 levels and high sea-levels (Fig. 1.8). Climate probably has long-term effects on the diversity of the biota and rapid climatic change may have been one of the causes of mass extinctions. Cyclic climatic change appears to affect the degree to which cyanobacteria are calcified and are preserved in stromatolites and other microbial carbonates. Periods of

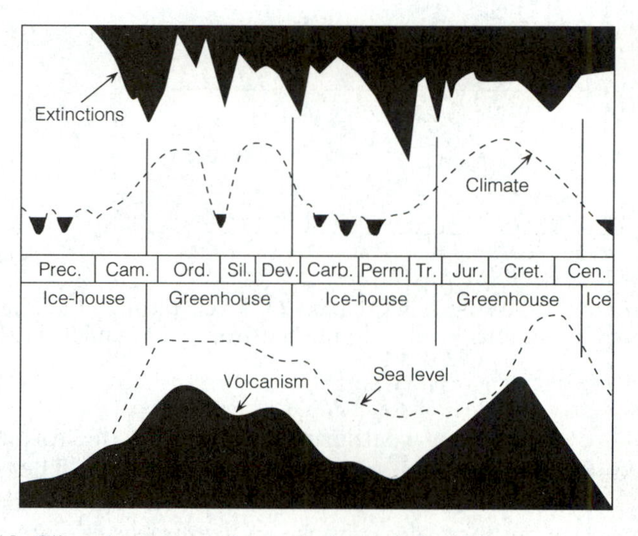

Figure 1.8 Climatic change through time, defined in terms of ice-house and greenhouse phases and related to fluctuations in sea-level and the intensity of volcanic activity. (Modified from Fischer, 1984.)

warm global climate apparently correlate well with extensive development of calcified cyanobacteria, while they are largely absent during cool climatic phases.

1.6.2 The Hydrosphere

The broad composition of the hydrosphere appears to have been established early in Earth history and to have retained a relatively stable condition that has allowed continuity of marine life, punctuated by extinctions of varied severity in response to local, regional or global perturbations of the system.

Most oceanic water had been generated by about 4 Ga. The initial molten crustal 'ocean' may have contained up to 20% water and conservative estimates suggest a depth of 50 km, adequate to provide sufficient water for the early Archaean hydrosphere. The composition of seawater is determined partly by gases from volcanic sources, but solid and dissolved material, including a large proportion of essential nutrients, enter the hydrosphere through river systems. There appears to have been a remarkable equilibrium between the supply of soluble compounds and their utilization and precipitation in the ocean, such that ocean salinity has remained constant throughout most of geological time. As a consequence the body fluids of marine organisms are unlikely to have changed significantly over the 3 Ga since they first developed. The main changes in the hydrosphere are driven by changes in the atmosphere or by biological changes within the ocean. Global changes in climate have important effects on the thermal structure of the oceans, which in turn causes radical changes in ocean circulation patterns. One consequence can be more vigorous upwelling associated with a rich supply of nutrients and phytoplankton blooms. The generation of organic matter and its storage on the ocean floor is part of the carbon cycle which appears to be intimately involved with the biological state of the ocean as indicated by the sharp changes in carbon cycling that coincide with many of the episodes of mass extinction.

The hydrosphere is also constantly interfacing with the atmosphere by diffusion of gases from one to the other. The changing levels of CO_2 in the atmosphere will have been linked with similar changes in dissolved CO_2 in the oceans. Correlated in time with these changes are cyclic alternations in the mineralogy of non-skeletal carbonate in ooids and cements from calcite at times of high CO_2 to aragonite at times of lowered CO_2 (Fig. 1.9).

At a smaller scale the hydrosphere interfaces directly with the lithosphere at deep-sea hot springs: here the huge volumes of water that circulate through the hot crust of mid-oceanic ridges vent to the seafloor carrying dissolved calcium, iron, manganese and silica together with CO_2. The emerging brine at temperatures in excess of 300° C then deposits manganese and iron hydroxides and hydroxides rich in impurities such as copper and silver. Some of the deep sea vents exuding abundant H_2S have a special community of organisms that do not rely

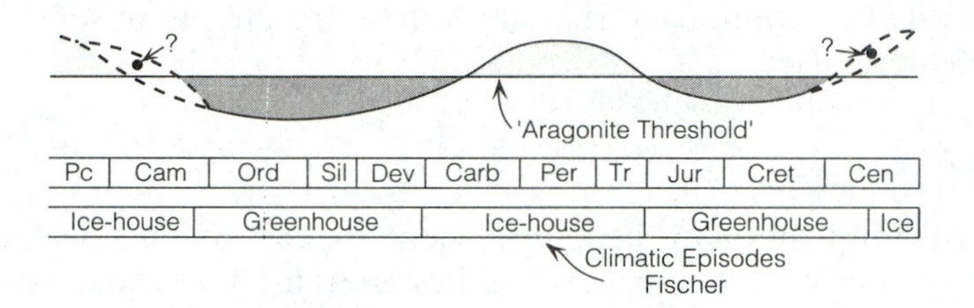

Figure 1.9 The aragonite–calcite cycle through time. (Modified from Sandberg, 1983.)

on solar energy but can utilize chemical energy at the base of the food chain.

On a broader scale there is an interaction between the hydrosphere and the organisms that inhabit it. The organisms rely on nutrients to metabolize, O_2 for respiration, CO_2 for photosynthesis and dissolved material to synthesize their shells. Their abundance and activity influences the amount of O_2 and CO_2 being withdrawn from the water and potentially being stored as organic carbon or carbonate. The diverse fauna and flora of a coral reef ecosystem, a marine analogue of the tropical rain forest, is a significant sink for CO_2 together with a large variety of other chemicals from seawater that precipitate in the carbonate skeletons.

1.6.3 The Lithosphere

The lithosphere is the main source of nutrients for living matter, it is a major source of carbon dioxide, it is the source of the main chemical constituents of seawater and its shape determines the landscape and seascape on which many organisms live. In spite of all the varied tectonic and magmatic events within the Earth's crust, the atmosphere and hydrosphere have remained remarkably stable throughout geological time. Once the crustal structure of the Earth was established in the early Archaean, changes in the lithosphere have been largely driven by plate tectonics. Changes in the global activity of plate movements have played an important role in determining CO_2 levels in the atmosphere. At times of rapid seafloor spreading, mid-ocean ridges are active and generate CO_2. On the other hand continental collision generates mountain belts, such as the Himalayas, in which the very active weathering of siliciclastic rocks tends to lower levels of CO_2, but the metamorphism of carbonates or carbonaceous sediments could raise the levels in the atmosphere. The estimated contribution and loss of CO_2 through the changing plate and palaeogeographic configurations of the Earth have formed the basis for estimating the history of CO_2 in the atmosphere (Fig. 1.6). On a shorter time-scale, periods of intense volcanicity, such as the explosive eruptions of Toba, in Sumatra, at 70,000 BP, or the immense basaltic eruptions of the Deccan

Traps at the end of the Cretaceous in India, inject vast quantities of volcanic gases into the atmosphere over a very short period of time, potentially with catastrophic effects.

Plate movements also change the distribution of continents and oceans, which affects currents within the oceans and the ability of organisms to migrate within the oceans or on land. Consequently the number of biogeographic provinces has changed throughout the Phanerozoic.

The changing nature of the lithosphere has had an important effect on the nature of the sedimentary record. During the early-mid Archaean, sedimentary facies were dominated by mafic-felsic volcaniclastics together with cherts, while shallow water clastics, turbidites and terrestrial alluvial facies have been recognized in late Archaean rocks; by the early Proterozoic more mature quartz sandstones were common in shelfal settings. Plate tectonic movements have influenced how much of the Earth's surface consisted of stable cratonic crust with slow subsidence rates and how much included tectonically active areas at subduction margins or zones of continental collision. Under marine conditions the former tends to be characterized by shallow marine shales, quartzose sandstones and limestones, while the latter commonly has deeper marine shales and turbidites deposited rapidly and forming thick sequences. Consequently, the plate tectonics of the globe influences the proportion of different facies and different environments that are available for exploitation by the existing biota. This proportion will be further moderated to an important degree by sea-level changes that determine the extent to which continental crust is covered by marine waters.

1.6.4 Biological change

Evolutionary change in organisms was first recognized in the mid-1800s (Darwin, 1859); although the mechanisms for evolutionary change have been widely debated (Dawkins, 1995) biological change at the species level is the fundamental feature of evolution. But groups of organisms such as coevolutionary partners, entire communities and even whole evolutionary faunas show similar patterns of change throughout geological time. These changes can be documented through palaeoecological studies (Boucot, 1983; Sheehan, 1992). In crude terms the appearance of major groups of faunas and floras can be correlated with changing environmental conditions. Most commonly the sequential appearances of the prokaryotes, eukaryotes and metazoans are plotted against the increasing partial pressure of atmospheric oxygen (Fig. 1.10).

1.7 GEOPHYSIOLOGY

In the search for an inclusive model for the operation and evolution of the planet, James Lovelock has presented Earth as a superorganism

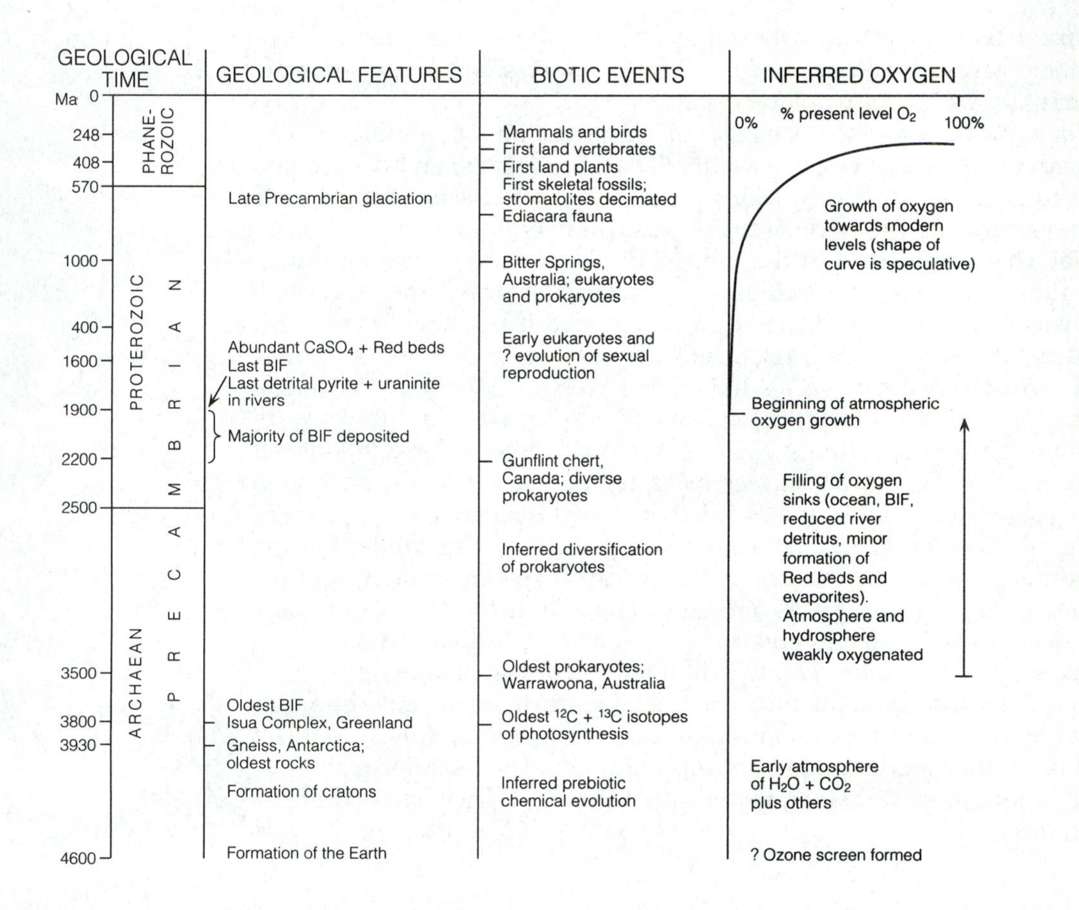

Figure 1.10 Some geological events together with the main evolutionary events in the planet's biota plotted against the exponential increase in atmospheric oxygen. (Redrawn from Kershaw, 1990.)

driven by its geophysiology. This model contrasts with the views of Robert Garrels who treated the Earth as primarily a chemical factory where carbon and sulphur are recycled; carbon is buried as limestone (calcium carbonate) and sulphur as gypsum (calcium sulphate). Both processes stabilize the level of oxygen in the atmosphere. The idea, however, of **Gaia** is not new. The Scottish scientist Sir James Hutton not only established the concepts of deep time and uniformitarianism during the late 1700s, he also modelled the Earth as a superorganism. The Gaia concept presents the planet as a living entity, regulating itself through interactions between the atmosphere, biosphere and the hydrosphere and to a lesser extent the lithosphere. So, for example, these interactions will control the planet's chemical composition and the climate through time. This planetary superorganism is called Gaia, the Greek name for the Earth.

Dead planets, such as Mars and Venus, have atmospheres in near chemical equilibrium. But a living planet, like Earth, has to cycle raw materials and waste products through its atmosphere, generating a disequilibrium amongst its chemical components. The atmosphere of Venus dominated by carbon dioxide but on Earth methane and oxygen coexist – two incompatible gases. Clearly another explanation must be found for this coexistence – the presence and functions of living organisms.

The planet, however, suffers a spectrum of climate conditions between two extremes: (a) **a greenhouse effect** – the heating up of the Earth's atmosphere by solar radiation trapped and re-emitted by greenhouse gases such as methane and carbon dioxide – and (b) an **ice-house effect** – cooling of the Earth with a loss of heat through the Earth's atmosphere. There are, however, many self-regulating processes. The control of **albedo** (reflectivity of solar radiation by the Earth) by variations in cloud cover is one such process. Clouds condense around the nuclei of sulphur gases generated by some marine algae. This cloud cover promotes a local cooling which can inhibit the spread of the algae. Eventually a balance is achieved between the optimum heat for the growth of the algae and amount of sulphur gas produced.

The **Daisyworld model** (Fig. 1.11) is a simple attempt to simulate Gaia (Lovelock, 1989). Two parameters are used – the growth of a single species of organism (a daisy) and the environment represented by temperature. Imagine planet Earth with few clouds and low concentrations of greenhouse gases; the albedo of the planet will determine

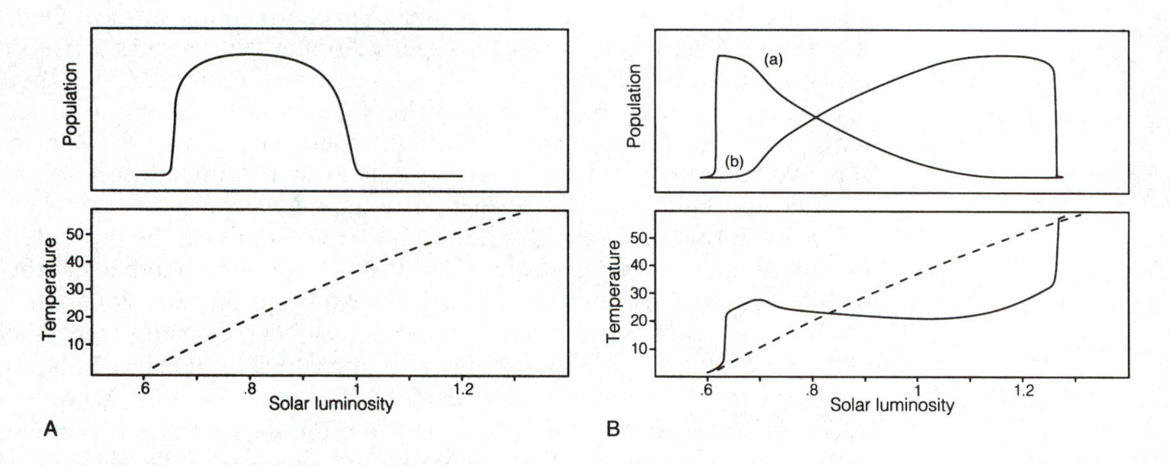

Figure 1.11 Evolution of Daisyworld based on A, conventional wisdom and B, geophysiology. The upper panels show the relative changes in daisy populations whereas the lower panels show temperature changes in degrees Celsius. Conventional models (A) predict that with increasing temperature the daisies first flourish and then die. A Gaian approach predicts that the competitive growth of the dark (a) and light (b) daisies will regulate the planet's temperature; the dashed line in the lower panel indicates increasing temperature without Gaian controls. (Replotted from Lovelock, 1989.)

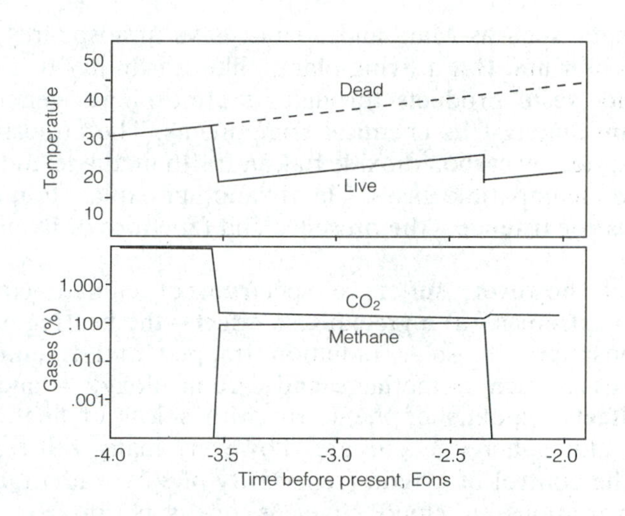

Figure 1.12 Transition from the Archaean to Proterozoic; model of a biosphere consisting of photosynthetic bacteria and methanogens. The upper panel shows the average global temperature, dashed line without life and the solid line with. The lower panel shows changes in abundance of carbon dioxide and methane; the abundance of these greenhouse gases decreases through time with the influx of oxygen from photosynthesizers. (Replotted from Lovelock, 1989.)

its temperature. Plenty of daisies grow on the planet's surface, well supplied by nutrients and water. Two types of daisy have evolved, both do not grow below 5°C or above 50°C but prefer temperatures of 22.5°C. During the first season the dark daisies prosper since they absorb heat and are generally warmer than the planet's surface. Over the next few seasons this morphotype multiplies but triggers a rise in the planet's temperature. But at temperatures above 22.5°C they decline and meet competition from the light daisies, seeking to lower temperature to those of more optimum conditions. So as the sun or star evolves, possibly heating up, so too must the mixed population of dark and light daisies to safeguard their own survival.

The evolution of the Earth's atmosphere and biotas can be presented in the so-called **ages of Gaia** (Lovelock, 1990): the Archaean, the middle ages and modern time. Early Precambrian life was dominated by bacteria; the cyanobacteria removed carbon dioxide from the planet's reserve of greenhouse gases tending to cool the Earth, analogous to the effect of the light daisies. The methanogens, however, added methane to the atmosphere and like the dark daisies promoted global warming. During the Archaean the sun was probably cooler and greenhouse gases were required to maintain the planet's temperature. By the mid-Proterozoic temperatures had approached an equilibrium with atmospheric carbon dioxide at about 1% (Fig. 1.12). Today if CO_2 is not removed by, for example, skeletal organisms, the planet would become too hot for life.

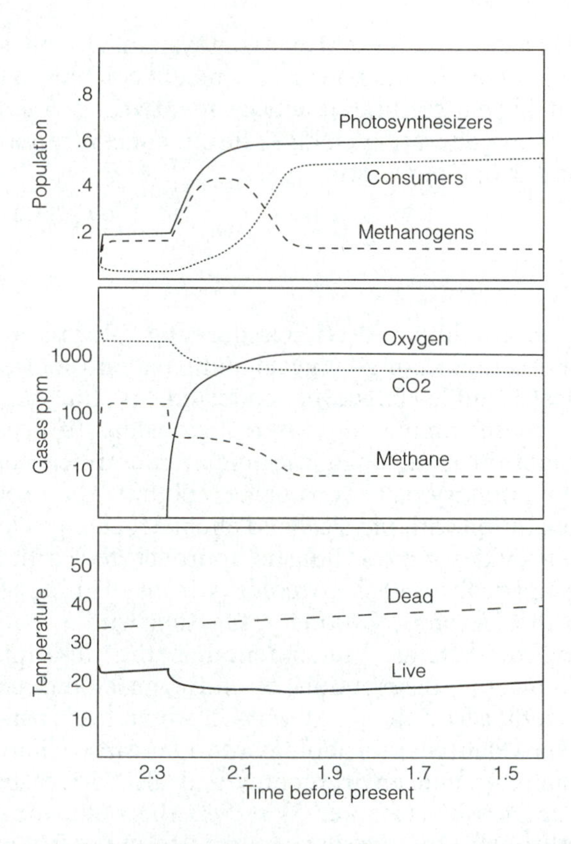

Figure 1.13 Evolution of the biosphere. The lower panel shows changes in climate in a live (solid) and lifeless (dashed) world; note the sharp fall in temperature when oxygen appears. The middle panel shows the changing abundance in atmospheric gases. The upper panel tracks changes in the population of ecosystems; both the photosynthesizers and methanogens increase initially when oxygen appears but the methanogens eventually decline to a much lower level of abundance. (Replotted from Lovelock, 1989.)

The planet's middle ages are signalled by the influx of large amounts of oxygen from cyanobacteria; initially most of the oxygen was removed during the oxidation of sediments, the burial of carbon in carbonate material and by solution in oceans. Nevertheless there was a gradual reduction of the anaerobic sector as atmospheric oxygen slowly built up and methane levels were reduced; the removal of this greenhouse gas promoted a drop in temperature. Although the temperature of the sun is currently increasing, the hike in atmospheric oxygen levels during the latest Precambrian and early Phanerozoic has helped regulate temperature, control climate and sustain the development of the planet's ecosystems (Fig. 1.13).

Gaia has helped focus biological research on a number of global issues. Life is in fact a planetary-scale phenomenon where living organisms must regulate their environment. The Gaia hypothesis complements Darwinian evolution; both organisms and their environment must be considered in a more integrated evolving system. Geophysiology may be a useful model to help understand very recent changes in the planet's systems forced by agricultural practices involving deforestation and industrial practices generating changes in the concentrations of greenhouse gases and ozone depletion.

1.8 STRATEGY

Palaeoecology is a field-based science deriving information from the collection and description of geological and palaeontological data. The palaeoecological sample, however collected, is the basic unit of analysis. From these samples the interrelationships of organisms and their environment through geological time are described and analysed. The field of evolutionary palaeoecology, combining the broader trends of biological evolution with physical and chemical changes in the planet through time, provides a more holistic approach to Earth history.

This book approaches palaeoecology from a time framework, accepting that there is an evolutionary, together with a spatial, component to ecological systems. Fundamental is the description of the factors that limit the distribution of individuals and communities (Chapter 2) together with the many ways in which fossils are preserved (Chapter 3). The adaptive morphology of organisms (Chapter 4) links evolution to habitats and environments and also drives major biotic radiations. Trace fossils (Chapter 5) record the behavioural patterns of organisms through time; fossils are thus useful environmental indicators (Chapter 6) and have evolved through time as integral parts of populations and communities (Chapter 7) with changing biogeographies (Chapter 8). Finally discussion of the simultaneous evolution of the marine (Chapter 9) and terrestrial (Chapter 10) biospheres is derived from the basic concepts of the book. The evolution of organisms and their characters together with community and large-scale biotic change may be linked to other planetary-scale changes involving climate and environment.

1.9 SUMMARY POINTS

- Palaeoecology is the study of the functions and relationships of ancient organisms and their communities.
- Fossil organisms, like their living counterparts, were probably restricted to particular habitats and environments – their 'addresses'.
- Understanding the life and trophic modes of fossil organisms – their 'occupations' – is an essential part of both aut- and synecology.
- The evolution and histories of the atmosphere, hydrosphere and lithosphere are linked to each other and to changes in the biosphere.

- The Gaia hypothesis suggests that the planet is a self-regulating organism involving the relationship, through time, between organisms and their environment.

1.10 FURTHER READING

Benton, M.J. 1990. *Vertebrate palaeontology*. Chapman & Hall, London. 377 pp.

Benton, M.J. and Harper, D.A.T. 1997. *Basic Palaeontology*. Addison, Wesley and Longman, Harlow.

Bosence, D.W.J. and Allison, P.A. (eds) 1995. *Marine palaeoenvironmental analysis from fossils*. Geological Society Special Publication 83, 272 pp.

Clarkson, E.N.K. 1993. *Invertebrate palaeontology and evolution*. 3rd Edition. Chapman & Hall, London. 434 pp.

Doyle, P. 1996. *Understanding fossils*. John Wiley & Sons, Chichester. 409 pp.

Goldring, R. 1991. *Fossils in the field: information potential and analysis*. Longman, Harlow. 218 pp.

Johnson, M.E. 1988. Hunting for ancient rocky shores. *Journal of Geological Education* 36, 147–154.

Kershaw, S. 1990. Evolution of the Earth's atmosphere and its geological impact. *Geology Today*, March/April 1990, 55–60.

Pianka, E.R. 1994. *Evolutionary ecology*. HarperCollins College Publishers, New York. 486 pp.

Van Andel, T.H. 1994. *New views on an old planet*. 2nd edition. Cambridge University Press, Cambridge. 439 pp.

Veizer, J. 1984. The evolving Earth: water tales. *Precambrian Research* 25, 5–12.

Westbroek, P. 1991. *Life as a geological force*. W.W. Norton & Co., New York, London. 240 pp.

1.11 REFERENCES

Ager, D.V. 1963. *Principles of paleoecology*. McGraw-Hill, New York. 371 pp.

Ager, D.V. 1965. The adaptations of Mesozoic brachiopods to different environments. *Palaeogeography, Palaeoclimatology, Palaeoecology* 1, 143–172.

Aronson, R.B. 1994. Scale-independent biological processes in the marine environment. *Oceanographic and Marine Biology: an Annual Review* 32, 435–460.

Boucot, A.J. 1981. *Principles of benthic marine ecology*. Academic Press, New York. 463 pp.

Boucot, A.J. 1983. Does evolution occur in an ecological vacuum?, II. *Journal of Paleontology* 57, 1–30.

Darwin, C. 1859. *On the origin of species by means of natural selection, or the preservation of favoured races in the struggle for life.* John Murray, London.

Dawkins, R. 1995. *River out of Eden. A Darwinian view of life.* Weidenfield & Nicholson, London. 172 pp.

Doyle, P., Bennett, M.R. and Baxter, A.N. 1994. *The key to Earth history. An introduction to stratigraphy.* John Wiley & Sons, Chichester. 231 pp.

Elias, M.K. 1937. Depth of deposition of the Big Blue (Late Paleozoic) sediments in Kansas. *Bulletin of the Geological Society of America* 48, 403–432.

Fischer, A.G. 1984. The two Phanerozoic supercycles. *In*: Berggren, W.A. and van Couvering, J.A. (eds), *Catastrophes and Earth history.* Princeton University Press, Princeton, 129–150.

Frakes, L.A., Frances, J.E. and Syktus, J. 1992. *Climate modes of the Phanerozoic.* Cambridge University Press, New York.

Hecker, R.F. 1965. *Introduction to paleoecology.* Elsevier, New York. 166 pp.

Holland, C.H. 1986. Does the golden spike still glitter? *Journal of the Geological Society of London* 143, 3–21.

Lamont, A. 1934. Lower Palaeozoic Brachiopoda of the Girvan district. *Annals and Magazine of Natural History* 14, 161–184.

Lovelock, J. 1979. *Gaia: A new look at life on Earth.* Oxford University Press, New York.

Lovelock, J.E. 1989. Geophysiology. *Transactions of the Royal Society of Edinburgh: Earth Sciences* 80, 169–175.

Lovelock, J. 1990. *The ages of Gaia. A biography of our living Earth.* Bantam Books, London and New York. 252 pp.

Lyell, C. 1850. Anniversary address of the President. *Quarterly Journal of the Geological Society of London* 6, xxvii–lxvi.

McKerrow, W.S. (ed.) 1978. *The ecology of fossils.* Duckworth Press, London. 383 pp.

Owen, R. 1854. *Geology and the inhabitants of the Ancient World.* Crystal Palace Library and Bradbury & Evans, London. 40 pp.

Raup, D.M. and Sepkoski, J.J. Jnr 1984. Periodicity of extinction in the geologic past. *Proceedings of the National Academy of Science USA* 81, 801–805

Richter, R. 1928. Aktuopaläontologie und Paläobiologie: eine Abgrenzung. *Senckenbergiana* 11, 285–292.

Sardeson, F.W. 1929. Ordovicic brachiopod habit. *Pan-American Geologist* 51, 23–40.

Schäfer, W. 1965. *Aktua-Palaeontologie nach studien in der Nordsee.* Waldemar Kramer, Frankfurt. 666 pp.

Schäfer, W. 1972. Ecology and palaeoecology of marine environments. University of Chicago Press, Chicago. 568 pp.

Sheehan, P.M. 1992. Patterns of synecology during the Phanerozoic. In *The Unity of Evolutionary Biology* 1, 103–118.

Stanley, S.M. 1975. A theory of evolution above the species level.

Proceedings of the National Academy of Science USA 72, 646–650.
Vrba, E. 1983. Macroevolutionary trends: New perspectives on the roles of adaptation and incidental effect. *Science* 221, 387–389.
Wilson, E.O. 1992. *The diversity of life*. Allen Lane, Penguin Press, London. 424 pp.
Windley, B.F. 1995. *The evolving continents*. 3rd edition. John Wiley & Sons, Chichester. 526 pp.
Ziegler, A.M. 1965. Silurian marine communities and their environmental significance. *Nature* 207, 27–272.

2 Environmental controls on biotic distribution

The biosphere has a hierarchical structure, with the individual animal or plant at the lowest level, then the population, the species, the community and the biogeographic province at the highest level. Each level of organization occupies a specific volume within ecospace; the volume occupied by a species is the niche, that by a community is the ecosystem. The way that physical and chemical factors such as light, temperature and oxygen determine the range of organisms is described and examples of how these controls can be recognized in the fossil record are discussed using specific examples.

2.1 THE STRUCTURE OF THE BIOSPHERE

The life of an organism is controlled by a wide range of physiological processes. These can generally operate only under a relatively narrow range of environmental conditions and each organism has its own tolerance of factors such as temperature and salinity. Organisms also attempt to maintain constant internal conditions (**homeostasis**), in the face of a fluctuating external environment, and their ability to do so sets limits to their environmental range.

The environment can be envisaged as multidimensional space, with each dimension representing a different environmental variable. An individual organism will have a specific range for each variable and when the ranges of all the environmental parameters are considered they will define a specific volume. Figure 2.1a shows the area occupied if only two parameters are considered and Figure 2.1b shows the volume if three parameters are considered. Further parameters cannot be represented diagrammatically, but the total volume could potentially be defined by multivariate mathematics. Most organisms do not occur singly but are part of an interbreeding population. The tolerance ranges of the individuals constituting a population are likely to vary a little so that the volume occupied by a population will be somewhat larger than that of any individual. The ecospace potentially occupied by a species is that volume occupied by all its constituent populations and is termed the **prospective niche.** Most species do not fully occupy their prospective niche, partly because they may be better adapted to part of the range, or because they are

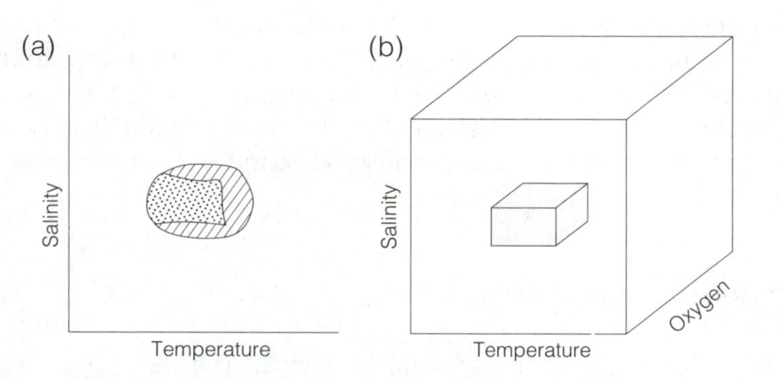

Figure 2.1(a) The prospective niche of a species defined by its temperature and salinity ranges. The cross-hatched area is that part of the niche which the species does not exploit and the stippled area is the realized niche.
Figure 2.1(b) The volume in ecological space occupied by a species, defined by its range of temperature, salinity and oxygen requirements.

in competition with other species, so they occupy a **realized niche** (Fig. 2.1a).

A hierarchical structure extends further upwards through the ecological scale (Fig. 2.2). Species that co-exist within a similar range of environmental parameters form **communities**, such as the *Littorina* community on rocky shores or the *Macoma* community on sandy beaches, each characterized by its own ecosystem, which in turn is mainly determined by the properties of the component niches (see Chapter 7).

When the distribution of communities is regarded on a global scale it is clear that groups of communities commonly occupy discrete parts of the globe. These community groups whose boundaries are commonly defined climatically or by physical barriers are **biogeographical provinces** and constitute the largest subdivision of the biosphere (see Chapter 8).

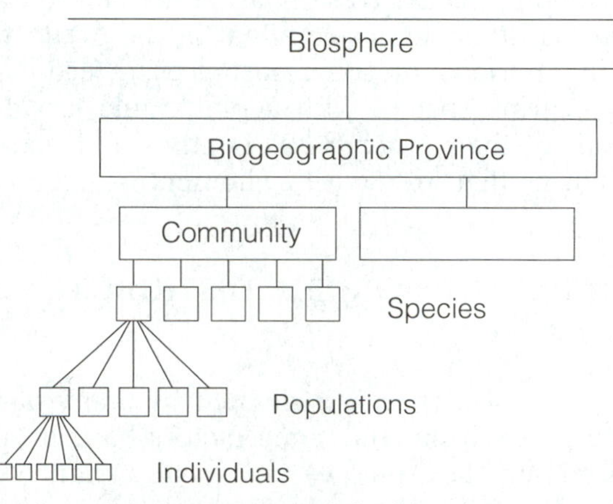

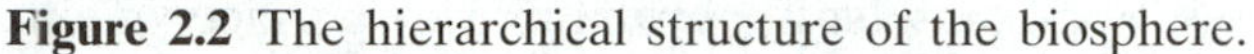

Figure 2.2 The hierarchical structure of the biosphere.

The palaeontological record helps to show how the number and range of niches have increased with time, it can show the change in community structure in response to environmental change and on a larger scale it records the changes in the number and distribution of biogeographic provinces in response to climatic change and plate movements.

2.2 DIVISIONS OF THE MARINE ENVIRONMENT

A profile from the shoreline to the bottom of the deep oceans shows the continental shelf, which is a nearshore region on average about 120 km wide sloping relatively gently down to a depth of about 200 m, then the continental slope extending down to about 4000 m and finally abyssal plains, deep-sea trenches and volcanic rises at greater depths (Fig. 1.2). Environments in the coastal zone include rocky shorelines, sandy or pebbly beaches, estuaries, lagoons and deltas and may show rapid lateral change along a stretch of coastline. Stretches of sandy coasts have a shoreface with intertidal beach environments on their landward side and a transition into more muddy shelf environments at about 10–20 m on their seaward side. Environments tend to become more uniform away from shore where mud forms the typical substrate, but even in deep-sea environments there may be considerable heterogeneity.

The environments that occur in waters on the continental shelf are referred to as **neritic**, whilst those further oceanwards are **pelagic.** The pelagic environments are divided by depth into epipelagic (<200 m), mesopelagic (<1000 m) bathypelagic(<4000 m) and abyssopelagic (<11,000 m). Greater depths are referred to as **hadal**.

Benthic organisms are bottom-living organisms and animals may live on the sea-floor as **epifauna** or below the sediment surface as **infauna**. The organisms may be immobile (sessile) or active (vagile). **Plankton** consists of organisms that float passively in the water, or at most, are active only in keeping themselves afloat. Holoplankton are organisms that live through all their life cycle afloat in the ocean waters, except for brief resting periods in some instances. Meroplankton, which includes many marine species with a planktonic larval stage, spend alternate periods of their life cycle as benthos and plankton. **Nekton** are those organisms that are active swimmers.

2.3 LIMITING FACTORS ON THE DISTRIBUTION OF ORGANISMS

The potential niche of a species is defined by many parameters, but, within a particular environment, some factors have a more profound effect in determining the range of the species than others, and are termed **limiting factors.** For example, in normal marine surface waters, salinity and oxygen levels are relatively constant and remain well within the tolerance range of most organisms, so do not determine their presence or absence. On the other hand, temperature in marine waters can vary considerably and may be an important limiting factor in determining the range of a species. In the following sections the various physical factors that influence the distribution of species are discussed.

2.4 LIGHT

Light is the fundamental source of energy for primary producers at the base of the food chain. In most photosynthetic plants light energy reacts with photosynthetic pigments, such as chlorophylls, and fixes carbon dioxide into organic compounds. Primary producers which create their own organic material in this way are termed **autotrophs,** as opposed to **heterotrophs** which utilize organic matter synthesized by other organisms. Diatoms, dinoflagellates, coccolithophorids and some cyanobacteria are typical of the phytoplankton that are dependent on light. Benthic algae are the main primary producers in nearshore areas. Primary productivity amongst plankton is influenced mainly by the intensity of light and the availability of nutrients (see section 2.5). The intensity of light decreases exponentially from the sea-surface downwards so that most photosynthetic activity has stopped at depths of about 150 m. However, the highest photosynthetic rates are commonly 10–20 m below the surface (Fig. 2.3), not at the

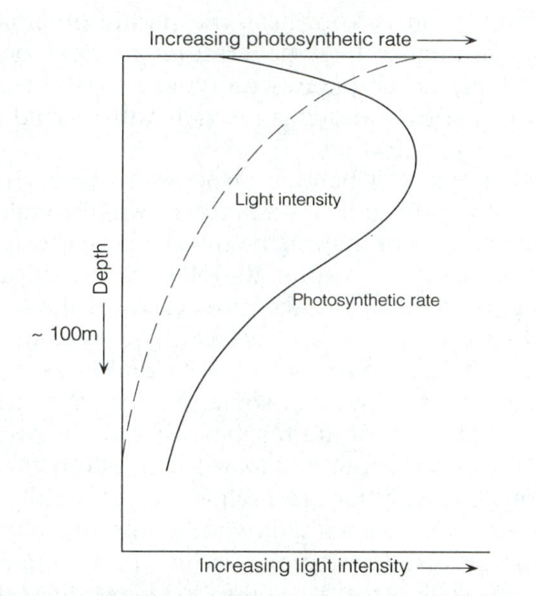

Figure 2.3 Generalised sketch of vertical profiles of light intensity and corresponding photosynthetic rate in a clear marine water column with uniform distribution of phytoplankton. (Modified from Barnes and Mann, 1991.)

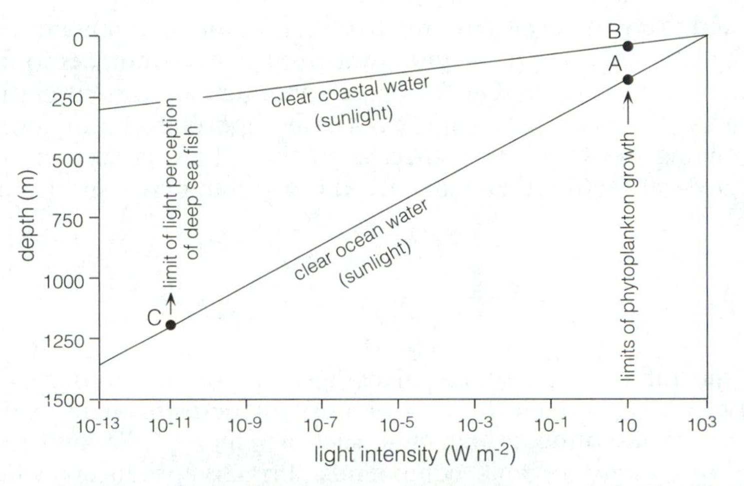

Figure 2.4 The relationship between illumination and depth. The limits to phytoplankton growth in clear ocean waters and clear coastal waters are shown by points A and B. The limit to light perception by deep-sea fish is shown by point C.

surface, because a high intensity of solar radiation can apparently inhibit photosynthesis by photo-oxidative reactions, and because of the destructive effects of high levels of ultra-violet light. The intensity of light penetrating a water surface varies with the angle of incidence of the light (light intensities will generally be lower in high latitudes) and with the amount of dissolved and particulate matter in the water. The depth range from the ocean surface through which there is net photosynthetic production is known as the **photic** or **euphotic zone**. Figure 2.4 shows the higher light penetration in clear ocean waters (down to about 200 m) as compared with typical coastal waters (about 40 m). Waters in estuaries and deltas are generally turbid and have a particularly low light penetration.

Marine phytoplankton and benthic algae show a very pronounced variation in both diversity and abundance down through the water column, with maxima several metres below the surface, decreasing to very low levels of productivity within 20–150 m of the surface depending on the intensity of the incident light, the clarity of the water and the degree of disturbance of the water surface. The maximum depth of the photic zone is estimated to be about 250 m, and this marks the limit to most algal productivity. However, some planktonic algae are found living at depths of thousands of metres, benthic red algae are recorded at 180 m, brown algae to a depth of 300 m and crustose coralline algae to 400 m. It is not clear whether these algae are especially adapted to very low light levels or are partially or wholly heterotrophic.

There is a zonation of benthic algae within the photic zone that is determined, at least partly, by their ability to absorb light. It was originally thought that it was the composition of their pigments that determined the depth to which the larger algae (macrophytes) could grow and that green algae would be expected in shallow waters and red algae in deeper waters. The zonation of algae with increasing depth from green to brown to red is unfortunately rarely clear, though green algae are particularly common in shallow water. Recent work has shown that it is the quantity of pigments rather than their composition that plays a dominant role in regulating algal growth at low light levels. Within the major groups of algae a distinct zonation of species has been recognized among red algae and in the genus of green algae, *Hallimeda*. A problem that affects the geologist in using algae as bathymetric indicators is that many disintegrate and are easily reworked. This is particularly true of the green, dasycladacean and codiacean algae, which in the case of *Penicillus* disintegrate into aragonite needles and in *Hallimeda* into gravel-sized particles. Crustose coralline algae which are robust and commonly firmly attached to the substratum may prove useful bathymetric indicators when more is known about their zonation in shallow clear water.

Algae also play a major part in the infestation of carbonate grains (endolithic algae) that are penetrated by minute borings that are commonly filled by microcrystalline calcite or aragonite (micrite). Micrite may rim the grain, forming a micrite envelope, or partially or wholly replace the grain. Many endolithic algae are photosynthetic and so are restricted to the photic zone, but other organisms such as sponges, fungi and bacteria, also micritize grains so that although extensive micritization is normally confined to the photic zone, it is not wholly diagnostic of shallow marine environments.

2.4.1 Palaeoecology

The vertical zonation of phytoplankton in response to decreasing levels of light is not recorded in the fossil record, but the distribution of benthic algae has been used to estimate bathymetry. As a general rule the presence of a varied benthic algal flora identifies the photic zone and commonly indicates the upper part of that zone, i.e., depths down to about 20 m, particularly if green algae are common. Similarly, the pervasive micritization of carbonate grains is generally a good indicator of shallow seas (see box 2.1).

2.5 NUTRIENTS

Nutrients are the inorganic and organic substances that are essential for the growth of plants. Together with light they exert a dominant control on productivity. The most important nutrient elements are nitrogen and phosphorus, but iron and silica may have a significant effect on productivity in some circumstances. Nitrogen is essential for the synthesis of amino-acids, proteins and nucleic acids and so plays an essential part in the growth of organisms. Phosphorus plays a particularly important role in the metabolism of an organism.

Box 2.1 A bathymetric analysis of Carboniferous carbonate mud mounds using algae

Carbonate mud mounds (also referred to in the Carboniferous as Waulsortian build-ups) are large structures, commonly a kilometre or more in diameter and 50–200 m high. Mounds result from the localized accumulation of carbonate mud which may enclose a very diverse benthic biota which lived on the mound. The mounds differ from reefs in that they are not constructed from a framework of rigid skeletons, such as those of corals and stromatoporoids. The thick accumulation of mud and the poor sorting of shelly material suggests that the mounds generally grew below storm wave-base, which in shelf seas is at a similar depth to the base of the photic zone (Fig. 2.6a). However, the presence of such large volumes of localized carbonate mud suggests the presence of highly productive algal or microbial populations, which, if they flourished below the photic zone, must have been heterotrophic. Similar highly productive heterotrophs are not known today but nor are large carbonate mud mounds whose abundance decreases following the end-Permian extinction. Did the postulated heterotrophs decrease or even disappear too? In this context an understanding of the depth at which carbonate mounds grew is particularly important.

In their study of Carboniferous mud mounds Lees and Miller (1985) used the distribution of calcareous algae and micritization within the mounds as bathymetric indicators. In the early stages of mound growth there are no calcareous algae or micritization. Micritization first affected the top of a mound and through time extended down the flank, in response to falling sea-level. At a later stage as sea-level continued to fall, the top of the mound was colonized by calcareous algae and these too migrated down the mound flank. At any one time the difference in height between the lowest limits of the calcareous algae and micritization was about 90 m (the height difference between A and B on Fig. 2.5), which suggests this was the difference in water depth between the two. When micritization reached the base of the mound it was 220 m high and there is evidence that

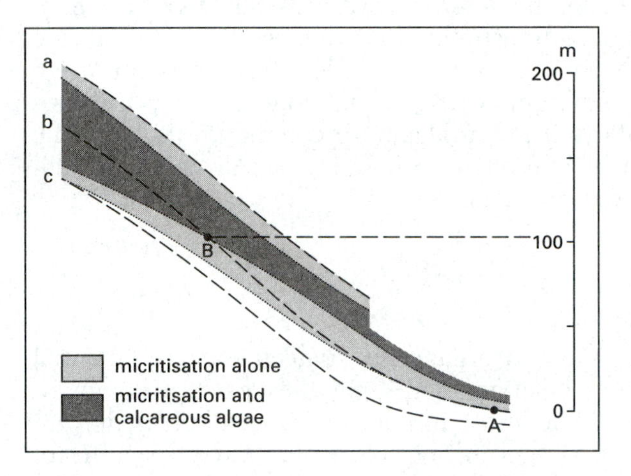

Figure 2.5 The flank of a carbonate mud mound. The dashed lines (a, b and c) are time-planes representing 3 stages of mound growth determined by conodonts. A marks the base of micritization on time plane b and B is the contemporaneous base of calcareous algae, 90 m higher. (Modified from Lees and Miller, 1985.)

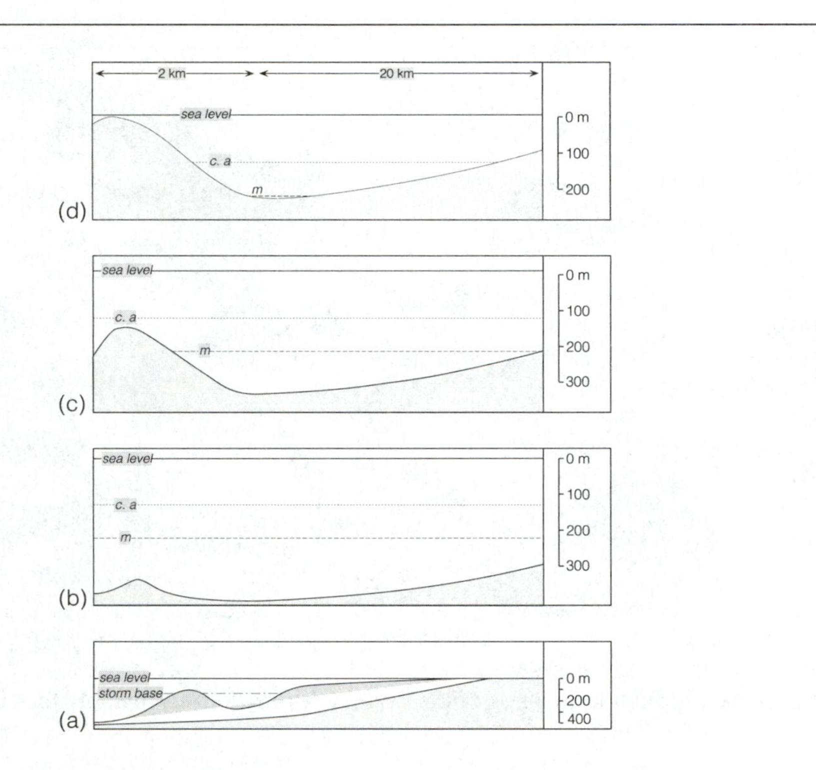

Figure 2.6 (a) Position of a typical carbonate mud mound low on a ramp. Figures b, c, and d show successive stages in the growth of a mound. (b) Start of the growth of the mound, (c) micritization affects the top of the mound, (d) micritization extends to the base of the mound, calcareous algae are 90 m higher and the top of the mound is at sea-level. (Adapted from Lees and Miller, 1985.

the crest was in very shallow water, so the mound apparently stood 220 m above the basin floor at this stage. The lower limit to calcareous algae would have been about 90 m above basin floor. The successive stages of mound growth can now be reconstructed (Fig. 2.6) and it can be seen that growth started at depths of more than 300 m (Fig. 2.6b) and grew upwards and outwards at shallower depths as sea-level fell (Fig. 2.6c and d). Nevertheless a substantial part of the mound was apparently constructed at depths below the productive levels of the photic zone suggesting that algal growth might have been heterotrophic not autotrophic. The main reservation about this conclusion is whether the limits of micritization are a reliable indicator of the base of the photic zone.

Although nitrogen is the most abundant gas in the atmosphere and is also dissolved in sea-water it is not available for organic metabolism until it is oxidized to form nitrate, nitrite or nitrous oxide. Nitrogen fixation is performed by certain bacteria that synthesize ammonia which is in turn modified to form nitrite and nitrate by oxidative steps within the food chain. Some additional nitrate is supplied

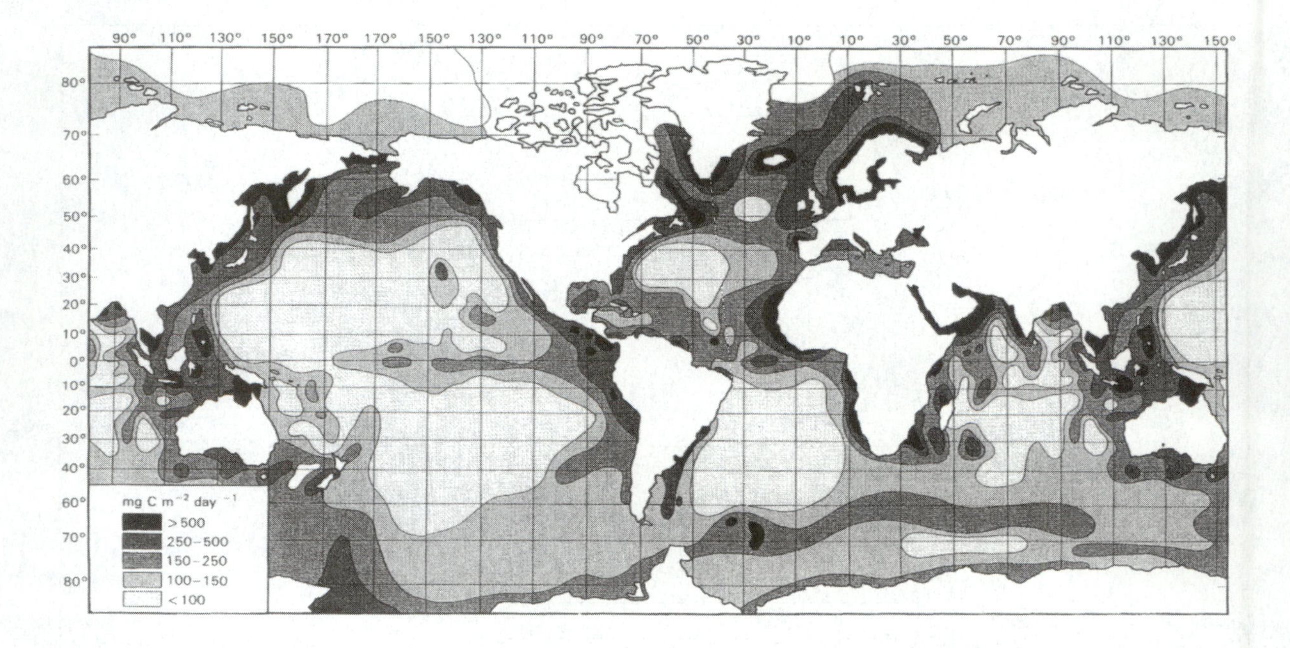

Figure 2.7 Primary production in the world's oceans. (From Barnes and Mann, 1991.)

to the ocean from the atmosphere and some from rivers. In contrast to nitrogen, most phosphorus is ultimately derived from continental weathering and much of it is transferred to the oceans bound to organic material. It is therefore thought that the supply of phosphorus to the oceans might have been considerably less prior to the evolution of terrestrial vegetation (Brasier, 1995).

Most organic production occurs in the surface layers of the ocean because light is necessary for photosynthesis (see previous section), but it is the availability of nitrogen and phosphorus that largely determines the geographic and temporal variations in productivity. A map of ocean productivity (Fig. 2.7) shows that high levels are found on the continental shelves where nutrient levels from continental run-off are high, and also in regions of the oceans where there are upwelling currents bringing nutrients from deeper waters.

In those parts of the ocean where the supply of nutrients is steady but relatively low, **primary production** (the fixation of carbon by photosynthetic organisms) creates **new production** in the form of phytoplankton. The organic matter is efficiently recycled through the food chain and nutrients released for further growth (**regenerated production**) (Fig. 2.8). Consequently, there is little fall-out of organic material from the euphotic zone (**export production**). In situations where there is turbulent mixing, or more commonly upwelling, nutrient supply is high and there are associated high levels of new production. As a result large amounts of organic material may be exported to greater depths. This can sink as

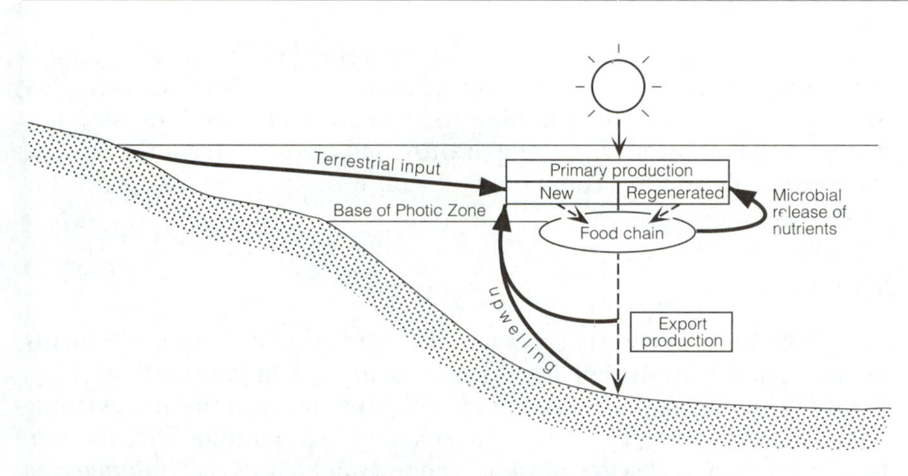
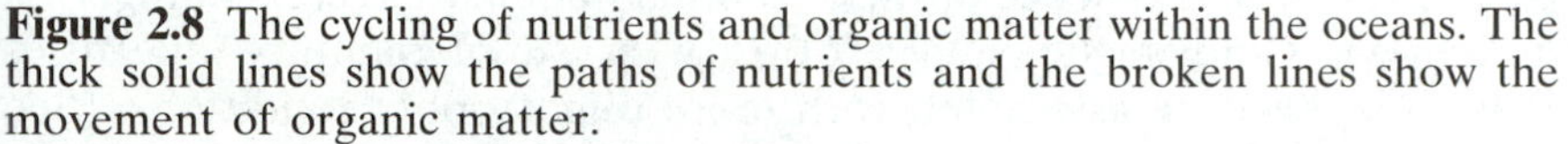

Figure 2.8 The cycling of nutrients and organic matter within the oceans. The thick solid lines show the paths of nutrients and the broken lines show the movement of organic matter.

dissolved organic material (**DOM**) or as particulate organic material (**POM**) which can be sedimented to form organic-rich muds. Some of the export production may ultimately be oxidized within the water column or at the sea-floor and be recycled from either site (Fig. 2.8).

Oceans, marine basins and lakes with low levels of nutrient supply are referred to as **oligotrophic.** Stable, nutrient-poor systems tend to produce ecosystems that have a high diversity of organisms but have a low abundance of individuals. Systems in which there is a rich supply of nutrients are referred to as **eutrophic** and are characterized by high productivity and abundant individuals of a relatively few species. High nutrient levels may be related to vigorous upwelling of nutrient-rich waters or to a high nutrient flux from land and increases in nutrient supply are generally related to climatic change, to new circulation patterns in the oceans or to changes in sea-level (Brasier, 1995a). Climatic change affects nutrient supply by determining the thermal regime in the oceans; stable, warm climates tend to produce oligotrophic oceans, while glacial periods favour eutrophic oceans. Climatic change can also influence the amount of run-off of fresh water from land.

Nutrient supply is thought to have a particularly strong affect on reef ecosystems (Hallock, 1988). The zooxanthellate organisms, particularly colonial corals, that contribute largely to the framework of reefs are adapted to low nutrient levels, while filamentous and fleshy algae and bioeroders (organisms that bore and destroy carbonate material) favour high nutrient levels. Consequently reefs generally flourish under oligotrophic conditions. Under somewhat more nutrient-rich, mesotrophic conditions growth tends to be suppressed and bioeroders create large amounts of carbonate mud, while under eutrophic conditions, carbonate production as a whole is reduced and bioeroded hardgrounds and carbonate non-sequences may develop.

Biological evidence for high productivity in areas of upwelling includes the abundance of certain species of foraminifera, planktonic gastropods, the particular types of dinoflagellate cysts present and the abundance of siliceous diatoms and radiolaria. Nutrients clearly play an important role in determining the biofacies in upwelling systems, but related factors, such as temperature and oxygen, may also play an important role and it is commonly difficult to isolate one limiting factor from others.

2.5.1 Palaeoecology

The abundance of certain species of fossil microplankton in sediments of late Cenozoic to Recent age has been used to identify upwelling areas. Planktonic foraminifera associated with the cold waters of upwelling zones may be recognised by their coastal biogeographic distribution, lack of spines and absence of supplementary apertures (e.g. *Globigerina bulloides*). Conversely, spiny forms with supplementary apertures (e.g. *G. ruber*) are associated with more oligotrophic conditions. (See Brasier, 1995, for a review.) Biogenic silica deposits formed from siliceous diatoms and radiolaria are commonly associated with upwelling. Prior to the proliferation of siliceous diatoms in the Cretaceous, reconstructions may be more difficult; nevertheless, the abundance of radiolaria, siliceous sponge spicules and assemblages with phosphatic skeletons (conodonts, inarticulate brachiopods, fishes' teeth) has been used to suggest high nutrient supply in rocks as old as the Cambrian. The presence of phosphatic deposits may be a reflection of high nutrient levels and high productivity in some situations, but they can accumulate under other circumstances. Dark, carbon-rich shales give an equivocal message because they may form under nutrient-rich upwelling areas but they may also form where the bottom waters were anoxic and productivity was only moderate (see section 2.6). A radical change in the cycling of nutrients associated with climatic change has been invoked as a major factor in major extinctions at the end of the Ordovician (Brenchley *et al.*, 1994), the middle to late Eocene (Brasier, 1995b) and in the Miocene (Arbry, 1992). In each case, cooling at high latitudes is thought to have promoted vigorous ocean circulation and upwelling and a switch from oligotrophic to eutrophic ecosystems with a consequent decrease in diversity. Changes in nutrient supply are believed to have played a major role in the growth and demise of reefs and may have affected the nature of carbonate systems in general. Brasier (1995) has made a more general case for nutrient supply and changes in the carbon cycle being involved in many of the major ecological revolutions, both extinctions and radiations, in earth history.

2.6 OXYGEN

Almost all eukaryotic organisms require oxygen for their metabolism. Oxygen requirements generally increase with the size and activity of an organism. Many very small-bodied organisms take up oxygen by

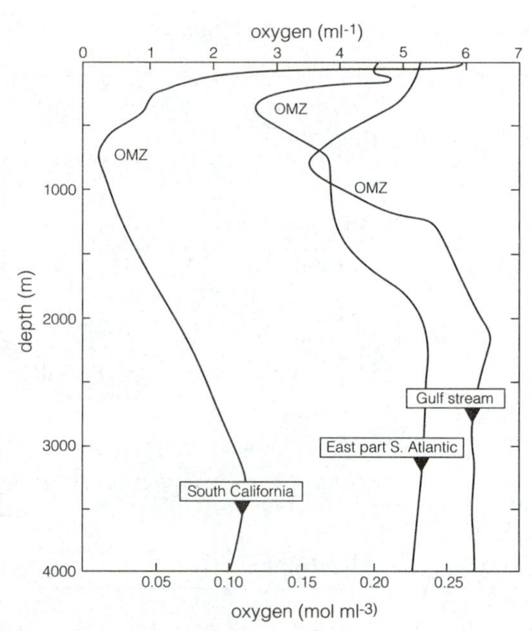

Figure 2.9 The vertical distribution of dissolved oxygen (concentration in ml/ l^{-1} and mol/ l^{-3}) in the ocean of south California, in the eastern part of the south Atlantic Ocean and in the Gulf Stream. OMZ = oxygen minimum zone.

diffusion, but most larger animals have respiratory devices, such as gills and a circulatory system and many have respiratory pigments that increase the carrying capacity of the blood. The oxygen content of sea-water varies between 8.5 and 0 ml/l (millilitres of oxygen per litre of water), but is mainly in the range of 6 to 1 ml/l, which is viable for the majority of marine organisms, though many live in the upper part of that range and some are 'uncomfortable' at lower oxygen levels. Oxygen starts to become an important limiting factor where concentrations fall below about 2 ml/l.

In the upper layers of the oceans the rate at which oxygen is added to marine waters by solution from the atmosphere and by the photosynthetic activity of phytoplankton is roughly balanced by that lost during respiration and oxidative processes, though fluctuations in productivity may result in fluctuating oxygen levels. Below the photic zone oxygen is not replenished by photosynthesis but is consumed by organic respiration and microbial degradation of organic material, leading to a fall in the levels of dissolved oxygen in the water. Consequently there is generally a well-developed oxygen profile in the ocean (Fig. 2.9) with the high levels of oxygen in the upper part of the ocean generally decreasing to depths of between 100 and several hundreds of metres, at which level the rate at which oxygen is consumed by the oxidation of organic matter exceeds the supply of oxygen by diffusion, giving rise to an **oxygen minimum zone (OMZ)** in which oxygen levels reach their lowest values. Below the OMZ,

Table 2.1 Classification of oxygen regimes, biofacies and physiological regimes according to different authors

OXYGEN ml per l	*Tyson and Pearson (1991)*	*Byers (1977)*	*Tyson and Pearson (1991)*	*Savrda and Bottjer (1991)*	*Wignall (1995)*	
OXYGEN ml per l.	*OXYGEN REGIME*		*BIOFACIES*			*PHYSIO-LOGICAL REGIME*
—8.0	Oxic	Anaerobic	Aerobic	Aerobic	Aerobic	Normoxic
—2.0						
—1.0	mod				Dysaerobic / Poikilo -aerobic	
—0.5	Dysoxic sev. extreme	Dysaerobic	Dysaerobic	Dysaerobic		Hypoxic
—0.2				Exaerobic		
—0.1	suboxic		Quasi -anaerobic	Quasi -anaerobic	Anaerobic	
—0.0		Anaerobic				
	Anoxic		Anaerobic	Anaerobic		Anoxic

oxygen levels may rise again in association with mid- to deep-water circulation of oxygenated waters. As a general rule oxygen levels within the OMZ are reduced by 25 to 50% of their initial level and this may not have a profound affect on organisms (Tyson and Pearson, 1991). However, where organic production in upwelling areas is particularly high, the depletion of oxygen in the underlying OMZ may produce dysoxic or even anoxic conditions (see South California, Fig. 2.9). Oxygen levels in deeper waters depend on the circulation structure of the oceans. In stable, well-stratified oceans, deep water circulation may be very sluggish and oxygen levels may be low, while in oceans with a vigorous thermohaline bottom circulation, oxygen levels will be relatively high.

Classifications of oxygen levels in the environment refer to the levels of dissolved oxygen using the suffix 'oxic' (Table 2.1). Other classifications are based on the biofacies present and use the suffix 'aerobic'. A classification based on biofacies, dividing environments into **anaerobic** (<0.1 ml/l O_2), **dysaerobic** (0.1–1.0 ml/l O_2) and **aerobic** (>1.0 ml/l O_2), according to their levels of dissolved oxygen (Byers, 1977), has been found useful in analysing the effects of oxygen on organic distribution. Three more recent classifications (Savrda and Bottjer, 1991; Tyson and Pearson, 1991; Oschmann, 1991 and Wignall, 1994) have redefined or further subdivided the categories of oxygen poor environments (Table 2.1), recognizing that the dysaerobic facies can be fully exploited, either permanently (exaerobic biofacies) or sporadically (poikiloaerobic biofacies).

The biotic response to low oxygen levels is varied. There is a marked decrease in faunal diversity when dissolved oxygen falls below about

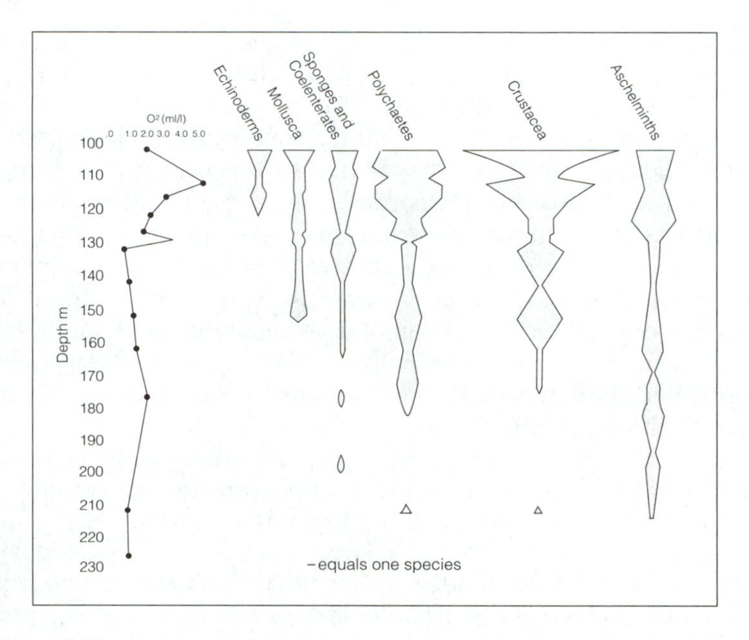

Figure 2.10 Change in species abundance with depth and dissolved oxygen in the Black Sea, where the sill depth is 100 m. Values above 130 m fluctuate from 2 ml/l upwards. Note the sharp decline in diversity at about 130 m where the dissolved oxygen level falls below 2 ml/l. (From Rhoads and Morse, 1971.)

2 ml/l. This is well illustrated by the faunal distribution in the stagnant waters of the Black Sea (Fig. 2.10). Shelf benthos appears to be more affected by oxygen levels in the range 2–1.4 ml/l than animals in the oceanic oxygen minimum zone. Those animals on the shelf that can survive at oxygen levels below 2 ml/l will commonly migrate away from the low oxygen area if they are active benthos or nekton, or will rise to the sediment surface if they are infaunal benthos. Adaptations to low oxygen levels can be morphological, physiological or behavioural (Oschmann, 1994). Morphological adaptation is shown by some bivalves (e.g. 'paper pectens') and a variety of 'flat worms' that have developed a flat body, which increases the surface area through which oxygen can be received by absorption or diffusion. Other animals adapt physiologically to periods of low oxygen lasting days or even weeks by drastically slowing down their metabolic rate, by having a particularly effective circulatory system and by having oxygen-binding pigments. During short periods of anoxia some organisms can metabolize via an anaerobic pathway. In most animals oxygen-binding pigments become less efficient at the lowered pH generated from the respiratory production of CO_2, but some polychaetes and 'blood cockles' have developed pigments that can tolerate these low pH levels. The problems caused by the toxic effects of the hydrogen sulphide that is generated in many anaerobic sediments is counteracted in some animals by having H_2S binding and transporting pigments. Alternatively, some organisms,

including 26 Recent bivalve species, belonging to five families, are chemosymbiotic and incorporate chemo-autotrophic bacteria, that oxidize H_2S and so prevent its toxic effects, but are ultimately ingested by their host. Chemosymbionts require sources of both sulphide and oxygen, which do not normally coexist, so the modern environments from which they have been recorded are generally rather specialized ones, such as deep-marine hydrothermal vents, petroleum seeps, the sea-floor close to the decaying carcasses of large marine animals and shallow marine areas colonized by seagrass where the abundant supply of decaying organic material promotes the generation of sulphide. The importance of chemosymbiosis in the colonization of oxygen-poor environments and their recognition in the fossil record is still a matter of debate (Oschmann, 1994; Wignall, 1994).

Where anoxia is seasonal some organisms adapt their life cycle to produce larvae at the start of the summer anoxic period and allow them to settle during the better-oxygenated autumn season. This behavioural adaptation allows a species to survive severe annual summer anoxia. More moderate levels of oxygen deficiency are survived by species with morphological or physiological adaptations (Oschmann, 1994).

In bottom waters with normal levels of dissolved oxygen the surface layers of the underlying sediment are well aerated and sustain a varied infauna, some of which burrow deep into the sediment while others exploit shallower levels, leading to a tiering of the animal traces (Fig. 2.11). With progressively lower levels of oxygen in the bottom waters, burrows penetrate less deeply and tiering is less pronounced. Anaerobic sediments with no bioturbation normally occur where there is a high flux of organic matter to the sediment surface and consequently the oxygen levels in the bottom waters are very low. Oxygen becomes exhausted in the near-surface layers, bioturbation is consequently suppressed and H_2S may be generated creating a toxic environment (see Chapter 6).

2.6.1 Palaeoecology

Shelled benthic faunas decline in diversity with reduced levels of oxygen and become rare in dysaerobic facies. In the Jurassic, for example, a few species of bivalves generally dominate in dark shale facies, though gastropods may be common at some horizons. In more oxygen-poor facies fossils are sparse except for exceptional bedding planes covered with fossils, commonly of a single species. Some of these fossils belonged to opportunistic benthic species capable of colonizing environments with low oxygen levels, others may have been infaunal benthos which emerged on to the sediment surface in response to a sudden lowering of oxygen availability, while others appear to have been species of epibenthos that lived a floating life attached to some object or were possibly pelagic forms. From his studies of the finely laminated dark grey shales of the upper Jurassic Kimmeridge

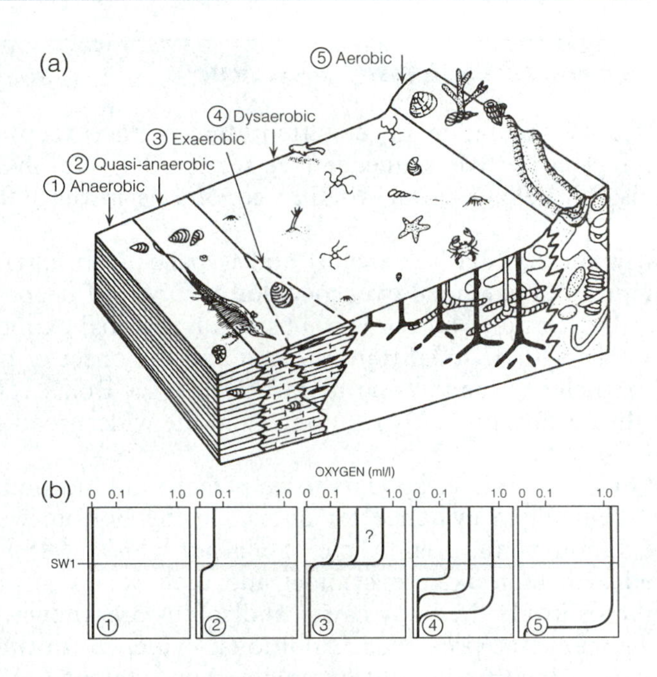

Figure 2.11 (a) Five trace fossil biofacies (labelled 1–5) determined by dissolved oxygen levels in the bottom waters and within the sediment (Savrda and Bottjer, 1991). Note decrease in burrow diversity, size and depth with decreasing oxygen availability. Note also the reduction in tiering in an offshore direction. (b) Concentrations of dissolve oxygen within bottom waters and below the sediment-surface interface. (From Savrda and Bottjer, 1991.)

Clay in England, Oschmann (1994) identified bivalve faunas which he interpreted as chemosymbionts and others that he regarded as adapted through their life cycle to seasonal oxygen depletion. The evidence for chemosymbiosis is that the lucinid *Mesomiltha concinna* and *Solemya* sp. grew to 4 cm, which is ten times larger than all the co-occurring endobenthic taxa. Comparison with Recent forms suggests that this is likely to be related to faster growth in chemosymbionts than in co-existing non-chemosymbiontic taxa. The evidence for a seasonal strategy for survival comes from those bivalve faunas and the inarticulate brachiopod *Lingula* that are continually present in small numbers through the sequence, suggesting they are adapted to low-oxygen conditions, but also occur in dense concentrations of articulated specimens on bedding planes. Such assemblages are typical of infaunal species that have reacted to a rise in the redox boundary by emerging to the surface and there suffered mass mortality. The recurrence of densely packed bedding planes suggests repeated mortalities, but the continual colonisation implies a fauna that can survive by its seasonal reproductive strategy. (See Chapter 6 for further discussion of oxygen poor biofacies.)

Box 2.2 The Posidonienschiefer – a controversial Jurassic black shale

The Posidonienschiefer is famous for its 'stagnation Lagerstätten', that is, the exceptional preservation of its marine fossils in the black shale sequence (see section 3.8). Yet in spite of decades of intense study there is still debate about whether conditions in the bottom waters were oxic or anoxic.

The Posidonienschiefer is of early Jurassic age (Toarcian) and is part of an extensive development of oxygen poor sediments that extended over much of northwest Europe and southwards to the margins of the Tethys Ocean. There was a substantial regional extinction in the early Toarcian, contemporaneous with a global transgression and the onset of black mud deposition. A positive carbon excursion in shallow marine carbonates (see Box 2.3) indicates a change in carbon cycling in the oceans, probably resulting from the widespread storage of carbon in bottom sediments.

Most of the fossil fauna in the Posidonienschiefer appears to be pelagic and indisputable benthos is rare. Some of the most compelling evidence for anoxia in the bottom waters comes from the exceptional preservation of the fossils (e.g., Seilacher *et al.*, 1985). For example, some fishes are preserved with their skeletons intact and their scales in place, ichthyosaurs are recorded with embryos inside the body cavity and squids are known with their tentacle hooks in place and their ink sac preserved. Additional evidence for anoxia comes from the sediments which have a high pyrite and organic carbon content (<15%), are commonly delicately laminated and undisturbed by burrowers and have a geochemistry which indicates oxygen deficiency.

An alternative view of the Posidonienschiefer sees the boundary between anoxic and oxic conditions as having been close to the sediment–water interface, with oxic bottom waters prevailing for long periods (Kauffman, 1981). The evidence on which this interpretation is based is as follows.

1. Investigations have been biased towards the exotic fossils, so that the more normal shelly layers, which may contain benthic bivalves, have been neglected. Some of the bivalve assemblages found attached to wood or ammonite shells were living attached to objects on the sea-floor and were not all pseudoplankton, living on a floating object. Benthic attachment is usually recognised by the fossils being attached to the upper surface of an object (Fig. 2.12a). Pseudoplankton is generally attached to the lower surface of floating wood and attaches to both flanks of a living ammonite, commonly with some organised pattern, e.g., clustered towards the aperture (Fig. 2.12b) (Wignall and Simms, 1990). Some of the dense bedding plane assemblages of bivalves, such as *Bositra*, (originally *Posidonia*) are reinterpreted as opportunistic benthic colonisation, not accumulations of pseudoplankton. A subsequent analysis of the life habits of *Bositra* noted that in addition to the accumulations of large specimens there are in some parts of the sequence accumulations of only small specimens. These are interpreted as pelagic larval forms which became precociously sexually mature and so remained pelagic and escaped the bottom conditions inimical to a benthic life (Oschmann, 1994).
2. There are thin limestone layers with bioclastic debris and current orientation of shell material is not uncommon.
3. The spectacular preservation of vertebrate skeletons is to some extent misleading, because the preparation of these skeletons has been made from the underside, which

is well preserved, while the upper surface is commonly poorly so. The implication drawn is that the skeletons lay in anoxic sediments but protruded into oxic or dysoxic bottom waters.

In response to the 'oxic-waters' interpretation of Kauffman, it has been suggested that the sea-floor might have been affected at intervals by storm-generated currents but the prevailing bottom conditions were anoxic. The geochemical and much of the biotic evidence suggest, on balance, that anoxia generally prevailed, but the environment was episodically disturbed by more dysaerobic events. Alternatively the environment could be regarded as poikiloaerobic.

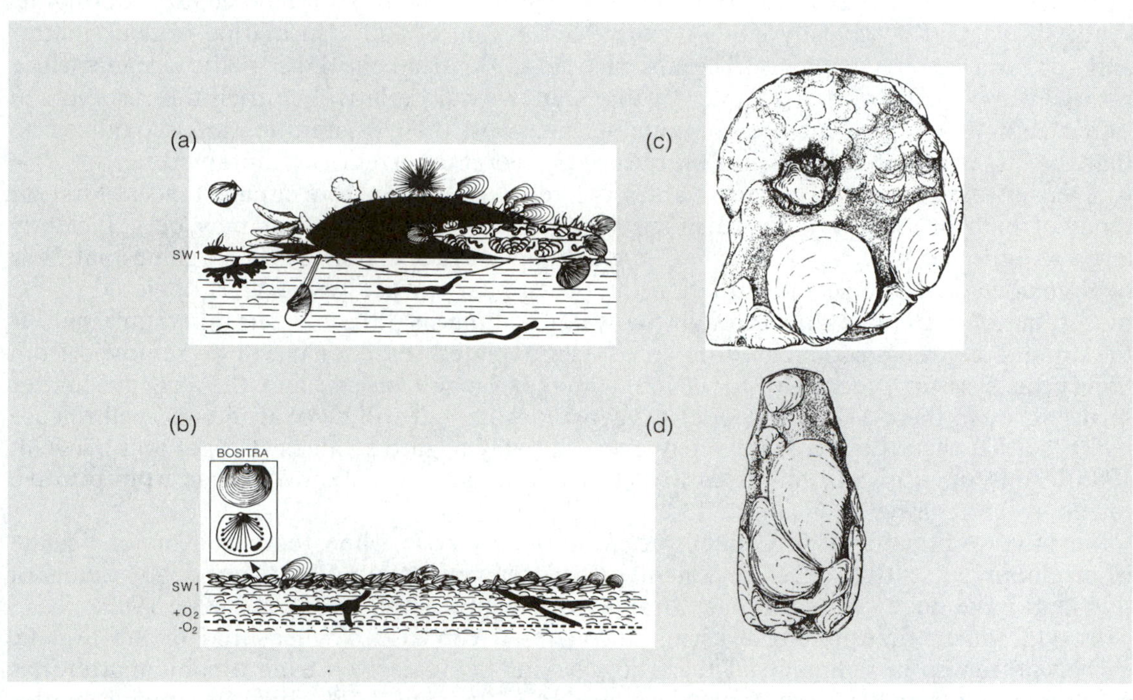

Figure 2.12 (a) '*Inoceramus*' palaeocommunity encrusting the shell surface of a dead ammonite; (b) Reconstruction of opportunistic colonisation of the seafloor by *Bositra*, which is here interpreted as being benthic (modified from Kauffman, 1981); (c) An example of an ammonite which was encrusted on the flanks and particularly towards the aperture by oysters while the animal was living; (d) shows the encrustation extends around the venter onto both flanks of the ammonite. The ammonite is *Buchiceras bilobatum* from the Cretaceous of Peru. (Modified from Seilacher, 1960.)

Box 2.3 Carbon cycling and carbon isotopes

Calcium carbonate in marine shells usually incorporates the carbon isotopes ^{13}C and ^{12}C in a ratio that is in equilibrium with their ratio in dissolved inorganic carbon (DIC) in seawater. The composition of calcite in a Jurassic belemnite is used as a standard (see also Box 2.4) and given the value zero. Positive and negative deviations in the ratio of ^{13}C to ^{12}C from zero recorded in shell carbonate are signified by the notation δ (for example as $\delta^{13}C = 3.0$ ‰ or $\delta\ ^{13}C = -3.0$‰) and reflect changes in the composition of seawater in terms of its dissolved inorganic carbon (DIC) unless the values have been modified by diagenesis or metamorphism.

^{12}C, the lighter isotope, is preferentially incorporated into organic matter during its synthesis, so that it typically has a negative $\delta^{13}C$ value of –20‰ in marine organic matter and –25‰ in terrestrial plants. The abstraction of ^{12}C from the DIC in the ocean-surface reservoir and its incorporation into organic matter would relatively enrich the reservoir in ^{13}C, except that organisms are continually dying and their organic matter is oxidized so that the ^{12}C is released back into the reservoir and equilibrium is maintained.

A positive excursion from zero values is likely to occur in two circumstances. First, at times of high productivity, the abstraction of ^{12}C into organic matter may be sufficiently large to produce positive $\delta^{13}C$ values even though a proportion of the organic matter is being oxidized. In the second situation the ocean near-surface reservoir is depleted of ^{12}C by settling of organic matter through the water column into the oxygen minimum zone. If the organic matter is sedimented on an oxygen-depleted sea-floor it will be removed from the ocean system, potentially for a long time. If on the other hand the organic matter settles through the OMZ into deeper oxygenated waters, it will be oxidized and will release its ^{12}C so that these deeper waters may give relatively negative values (Berger and Vincent, 1986). Consequently, a profile from the surface to depths may show a change from positive to relatively negative $\delta^{13}C$ (Fig. 2.13).

Negative excursions of $\delta^{13}C$ in surface waters may arise when there is a major decline in productivity, so that ^{12}C is returned to the ocean reservoir. Relatively dead oceans of this kind have been referred to as 'Strangelove oceans' (Hsü and McKenzie, 1985).

In addition to determining $\delta^{13}C$ in carbonates it can also be measured in the organic carbon preserved in sediments. The carbon, having being derived from organic matter, has a negative value that is commonly between $\delta^{13}C$–20‰ and –30‰, but the organic matter forms in equilibrium with the near-surface DIC, so changes in ^{13}C values reflect the changes in that reservoir and move in a similar manner to those recorded in carbonates. However, because the $\delta^{13}C$ ratio is attained in the surface waters it will be retained as the organic matter sinks through the water column, so a $\delta^{13}C$ profile like that in carbonates is not developed.

Palaeontological studies have used carbon isotopes to identify periods when there was exceptionally high or low productivity and times when storage of organic matter was particularly enhanced as in the Mesozoic anoxic events. Carbon isotopes have also been used, in a different way to that described above, to interpret the hydrological regime in which a range of palaeosalinities might arise (section 2.7).

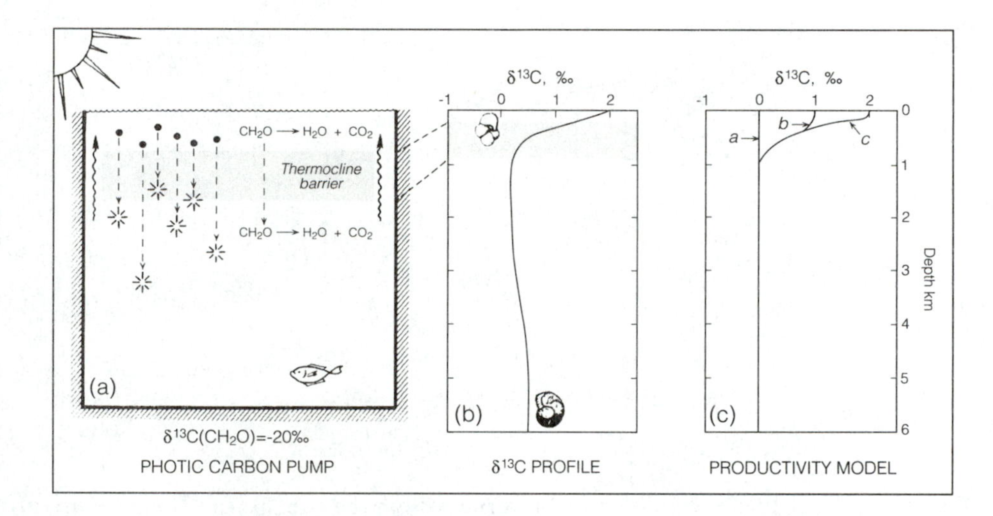

Figure 2.13 Principle of $^{13}C/^{12}C$ fractionation between surface and deep waters. (A) Particulate organic matter sinks below the thermocline, where it is oxidized and releases dissolved carbon that migrates upwards and downwards by diffusion. (B) World-ocean average $\delta^{13}C$ profile. High productivity in the photic zone removes ^{12}C from the surface reservoir. ^{12}C is released as organic matter is oxidized in the thermocline, producing progressively more negative $\delta^{13}C$ values to the base of the thermocline. (C) Conceptual model of changing $\delta^{13}C$ profile in response to increasing productivity ($a<b<c$). (Modified from Berger and Vincent, 1986.)

2.7 TEMPERATURE

Temperature is one of the most pervasive influences on the distribution of organisms. It is probably a limiting factor in establishing the range of many organisms within the ocean and plays a major role in determining the latitudinal partitioning of the biosphere into biogeographic provinces (see Chapter 8).

Most marine invertebrates and fishes are **poikilotherms** that do not regulate their body temperature to a constant level but have the same body temperature as seawater. On the other hand, marine mammals are **homoeotherms** that maintain a constant body temperature. Temperature has a major influence on body metabolism. The rate of metabolism, measured by the rate of oxygen consumption and referred to as Q_{10}, generally increases two or three times in poikilotherms for every 10°C rise in temperature. The majority of marine invertebrates live within a temperature range from –1.7°C to 30°C. A few animals are recorded from hot springs with temperatures up to 55°C and bacteria are known to live in waters of 100°C. Most marine animals apparently live within a relatively narrow range of temperature that is commonly determined by the optimal temperatures for reproduction and early growth. However, many organisms have some ability to adapt to fluctuating temperatures.

Temperature in the oceans in mid- and low latitudes decreases downwards along a sharp temperature gradient to the base of the

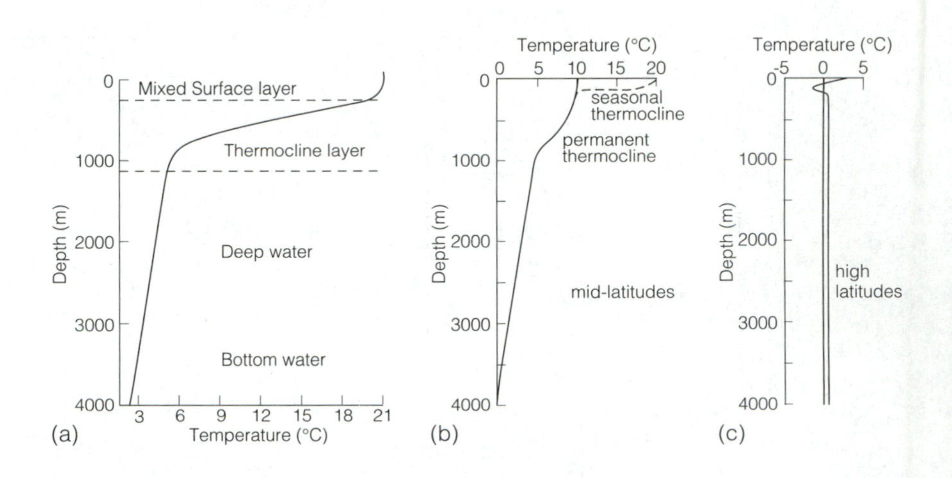

Figure 2.14 Thermal layering within the oceans. (a) Temperature profile with depth in the open tropical ocean (b) Profile in mid-latitudes with a summer season thermocline, (c) Almost uniform temperatures in the high latitude oceans, except for a cold layer at about 50 m and a seasonal summer thermocline above.

permanent thermocline at about 1000 m in low latitudes and then very gradually below that level, reaching temperatures of about 2°C on the ocean floor (Fig. 2.14a). In mid-latitudes, where there are big differences between summer and winter temperatures there is also a seasonal thermocline which develops at a depth of a few tens of metres during the spring and summer months but is lowered to a depth of 200–300 m at a winter maximum, when autumn and winter storms increase the depth of the mixed surface layer (Fig. 2.14b). At high latitudes temperatures are uniformly low below the surface layers except for a particularly cold layer of relatively less saline water at about 50–100 m and a shallow summer thermocline above (Fig. 2.14c). The influence of temperature on animals is likely to be most clearly shown in their depth distribution in the permanent or seasonal thermoclines.

Temperature also changes with latitude (Fig. 2.15) and plays a large part in determining the broad geographic distribution of organisms. Because the circulation of surface ocean waters is organized into discrete currents or gyres, each with its own temperature range, the distribution of planktonic and benthic communities often shows sharp discontinuities at the boundaries of these water masses. This is well illustrated by the bivalve communities along the west coast of North America whose overall latitudinal distribution is probably determined by the regional temperature gradient, but the sharp boundaries between one group of communities and the next is related to the boundaries between water masses impinging on the coast (see Chapter 8).

The stability of the thermal regime is also a factor influencing biological adaptation to temperature (Clarke, 1993). The tropics, polar regions and the deep sea, all have stable thermal regimes, while in temperate regions the thermal regime is unstable, with seasonal surface

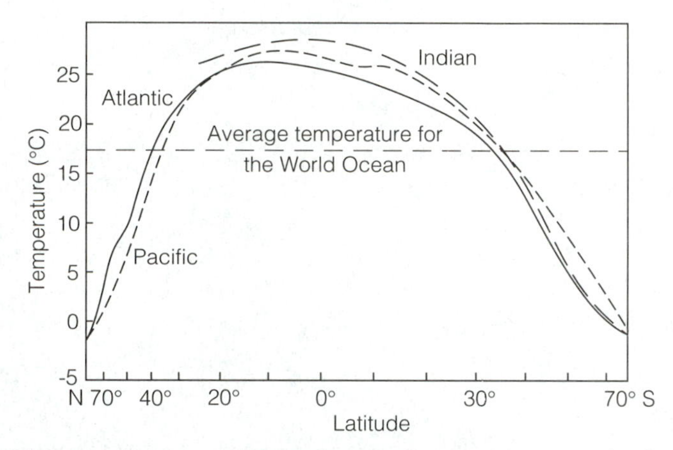

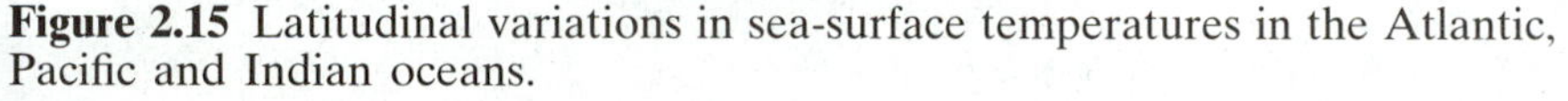

Figure 2.15 Latitudinal variations in sea-surface temperatures in the Atlantic, Pacific and Indian oceans.

temperatures varying by as much as 20°C. High rates of temperature change may have a strong affect on faunas, particularly those adapted to a stable thermal regime. For example, the El Nino Southern Oscillilation of 1982–1983 brought warm waters to the Pacific coast of South America causing the total extinction of the bivalve *Mesodesma donacium*, which formerly constituted over 90% of individuals in the fauna, and its replacement by *Donax peruvianus*, which was previously rare.

2.7.1 Palaeoecology

Although temperature is almost certainly a limiting factor in the distribution of many species it is difficult to demonstrate its effects on most fossil faunas. However, the boundaries of many palaeobiogeographic provinces was probably determined largely by temperature (see Chapter 8). Some of the most fruitful palaeontological studies have been those on Quaternary microfossils from ocean cores. Phleger *et al.* (1953), using cores from the Atlantic, identified low-, mid- and high-latitude assemblages by comparison with Modern faunas (**taxonomic uniformitarianism**) and plotted warmer-than-present and colder-than-present intervals through the sequence. Subsequently the details of the climatic record from ocean cores has been greatly enhanced by correlating the cyclical fluctuations in warm and cold water microfossil assemblages with the oxygen isotope record (see Box 2.4). This has convincingly demonstrated that climate varied on a time-scale of about 20,000, 40,000 and 100,000 yrs, which are the Milankovitch cycles of solar insolation.

The most comprehensive reconstructions of climates for time-slices in the geological past are those of CLIMAP (CLIMAP, 1976, 1981, 1984) which reconstructed global climate at the Last Glacial Maximum (18,000 BP) and ocean temperatures for the Last Interglacial (~122,000 BP). The reconstructions of the oceans were based on the assemblages

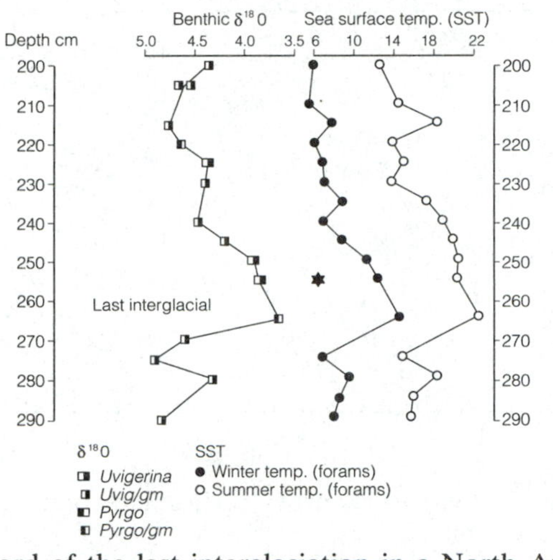

Figure 2.16 Record of the last interglaciation in a North Atlantic core. Sea-surface temperatures (SST) for summer and winter were estimated using assemblages of Foraminifera. $\delta^{18}O$ values measured from benthic foraminifera reflect changes in ice volume and are used to identify the interglacial interval and to correlate between cores. The star marks the estimated last interglacial level. (Modified from CLIMAP, 1984.)

of foraminifera, radiolaria or coccoliths in numerous deep-sea cores with the $\delta^{18}O$ record (Fig. 2.16). Estimates of sea-surface temperatures (SST) were made using a powerful analytical technique, known as the transfer function. This provides a means of translating numerical descriptions of the planktonic biota in the deep-sea cores into past seasonal sea-surface temperatures (CLIMAP, 1976), based on a knowledge of biotic assemblages found in modern deep-sea sediments.

The results for the Last Interglacial showed that sea-surface temperatures were similar to those at the present day. The results for the Last Glacial Maximum showed that while the tropical and sub-tropical sea-surface temperatures were similar to that at the present day, cooler water migrated towards the equator and the subpolar and transitional zones were narrowly constricted, indicating a high latitudinal temperature gradient (Fig. 2.17).

Comparison of fossil assemblages with modern forms and the application of transfer functions becomes progressively less reliable with pre-Pleistocene assemblages as the proportion of extant species declines. Nevertheless, it is still commonly possible to recognise assemblages of microplankton that are confined to distinct biogeographic regions and probably reflect bodies of ocean water with a particular temperature regime and circulation (see Murray, 1995 for a brief review). In a general way the diversity of foraminiferal assemblages decreases towards high latitudes, but diversity is affected by other factors such as nutrient supply, so that the pattern is complex in detail.

Reliable indicators of past temperature levels in macrofossils are not common. As a general rule the abundance of shell-bearing animals

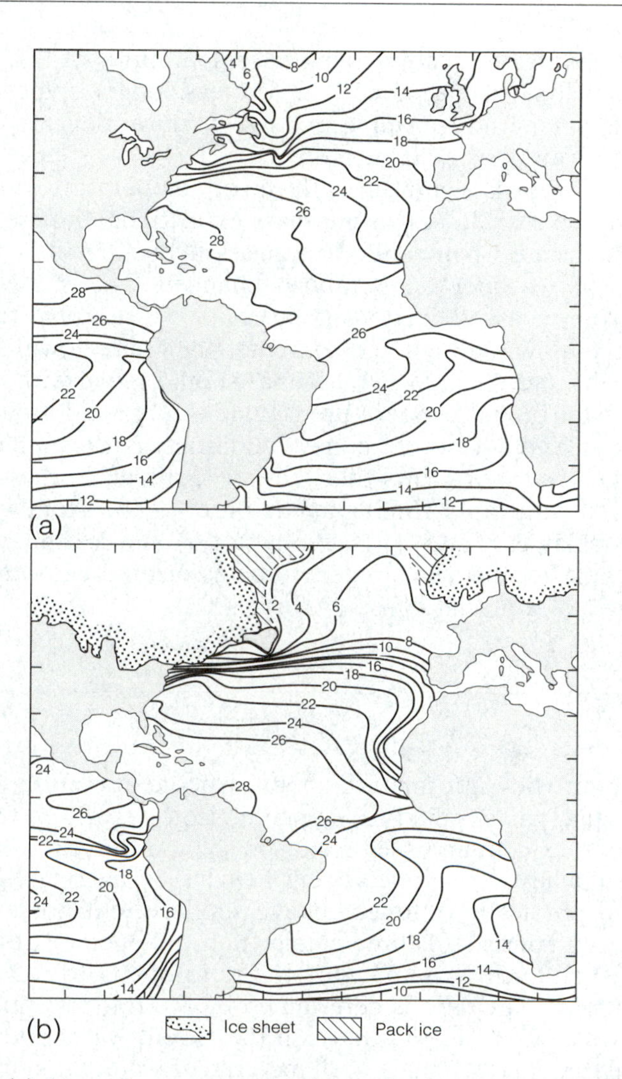

Figure 2.17 (a) Map of sea-surface isotherms in the Modern Ocean; (b) map of isotherms at the Last Glacial Maximum (18, 000BP). Note the closely spaced isotherms and contraction of climatic belts in the northern Atlantic, but the similarity of temperatures in the subtropics. (From CLIMAP, 1981.)

increases towards equatorial regions so that the presence of thick, shelly limestones is a good indicator of low latitudes, though there are some exceptions. The morphology of macrofossils can be used as a temperature indicator in some circumstances. Polar bivalve faunas are thin-shelled, have little or no ornamentation, are generally of small size and growth rings show that they grew slowly; forms that are cemented to the substrate, such as oysters, are generally absent. Many of the features of polar faunas are also found among the cold water abyssal bivalves. With increasing water temperatures shells tend to be more strongly ornamented. This is exemplified by the nature of spines which are absent on bivalves living in temperatures of less than 10°C, are <10 mm long in waters <20°C, whereas longer spines are found above this temperature and the longest, most delicate spines are confined to waters of 25°C or more.

One of the most significant effects of temperature on the fossil record may be its role in causing mass extinctions, though this is still controversial (see Chapter 9). It appears likely that it is the rate at which global seawater temperatures changed that is the significant factor in some mass extinctions (Clarke, 1993). If global temperature changes are slow and not too extreme, their effect will be to shift climatic belts and the associated faunas should move with them. If on the other hand global temperature change is large and rapid, the biota may be unable to adjust to the new conditions. For example, evidence from oxygen isotopes shows that the late Ordovician mass extinction is associated with rapid cooling at the onset of the Gondwana glaciation (Brenchley *et al.,* 1994) (see Chapter 9) and a similar extinction event in the Miocene also appears to be associated with rapid cooling at the onset of a glacial phase (Arbry, 1992).

Box 2.4 Oxygen-stable isotopes

Stable oxygen isotopes potentially provide information about palaeotemperature, the fluctuations in the size of icecaps, and palaeosalinity. Elemental isotopes differ from one another in the number of neutrons they contain which results in slight differences in weight, hence it is common to talk colloquially of isotopes being heavier or lighter, e.g. ^{18}O is heavier than ^{16}O. Chemically and physically isotopes behave similarly but not exactly in the same ways. Atmospheric oxygen consists of three stable (non-radiogenic) isotopes, of which 99.76% is ^{16}O, 0.04% is ^{17}O and 0.20% is ^{18}O. It is the ratio of ^{18}O to ^{16}O which is used in most geological investigations. This ratio is determined by the fractionation factor, that is, the change in the proportions of ^{18}O to ^{16}O between the parent water and another phase, such as a marine carbonate cement, carbonate shell material or water vapour. During precipitation of calcite from seawater the degree of fractionation is determined by the temperature such that the $^{18}O/^{16}O$ ratio increases with decreasing temperature, i.e., a relatively positive value, such as +2, reflects a low temperature (~8°C) (Fig. 2.18). The isotopic ratio is given relative to a natural standard that can be ‘standard mean ocean water’ (SMOW), which is commonly used in oceanographic studies, or it can be the Peedee belemnite standard (PDB), which is most commonly used in geological studies. The PDB standard is based on samples taken from *Belemnitella americana,* from the Peedee Formation, South Carolina, but is now calibrated against further material after the original material became exhausted. The values of $^{18}O/^{16}O$ are generally reported in the literature as parts per thousand (‰) deviation from a standard. A 1‰ shift in $\delta^{18}O$ values reflects approximately a 4–5°C change in marine water temperature.

Material that is suitable for palaeotemperature determinations must have precipitated in equilibrium with seawater. The shells of fossil brachiopods, bivalves and foraminifera are the most commonly used, though early marine cements can also yield good results. There

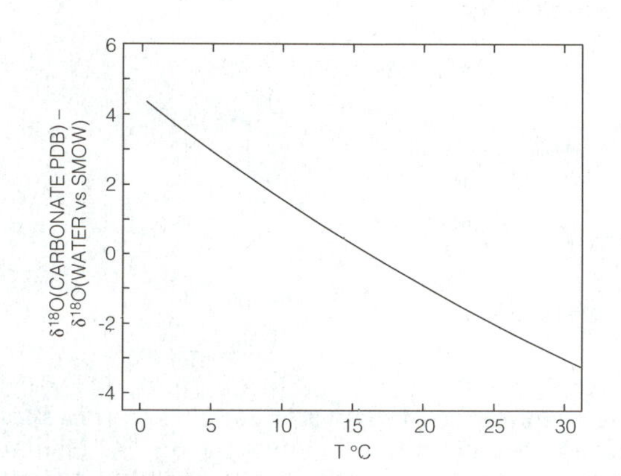

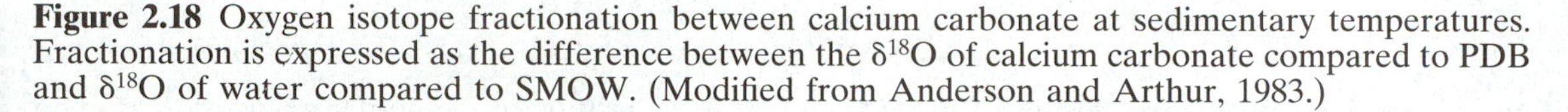
Figure 2.18 Oxygen isotope fractionation between calcium carbonate at sedimentary temperatures. Fractionation is expressed as the difference between the $\delta^{18}O$ of calcium carbonate compared to PDB and $\delta^{18}O$ of water compared to SMOW. (Modified from Anderson and Arthur, 1983.)

are, however, several reasons why temperature estimates can be wrong (Marshall, 1992). Not all organisms precipitate their shells in equilibrium with seawater, instead their physiology produces quite different fractionation effects (vital effects); corals, calcareous algae and echinoderms are known to produce poor results for this reason. Many shells are affected by diagenesis and the $\delta^{18}O$ values recorded are commonly derived from both shell material and cement. Cements generally have lighter values than shell material deposited in equilibrium with seawater so that the 'mixed' value is commonly light, falsely implying higher marine temperatures than really existed. Additionally, estimates of palaeotemperatures are made on the assumption that the seawater from which they were precipitated was the same as normal recent ocean waters. In general this assumption is valid for sub-Recent open marine waters which are well mixed and relatively homogeneous, but marginal marine waters may be brackish or hypersaline and these have abnormal ^{18}O values (see below). There is a greater problem about the changes in isotopic composition of ocean waters with time. It is known that the composition varies between non-glacial and glacial periods, like those of the present day, when it reflects the volume of ice caps. Isotopic composition may also have varied through other causes throughout the Phanerozoic, so relative changes in $\delta^{18}O$ values are commonly more useful as an indication of temperature or water-composition change than as absolute values.

Deviations in the isotopic values of water, away from seawater, arise through fractionation during vapourization and condensation. When vapour forms from seawater the lighter isotope ^{16}O is preferentially fractionated and the vapour which rises into the atmosphere has a light isotopic value. However, the reverse happens during condensation of water vapour as rainfall which is then relatively enriched in ^{18}O and is relatively heavy. As the water vapour in the atmosphere moves inland it progessively loses heavy oxygen by condensation and precipitation so that the rainfall becomes isotopically lighter (Fig. 2.19). As a consequence, inland rain and river water tend to be isotopically light. This effect is reinforced by condensation at low temperatures, e.g. over high mountains and towards low latitudes. When isotopically light vapour is precipitated as snow and stored in polar ice

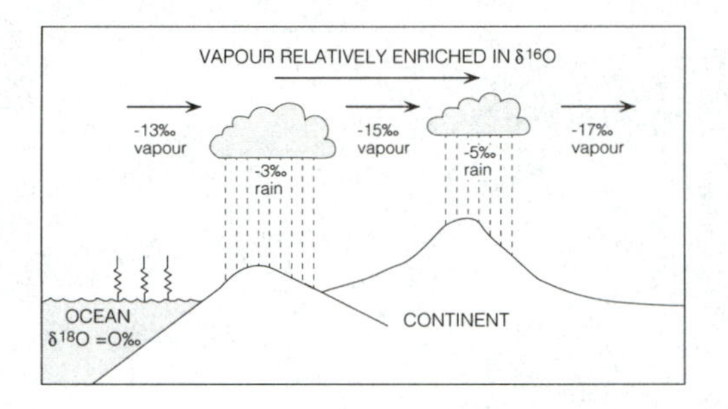

Figure 2.19 Fractionation during evaporation of sea-water enriches water vapour in ^{16}O. Fractionation of water vapour during condensation to rain preferentially removes ^{18}O leading to progressive enrichment of ^{16}O in clouds with distance from their ocean source. (Modified from Siegenthaler, 1979.)

caps it leaves the oceans depleted in ^{16}O. Hence $\delta^{18}O$ values in seawater are higher during glacial periods and lower during interglacials and in non-glacial times. The isotopic shift caused by ice-volume change from the last glaciation to the present interglacial is thought to have been about –1.5‰ and from the present to a non-glacial state would be ~1.0–1.2‰. Note that both the increased volume of ice and the lowered temperatures during a glaciation cause the $\delta^{18}O$ values to shift positively, and it difficult to isolate one effect from the other.

The fractionation of oxygen isotopes into water vapour affects the hydrological cycle in ways that have been widely used to estimate palaeosalinities. In enclosed marine areas, such as hypersaline lagoons, where evaporation exceeds precipitation ^{18}O becomes enriched and shell material gives heavy values. Conversely, in fresh and brackish waters fed from isotopically light rain or river water, $\delta^{18}O$ values tend to be light (see box 2.5).

2.8 SALINITY

Most ocean waters have relatively uniform levels of salinity to which the majority of marine animals are well adapted. Salinity generally only becomes an important limiting factor in marginal marine environments that are brackish or hypersaline.

Animals normally have a constant concentration of dissolved ions within their cell fluids. Water will enter a cell across its semi-permeable membrane by **osmosis** if the internal solution is more concentrated than that outside and will exit a cell if the difference in concentration is reversed. Most marine animals are isotonic (isosmotic) with seawater and live within a narrow range of salinities (they are **stenohaline**). Some other organisms are isotonic with seawater, but can tolerate a broader range of salinities (they are **euryhaline**) while others, particularly Crustacea, can keep their internal solutions constant in spite of fluctuating levels of salinity by a variety of methods of osmoregulation.

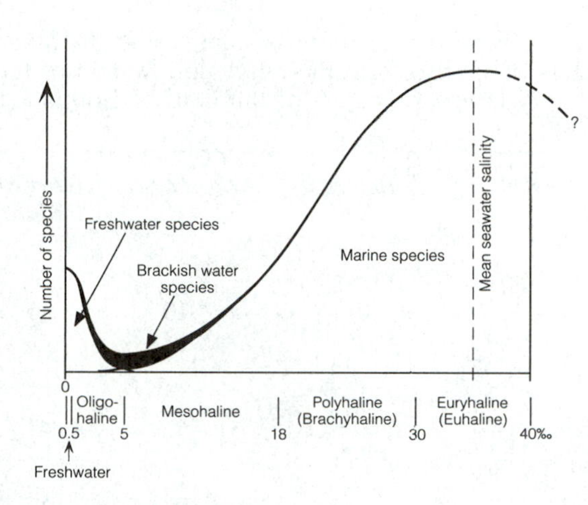

Figure 2.20 Classification of salinity levels and generalized relationship of species diversity with respect to salinity. Note that brackish and freshwater faunas are of low taxonomic diversity and the diversity of marine species increases as salinity increases and peaks at average seawater salinity levels. (From Hudson, 1990.)

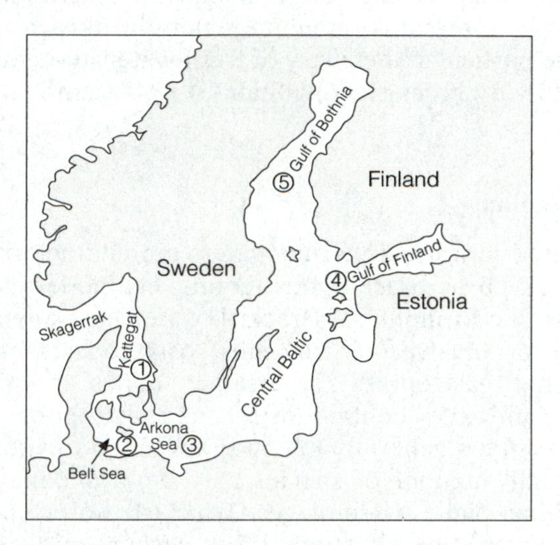

Figure 2.21 Map of the Baltic Sea. The Skagerrak opens into the North Sea with normal marine waters. (Modified from Robertson, 1989.)

Most oceans and seas have seawater with a salinity within the range 30–40‰ (parts per thousand of dissolved salts in seawater). Faunal diversity is generally highest within this range and decreases with higher or lower salinities (Fig. 2.20). The decrease in diversity as waters become more brackish is well exemplified by the faunas of the Baltic Sea (Fig. 2.21, Table 2.2). Faunas that can tolerate brackish waters (in the range 30–0.5‰) appear to be composed mainly of euryhaline marine species together with some osmoregulators that are specially adapted to low salinities. The latter group of organisms are also

Table 2.2 Table showing the decrease in salinity and the associated decline in numbers of species from the euryhaline waters of the Kattegat to the oligohaline (very brackish) waters of the Gulf of Bothnia (from Segerstrale, 1957)

	Kattegat	*Belt Sea*	*Arkona Sea*	*Gulf of Finland*	*Gulf of Bothnia*
Salinity (%)					
bottom	30–34	14–30	13–17	8–9	4–7
surface	20–30	10–16	7–10	c.5	3–6
Marine groups					
Hydroids	c.47	36	21	1	–
Polychaetes	193	147	15	3	–
Bivalves	92	34	24	4	–
Amphipods	132	c.52	17	5	–
Decapods	64	c.25	5	2	–
Fishes	75	55	30	22	–

commonly well adapted to cope with the fluctuations in salinity which are particularly common in marginal marine environments, such as estuaries. Fresh water environments generally have a distinct fauna adapted to the particular chemistry of fresh water, though a few species may penetrate very brackish (oligohaline) waters and so slightly raise the diversity.

2.8.1 Palaeoecology

Facies deposited in normal marine waters are characterized by stenohaline groups such as brachiopods, corals, echinoderms, ammonoids and larger benthic forminifera. Brackish water facies generally have a low diversity of bivalves, crustaceans, ostracodes, smaller benthic foraminifera and gastropods. Hypersaline faunas are of low to very low diversity and may contain ostracods, gastropods, bivalves etc. Although diversity is generally low in brackish and hypersaline facies, fossils of a small number of species may be abundant and occur as crowded bedding-plane assemblages. Brackish water species tend to be small and are commonly stunted, but certain animals, such as the bivalves *Mya* and *Scrobicularia* and the gastropod *Hydrobia* are well adapted to brackish conditions and may actually be larger there.

Detailed studies of salinity changes within Mesozoic and Cenozoic rocks have generally interpreted the salinity levels of biofacies by comparison with modern faunas. A study of the Great Estuarine Group (M. Jurassic) in Scotland (Hudson, 1963) identified broadly brackish environments using sedimentary evidence and faunal evidence, such as:

1. the low diversity of the fauna;
2. absence of stenohaline forms;
3. presence of euryhaline forms.

Box 2.5 Paleosalinities in Jurassic rocks of the Great Estuarine Group

In a series of papers spanning more than 30 years John Hudson and his co-workers have made detailed studies of palaeosalinities in the Great Estuarine Group and have shown that it is valid to reconstruct salinity conditions in rocks as old as the Jurassic by using the principles of taxonomic uniformitarianism and identifying the modern descendants of the fossil biota, whose salinity ranges are known. The Bathonian rocks of the Great Estuarine Group include mudrocks, with two large intercalated sand bodies that are interpreted as fluvial deltas. The initial work (Hudson, 1963) was mainly concerned with showing how the bivalves and gastropods, when compared with modern descendants, appeared to reflect changes in salinity through the sequence. These initial conclusions have subsequently been greatly strengthened by a detailed study of the Kildonnan Member which demonstrates that other parts of the biota, such as ostracodes, concostracans, dinoflagellate cysts, acritarchs and the xanthophyte alga, *Botryococcus*, show a pattern consistent with a salinity control of the assemblages (Hudson *et al.,* 1995). The microplankton (the dinoflagellate cysts, acritarchs and *Botryococcus*) are particularly informative because they give information about the water mass in which they lived, independent of substrate which is another variable that might have had a major influence on the benthic biota. *Botryococcus* shows an inverse correlation with marine/brackish dinoflagellate cysts and is a particularly good indicator of low salinity (Fig. 2.22).

Many of the bivalve shells in the Great Estuarine Group retain their original shell structure and have suffered little diagenesis, which makes them ideally suited for isotopic analysis. Determinations of both $\delta^{13}C$ and $\delta^{18}O$ support the salinity variations shown by the faunas and floras. Low salinity is generally associated with negative $\delta^{18}O$ values when the water body was fed by riverine water (see Box 2.4). $\delta^{13}C$ values also tend to be low because river water generally derives its dissolved carbon from organic matter in soils. However, they may be higher in standing water where dissolved bicarbonate tends to equilibrate with the atmosphere. Samples taken at close intervals through parts of the sequence showed that isotopic values became relatively more positive or negative in phase with biotic changes and suggest that some of the salinity changes are reflecting hydrological changes to a brackish lagoon. Using estimates of average sedimentation rates it is suggested that overall the lagoon was relatively stable but that there were some short-term changes in lagoonal salinity that occurred on a decadal scale or less.

Analysis of a brackish lagoon higher in the sequence (the Duntulm Formation – Andrews and Walton, 1990) reveals a wider range of salinities. Oyster banks composed of *Praeexogyra hebridica* reflect relatively marine conditions at some levels, whereas varieties of algal limestones reflect supralittoral marsh to littoral stromatolitic environments. High in the sequence the presence of *Unio, Neomiodon* and *Botryococcus* associated with terrestrial pollen and spores indicate a fresh-water environment.

Changing salinity levels through the sequence were estimated by making comparisons with similar taxa in the Recent (see Box 2.5). Using a similar approach, detailed salinity profiles were constructed for Upper Eocene marginal marine and fresh water facies of the Hampshire Basin, England, using ostracode assemblages. More contro-

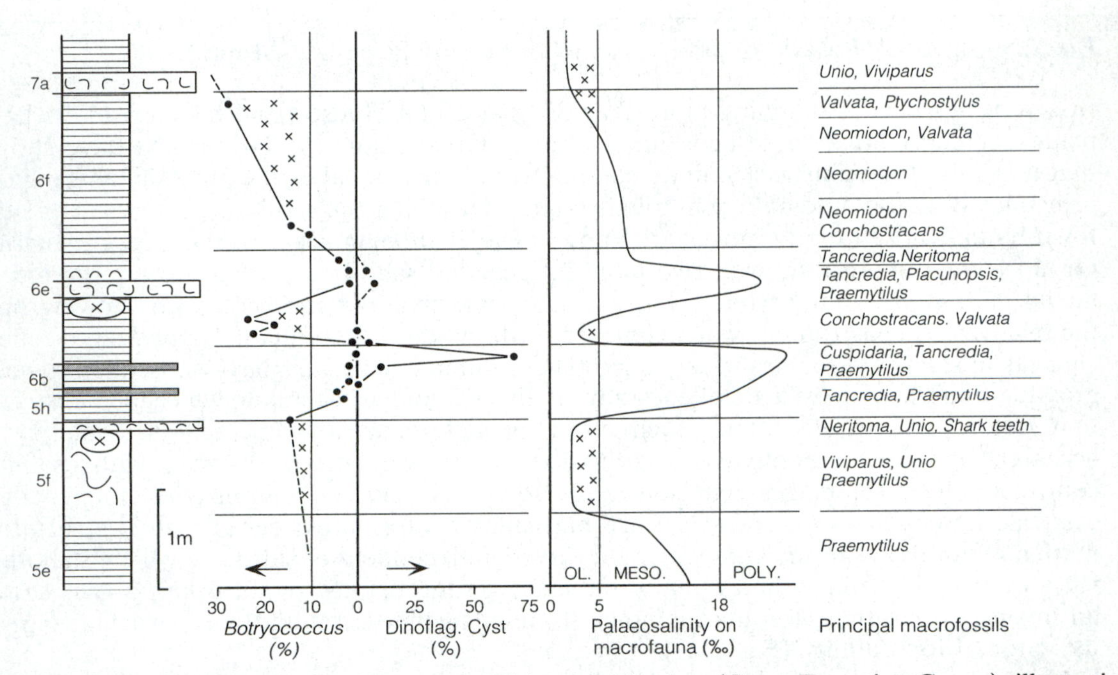

Figure 2.22 Detailed log of part of the Jurassic Kildonnan Member (Great Estuarine Group), illustrating the variations in macrofauna and palynoflora and inferred palaeosalinity. Crosses illustrate oligohaline biotas; stipple indicates polyhaline biotas. Note the co-occurrence of dinoflagellate cysts with polyhaline bivalves and *Botryococcus* with an oligohaline-low mesohaline macrofauna. (Modified from Hudson *et al.*, 1995.)

versially it has been suggested that extensive areas of epeiric seas may have had lowered salinities at times in the past. A study of end-Triassic beds in the Alpine region and northwestern Europe during the Rhaetian transgression (Hallam and El Shaarawy, 1982) showed that the diversity of bivalves, foraminifera and ostracodes decreased and stenohaline forms, such as ammonites, disappeared northwestwards away from the fully marine Tethyan area. This was interpreted as being related to a regional salinity gradient in this shallow epeiric sea.

2.9 SUBSTRATE COMPOSITION

The grain size of sediment is one of the most easily assessed physical parameters that palaeontologists can relate to faunal distribution. However, the role of substrate as a limiting factor is a complex one and grain size is one amongst many contributing aspects of substrate. Although a new population of marine animals may become established by the random settlement of larvae on the sea-floor, more commonly larvae select the site where they will settle. The choice of substrate is made according to a wide variety of criteria. Some do appear to select according to sediment grain size, but current movement and turbulence are important to many larvae and others select according to such factors as the organic covering of the grains, the roughness of a surface, light levels, and the chemistry of the sediment. For example, the wood-boring larvae of the 'ship worm' (a highly specialized bivalve, *Teredo*), are responsive to the chemistry of lignin and so actively select a wood substrate (they are chemotaxic), while the larvae of the Recent polychaete worm *Ophelia bicornis* select clean sand guided by the organic coating to the particles.

Studies of modern communities have generally shown that there is a broad correlation between community distribution and the grain size of the substrate, though there are a few studies in which this relationship is seen to be weak. In nearshore environments the species diversity of animals is generally highest in muddy sands, moderate in sandy muds, low in pure sands and may be near zero in soft muds. In many instances it is the firmness of the sediment that is an important factor in influencing faunal distribution. Sediments vary from soupy muds with a very high water content, to soft loose sediments, to firmgrounds and hardgrounds, and both the shelled fauna and soft bodied infauna vary accordingly (see Chapter 6). Bioturbation of sediment by deposit-feeders also affects faunal distribution; intensive reworking of the surface layers produces a soft, faecal-rich sediment that is prone to resuspension and the resultant turbidity appears to inhibit colonization of the area by suspension-feeders. In general, suspension-feeders favour the less muddy sediments, while deposit-feeders are more common in the more organic-rich muddy sediments. A more specific relationship between organisms and their substrate is seen among those that require a hard surface on which to attach themselves. For example, the majority of large seaweeds live on rocky or stony shores which provide an anchor-site for the algal holdfast. Among epifaunal animals, barnacles, most large colonial corals and some bivalves require a hard substrate for attachment, while some sponges, bryozoans and bivalves live infaunally within the borings they have made.

The relationship between animals and the substrate is not simply one way, because a fauna can modify the substrate on or in which it lives. The effects of bioturbating organisms has already been described, but the shell-bearing animals in a community can also modify the substrate by generating dead shells and shell fragments that can play an important role in providing varied settlement sites in an otherwise fine-grained sediment (this process is referred to as **taphonomic feed-back**). Taphonomic feed-back is a sporadic feature of siliciclastic environments, but is prevalent in many carbonate environments where the substrate is largely of biogenic origin. For example, it is common for disaggregated crinoids to form loose coarse sand to granule sediments, shells of brachiopods or bivalves form larger clasts for attachment, while corals may form even more stable attachment sites (see Brett, 1991 for an excellent description of biogenic modification of sediments in the Silurian).

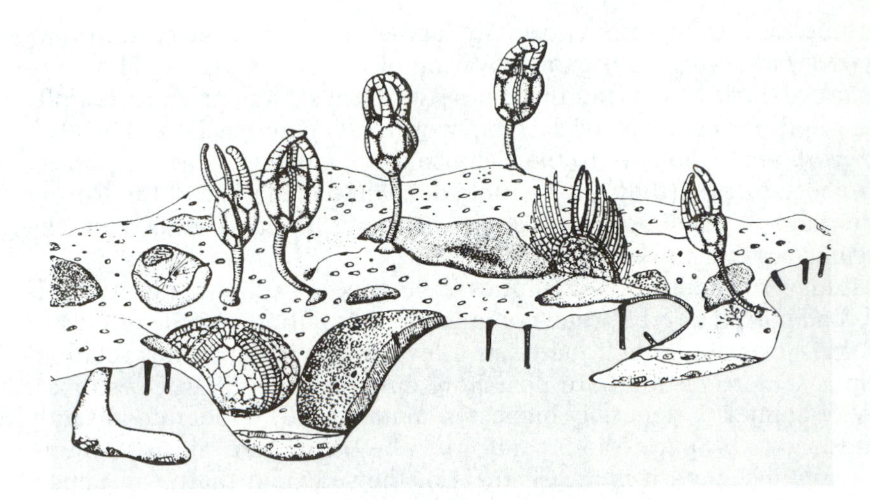

Figure 2.23 Hardground surface bored and colonised by epifaunal animals some of which are moulded to the surface. (From Brett and Liddell, 1978.)

2.9.1 Palaeoecology

Important information about the nature of the substrate is provided by the types of trace fossils present and their preservation. Trace fossils record the activities of animals which may in turn reflect the nature of the sediment in which they lived. For example, animals living in loose silt or fine sand commonly line their burrows, as with the pellet-lined walls of *Ophiomorpha*, while animals living in firmgrounds may leave scratch marks as clear, sharp traces. Preservation of traces becomes more distinct as the sediment becomes firmer; thus burrows in soupy sediment become compressed and have indistinct boundaries, while in contrast burrows in firmgrounds are uncompressed and have sharp outlines.

The morphology of both epifaunal and infaunal animals can reflect the nature of the substrate where they lived (Chapter 6). Further, a careful study of the attachment area of sessile organisms can show the nature of the substrate to which the animal was attached. The identification of rocky shorelines and **hardgrounds** at which carbonate sediment had become at least partially cemented on the sea-floor, depends largely on fossil evidence of a hard surface (Fig. 2.23). For example, oysters have an attachment area that is moulded to the rock surface so that the shape of the lower, attached, shell is considerably modified according to the configuration of the substrate, and the upper shell too may reflect the shape, though less precisely. Attached crinoids too commonly provide valuable evidence of their former substrate; those that were attached to firm or hard surfaces have a thickened base to the stem and an attachment disk shaped to the surface on which the crinoid lived. Those crinoids that lived on soft sediment have branching rooting systems. In addition, the presence of rock borings penetrating both bioclasts and cement, made particularly by bivalves and sponges, is important evidence of a hard substrate.

In some fossils the shape of the attachment area may reflect a host organism to which the animal had been attached. Examples of this are known from oysters that had encrusted ammonites, belemnites, gastropods, echinoids and corals. Studies of ammonites have shown examples of encrustation on both flanks of the ammonite, suggesting the ammonite was still living and of encrustation on one side only, indicating that the ammonite was dead and lying on its side (Box 2.2). Particularly interesting are the examples of attachment areas that reflect the shape of soft-tissued organisms which had little chance of survival as fossils. Oysters have been shown to record the impressions of the roots of Eocene mangrove roots, and bryozoans the impressions of Cretaceous sea-grass.

A more general correlation between fossil faunas and substrate type has been found in several studies. Benthic bivalve communities in the Corallian (Jurassic) have been shown to vary in overall diversity and in the relative proportions of epifauna and infauna according to the associated substrate (Fürsich, 1976), and, in the Ordovician, benthic communities largely composed of brachiopods have been shown to be depth-related, but also influenced significantly by sediment type (Pickerill and Brenchley, 1979).

There are, however, pitfalls in inferring an ecological association between fossils and the sediment in which they are found. In many instances fossil assemblages are formed by current winnowing from the underlying sediment (see Chapter 3), so that it is not uncommon for an animal that lived on a mud substrate to be found in a storm-sandstone bed. It is important that evidence of sediment reworking is taken into account in reconstructing animal/sediment relationships. This problem is of course absent in dealing with trace fossils (see Chapter 5).

2.10 SUBSTRATE MOBILITY, SEDIMENTATION RATES AND TURBIDITY

Sediment mobility and the turbidity of the water are important limiting factors in certain nearshore environments. Epifaunal animals in particular are seriously affected by mobile substrates, but most animals have difficulties in surviving on very mobile sand. Some infaunal animals have adapted to sediment instability by rapid and/or deep burrowing. Animals may burrow rapidly downwards if they are in danger of being eroded out of the sediment, or may burrow upwards if they are in danger of being deeply buried by rapid sedimentation. Cockles (the bivalve *Cardium*) are shallow burrowers and may be excavated in large numbers from beach sediments during storms, but can re-establish themselves rapidly afterwards. The razor shells (*Ensis* and *Solen*) are both deep-burrowing and burrow rapidly if threatened with exposure.

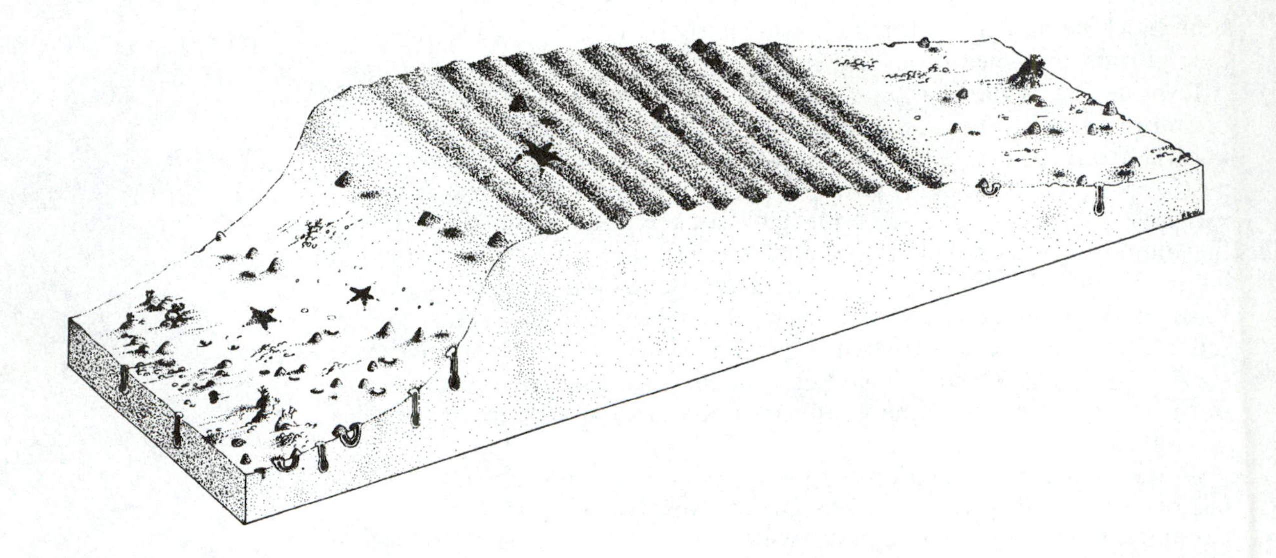

Figure 2.24 Block diagram of a stable sand wave in the Kiel Bight showing the colonization of the trough between the sand waves by an infauna of polychaetes and bivalves. (From Wilson, 1986.)

Few animals can survive on very mobile substrates such as rapidly migrating sand waves and only a relatively few epifaunal animals can survive the high-energy conditions on a beach or upper shoreface. Observations on the tidal shelf around the British Isles show that the tidally scoured shelf and fields of rapidly migrating sand waves have sparse to very sparse faunas. Where sand waves become stabilized for a short period they become colonized, particularly in the troughs (Fig. 2.24); in lower-energy areas of rippled sand where tidal currents are generally less than 50 cm/sec, a rich fauna is commonly maintained. Unusually high rates of sedimentation during storms or floods can cause mass mortalities. For example, the deposition during a flood of ~10 cm silt and mud in the low-energy areas of a lagoon on the coast of California caused extensive but selective mortality; some bivalves were killed and buried at the base of the deposit, but others migrated upwards and survived. Mortality was greatest among suspension-feeding species, individuals of large size and populations that were densely packed (Peterson, 1985).

Muds that lack cohesion and are easily resuspended to create high turbidity have a serious effect on suspension-feeders and some deposit-feeders. Bioturbation tends to promote resuspension of mud, so favouring deposit-feeders at the expense of suspension feeders, whose gills or other feeding organs tend to become clogged. Turbid environments are common in estuaries and deltas and may temporarily occur more offshore through the activities of storms.

2.10.1 Palaeoecology

Ancient nearshore mobile-sand environments include those formed in strongly tidal seas, a variety of channel settings (deltas, estuaries) and those of the upper shoreface and beach. Generally the sandstones in these situations have a low faunal diversity and low abundance of fossils, though under certain conditions one or two species may occur in abundance. The general absence of fossils in mobile-sand facies is particularly pronounced in Palaeozoic rocks because the predominant shallow marine fauna consisted of epifaunal brachiopods that were ill-suited to such substrates. Very sparsely fossiliferous Palaeozoic sandstone facies are exemplified by the Grès Armoricain, which is a thick Lower Ordovician tidal and storm-influenced sand body spreading thousands of square kilometres across Gondwanaland (Spain, France, N. Africa, Jordan, Arabia) and containing little else but trace fossils, such as *Skolithos* and *Monocraterion*. It was not until the Mesozoic and even more the Cenozoic that the active deep-burrowing mode of life developed more widely amongst bivalves and fitted them for a life in nearshore sands.

In shelf environments the effects of high sediment mobility are generally related to short-lived storm events. Storm-event sandstones rarely contain an indigenous fauna but may have shell concentrations at the base of a bed, winnowed from the underlying sediment. Rapid rates of sedimentation associated with storms, floods and turbidity currents may in some instances be recognized by escape burrows adjacent to which the sedimentary laminae are disturbed by the movement of the burrowing organism as it attempts to escape being buried. Some event beds have been shown to have a layer of mud several centimetres thick at the top, which is thought to have been deposited from suspension at a late stage in a storm and must have had deleterious effects on any benthos that had survived the turbulent phase of the storm. Exceptional preservation of whole crinoids in their vertical life-orientation is thought to reflect rapid burial by a mud blanket.

2.11 DEPTH

The effects of depth of water on an organism are experienced through the increase in hydrostatic pressure. This can have both physical and chemical consequences. Water does not undergo significant distortion with increasing pressure, nor does the the soft tissue or shells of organisms. Gas on the other hand does suffer marked changes with the increasing pressure felt with increasing water depth, so the physical consequences are mostly experienced by those animals that have enclosed pockets of gas in their bodies. For example, at 10 m depth there is an increase of 1 atmosphere pressure, and the swim bladder in a fish may be reduced to a half its volume. This generally has severe limiting effects on the depths to which animals with swim bladders, lungs, and gas-filled shells can go. For example, *Nautilus pompilius* lives abundantly at depths of about 600 m around Fiji, but was found to implode at depths between 730–900 m when lowered in a cage (Westermann and Ward, 1980). It appears that in nautiloids the curvature and thickness of the septa are designed to optimize strength and buoyancy. In spite of the effects of hydrostatic pressure, animals such as whales, some sharks and diving birds of the auk family dive to a depth of a few hundred metres, and sperm whales to a depth of over 1000 m, having adaptations that allow them to dive deeper and stay down longer than would be expected.

The second major aspect of depth as a limiting factor is its effect on the solubility of calcium carbonate. Surface ocean water is supersaturated with respect to $CaCO_3$, but becomes undersaturated with increasing depth. The depth at which significant dissolution of calcium carbonate begins is the **lysocline**, and the depth at which carbonate dissolution exceeds supply and all carbonate is dissolved is the **carbonate compensation depth** (**CCD**). The depth of the CCD varies considerably with carbonate production but is generally about 3–4 km, but it is shallower for aragonite (aragonite compensation depth, ACD) which is more soluble so that the ACD is rarely more than 1–2 km. Benthic animals find it more difficult to construct large thick shells as the lysocline is approached and neither calcareous benthic nor pelagic faunas are preserved below the CCD. This knowledge has been effectively used in the reconstruction of the depths of Tethyan basins in the Alpine region, where the shallower basins preserve calcareous oozes whereas the deeper basins have only terrigenous mudstones.

Apart from the effects of hydrostatic pressure and the solubility of carbonate, depth alone has relatively little effect on biotic distribution. However, many of the other limiting factors, such as temperature, substrate and turbulence, discussed earlier in this chapter are depth-related and are commonly used to infer relative depth in environmental reconstructions (see Chapter 6).

2.12 THE EFFECTS OF LIMITING FACTORS IN DIFFERENT ENVIRONMENTS

Several physical factors have a strong limiting effect on the distribution of animals in nearshore environments and others have an influence on the faunas in deep marine situations. In between these two ends of the depth spectrum the faunas of open marine shelves appear to be more subtly controlled (Fig. 2.25).

Light and nutrients are essential to primary productivity in the surface layers of marine waters. Nutrients are generally most abundant in shallow marine regions and areas of upwelling, but food is generally amply available in most neritic ecosystems and only becomes a seriously limiting factor on biomass production in deep marine

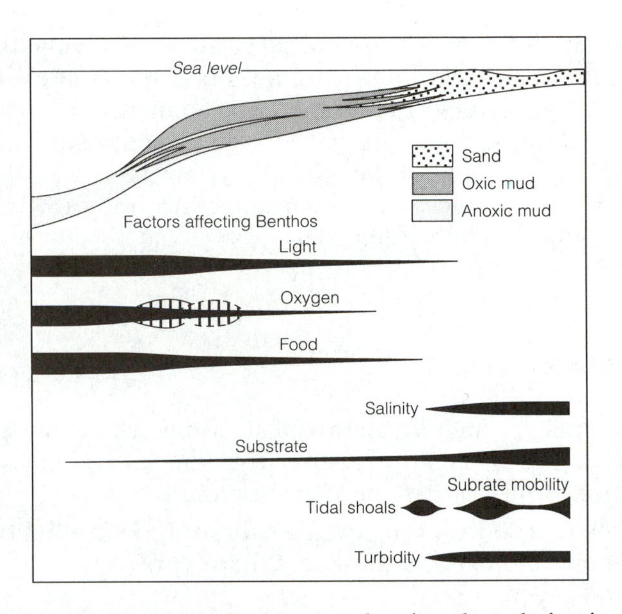

Figure 2.25 A shoreline to basin transect showing the relative importance of different physical factors in determining the distribution of faunas. The dashed oxygen line indicates episodes of shelf anoxia. (Modified from Brenchley and Pickerill, 1993.)

waters. However, temporal variations in the amount of nutrients reaching the surface have an important role in determining the nature of ecosystems, i.e., whether they are eutrophic systems or oligotrophic systems.

Many nearshore environments are subject to large diurnal and seasonal changes in temperature and the effects of salinity can be important in lagoonal, estuarine and deltaic situations; substrate type is commonly very varied, ranging from the hard substrates of the rocky shore, to the sands of the shoreface and the muds, silts and sands in estuaries and deltas. Nearshore substrates are commonly mobile and the overlying waters may be turbid.

Towards the other end of the depth spectrum, i.e., in deep water, it is light, food and in some places oxygen that tend to be limiting factors. Primary productivity is largely confined to the photic zone, so primary producers are excluded from deep water environments. Much of the organic material generated in the photic zone is consumed in the water column so that the food available at depth is limited. Oxygen levels are generally low in the oxygen minimum zone which is most likely to impinge on the sea-floor on the continental slope. Very low levels of oxygen may also be associated with stagnant parts of deep oceans and in some circumstances the central parts of cratonic basins (see Chapter 6). In the mid-part of a shoreline to basin transect, marine conditions are at their most constant and the environment is not strongly stressed. Salinity variations are small, turbidity and turbulence are low except during major storms, substrate is relatively uniform and temperature variations are not extreme. Food levels are sufficient for an abundant and diverse fauna and the sediment–water interface is generally well oxygenated except during exceptional transgressive episodes. In such a situation the physical controls on distribution may be quite subtle differences of temperature, substrate, or depth, which possibly influence larval settlement.

2.13 SUMMARY POINTS

- The biosphere has a hierarchical structure with each level contributing to a larger structure, e.g., the individual is part of a population, which is part of a species and so on.
- The niche of a species is its ecological range, controlled by multiple physical and biological parameters that define a specific volume in ecological space.
- Some factors play a particularly important role in determining the limits to a niche. These are referred to as limiting factors.
- Light is a limiting factor for all photosynthetic organisms, confining most to the photic zone.
- Nutrients are limiting in that they affect the structure of an ecosystem and determine whether it is oligotrophic or eutrophic.
- Oxygen becomes limiting where the loss of oxygen through oxidation of organic material exceeds the oxygen dissolved in the seawater. This may occur in the oxygen minimum zone or in basins with a stratified water column. Oxygen depletion occurs in the upper layers of sediment and this is commonly reflected in the trace fossil assemblages.
- Temperature is a pervasive control on faunal distribution, but it cannot generally be demonstrated in the fossil record, except in the broad distribution of biogeographic provinces according to latitude.
- Salinity can be an important limiting factor in nearshore brackish and hypersaline environments.
- Substrate clearly has an influence on faunal distribution but the control may be grain size, firmness, chemistry etc.
- Sediment mobility is an important limiting factor in high energy, sandy nearshore environments. Turbidity may be important in more muddy environments.
- Salinity, substrate, substrate mobility and turbidity are important factors in nearshore environments. Temperature is particularly important in shallow waters generally. Light, oxygen and food are commonly limiting in deeper waters.

2.14 FURTHER READING

Barnes, R.S.K. and Mann, K.H., 1991. *Fundamentals of Aquatic Ecology*. Blackwell Scientific Publications, Oxford. 270 pp.

Biology: Function, Biodiversity, Ecology. 1995. Oxford University Press, 420 pp.

Bosence, D.W.J. and Allison, P.A. (eds) 1995. *Marine Palaeoenvironmental Analysis from Fossils*, Geological Society Special Publication, 83. 272 pp.

A broad modern review of many aspects covered in this chapter. Levinton, J.S., 1982. *Marine Ecology*, Prentice Hall. 52 pp.

There are many good books on marine ecology; these have been selected because they cover many aspects which are relevant to palaeoecologists.

Environments and Physiology of Marine Organisms. *Transactions of the Royal Society of Edinburgh, Earth Sciences*, 1989, 80, parts 3 and 4. A collection of papers aimed at integrating the experience of biologists and palaeontologists.

2.15 REFERENCES

Andrews, J.E. and Walton, W. 1990. Depositional environments within Middle Jurassic oyster-dominated lagoons: an integrated litho-, bio-, and palynofacies study of the Duntulm Formation (Great Estuarine Group, Inner Hebrides). *Transactions of the Royal Society of Edinburgh: Earth Sciences* 81, 1–22.

Arbry, M.-P. 1992. Late Paleogene calcareous nannoplankton evolution: A tale of climatic deterioration. *In:* Prothero, D.R. and Berggren, W.A. (eds) *Eocene-Oligocene Climatic and Biotic Evolution.* Princeton University Press, Princeton, 272–309.

Berger, W.H. and Vincent, E. 1986. Deep-sea carbonates: Reading the carbon-isotope signal. *Geologische Rundschau*, 75, 249–269.

Brasier, M.D. 1995a. Fossil indicators of nutrient levels. 1: Eutrophication and climate change. *In:* Bosence, D.W.J. and Allison, P.A. (eds) *Marine Palaeoenvironmental Analysis from Fossils.* Geological Society Special Publication. 83, The Geological Society, Bath, 113–132.

Brasier, M.D. 1995b. Fossil indicators of nutrient levels. 2: Evolution and extinction in relation to oligotrophy. *In:* Bosence, D.W.J. and Allison, P.A. (eds) *Marine Palaeoenvironmental Analysis from Fossils.* Geological Society Special Publication, 83, Geological Society, Bath, 133–150.

Brenchley, P.J. and Pickerill, R.K. 1993. Animal-sediment relationships in the Ordovician and Silurian of the Welsh Basin. *Proceedings of the Geologists' Association* 104, 81–93.

Brenchley, P.J., Marshall, J.D., Carden, G.A.F., Robertson, D.B.R., Long, D.G.F., Meidla, T., Hints, L. and Anderson, T. 1994.

Bathymetric and isotopic evidence for a short–lived Late Ordovician glaciation in a greenhouse period. *Geology* 22, 295–298.

Brett, C.E. 1991. Organism-sediment relations in Silurian marine environments. *Special Papers in Palaeontology* 44, 301–344.

Brett, C.E., Boucot, A.J. and Jones, B. 1993. Absolute depth of Silurian benthic assemblages. *Lethaia* 26, 23–40.

Byers, C.W. 1977. Biofacies pattern in euxinic basins: a general model. *In:* Cook, H.E. and Enos, P. (eds) *Deep-water carbonates*. SEPM Special Publication, 25, Society of Economic Paleontologists and Mineralogists, Tulsa, 121–138.

Clarke, A. 1993. Temperature and extinction in the sea: a physiologists view. *Paleobiology* 19, 499–518.

CLIMAP, 1976. The surface of the Ice-Age Earth. *Science* 191, 1131–1137.

CLIMAP, 1981. Seasonal reconstructions of the Earth's surface at the last glacial maximim. *Geological Society of America Map and Chart Series* MC-36, 1–18.

CLIMAP, 1984. The last interglacial. *Quaternary Research* 21, 123–224.

Fürsich, F.T. 1976. Fauna-substrate relationships in the Corallian of England and Normandy. *Lethaia* 9, 343–356.

Hallam, A. and El Shaarawy, Z. 1982. Salinity reduction of the end-Triassic sea from the Alpine region into northwestern Europe. *Lethaia* 15, 169–178.

Hallock, P. 1988. The role of nutrient availability in bioerosion: consequences to carbonate buildups. *Palaeogeography, Palaeoclimatology, Palaeoecology* 63, 275–291.

Hsü, K.J. and McKenzie, J.A. 1985. A 'strangelove' ocean in the earliest Tertiary. *In:* Sundquist, E.T. and Broecker, W.S. (eds) *Natural Variations; Archaean to Present.* American Geophysical Union Geophysical Monograph, 32, 487–492.

Hudson, J.D. 1963. The ecology and stratigraphic distribution of the invertebrate fauna of the Great Estuarine Series. *Palaeontology* 6, 327–348.

Hudson, J.D., Clements, R.G., Riding, J.B., Wakefield, M.I. and Walton, W. 1995. Jurassic palaeosalinities and brackish-water communities – a case study. *Palaios* 10, 392–407.

Kauffman, E.G. 1981. Ecological Reappraisal of the German Posidonienschiefer. *In:* Gray, J.B. and Berry, W.B.N (eds) *Communities of the Past.* Hutchinson Ross Publishing Company, Stroudsburg, 311–382.

Lees, A. and Miller, J. 1985. Facies variations in Waulsortian buildups, Part 2: Mid-Dinantian buildups from Europe and North America. *Geological Journal* 20, 159–180.

Marshall, J.D. 1992. Climatic and oceanographic isotopic signals from the carbonate rock record and their preservation. *Geological Magazine* 129, 143–160.

Murray, J.W. 1995. Microfossil indicators of ocean water masses, circulation and climate. *In*: Bosence, D.W.J. and Allison, P.A. (eds) *Marine Palaeoenvironmental Indicators from Fossils*. Geological Society Special Publications 83, The Geological Society, Bath, 245–264.

Oschmann, W. 1991. Anaerobic-poikiloaerobic-aerobic: a new facies zonation for modern and ancient neritic redox facies. *In:* Einsele, G., Ricken, W. and Seilacher, A. (eds) *Cycles and Events in Stratigraphy*. Springer-Verlag, Berlin, 565–571.

Oschmann, W. 1994. Adaptive pathway of benthic organisms in marine oxygen-controlled environments. *Neues Jahrbuch für Paläontologie* 191, 393–444.

Peterson, C.H. 1985. Patterns of lagoonal bivalve mortality after heavy sedimentation and their paleontological significance. *Paleobiology* 11, 139–153.

Phleger, F.P., Parker, F.L. and Peirson, J.F. 1953. North Atlantic Foraminifera. *Reports of the Swedish Deep-Sea Expedition, 1947–1948* 7, 1–122.

Pickerill, R.K. and Brenchley, P.J. 1979. Caradoc marine communities of the south Berwyn Hills, North Wales. *Palaeontology* 22, 229–264.

Savrda, C.E. and Bottjer, D.J. 1991. Oxygen-related biofacies in marine strata: an overview and update. *In:* Tyson, R.V. and Pearson, T.H. (eds), *Modern and Ancient Continental Shelf Anoxia.* Geological Society Special Publications, 58, Geological Society, Bath, 201–219.

Seilacher, A. 1960. Epizoans as a key to ammonoid ecology. *Journal of Paleontology* 34, 189–193.

Seilacher, A., Reif, W.-E. and Westphal, F. 1985. Extraordinary Fossil Biotas: Their Ecological and Evolutionary significance. *Philosophical Transactions of the Royal Society, London* B 311, 5–26.

Tyson, R.V. and Pearson, T.H. 1991. *Modern and Ancient Continental Shelf Anoxia*. Geological Society Special Publications, 58, Geological Society, Bath, 470 pp.

Westermann, G.E.G. and Ward, P., 1980, Septum morphology and bathymetry in cephalopods: *Paleobiology* 6, 48–50.

Wignall, P.B. 1994. *Black Shales*. Clarendon Press, Oxford, 127 pp.

Wignall, P.B. and Simms, M., 1990, Pseudoplankton: *Palaeontology* 33, 359–378.

3 Taphonomy

The aim of this chapter is, first, to discuss the extent to which the fossil record is a true record of past biotas, and secondly, to show that the manner in which fossils are preserved can contribute to an understanding of past ecology and environments.

3.1 THE STUDY OF FOSSIL PRESERVATION

The fossil record provides a striking but incomplete history of the changing biology of the Earth through geological time. Although fossil assemblages collected in the field can provide important information about past ecology and evolutionary history, it is important to understand the degree to which that information is incomplete and may have been biased by the preservational history of the fauna. The 'science of the laws of burial' is termed **taphonomy**. Although important information is lost from the record during burial, there can also be a gain in information because the style of preservation of many fossil assemblages can provide important insights into the depositional environment in which the shells accumulated and the shallow diagenetic environment in which they were preserved.

3.2 PRESERVATION POTENTIAL AMONGST BIOLOGICAL COMMUNITIES

Living marine benthic communities are composed of both plants and animals, some of which are soft bodied or only weakly skeletonized, but some have mineralized skeletons which have a good prospect of being preserved. Non-biomineralized plants largely disappear without trace, but the soft-bodied fauna is partially recorded by the tracks, burrows and trails (**trace** or **ichno-fossils**; see Chapter 5) which record the activity of this part of the biota. Some groups, such as the annelids (worms), are very rarely preserved as body fossils but have left a variety of trace fossils; other groups, such as the Crustacea, have a patchy body and trace fossil record and some groups such as the ophiuroids (brittle stars) have a very limited fossil record, though they have been part of the marine biota since the Palaeozoic.

Studies of living communities have shown that the proportion of shelled species that stands a good chance of being preserved as fossils varies from 7 to 70% or more (Lawrence, 1968; Kidwell and Bosence, 1991). Studies of shelf faunas mainly consisting of polychaetes, crustaceans, molluscs and echinoderms have shown that the likely preservable fauna off Southern California was 33% but was only 21% on the Georgia shelf. Individuals in the shelled portion of a community constitute, on average, about 40% of those in the total fauna, and form about 50% of its biomass. The loss of the soft-bodied components of faunas has serious implications for attempts to reconstruct the biological relationships within a community. For example, most of the algae, which are the primary producers at the bottom of the food chain, are lost and so are many important predators higher in the food chain, such as polychaetes and starfish, so that it is difficult to reconstruct the feeding structure of most past communities (see also Chapter 7).

3.3 THE FIDELITY OF FOSSIL ASSEMBLAGES

Accepting the limitations imposed on palaeoecology by the general absence of fossils of soft-bodied organisms, a further concern in many palaeoecological studies is the extent to which the fossil shell assemblages were transported from their place of life and do not reflect the live shelled population of the area. However, the results from studies of Recent environments are generally encouraging for the palaeoecologist. Where dead shell assemblages have been compared with the living populations of the same environment, the assemblages have been broadly comparable in most cases, though there is commonly a loss of small or fragile forms. The similarity of the death assemblages to the live populations (its **fidelity**) can be measured in different ways. In a study of live and dead assemblages in Texas Bays, Staff *et al.* (1986) found that, of six parameters studied, taxonomic composition and the biomass of the assemblages showed the closest similarities. In other studies it has been shown that when a census of live species is compared with the dead species from the same site, the agreement is generally good: mean fidelities are 83–95% if data are related to the study area (Table 3.1 column, A) and are 75–98% if the focus is on individual habitats or facies (Table 3.1, column, B). In most studies, only a small number of individuals (<10%) in the dead assemblages are 'out-of-habitat', exotic species. It is only in particular environmental situations, such as in turbidites, tidal channels, beaches or washover fans that a high proportion of the individuals are exotic. When, however, the census of live species, obtained by single 'grab' samples, is compared with the census of dead species, it is the live species that appear to be missing (Table 3.1, column C). This is partly a consequence of grab sampling, which commonly captures only a part of the living material in marine waters, but it also reflects an important differ-

Table 3.1 The fidelity of shell assemblages in terms of A. the % of live species found dead in the study area and B. the % live species found dead within the same facies. C. shows the % of dead species found live within the study area (from Kidwell and Bosence, 1991)

	A *What % of live species are found dead within the study area?*		*B* *What % of live species are found dead within the same facies?*		*C* *What % of dead species are found alive within the study area?*	
Setting	*Mean*	*Range*	*Mean*	*Range*	*Mean*	*Range*
Intertidal	83%	45–100%	90%	62–100%	54%	27–100%
Coastal subtidal	95%	87–100%	98%	82–100%	33%	10–58%
Open Marine	84%	54–97%	75%	30–100%	45%	38–64%
Grand Means	87%		88%		44%	

ence in the nature of death assemblages. An assemblage of dead shells commonly represents the accumulation of material over a period of time and is likely to be a mixture of many successive populations. The assemblage is referred to as '**time averaged**' and it is a commonly a better measure of the 'typical' community within a habitat than that made by grab sampling the live population at any one time. If samples of the living population are made over a long period the total species census progressively resembles the dead census more closely.

Shells, because of their large size and relatively stable production levels, reflect an important part of any community in terms of biomass. Most dead shells recorded in modern environments belong to species living in the same area. Preserved shell accumulations appear to record a time-averaged species census of the shelly portion of the living communities with relatively little dispersal between habitats, though small patchy populations may become mixed. The study of fossil accumulations is therefore a realistic basis for reconstructing the shelly part of palaeocommunities.

When tracking the fate of a biota on its path to being fossilized, it is useful to identify the successive stages through which it passes. The living assemblage is a **biocoenosis**, which is transformed into a **thanatocoenosis** after death and decay. Various taphonomic processes act on the remaining skeletons to create a **taphocoenosis**, which is the fossil assemblage that is finally preserved. Many palaeontologists alternatively refer to assemblages preserved in their life position as **life assemblages** (= autochthonous thanatocoenoses) and reworked assemblages as **death assemblages** (= allochthonous taphocoenoses). Before fossil assemblages can be used with confidence in community reconstructions, some estimate must be made of their fidelity, based on evidence of the degree to which it was transported from its original habitat or mixed with members of other communities. It is possible to characterize assemblages into three categories according to their degree of dislocation, as follows.

1. **Life assemblages** are fossil assemblages which retain life orientations of some of the organisms. The shells of bivalved animals are commonly still joined and skeletons composed of plates or weakly joined parts may be complete. Shells may commonly be in clusters and in the orientation in which they lived (Fig. 3.1).
2. **Neighbourhood assemblages** (Fig. 3.2) are composed of shells which have been disturbed and displaced from their original living position but are believed to be in their original habitat. Bivalved shells are commonly separated, but breakage and abrasion is rare. Assemblages from adjacent beds commonly have a similar species composition, suggesting that they have not been mixed with other communities.
3. **Transported assemblages** (Fig. 3.3) commonly contain broken and sometimes abraded shells and the species composition of adjacent assemblages may be variable.

The distinction between neighbourhood and transported assemblages is not always clear, partly because the two categories are intergrading.

3.4 POST-MORTEM PRESERVATION AND LOSS

The fossils that are preserved in sedimentary sequences probably represent only a small proportion of the original shelled fauna. Although there are no reliable estimates of the proportion that survive, Raup and Stanley (1971, p. 15) have noted that the number of shells

Figure 3.1 An assemblage of the Lower Silurian brachiopod, *Pentamerus*, in life position (the block is upside down X0.75). The brachiopods lived in clusters with their umbones buried in the sediment.

Figure 3.2 A neighbourhood assemblage of bryozoans and brachiopods on the surface of a bioclastic limestone bed from the Wenlock Limestone, Dudley, England (Silurian). Some of the brachiopods are articulated and some bryozoans with delicate branches are present, but the fossils are clearly not in life orientations.

that would have occupied a quarter square metre of the sea floor over a period of a million years could be in excess of 100 million, more than enough to fill the museums of the world if none were lost. Some shelled communities are preserved almost entire, others are affected by mechanical or biological degradation and/or chemical dissolution and may disappear completely. Shells on the sediment surface in shallow marine waters are particularly likely to be affected by mechanical destruction or bioerosion (boring or predation), while they are more likely to suffer dissolution in deep marine waters. In general, the resistance of shells to destruction is related to their mineralogy, their internal structure and the amount of organic matter distributed within the shell. Calcite is more stable than aragonite, though calcite solubility increases with magnesium content. Consequently shells that are composed of calcite stand the best chance of survival, shells composed of high Mg calcite are more prone to dissolution and aragonite shells are the least resistant (Table 3.2). Palaeozoic faunas usually have more taxa with preserved calcite than later faunas. The

Figure 3.3 A transported assemblage of the bones of small amphibious reptiles, fish bones and teeth and coprolites from the base of a thin sandstone bed (the Rhaetic Bone Bed, Jurassic), Westbury on Severn, England. The organic remains are preserved in phosphate and the matrix is largely pyrite.

preservation of aragonitic shells becomes progressively rare with age amongst Cenozoic and Mesozoic rocks, though fossils of aragonitic shells are commonly preserved as moulds or calcite casts.

Table 3.2 Mineralogical composition of the skeletons of major groups of organisms

low Mg calcite	*high Mg calcite*	*aragonite*
brachiopods, trilobites, bryozoans, belemnites many rugose and tabulate corals, pelagic foraminifera coccolithophorid and charophytic algae some bivalves (parts of oysters), sponges ostracodes	echinoderms most benthic foraminifera some sponges, tabulate corals, bryozoans, ostracodes and rhodophytic algae	molluscs scleractinian corals some sponges, bryozoans and rhodophytic and chlorophytic algae

3.5 DESTRUCTION BY PHYSICAL, BIOLOGICAL AND CHEMICAL PROCESSES AT THE SEDIMENT SURFACE

Post-mortem destruction of shells and skeletons on the sediment surface commonly occurs through physical and biological processes. The component parts of many skeletons, such as the plates forming an echinoid test, or the two halves of a bivalve, are held together by organic material that progressively decays on death so that the parts become disarticulated and commonly dispersed. Many shells are robust when the animal is alive but rapidly loose strength by decay of organic matter after death. Shells that have a microstructure that incorporates relatively large amounts of organic matrix disintegrate into crystallites relatively rapidly after death (Glover and Kidwell, 1993). Mechanical destruction of shells is most common where waves and currents batter the shells against a hard substrate or where the shells are transported among mobile gravel. Away from the shoreline shells are locally reworked and redistributed by waves and currents, but it is unlikely that they are commonly fragmented unless weakened by bioerosion.

Bioerosion and biofragmentation has been shown to be one of the most important processes in shell degradation in modern environments. In carbonate habitats the measured rates of sediment production resulting from bioerosion may be as high as the rate at which shell material is being generated by organisms. This helps to explain the huge amount of shell debris accumulated in some thick limestone formations. Bioerosion is generally caused by organisms such as fungi, algae, sponges, barnacles, bivalves, echinoids, worms, and bryozoans that bore or excavate shells and so weaken them. In quiet, offshore environments it has been shown that fragmentation is correlated with the degree of bioturbation, suggesting that there it is related to biological rather than hydrodynamic processes.

Biofragmentation can be caused by predators and scavengers, such as fishes and large crustaceans, that crush shells. In some instances, the shell material is abandoned on the sea-floor, in other cases it is ingested and may be preserved as faecal accumulations. It is commonly difficult to distinguish between particles formed by physical and biological processes. However, it seems likely that bioerosion forms mainly fine-grained material, while physical fragmentation and biofragmentation forms a wide spectrum of grain sizes. The shape of the debris formed by the crushing of shells resembles that formed by physical fragmentation, but some fragments produced by shell crushing have a distinctive morphology (see Fig. 7.27).

The likelihood of shells being dissolved on the sediment surface depends largely on the degree of saturation of water with respect to $CaCO_3$. Most Recent shallow-marine waters are saturated or oversaturated with respect to $CaCO_3$, so there is little dissolution of shell material. On the other hand, some brackish and fresh waters are acidic and shells may suffer corrosion or total dissolution. In marine environments there appear to have been periods in the past when the degree of saturation was reduced and aragonite dissolution may have occurred in relatively shallow marine waters. In deep oceanic waters undersaturation increases with depth, partly as a result of the oxidation of organic matter and consequent increase in CO_2 but due also to a lesser degree to the effects of increasing pressure and decreasing temperature on calcite solubility. Where dissolution equals the supply of carbonate (the carbonate compensation depth or CCD), only small amounts of carbonate sediment will be preserved. In modern oceans the CCD is usually between 3 and 4 km although if the predominant carbonate shell material is aragonite it can be as little as 1 km.

3.6 PRESERVATION AND DESTRUCTION BELOW THE SEDIMENT SURFACE

Once biological material is buried, its preservation potential is related to the nature of the enclosing sediment and the associated chemical environments.

In most sandy sediments organic carbon in soft tissue is usually oxidized in the well-irrigated surface layers, but the fate of shell material depends largely on the nature of the enclosing pore fluids during burial. If at any stage the fluids become acidic or undersaturated with respect to calcite, then dissolution is likely to occur. However, if there has already been some cementation of the sandstone, shells may be preserved as moulds.

The preservation potential of shell material, particularly calcite shells, in carbonate sediments is generally relatively good and the detailed morphology of shells is commonly well preserved. In shallow marine carbonates the sediment commonly becomes lithified at an early stage, prior to compaction. The early cements can be marine, usually occurring as a fringe of aragonite or high magnesium-calcite needles perpendicular to the surface of grains and replaced by calcite in ancient limestones. Most commonly the main pore-filling cement is a clear, drusy cement (cement with an increase in crystal size away from the substrate) which may succeed the early fibrous cement or itself be the first cement. Many drusy cements are thought to be deposited from meteoric waters (fresh waters) in the phreatic zone (below the water table) though where there is evidence of fracturing of grains and cements they may have been formed at substantially greater depths. Calcite shells in limestones commonly retain their original shell structure, but aragonite shells are generally dissolved away and the resulting cavity filled with a drusy cement or the aragonite is neomorphically replaced by calcite. In mudrocks the fate of organic material is determined largely by the geochemical environment in the top few metres of sediment. Some shells pass through these sediments unaltered except for the loss of organic tissue, others are enclosed within carbonate concretions, others are coated, filled or replaced by phosphate, pyrite, or silica, whilst others are lost by dissolution. Shells

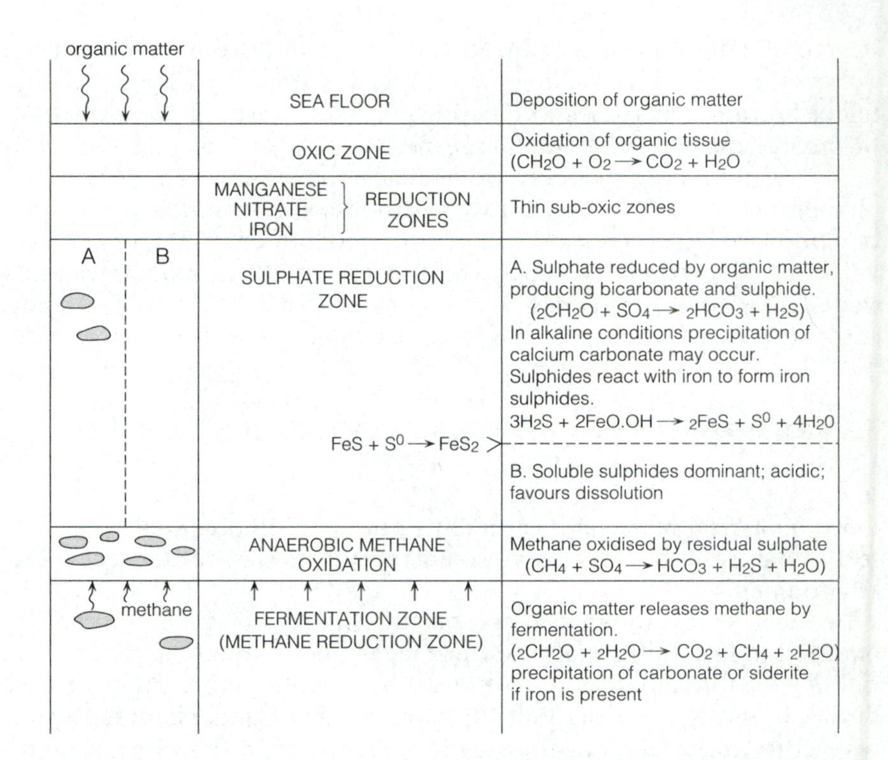

Figure 3.4 Diagenetic zones in the near-surface layers of marine muds. These are the favoured locations for concretion growth and pyrite formation.

that survive through the surface layers commonly suffer compaction with further burial and may be subject to further mineralization or dissolution.

3.6.1 The shallow diagenetic environment

In the surface layers of muddy sediments there is a zonation of chemical reactions, promoted by bacterial activity, from the sediment surface downwards (Fig. 3.4). In the thin near-surface layer the muds are continuously oxygenated by diffusion from the overlying seawater. Organic material is oxidized with the mediation of bacteria. In the underlying anoxic layers microbes utilize manganese oxides, nitrates, iron compounds and sulphates as alternative reducing agents of organic matter. The suboxic zones of manganese, nitrate and iron reduction are generally thin, and are important sites for the growth of phosphate and glauconite, but it is the sulphate reduction zone that appears to play a dominant role in determining the fate of most potential fossil material and in the growth of pyrite and chert. Within this zone soluble sulphide can produce acidic, undersaturated pore waters favouring dissolution of shell material. On the other hand, when the effects of sulphides are diminished by reactions with iron and the sulphate is rapidly exhausted, the resultant high alkalinity favours precipitation of carbonate and the preservation of shells. The state of the sulphate reduction zone generally determines the fate of shell material as it becomes buried. Where there is pervasive bioturbation, oxic pore waters penetrate deeper into the sediment and the development of alkalinity is suppressed and acidity, caused by oxidation of organic matter, leads to the dissolution of shells. Conversely, where bioturbation is weak the sulphate zone is commonly well developed and is associated with high alkalinity and shells are more likely to be preserved. The availability of iron also affects the preservation potential of shells. In sediments where the iron content is low, acidity develops in the absence of iron reduction, while alkalinity and preservation of shells is favoured in more iron-rich sediments.

Below the sulphate reduction zone, where sulphate has become exhausted, any surviving organic material is degraded by microbial processes to produce methane and CO_2 (zone of methanogenesis). Preservation and precipitation of $CaCO_3$ is generally favoured in this zone and some of the methane generated can diffuse upwards into the sulphate reduction zone where it is oxidized (zone of methane oxidation) and may contribute to further deposition of carbonate (Raiswell, 1988). Within the zone of methanogenesis, precipitation of carbonate in association with iron, already reduced to the ferrous form, may result in the formation of iron carbonates, such as ferroan calcite and siderite.

3.6.2 Preservation in carbonate concretions

Many of the best-preserved fossilized shells and skeletons have been recovered from concretions. Examples include uncompacted ammonites from the Lower Jurassic of the Dorset coast, the Old Red Sandstone (Devonian) 'Fish Beds' of Caithness, Scotland, and the amazingly varied Pennsylvanian fauna of Mazon Creek, Illinois. In all these examples the concretions grew in the near-surface sediment, prior to compaction.

Concretions reflect carbonate deposition that was focussed around a nucleus, such as carcass or shell, or along particular horizons to form spherical or ellipsoid bodies, or in some situations, continuous carbonate beds. The formation of carbonate concretions in the sulphate reduction zone is favoured by rapid burial of organic matter, restricted bioturbation, the presence of iron and possibly skeletal debris and, importantly, a pause in sedimentation that allows time for concretions to grow. In fresh waters the conditions for concretion formation are rather different. Concentrations of sulphate ions are much lower in fresh water, consequently the sulphate reduction zone is virtually absent. In the absence of high levels of sulphide production but where carbonate activities remain high and there is a supply of iron, siderite nodules form. Siderite nodules that may yield well-preserved fossils are widespread in various fresh water mudrocks, particularly in lacustrine sediments.

3.6.3 Pyritic preservation

Most pyritized fossils are found in mudrocks because their formation requires organic carbon, iron and sulphate and these constituents are commonly available in marine muds, but one or more is deficient in other sediments. Limestones, for example, generally lack pyrite because there is a deficiency of sedimentary iron in most carbonate environments and pyrite is generally absent in sandstones because organic carbon is destroyed by oxidation in the well-irrigated surface sediments.

Most pyritization occurs within the sulphate reduction zone in mudrocks within a few metres of the sediment surface. There, H_2S, generated by sulphate reduction, combines with reactive iron minerals to produce finely particulate iron monosulphides (FeS). These in turn react with sulphur, formed by the bacterial breakdown of H_2S, to produce pyrite (FeS_2) (see Canfield and Raiswell, 1991, for a detailed discussion of pyrite formation). Formation of pyrite is favoured in sulphidic sediments in environments that are neither anoxic nor strongly oxic at the sediment surface but in which anoxia develops at shallow depths (Fisher and Hudson, 1987). The form of the pyrite is related to the diagenetic environment in which it grows, determined largely by the degree of oxygenation in the surface layers of sediment and by the rate of sedimentation (Brett and Baird, 1986). In the well-oxygenated muds, pyrite is sparse because most organic matter is destroyed by oxidation (Fig. 3.5, field A.). In more weakly oxygenated environments that are nevertheless bioturbated, pyrite will not generally form in the body of the sediment but forms in anaerobic micro-environments, commonly within shells, where it forms internal moulds (Fig. 3.5, fields B and C). The early pyrite is generally framboidal, but later pyrite has equant crystals and may create stalactitic growths. In anaerobic environments, associated with dark grey to black mudstones, there is commonly a plentiful supply of dispersed organic matter, iron and sulphate, so diffusion gradients are low and pyrite forms framboids dispersed throughout the sediment (Fig. 3.5, field D). More rarely pyrite occurs as a shell replacement, or it may form a coating on soft tissue or be disseminated within the tissue. In addition to oxygenation levels, rate of sedimentation can also influence pyrite formation. Where sedimentation rates are low, organic matter is commonly destroyed before it reaches the zone of sulphate reduction (Fig. 3.5, field E). Where rates of sedimentation are high, sulphate reduction may be restricted because there is a dilution of organic matter and also because sediment may be buried so quickly that it passes rapidly through the sulphate reduction zone (Fig. 3.5, field F).

3.6.4 Preservation in phosphate

Many of the most exquisite fossils are preserved in phosphate. Replacement of shells or organic tissue may occur at such an early

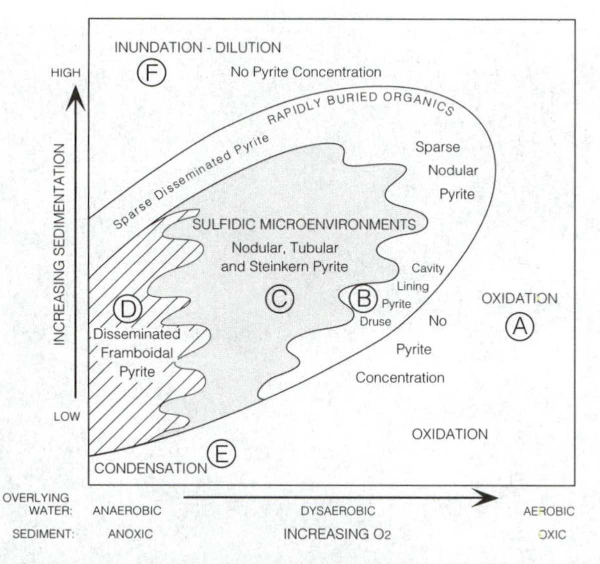

Figure 3.5 Pyrite growth in sediments in relationship to levels of oxygenation and sedimentation rate. The fields A to F are explained in the text. (Modified from Brett and Baird, 1986.)

stage that the original fine detail is preserved. Phosphate also cements and replaces bone, preserves faecal material as coprolites and replaces or encrusts limestones. Most commonly the phosphate occurs as concretions, sometimes preserving the delicate anatomy of soft-bodied animals (Fig. 3.6), but it may replace small fossils which are then encased within a carbonate concretion.

Phosphatization involves the replication of organic material by carbonate-fluorapatite. The phosphorus required for the formation of phosphates is derived from organic matter that has been sedimented from the water column plus some detrital phosphate. The phosphorus is released by microbial activity and potentially diffuses back into the water column, unless burial or changes in the redox of the surface layers allows further microbial activity in an anoxic environment to fix the dissolved phosphate. This process appears to be one of the earliest phases of early diagenetic mineralization, occurring in the suboxic zones, which accounts for the instances of soft tissue preservation. Conditions favouring the growth of phosphate are low rates of sedimentation coupled with high productivity of phytoplankton leading to rapid sedimentation of organic matter, bottom sediments that are anaerobic and rarely bioturbated and low levels of CO_2 that would otherwise inhibit the growth of apatite. Upwelling systems in the oceans are particularly favourable sites for the formation of primary phosphate, but phosphatized fossils are recorded from a wide range of environmental settings including both shelf and marginal marine environments.

Figure 3.6 The shrimp *Pseudogalatea ornatissima* preserved in phosphate, from the Granton Shrimp Bed, upper Carboniferous, Granton, Scotland. (Photograph by courtesy of E.N.K. Clarkson.)

3.6.5 Preservation in silica

Silica can replace calcite and aragonite shells, permineralize (permeate the pore spaces) of wood or peat or pervasively fill burrow systems with cement. It can also form nodules and layers of chert, replacing carbonate sediment and it can surround or fill fossils, so that if the fossil is later dissolved it will be preserved as an external or internal mould. Silica has been particularly important in preserving algal cells and filaments in Precambrian rocks, so providing evidence of the early stages of evolution and it has also preserved the plants in the Devonian Rhynie Chert, which is one of the earliest records of an *in situ* terrestrial flora.

There are three common modes of shell replacement:

1. as a granular white crust;
2. as finely granular replacement; and
3. as concentric rings of silica (beekite rings).

All internal shell structure is destroyed during silification, and some details of the surface morphology can be obscured, particularly with beekite replacement. Nevertheless, in some circumstances details of the external morphology are perfectly preserved.

Deposition of silica appears to depend on an adequate supply of silica and suitably acidic conditions in the diagenetic environment. Sources of the silica are mainly biogenic (radiolarians from the Ordovician onwards, sponges from the Cambrian onwards and siliceous Crysophyta (diatoms) from the Cretaceous onwards), or volcanogenic, though the alteration of clays also releases some silica too. Most silicification appears to occur during early burial in a wide variety of situations, but some cherts are formed at greater burial depths.

Precipitation of silica in shallow marine sediments mainly occurs in the sulphate reduction zone where reactions that produce acidity may favour the dissolution of carbonate and the deposition silica. The process is strongly influenced by the solubility of the available biogenic silica. The conditions favouring the growth of siliceous nodules are similar to those for carbonate concretions, namely, a sufficiently rich source of organic matter and a pause in sedimentation that allows time for nodules to grow. Environments where silification appears to be favoured are shallow marine basins and lakes, in some instances with evaporites, where evaporation increases the pH and mobilizes silica. Silicification may also occur in particularly acidic conditions, such as peat bogs, provided there is a sufficient supply of silica.

Chert is not exclusively the product of the diagenetic replacement of carbonate and evaporitic rocks. Bedded chert, formed by the diagenetic transformation of biogenic silica (mainly radiolarian oozes), is widely recorded from deep marine environments and it is claimed that much of the chert in the banded ironstones of the Proterozoic, Gunflint Formation was deposited directly from seawater.

3.7 PRESERVATION OF SOFT TISSUE

The rare instances in which evidence of soft tissue is preserved provide important information about the morphology of the organisms and, where a whole biota is preserved, about their ecology (see Allison and Briggs, 1991a).

Non-mineralized animals, such as worms, can survive considerable transport immediately after death but disaggregate rapidly as soon as decay starts (Allison, 1986). The putrefaction of some larger organisms on the sediment surface can influence the state of their final preservation. Most organisms, such as fishes and marine reptiles, have negative buoyancy on death and sink to the sea-floor, where putrefaction will begin. After a certain period of decay, gases build up within the organs and tissues and the carcass will 'bloat and float'. If disintegration occurs while the carcass is floating, various parts of the body can become dispersed and the bones will be deposited as isolated parts. If the carcass loses buoyancy while it is entire, the body will be deposited as a whole skeleton.

In most situations scavenging of organic material begins immediately after death and putrefaction begins within a few days. Destruction of organic matter is mediated by microbial activity, which is initiated in the near-surface oxic environment and continues with shallow burial in the anoxic sulphate reduction zone. Any surviving organic material will normally be consumed in the fermentation zone. The extent to which organic tissue is destroyed depends partly on the rate at which it is oxidized or attacked by microbial activity and also on the resistance of the organic matter to decay; the macromolecules of some tissue are more resistant than others. Very labile tissue is most easily destroyed, followed by chitin, cellulose, and then lignified cellulose. Although the more refractory materials normally suffer biodegradation and chemical transformation to more resistant molecular structures, the original morphology of the tissue can be preserved.

3.8 FOSSIL LAGERSTÄTTEN – THE EXCEPTIONAL PRESERVATION OF FOSSILS

Exceptionally well preserved individual fossils and fossil assemblages yield a disproportionately large amount of important palaeontological information. **Lagerstätten** is a German term for economic bodies of ores or minerals, and was applied to fossils by Seilacher for 'bodies of rock unusually rich in palaeontological information, either in a qualitative or quantitative sense' (Seilacher *et al.*, 1985). Lagerstätten include fossil discoveries ranging from particularly thick accumulations of shells, life assemblages overwhelmed *in situ*, preserved soft-bodied faunas, and soft-bodied organisms exquisitely preserved in phosphate. Although the preservation of Lagerstätten represents special circumstances, they are one end of a spectrum of normal sedimentary and preservational processes.

Lagerstätten can be usefully divided into two main types (Fig. 3.7): first, **concentration deposits**, which are notable for the close-packing and abundance of shells they contain, e.g. oyster beds, bone beds and varied coquinas (section 3.9); and secondly, **conservation deposits**, which are notable for the quality of preservation, particularly of organisms with soft tissues. Conservation deposits arise in a variety of ways associated with differences in rates of deposition, oxygenation of bottom waters and the diagenetic environment in the near-surface sediments. Nevertheless, two broad types, 'obrution deposits' and 'stagnation deposits', can be distinguished.

Obrution deposits are formed by short-lived events such as storms or turbidity currents that lead to the rapid burial of a fauna. Preservation of soft-bodied faunas is favoured where the bottom waters in the depositional area are anaerobic or the sediments are anoxic, so inhibiting bioturbation and slowing bacterial decay. Obrution deposits include smothered hardground faunas, clustered fossils, such as trilobites, indicating gregarious behaviour and faunas

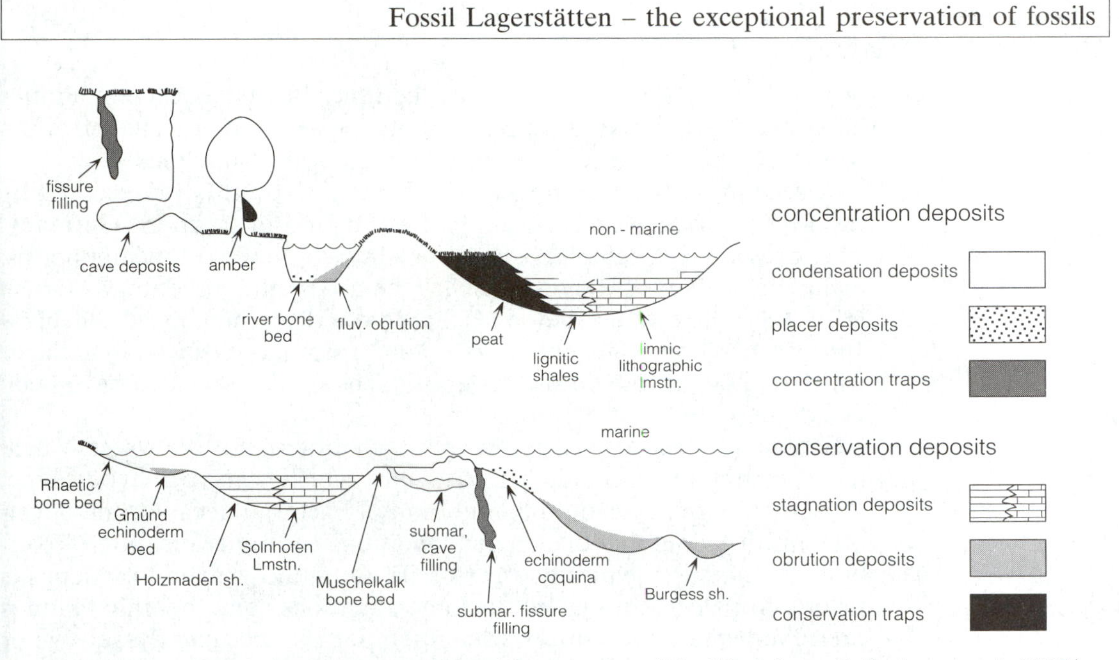

Figure 3.7 Classification of Lagerstätten, with examples. (Modified from Seilacher, *et al.* 1985.)

transported and buried in turbidites or storm flows. An example of the last type of preservation is the Lagerstätten in the Middle Cambrian Burgess Shale in the Rocky Mountains of Canada. A rich and varied soft-bodied fauna was carried offshore and downslope by turbidity currents or storm flows and rapidly deposited in a mainly anaerobic environment. Most of the fauna is now seen as silvery films which commonly outline even the most delicate structures on the fossilized animal. The films appear to be formed of phyllosilicates on a coating of kerogenized carbon. Fossils come from two main levels, the lower of which, the Phyllopod Bed, has alone yielded about 65,000 specimens belonging to more than 100 species. The fauna is particularly rich in arthropods, but has a good variety of sponges, echinoderms, priapulid worms and species belonging to several other known groups. In addition there is a variety of forms that cannot be easily assigned to established taxa (Conway Morris, 1986). More than 90% of the living fauna was soft-bodied and is not normally preserved, though Lagerstätten similar to the Burgess Shale fauna have now been discovered in the USA, Greenland and China. The Burgess Shale-type Lagerstätten offer an exceptional opportunity to record a wide range of rarely preserved Cambrian biota and have been the basis for important discussions of the origins and evolutionary divergences of major taxa. The exceptionally wide range of organisms preserved has also offered an opportunity to reconstruct the ecosystem of the fossil assemblage and identify both predators and prey (Chapter 7, section 7.11). Some caution has to be applied to the ecological interpretations because the assemblage is transported and potentially derived from different sites. Representatives of benthic and pelagic communities have been differentiated, but it is unfortunate that the site of origin of the benthic faunas cannot be identified with confidence.

A second, very different type of obrution deposit is represented by the Lower Jurassic echinoderm bed at Gmund in southern Germany. The preserved assemblage consists almost wholly of echinoderms, including stalked crinoids, grazing echinoids and predatory starfish (Fig. 3.8a). The fauna lies on a hardground at the top of the basal Jurassic conglomerate and is overlain (smothered) by a mud layer which is thought to be a storm deposit. The muds appear to have been anoxic and inhibited bioturbation.

Stagnation deposits arise from the stagnation of bottom waters which most commonly become anoxic, but which can be hypersaline. Preservation depends on the inhibition of bacterial decay or the development of favourable diagenetic conditions for mineralization of soft-bodied animals. Typically the fossils found in stagnation deposits belong to pelagic groups but in some situations some benthic fauna is preserved where the bottom waters temporarily became dysaerobic or even aerobic. The Posidonienschiefer in the Toarcian (Jurassic) of Holzmaden, Germany, is a classic stagnation deposit. The dark bituminous marls preserve a fauna including nektic ammonites, belemnites and ichthyosaurs, epiplanktonic crinoids and benthic bivalves that are concentrated at discrete horizons, marking intervals when the sea-floor was oxygenated (Fig. 3.8b) (see Chapter 2). The fossils recovered from the Posidonienschiefer offer a time-averaged view of a substantial part of the ecosystem of a stagnant basin, together with a terrestrial component that was washed in (see Chapter 10).

The Solnhofen Limestone (Upper Jurassic of Bavarian Germany) represents a different type of stagnation deposit. The fauna in the thin bedded micritic limestones is famous for the presence of *Archaeopteryx*, but consists mainly of a mixture of pelagic forms and washed-in benthos (Fig. 3.8c), though some of the crinoids are thought to be *in-situ* benthos. Evidence that much of the benthos is washed-in comes from the death trails of some of the animals. Most spectacular are the irregular spiral tracks of *Mesolimulus* with the dead animal preserved at the end of the track. Some fossil fishes show a dorsal curvature of the vertebral column indicative of desiccation that might indicate highly saline bottom waters in a well-stratified sea which would also have been deficient in oxygen. Under these conditions bioturbation was rare, predators and scavengers absent and levels of bacterial activity low. Fossils of soft-bodied animals were formed as impressions of the carcass on the carbonate-mud surface, though this was sometimes enhanced by the development of a carbon film. Ammonites are crushed and in some cases were deformed within soft sediment, suggesting that aragonite dissolution occurred within the upper layers of the sediment. The early growth of phosphate has preserved coprolites and the ink sacs of coleoid cephalopods. There is evidence of wrinkling of the surface mud that suggests there was a

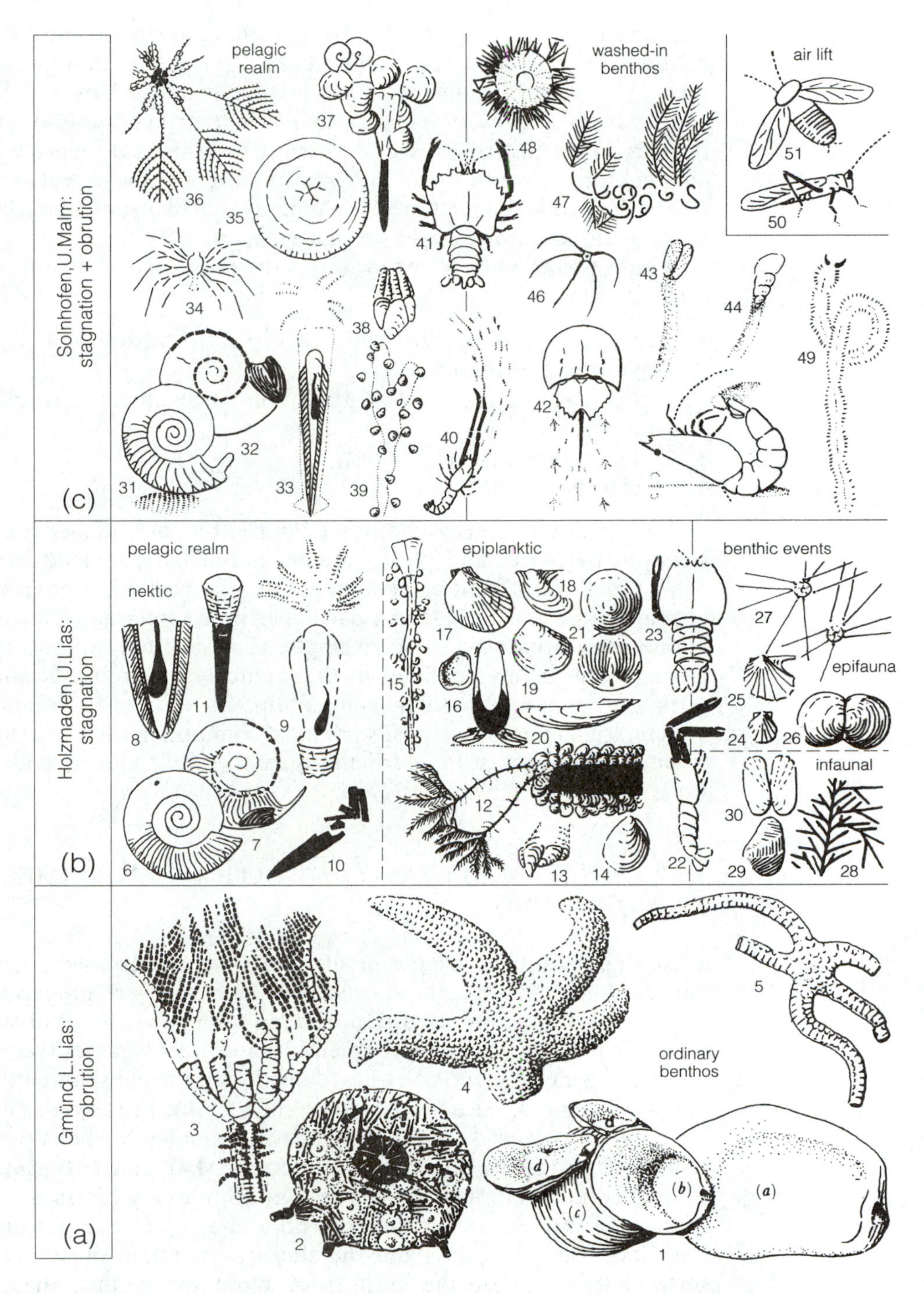

Figure 3.8 Three faunas preserved in Lagerstätten. (a) The echinoderm fauna in the Lower Lias (Jurassic) of Gmund, Germany, preserved as an obrution deposit. (b) The mixed pelagic, epiplanktonic and benthic fauna preserved as a stagnation deposit in the black shales of the Posidonienschiefer (U.Lias, Jurassic) at Holtzmaden, Germany. (c) The pelagic and washed-in benthos in the stagnation deposit in the Solnhofen Limestone (Late Jurassic) of Solnhofen, Germany. (From Seilacher *et al.*, 1985.)

cyanobacterial mat that might have played a part in the special preservation of the fauna. The Solnhofen Limestone is thought to have formed in a hypersaline stagnant lagoon protected by a barrier of sponge reefs. It preserves a very sparse but very varied fauna ranging from flying reptiles and insects, a variety of fishes, rare marine reptiles and nektonic invertebrates and a very rare benthos (Barthel *et al.*, 1990). The total Lagerstätte is a time-averaged hypersaline lagoonal fauna with an admixture of foreign elements.

Situations particularly favourable to the preservation of conservation Lagerstätten are:

1. deposits from dilute turbidites or distal storm deposits on a basin slope or distal ramp;
2. local areas of anoxic sediments within basins or more rarely basin-wide anoxic muds;
3. deltaic sediments;
4. lacustrine sediments.

Preservation of Lagerstätten is favoured by early diagenetic mineralization by pyrite and phosphate or within early-formed carbonate concretions. Preservation of Lagerstätten in particular environments appears to be concentrated in particular time intervals (referred to as **taphonomic windows**). For example, Lagerstätten in basinal-slope deposits are mainly confined to the Cambrian, possibly because bioturbation penetrated shallower depths in the early Phanerozoic. Lagerstätten in deltaic deposits are most common in the Carboniferous, when delta plains were particularly extensive (Allison and Briggs, 1991b).

3.9 TYPES OF SHELL CONCENTRATIONS AND TIME-AVERAGING

Most shell-bearing marine animals live at relatively low densities on the sea-floor, though some species such as oysters are gregarious, and reefs are a prominent exception. During intervals of normal sedimentation, particularly in environments below fair-weather wave base, shells are generally buried at low densities and are dispersed throughout the sediment, except where life clusters are preserved. However, the geological record contains abundant examples of shells that occur at high densities; these deposits are termed **shell concentrations**.

Shell concentrations can arise in a multitude of situations; they are commonly formed by current transport, they can arise because sedimentation rates are slow and the shells accumulate on the sediment surface, they can be life clusters or more rarely they may be the stomach contents of large vertebrates.

A classification of shell concentrations into four categories groups them according to their rate of formation that broadly reflects rates of sedimentation (Fig. 3.9) (Kidwell, 1991).

1. **Event concentrations** are those formed over a short period, generally a few minutes to a few days. They are typically formed by the action of short-term events such as storms or turbidity currents, but would include the stomach contents of predators. Event concentrations commonly occur as coquinas at the base of sandstone beds or as shell beds of shell pavements within shales, though more rarely they may occur as isolated lenticular accumulations. The taxonomic composition, and preservational characters are normally uniform throughout the thickness of a bed, reflecting rapid deposition of a well-mixed population of shell material.
2. **Composite concentrations** are beds which show changes in the taxonomic composition and/or state of preservation of shells in successive layers. This is thought to reflect the stacking of event concentrations that were deposited one on another either in rapid succession or over a longer time span. The common occurrence of composite concentrations within a sequence is generally taken as an indication of relatively slow sedimentation rates.
3. **Hiatal concentrations** are heterogeneous beds with varied taxonomic composition and preservation. The internal stratification is commonly a complex arrangement of irregular, discontinuous partings. Fragmentation and boring of shells is common and there is a common association with authigenic minerals such as chamosite and glauconite. Some of the heterogeneity of hiatal beds is the result of varied diagenesis within the bed. Some hiatal concentrations can be shown on biostratigraphic evidence to be condensed horizons or disconformities at which biozones are missing and most are believed to reflect long periods (hundreds to thousands of years) during which sedimentation was minimal.
4. **Lag concentrations** form by erosion of pre-existing fossiliferous deposits. The term is restricted to those deposits which were derived from previously consolidated, usually lithified, sediment. It is not applied to lag deposits formed by excavating nearly contemporaneous shells such as those formed by migration of a tidal channel; these would be regarded as event beds or composite beds, depending on their internal stratigraphy. Lag concentrations typically occur as patches or more laterally extensive linings to nearly planar erosion surfaces or as lenses or linings on a channel floor. Fossils in a lag deposit may come from different stratigraphic horizons, may have adhering lithified sediment and internal cement, and commonly occur in association with lithoclasts.

Shell concentrations can be formed from the remains of a single contemporaneous association of animals but more commonly they are time-averaged and formed from a mixture of biological material from a succession of time horizons (see Kidwell and Bosence, 1991). Shell concentrations are generally an average of minor fluctuations in community composition over time, but they can also be a mixture of very different communities drawn from different environments. The

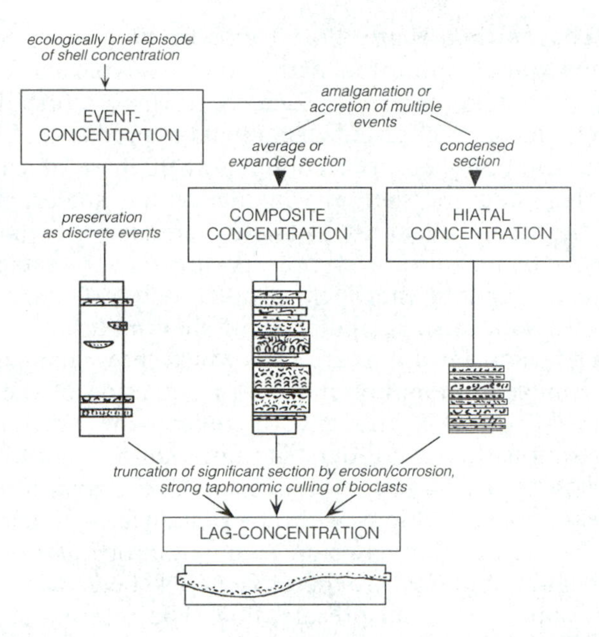

Figure 3.9 Classification of shell concentrations into four different types, arising from different processes of concentration. (From Kidwell, 1991.)

mixing of faunas from a single habitat obscures details of the palaeoecological record, such as stages in community succession, but it generally increases species diversity and may give a more complete picture of the shelly portion of the 'normal' community. The chances that an assemblage of shells from a single community will be preserved is greatest when sedimentation rates are high and the shells are rapidly buried as single event beds. Slow sedimentation produces a more condensed stratigraphy so that concentrations are likely to be time-averaged, event beds will be more closely spaced in the sequence and composite beds more common. Hiatal horizons at which shell material has accumulated and been mixed in the absence of appreciable sediment accumulation may record communities time-averaged over a long time-span. Time-averaged concentrations may be recognized by the presence of shells from distinctly different habitats (e.g. fresh water taxa mixed with marine forms) and mixtures of shells showing different taphonomic states, including different degrees of boring, encrustation, abrasion, breakage. Differences in the diagenetic state of bioclasts, such as different sediment fills, internal cements, or the presence of authigenic minerals such as chamosite or glauconite, may identify long-term time-averaged faunas, as will a mixture of taxa of different stratigraphic age (Fürsich and Aberhan, 1990).

3.10 INFORMATION FROM SHELL CONCENTRATIONS IN THE MARINE ENVIRONMENT

A taphonomic understanding of shell concentrations is not only important in the reconstruction of past communities, it can yield important information about the environment in which the beds accumulated. Concentrations can convey important information about the hydrodynamic regime in which they were deposited, rates of sedimentation and the degree of time averaging. They may also be important in marking hiatal surfaces of particular significance in sequence stratigraphy.

The state of preservation of shells and bioclasts provides information about their post-mortem history, i.e. whether they have been transported, the severity of the hydrodynamic regime that they have encountered and the duration that they have spent on the sediment surface. Important aspects of preservation that can yield relevant information include the following.

1. Articulation ratio of bivalved shells such as brachiopods and bivalves (Fig. 3.10). Most shells that are articulated during life separate soon after death as the muscles or ligaments that hold the shells together decay. It follows that most shell concentrations with a high articulation ratio have been preserved more or less as they lived in life (**life assemblage**). This is thought to arise when living communities are rapidly buried, particularly when they are living in dense clusters and the shells are supporting one another. Bivalved animals living within the sediment are also more likely to be preserved as articulated shells but are unlikely to form concentrations. There are exceptions to the general rule that articulated shells indicate *in situ* assemblages. On rare occasions living animals can be transported in strong currents (turbidity currents or storm currents) and rapidly deposited and buried. These assemblages are usually found at the base of sandstone beds and are easily identifiable as being transported. The ease with which shells become disarticulated varies considerably. Bivalves with a strong ligament resist disarticulation longer than those with a weak ligament and brachiopods with a more retentive articulation of teeth and sockets (atrypids and rhynchonellids) are more likely to remain articulated. Amongst animals with an exoskeleton composed of plates, such as the echinoids, disaggregation occurs soon after decay of organic matter is initiated. It is not possible to assess the degree of transport if the shells are disarticulated. Shells can disarticulate *in situ* without any disturbance, or they may be disarticulated by current action.
2. Valve ratios in bivalved shells (Fig. 3.10). The substantial difference in shape of the opposed valves of brachiopods means that they have a different response to a current and will become separated during transport. Even the left and right valves of bivalves that are near mirror images of each other have been shown to behave differently in response to currents and so follow different transport paths

(Frey and Henderson, 1987). It follows that shell assemblages that have an equal proportion of opposing valves are unlikely to have been transported far, but a greater degree of transport is indicated as the ratio of one valve to the other departs from unity.

3. Degree of fragmentation (Fig. 3.11). Fragmented shell material is good evidence of transport or a long post-mortem residence time on the sediment surface. Mechanical fracturing and fragmentation of fresh shells requires vigorous current activity and is most commonly the product of nearshore waves. The effect of transitory events, such as storms, does not normally cause fragmentation except in nearshore environments. Biological activity is also a major cause of destruction. Shells are fragmented by predators, such as cephalopods, crustaceans and fish, that crush shells (Carter, 1968) and they are also destroyed by boring organisms such as algae, sponges and gastropods, which weaken shells until they disintegrate with even feeble current activity. As a general rule fragmented shell material is common in two very different contexts. It is common in some highly energetic nearshore environments where the fragmentation is mechanical but it also occurs in quieter offshore environments where the shell material lies for a relatively long period on the sea-floor and is subject to bioerosion.

The sorting and fabric of shell concentrations also provide evidence of the hydrodynamic regime in which they accumulated. Shells can generally be regarded as sedimentary particles and their sorting characteristics can be interpreted in the same way as other sedimentary clasts (Fig. 3.11). In general, good sorting reflects sustained transport by a steady current or winnowing by waves. The exceptions are life assemblages formed of shells of one generation that have a narrow size range. Less well-sorted shell concentrations may partly reflect the size distribution of the animal populations that were the source of

			ASSEMBLAGE	RATIO
(a)	ARTICULATION RATIO	brachial valve / pedicle valve		2 : 1
(b)	RATIO OF VALVES	bv / pv		PV :BV 3 :2
(c)	FRAGMENTATION	A	B	C

Figure 3.10 Some preservational features of shell assemblages which may be a guide to the degree of post-mortem transport.. (a) Articulation ratio; (b) Ratio of valves; (c) Fragmentation, characterized in terms of size and shape. Large fragments (a) are commonly the result of shell crushing by predators. Shell gravels and sand (b and c) commonly result from biogenic weakening by boring followed by hydrodynamic fragmentation.

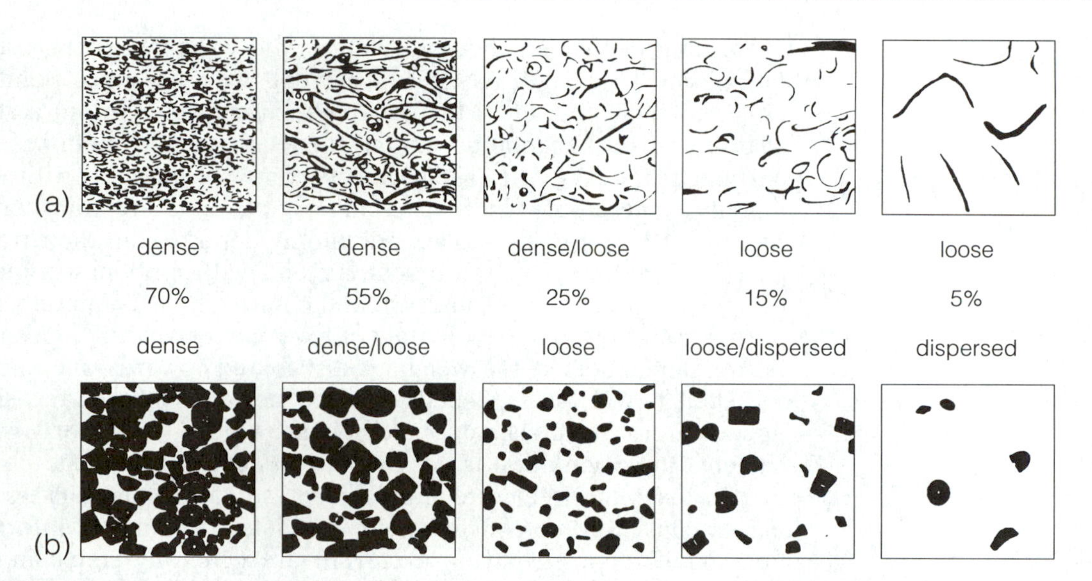

Figure 3.11 Different degrees of close-packing are shown. (a) Typical shells of bivalved organisms. (b) Crinoid columnals. Note that bioclasts of different shapes may constitute the same volume-percent of the rock but exhibit significantly different degrees of close packing. (From Kidwell and Holland, 1991.)

shells and partly reflect later sorting, but the separate effects on the size frequency distribution can rarely be evaluated (see also Chapter 7). The density of packing within a concentration (Fig. 3.11) can provide further evidence of the depositional regime. In general, densely packed concentrations are *in situ*, life assemblages or well winnowed transported assemblages. More loosely packed concentrations are more varied in origin and can be the result of poor sorting of an initially mixed grain size population, or the result of bioturbation.

The orientation of shells is another source of preservational information. Shells can be orientated in a range of attitudes relative to the bedding plane, varying from concordant, to oblique to vertical (Fig. 3.12). They may also be clustered and stacked in a variety of manners (Fig. 3.12). Shells may also show different styles of orientation when viewed on a bedding plane. Some types of orientation are particularly informative.

1. Imbrication: the imbrication of shells provides good evidence of transport by a unidirectional current and can be used to determine the direction of the depositional current.
2. Vertical stacking: this occurs under conditions of sustained oscillatory currents typical of some wave activity.
3. Preferred orientations on bedding planes: strong unimodal orientations usually indicate unidirectional currents (Fig. 3.12). Bimodal, opposed orientations commonly reflect oscillatory (wave) currents. But the origins of shell orientations are not simple. In a unidirectional current shells are transported with their length across current

but during deposition they reorientate into a stable resting orientation which is length parallel to current, with umbones pointing either to right and left. But the weight distribution within a shell also affects the orientation and if, for example, the umbones are particularly heavy the shell can become orientated with the umbones upcurrent. Shells in oscillatory currents are transported rapidly back and forth and are commonly deposited in their transport orientation i.e. length across current, with umbones pointing in opposed directions. Cylindrical and conical shells commonly roll during transport and reorientate into a current-parallel position during deposition. If the weight is distributed towards one end of the shell it will be deposited with the heaviest part of the shell upcurrent (gastropods and belemnites). Under oscillatory wave currents the orientation is normally across current with the points of the cones in either direction. The preservation of shell assemblages convex-up or convex-down can yield some useful information. Shells deposited from a current are generally stable in the convex-up position, but can be deposited concave-up in the lee of a wave ripple after having been flipped over at the ripple crest (Clifton and Boggs, 1970). In fine-grained sediments, unaffected by currents, shells commonly lie concave-up as a consequence of the activities of predators and scavengers and of bioturbation which rotates the shells.

Preservational characters similar to those described above have been used to characterize fossil assemblages with different preservational styles (**taphofacies**) (Speyer and Brett, 1988). A study of taphofacies in a Recent intertidal/shallow shelf environment (Meldahl and Flessa,

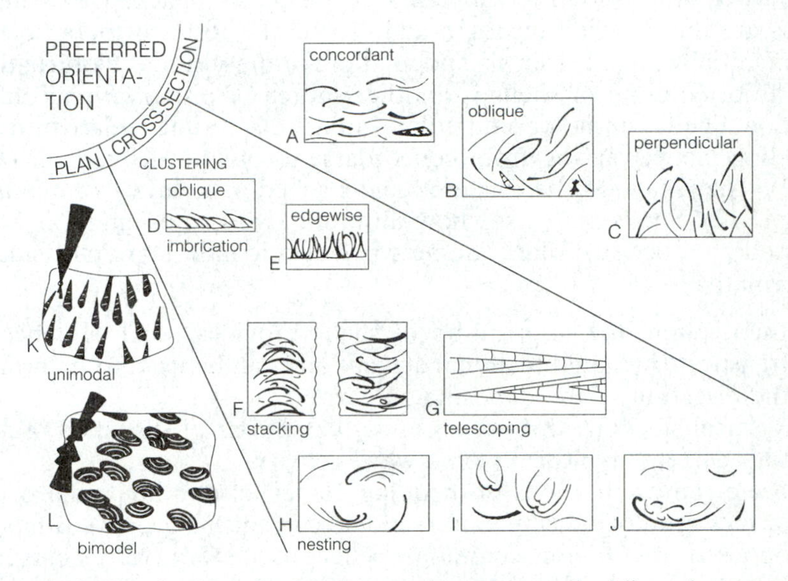

Figure 3.12 Terminology for hardpart orientation on bedding planes and in cross-section of a bed. (From Kidwell *et al.* 1986.)

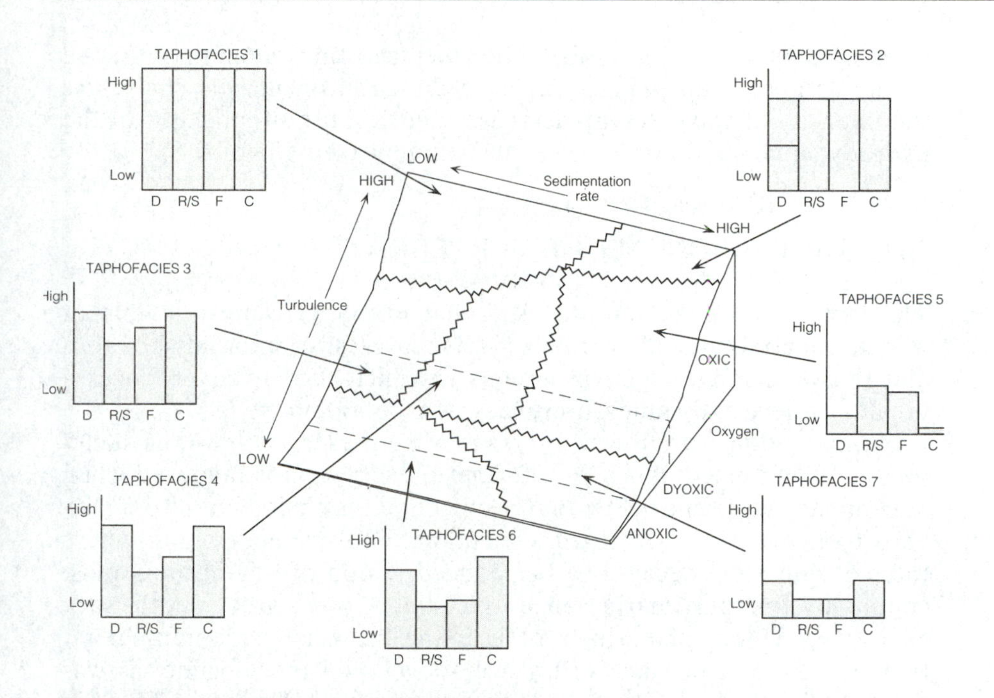

Figure 3.13 Taphofacies for Palaeozoic epeiric seas. Seven distinct taphofacies are recognized on the basis of differences in four taphonomic properties: D=disarticulation, R/S=reorientation and sorting, F=fragmentation and C=corrasion (Corrosion/abrasion). The seven taphofacies in turn, reflect environmental conditions in corresponding fields of the block diagram. Environmental conditions are broadly represented by the three indicated parameters (sedimentation rate, turbulence and oxygenation). (From Speyer and Brett, 1988.)

1990) showed that each lithofacies reflected the prevailing energy in the environment and the preservational characteristics of the shell assemblages reflected their post-mortem history. Those taphofacies with an abrasion-dominated history were found in high-energy low-intertidal and upper-subtidal environments while those with an encrustation-bioerosion history were found in low-energy environments. Seven taphofacies have been recognized by Speyer and Brett (1988) on the basis of:

1. the degree of disarticulation of shells;
2. the amount of reorientation and sorting;
3. the degree of fragmentation; and
4. degree of corrasion.

Each taphofacies was related to differences in turbulence and sedimentation rate in the environment (Fig. 3.13). High levels of disarticulation, good sorting, widespread fragmentation and abrasion reflect high turbulence and low sedimentation rate. Slightly lower scores on these parameters may reflect higher sedimentation rates with more rapid burial, while low scores reflect quiet environments with high rates of deposition (Fig. 3.13). The thickness and geometry of shell accumulations changes in an offshore direction on marine shelves as the intensity of wave energy decreases and is a useful indicator of the hydrodynamic conditions in the environment (see Chapter 6).

3.11 THE TAPHONOMY OF VERTEBRATES

The taphonomy of vertebrates is similar in many respects to that of marine invertebrates. Nevertheless their large size, internal skeletons and the wide range of environments in which they lived produces a greater variety and some differences in taphonomic style.

Marine and lacustrine vertebrates may be preserved as whole skeletons or as dispersed bones. The former preservation is favoured when animal carcasses sink below the thermocline and halocline into a still, anoxic environment which lacks predators and scavengers and where rates of microbial decay may be reduced. Good preservation is most commonly found in marine anaerobic shales of no great depth, such as the varied fauna of marine reptiles in the Jurassic Posidonienschiefer (Chapter 2), and in lakes with a well-developed thermocline, as was the case for the rich Devonian fish faunas of the Caithness Flags in northern Scotland. After death carcasses may suffer various degrees of disaggregation and dispersal. In his study of the varied marine vertebrate faunas of the Jurassic Oxford Clay, Martill (1985) recognized five preservational styles, each modified by diagenetic and compactional processes (Fig. 3.14).

1. Articulated skeletons with intact bone-to-bone relationships. A few of these had coprolitic material in the position of the gut and very rare specimens had evidence of soft tissue. The underside of some remains was better preserved than the upper side because the upper surface is exposed to scavengers, bacterial decay and disruption by escaping gas whereas the lower surface is protected and commonly lies within anoxic sediments where scavenging is inhibited. A similar preservation of the undersurface of vertebrates is seen in the Jurassic Posidonienschiefer (see Chapter 2).
2. Disarticulated skeletons with the bones disaggregated, but a substantial number present. A limited degree of disaggregation is generally the result of scavenging and predation or disturbance by storm waves.
3. Isolated bones and teeth that are mainly the product of skeletons decomposing in the water column or dispersal by scavengers. Teeth may have dropped from living animals, which happens with some living reptiles.
4. Worn bones that reflect a long taphonomic history and are commonly associated with hiatal surfaces.
5. Coprolitic material that may give an insight into the prey that was devoured but cannot always be associated with a particular predator.

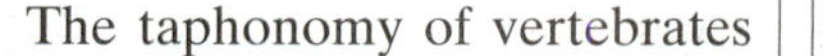

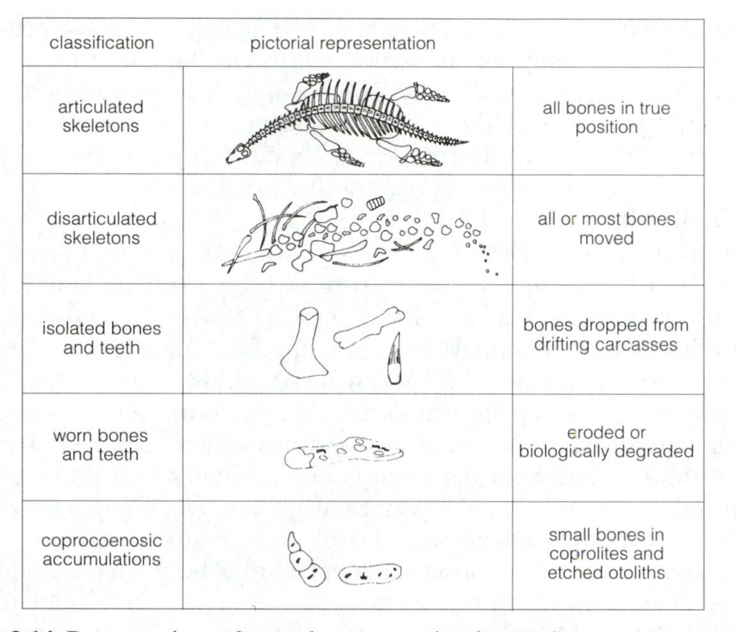

Figure 3.14 Preservation of vertebrate remains in marine settings as shown by the record in the Oxford Clay, Jurassic, England. (From Martill, 1985.)

Preservation of vertebrate material in terrestrial environments is very varied, being associated with a wide range of environments. Whole-body preservation is generally rare and is most commonly associated with Pleistocene and Holocene deposits. The most dramatic of these remains are the faunas preserved by freezing. In September, 1991, a frozen body emerged from a glacier below the Hauslabjoch close to the Austrian-Italian border. The corpse appeared to be that of a mountaineer, but as an investigation of the remains proceeded they proved to be those of a Neolithic hunter, preserved with his clothes and possessions, providing remarkable evidence of the life of a late Stone Age man (Spindler, 1995). Mammoths are the most famous representatives of the faunas that lived during the Last Glacial but a variety of other animals, including horses, woolly rhinoceroses stag-moose and bison are also known. Freezing, which is akin to desiccation, preserves an exceptionally full record of the animals. Soft tissue retains histological details and the nature of wounds and disease. Gut contents reveal the nature of the animal's diet. The remarkable detail of reconstruction possible and the controversies that nevertheless still remain are described in Dale Guthrie's fascinating reconstruction of an exhumed bison from Alaska (Guthrie, 1990). Although discussion of frozen carcasses has often centred on the cause of death, their preservation probably depends much more on special conditions after death. The key to preservation appears to commonly entail entrapment in mud, rather than drowning in water, where the carcass would float to the surface, followed by freezing and then rapid sedimentation and burial below the level of summer thaw.

Whole-body remains are also preserved in peat. Carcasses that sink rapidly into peat may be preserved with skin, hair and other tissue, though bone may be destroyed. The strongly acidic conditions within peat inhibits microbial decay so favouring the preservation of soft tissue, but causing the destruction of bone. Some of the most spectacular remains preserved in peat are those of Iron-Age man preserved in a Danish bog (Glob, 1971).

The tar pits of La Brea are another striking example of vertebrate preservation. The tar pits were formed by a seep of heavy hydrocarbons accumulating as a viscous pool at the surface. A variety of large Pleistocene mammals became engulfed in the tar (see also Chapter 10), presumably as they were drinking from surface water, and their bones have been preserved. The skeletons are almost always disarticulated, either by predators and scavengers or by the circulation of the tar. The bone material is exceptionally well preserved and skeletons, representative of a variety of species, can be assembled from disparate parts that are not necessarily from a single carcass.

In many terrestrial environments vertebrate bones have undergone various degrees of transport. In these circumstances an understanding of the taphonomic history of a bone deposit is vital in determining whether the remains belong to the environment in which they are found. The dispersal of bones may occur through the activities of predators or scavengers, or by physical processes (see Behrensmeyer, 1991). The transport of bones in running water follows the same laws that govern other sedimentary particles. However, bones are complex particles with different densities, complex shapes and irregularly distributed pore space. The overall density of fresh bone varies from less than 1 to 1.7, which means that some float and some sink. Teeth on the other hand have a density of about 2 and are less readily transported and may be concentrated in lag deposits. Bones from different parts of a carcass have a different potential for being transported, so become separated during transport. Amongst medium to large mammals it has been found that vertebrae, ribs and the sternum are most easily transported, then limb parts and finally skulls, mandibles and teeth.

Transported bones are recognized by their dispersed preservation, by the sorting of particles, their current orientation and their association with fluvial facies. The degree of transport may be assessed by the degree of separation of different elements of a skeleton, the amount of breakage and the degree of abrasion. Fresh bone does not fragment or abrade easily in sand and gravel rivers, though weathered bone is much more rapidly affected. Generally, abraded bone signifies substantial downstream displacement. Bones deposited in terrestrial environments, particularly fluvial ones, are subject to repeated burial and exhumation so bone accumulations are commonly a time-averaged association of remains of different ages and from different environments.

Cave and fissure remains form one special category of terrestrial remains that are volumetrically small but have yielded a disproportionately large body of information about vertebrate faunas. Vertebrate remains may be concentrated in surficial karst, in a variety of other fissures commonly resulting from downslope movement of blocks or extensional tectonic movements, in lava tubes and in cavities amongst accumulations of large boulders. There are three main situations in which an accumulation may develop (Simms, 1994):

1. The animals lived and died in an enclosed environment. This situation includes animals such as cave fish, but preservation of such faunas is likely to be very rare.
2. Faunal remains were transported into the environment by predators and scavengers. Predators that used the cave as a dwelling can sometimes be identified by the association of bones of carnivores mixed with a variety of herbivores. The herbivore bones generally show signs of crushing and gnawing. The bone accumulations are often concentrated at the cave entrance (Fig. 3.15a). Roosting and hibernation sites may be identified by being low-diversity deposits on the floor of a cave (Fig. 3.15b). The remains of bats and birds, commonly associated with phosphate deposits derived from accumulations of guano, suggest a roosting site, whereas monospecific accumulations of mammals such as bears suggest a

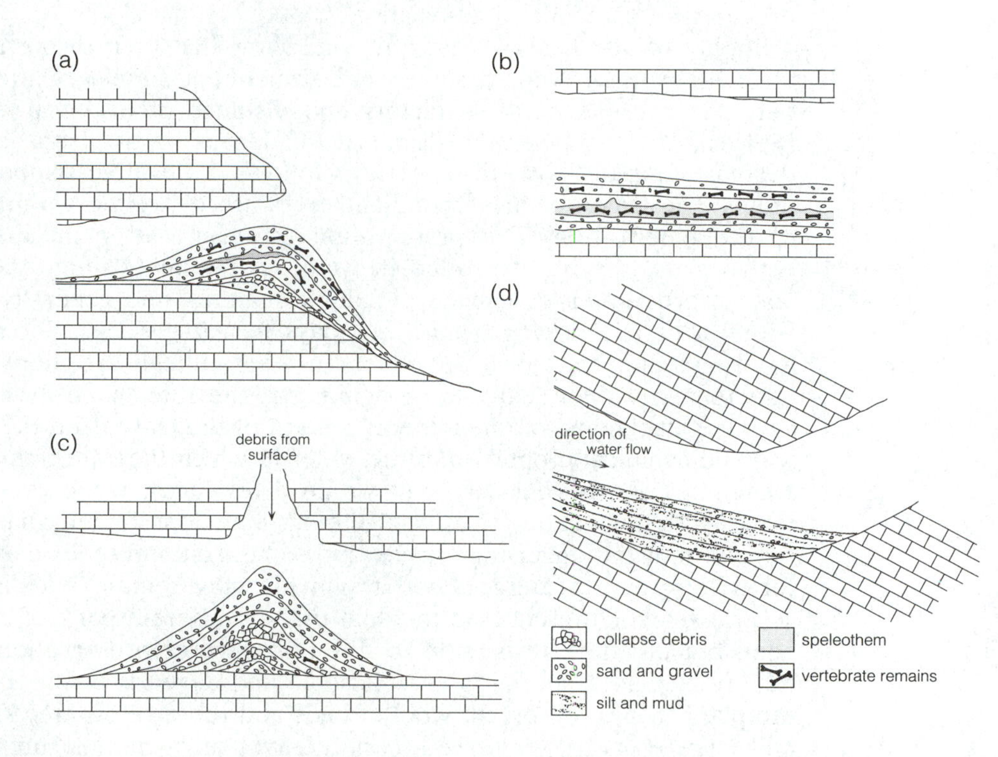

Figure 3.15 Bone accumulations in caves. (a) Cave entrance accumulations: commonly with bones brought to the cave by predators together with the remains of the cave dwellers. (b) Low diversity bone-rich layers interbedded with barren layers, commonly reflecting intermittent roosting or hibernation sites. (c) Pitfall trap with talus-cone of debris from a surface pit. (d) Flood deposits with fining upward cycles and coarse lags of vertebrate remains. (Modified from Simms, 1994.)

hibernation site. The accumulation of many thousands of skeletons of the bear *Ursus spelaeus* in a single cave at Mixnitz in Austria is a particularly spectacular example.

3. Remains were transported into the environment by physical means. Gravity and flowing water transport sediment together with bones into caves and fissures. Debris falling through the roof of a cave from a karstic pit above may produce piles of debris on the cave floor and may potentially contain remains that accumulated in the surface pit (Fig. 3.15c). Commonly sediment and bone will be deposited in caves by running water, either during normal flow conditions or at times of flood (Fig. 3.15d). These remains are likely to be derived from the land upstream.

3.12 THE PRESERVATION OF PLANTS

Plants are composed of a variety of parts (branches, roots, leaves, pollen, fruit, seeds), some of which become separated during life while others become disaggregated after death. An understanding of dispersal processes affecting the various parts is important to the interpretation of palaeofloral associations.

Studies of dispersal of leaves by wind show that their dispersal was affected by their weight, shape and size, but a simple relationship between any one of these factors and distance of dispersal is rare (Spicer, 1991). Generally dispersal of leaves away from source decreases rapidly with distance and follows a negative exponential mode. The leaves of most woodland trees are deposited within 50 m of the parent tree. The preservation of nearly all plant material depends on it being deposited in water. After being immersed, the leaf absorbs water, is leached of soluble material and generally sinks within a period varying from a few days to a few weeks. Soon after entering water, microbial decay sets in. Leaf cuticles are more resistant to decay than cellulose or lignin, but the rate of decay is very variable, depending on the mineral content of the leaves and the physical and chemical properties of the water in which they are immersed. Deciduous leaves generally sink within a few days, while evergreen leaves with thick cuticular or wax coatings may float longer. Diaspores remain buoyant longer than leaves and some fruit and seeds may float for a long period. Large wood fragments may remain buoyant for several years. Differences in the float time of different parts of a plant contribute to their dispersion to different sites of preservation. The fidelity of plant material is very variable and depends on the part of the plant that is preserved. Rootlet beds and more rarely tree stumps and large roots are preserved *in situ*, leaves are generally preserved near their origin, but wood is commonly transported large distances and may be found in marine environments far from shore. Pollen and spores have walls composed of sporopollenin, which is particularly resistant to decay, so they too are commonly widely dispersed by wind and water, though a recurrently recorded association of pollen or spores is usually a good record of the surrounding vegetation.

Plant material may be preserved in a variety of forms including:

1. preservation of original material
2. permineralization
3. coalification and
4. moulds and casts, impressions and compressions.

Preservation of original plant material is mainly in the form of resistant spores and pollen that may retain all the surface detail of the cell wall. Permineralization preserves material in three dimensions, commonly retaining the detailed cellular structure. The preservation can be in silica, carbonate or pyrite. Silica may permeate the interior of the plant cells, leaving the cell wall as an organic coat, or it may permeate the whole plant structure. Silicification is related in some instances to highly silicic volcanic environments, as is traditionally envisaged for the Devonian Rhynie Chert. However, it also occurs in silica-poor environments where it is thought to be related to microbial activity in the early stages of organic decay, but the exact conditions that promote the silification are far from clear. Permineralization of plant material by calcium carbonate has produced the superb preservation of woody material found in coal balls, which are carbonate concretions formed in mainly uncompacted peaty material in Carboniferous Coal Measure bogs. The growth of the concretions appears to require a marine source of sulphate to mediate sulphate reduction of organic material. Although some of the coal horizons bearing concretions are overlain by a marine band, this is not always the case. A solution for this dilemma has been offered by Spicer (1989), who suggested that rising sea-level may cause an intrusion of a sub-surface saline wedge within a delta plain without necessarily inundating the whole delta. Permineralization by pyrite is mainly confined to marine environments where it preserves woody tissue and seeds as in the terrestrial plant flora found in the marine Tertiary London Clay. Coalified remains are common in a wide variety of facies; with this preservation the external details may be exquisitely preserved but the internal structure is lost.

A large proportion of plant fossils are preserved as moulds and casts. Leaves retain enough rigidity to leave an impression on a soft sediment surface, even though they may be destroyed before burial. More commonly the leaves are buried and preserved as internal or external moulds, commonly with a carbonaceous veneer. Plant stems and roots undergo various degrees of compression during burial which may affect their morphology, so it is important to appreciate the processes that may be involved (Rex, 1986). Figure 3.16a shows how a hollow *Calamites* stem could be preserved as an undistorted external mould (em) or internal mould (im). Alternatively it is likely to be compressed during burial to give an oval cross-section (Fig. 3.16b to e). In some instances compression of the coaly vertical sides to the plant stem or

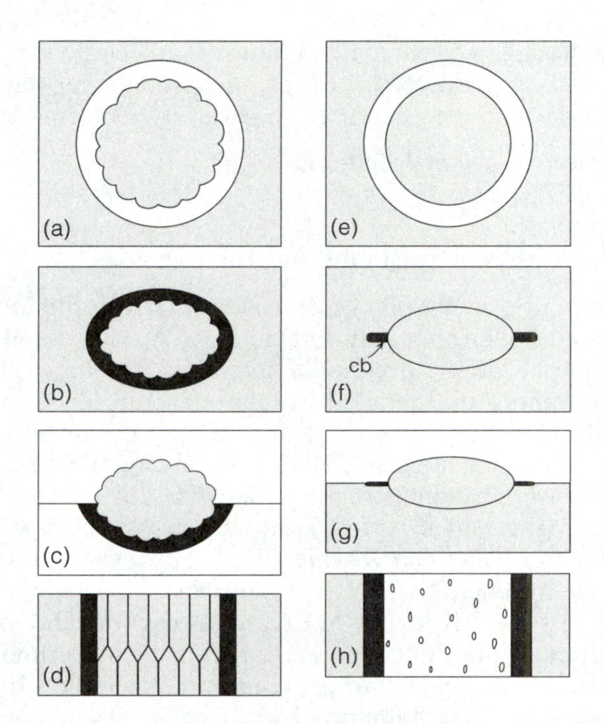

Figure 3.16 Preservation of plant material and the formation of 'compression borders'. (a) A hollow *Calamites* stem is filled with sediment. The stem has internal ribbing. (b) After compression the stem is deformed into an ellipse and the plant tissue is coalified. (c) Pathway that a fracture plane might take to expose the fossil. (d) View of the pith cast and coalified border seen from above. (e) Hollow stem filled and buried in sediment. (f) Compressed stem and the deformed tissue forms a compression border(cb). (g) The pathway taken to expose the cast. (h) View of the pith cast from above, showing prominent compression borders. (Modified from Rex, 1986.)

root squeeze it laterally to give a compression border (Fig. 3.16 f to h). Lateral branches and leaf bases are particularly prone to deformation and distortion and may produce many misleading shapes.

3.13 SUMMARY POINTS

- Except in rare instances, all biota lacking bio-mineralized parts is lost. This means that both the algae at the base of the food chain and many of the soft-bodied animals which are an integral part of the trophic structure are unrecorded in the fossil record, except sporadically as trace fossils.
- Generally less than 50% of a living community consists of animals with a mineralized skeleton.
- The shelled fossils that are preserved represent only a small proportion of the living populations that had shells, most shells having been destroyed by physical fragmentation, bioerosion, and most

importantly by chemical dissolution. The absent fossil biota represents an important 'information loss' that should be appreciated in any palaeobiological investigation.

- The fauna that is preserved is fortunately a rich source of biological and geological information which is revealed by careful taphonomic analysis. Shell accumulations are time-averaged to varying degrees but generally are representative of the habitats in which they lived and died.
- Shells are destroyed on the sediment surface by physical, chemical and particularly biological processes. They commonly suffer extensive dissolution after burial.
- Shells are preserved whole, as moulds and casts, and may be particularly well preserved in pyrite, phosphate and silica and within carbonate concretions.
- Organisms with soft tissue normally require early mineralization in order to be fossilized.
- Lagerstätten are deposits with exceptional preservation of fossils. They may be concentration deposits or conservation deposits. The latter include obrution deposits that commonly represent populations and communities that have been instantaneously overwhelmed and buried by sediment. Stagnation deposits accumulate in anoxic or hypersaline conditions favouring special preservation. They present a time-averaged view of the biota of a particular environment.
- Shell concentrations are a major source of fossil data. An understanding of their taphonomy is important in assessing the palaeoecological information they yield. They also provide important information about the hydrodynamic environment where they accumulated.
- Different styles of concentration (event, composite, hiatal and lag concentrations) are related to different rates of accumulation and different rates of sedimentation.
- The taphonomy of vertebrates is similar to that of invertebrates, except that a greater proportion of marine vertebrate faunas are nektonic and liable to dispersal by floating; the terrestrial faunas are more likely to be dispersed in fluvial systems.
- Plant taphonomy is influenced by the different potential for dispersal and preservation of different parts of the same plant.

So how useful is the palaeontological record, considering the patchy preservation? Well, it depends on which questions are being asked. The record is generally inadequate to answer questions about detailed biological relationships in the past, but it is adequate to answer questions about the broad composition of the shelled part of ancient communities, the relationship of those communities to environment and how those communities changed with time. Lagerstätten are important because they preserve taxa that otherwise go unrecorded, they provide more detailed information on a whole fauna and allow reconstructions of ecosystems and they help to correct perspectives based on shelly faunas alone. But Lagerstätten are rare, so the information gained from them is limited to a few points in time. In palaeoecology it is important to investigate problems that are capable of being solved with the information available.

3.14 FURTHER READING

Allison, P.A. and Briggs, D.E.G. (eds) 1991. *Taphonomy: Releasing the Data Locked in the Fossil Record*, Plenum Press, New York, 560 pp. A comprehensive and authoritative account of most aspects of taphonomy at a fairly advanced level.

Donovan, S.K. (ed.) 1991. *The Processes of Fossilization*, Belhaven Press, London, 303 pp. Covers some of the same ground as Allison and Briggs, but has particularly useful chapters on the taphonomy of different groups (trilobites, ammonites, echinoids).

3.15 REFERENCES

Allison, P.A. 1986. Soft-bodied animals in the fossil record: The role of decay in fragmentation during transport. *Geology* 14, 979–981.

Allison, P.A. and Briggs, D.E.G. 1991a. The taphonomy of soft-bodied animals. *In:* Donovan, S.K. (ed.), *The Processes of Fossilisation.* Belhaven Press, London, 120–140.

Allison, P.A. and Briggs, D.E.G. 1991b. Taphonomy of non-mineralized tissues. *In:* Allison, P.A. and Briggs, D.E.G. (eds) *Taphonomy: Releasing the Data from the Fossil Record*. Plenum Press, New York, 26–71.

Barthel, K.W., Swinburne, N.H.M. and Conway Morris, S. 1990. *Solnhofen: A Study in Mesozoic Palaeontology*. Cambridge University Press, Cambridge, 236 pp.

Behrensmeyer, A.K. 1991. Terrestrial vertebrate accumulations. *In:* Allison, P.A. (ed.) *Taphonomy: releasing the data locked in the fossil record.* Plenum Press, New York, 229–335.

Brett, C.E. and Baird, G.C. 1986. Comparative taphonomy: A key to palaeoenvironmental interpretation based on fossil preservation. *Palaios* 1, 207–227.

Canfield, D.E. and Raiswell, R. 1991. Pyrite formation and fossil preservation. *In:* Allison, P.A. and Briggs, D.E.G. (eds) *Taphonomy: Releasing the Data Locked in the Fossil record.* Plenum Press, New York, 337–387.

Carter, R.M. 1968. On the biology and palaeontology of some predators of bivalved molluscs. *Palaeogeography, Palaeoclimatology, Palaeoecology* 4, 29–65.

Clifton, H.E. and Boggs, S.J. 1970. Concave-up pelecypod (*Psephidia*) shells in shallow marine sand, Elk River Beds, southwestern Oregon. *Journal of Sedimentary Petrology* 40, 888–897.

Conway Morris, S. 1986. The community structure of the Middle Cambrian Phyllopod bed (Burgess Shale). *Palaeontology* 29, 423–467.

Fisher, I.S.J. and Hudson, J.D. 1987. Pyrite formation in Jurassic shales of contrasting biofacies. *In:* Brooks, J. and Fleet, A.S. (eds) *Marine Petroleum Source Rocks.* Geological Society Special Publication, 26, Geological Society of London, London, 69–78.

Frey, R.W. and Henderson, S.W. 1987. Left-right phenomena amongst bivalve shells: Examples from the Georgia coast. *Senckenbergia Maritima* 19, 223–247.

Fürsich, F.T. and Aberhan, M. 1990. Significance of time-averaging for palaeocommunity analysis. *Lethaia*. 23, 143–152.

Fürsich, F.T. and Oschmann, W. 1993. Shell beds as tools in basin analysis: the Jurassic of Kachchh, western India. *Journal of the Geological Society, London* 150, 169–185.

Glob, P.V. 1971. *The Bog People*. Paladin, London, 137 pp.

Glover, C.P. and Kidwell, S.M. 1993. Influence of organic matrix on the post-mortem destruction of molluscan shells. *The Journal of Geology* 101, 729–747.

Guthrie, R.D. 1990. *Frozen fauna of the mammoth steppe: the story of Blue Babe*. The University of Chicago Press, Chicago, 323 pp.

Hudson, J.D. 1982. Pyrite in ammonite-bearing shales from the Jurassic of England and Germany. *Sedimentology* 29, 639–677.

Kidwell, S.M. 1991. The stratigraphy of shell concentrations. *In:* Allison, P.A. and Briggs, D.E.G. (eds) *Taphonomy: releasing the data locked in the fossil record.* Plenum Press, New York, 115–209.

Kidwell, S.M. and Bosence, D.W.J. 1991. Taphonomy and time-averaging of marine shelly faunas. *In:* Allison, P.A. and Briggs, D.E.G. (eds) *Taphonomy: Releasing the Data Locked in the Fossil Record.* Plenum Press, New York, 115–209.

Lawrence, D.R. 1968. Taphonomy and information losses in fossil communities. *Geological Society of America Bulletin* 79, 1315–1330.

Martill, D.M. 1985. The preservation of marine vertebrates in the Lower Oxford Clay (Jurassic) of central England. *In:* Whittington, H.B. and Conway Morris, S. (eds) *Extraordinary fossil biotas: their ecological and evolutionary significance.* Philosophical Transactions of the Royal Society of London, B311, The Royal Society, London, 155–165.

Meldahl, K.H. and Flessa, K.W. 1990. Taphonomic pathways and comparative biofacies and taphofacies in a Recent intertidal/shallow shelf environment. *Lethaia* 23, 43–60.

Raiswell, R. 1988. Chemical model for the origin of limestone-shale cycles by anaerobic methane oxidation. *Geology* 16, 641–644.

Raup, D. and Stanley S.M. 1971. *Principles of Paleontology.* Freeman and Company, San Francisco. 481pp.

Rex, G. M. 1986. Further experimental investigations on the formation of plant compression fossils. *Lethaia* 19, 143–159.

Seilacher, A., Reif, W.-E. and Westphal, F. 1985. Sedimentological, ecological and temporal patterns of fossil Lagerstätten. *In*: Whittington, H.B. and Conway Morris, S. (eds) *Extraordinary Fossil Biotas: Their Ecological and Evolutionary Significance. Philosophical Transactions of the Royal Society, London* B 311, 5–26.

Simms, M.J. 1994. Emplacement and preservation of vertebrates in caves and fissures. *Zoological Journal of the Linnaean Society* 112, 261–283.

Speyer, S.E. and Brett, C.E. 1988. Taphofacies models for epeiric sea environments: Middle Paleozoic examples. *Palaeogeography. Palaeoclimatology, Paleoecology* 63, 225–262.

Spicer, R.A. 1989. The formation and interpretation of plant fossil assemblages. *Advances in Botanical Research* 16, 95–191.

Spicer, R.A. 1991. Plant taphonomic processes. *In:* Allison, P.A. and Briggs, D.E.G. (eds) *Taphonomy: releasing the data locked in the fossil record.* Plenum Press, New York, 71–113.

Spindler, K. 1995. *The man in the ice*. Phoenix, London, 305 pp.

Staff, G.M., Stanton, R.J.J., Powell, E.N. and Cummins, H. 1986. Time averaging, taphonomy, and their impact on paleocommunity reconstruction: Death assemblages in Texas bays. *Geological Society of America Bulletin* 97, 428–443.

Adaptive morphology 4

Autecology is the study of the life modes of organisms and the relationship between individuals and the environment. It focusses on the growth and shapes of organisms, whether individuals or colonies, and the correspondence of morphology to both life strategies and habitats. Every organism is adapted to a specific life mode and is contained by environmental limitations. The investigation of adaptive morphology of fossil animals and plants has many angles; many functional studies have been published. Dodd and Stanton (1981) and Pickerill and Brenchley (1991) have discussed and listed a large selection. Another *modus operandi* involves the analysis of autecology against a framework of long-term biological change. With this strategy the importance and influence of functional changes at the individual level is emphasized. These changes can be transmitted and amplified into evolution at the community and ecosytem levels and are thus building blocks of large-scale biotic and planetary change.

4.1 INTRODUCTION

Many morphologies and structures have been developed, as adaptations, in response to a particular life style. Evolutionary convergence is thus rife in the fossil record. For example, dolphins, sharks and swordfish together with the fossil ichthyosaurs all have similar exteriors, although they are only distantly related. Their spindle-shaped bodies are streamlined for high-velocity movement through water. More remarkable is the development of coral-like external morphologies not only in the solitary rugose and scleractinian corals but also during the Permian in the richthofeniid brachiopods and during the Cretaceous in the rudist bivalves. Clearly the relationship of organisms to their environments and life styles can in many cases be interpreted from morphology. Adaptation is broadly defined as the goodness of fit of an organism to its environment, modified by natural selection. This concept is derived from the core of Darwinian evolutionary theory.

Organisms are functionally constrained by their overall morphology, inherited from ancestors; moreover, biological structures may have multiple functions. Some structures apparently have a nonadaptive significance, such as male nipples, and other more neutral structures

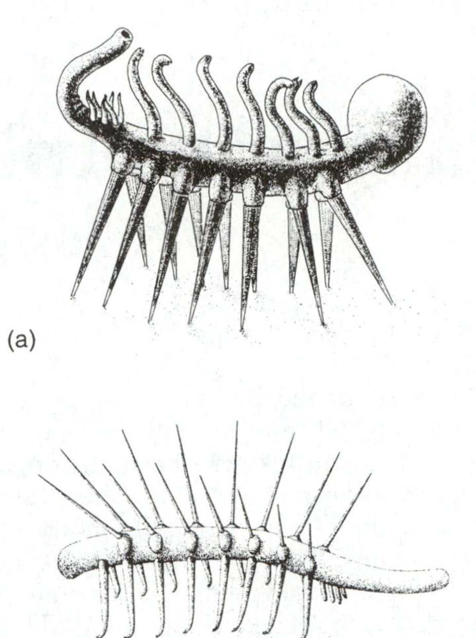

Figure 4.1 Reconstructions of the mid-Cambrian armoured onychophoran *Hallucigenia* : a. Mid 1970s reconstruction, b. Current reconstruction, early 1990s. (Redrawn from various sources.)

may remain in evolving populations through genetic drift. But the major environmental changes driving mass mortalities and extinction events disrupt the process of fine-tuning through selection and can dramatically reset macroevolutionary clocks. The process then starts again.

Unfortunately fossil organisms cannot be observed in action. Most analyses of adaptations and functions involve at least some speculation which may be subjected to tests and subsequent revision. *Hallucigenia*, a spiny worm-like animal described from the mid-Cambrian Burgess Shale was initially reconstructed, on the then available evidence (Conway Morris, 1978) as a cylindrical, flexible tube supported by pairs of spines and with a series of tube-like extensions developed along the upper surface of the animal. New discoveries suggested that the animal had been reconstructed upside-down (Ramsköld and Hou, 1991). The spines were in fact developed in pairs along the back of the animal, probably for protection; the tubes were the legs. *Hallucigenia* was not so odd after all; it was merely an armoured onychophoran (Fig 4.1).

4.2 TERMINOLOGY

Studies of adaptation have evolved a range of terms. Adaptation, strictly speaking, is the fit of an organism to its environment; fundamental, however, is the role of natural selection in shaping adaptations.

Exaptations are functionally useful structures that apparently were not shaped by natural selection, whereas aptations are any structures fitted to a particular function; **aptations** thus include both adaptations and exaptations. Moreover exaptations may be transferred from pre-existing adaptations used in other functions (**coaptations**) or from structures apparently without a previous functional significance (**non-aptations**). **Preadaptation** involves structures or groups of structures already functional but available and suited by chance for a more innovative function; this, subsequently, becomes the main function of the structure in future generations (**postadaptation**).

Although there have been a number of philosophical studies of the definition and use of these terms (Gould and Vrba, 1982), in practice the terms adaptation and preadaptation are most widely used and sufficient to deal with the majority of palaeontological cases (Skelton, 1990).

4.3 INFLUENTIAL FACTORS

The conventional view that genome information controls the development of structures and their functions must now compete with a more holistic concept, suggesting structures may arise randomly, through chance and contingency, as part of a more inclusive package; subsequent selection and fine-tuning then develops their functions. This reflects a current move from a deterministic world, governed by set rules, to more chaotic systems where self-organization amongst organisms and their structures is possible.

The evolution of form can be considered in terms of three main constraining factors: the genome, the development and the function of the organism. These factors have been graphically displayed as a triangle (Seilacher, 1989). The three apices of the triangle are combined to define the overall constructional morphology of an organism. This has been modified more recently to include environmental factors; the original triangle is now an inverted pyramid with four apices and combines an inclusive organism together with its constructional morphology (Fig. 4.2). Studies of construction morphology have been applied to a number of groups such as the Ediacara animals (Seilacher, 1989), and the Brachiopoda (Vogel, 1986).

The development of a set body plan (Fig. 4.2) is mainly controlled by its evolutionary history or genetic heritage. The phylogenetic apex thus marks morphology arising from genetic signals. The fabricational factor involves mainly the growth programme of the organism. The constraints on organisms with accretionary growth programmes, such as the brachiopods and molluscs, are quite different from those that grow by ecdysis or moulting, such as the arthropods, although these animal groups increase steadily in size throughout ontogeny. The growth of vertebrate bone, however, involves a different set of problems. The internal skeleton of a vertebrate must support a changing

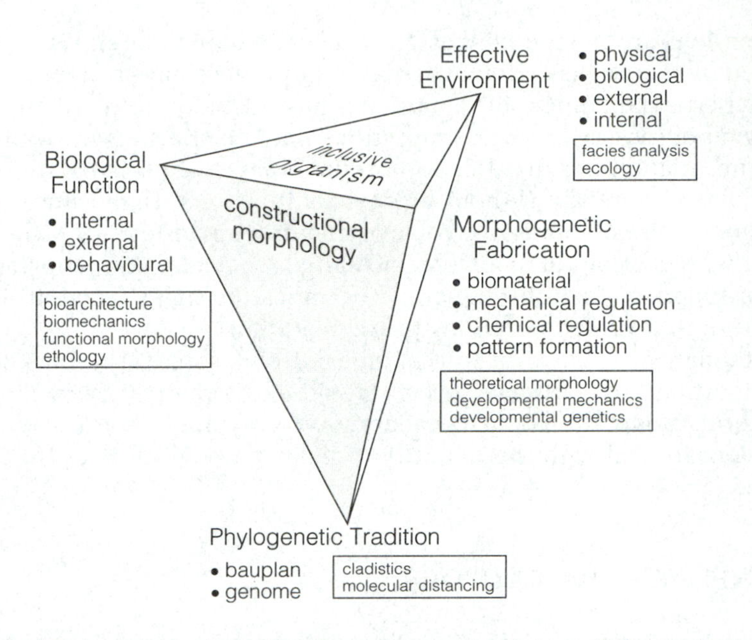

Figure 4.2 A morphodynamic view of functional morphology: changes in biological function, effective environment and morphogenetic fabrication can trigger evolutionary changes. (After Seilacher, 1992.)

body weight throughout its life span. Simple isometric growth of bone is in theory matched by a cubic expansion of body volume; if this process continued, soon the increased weight of the animal would no longer be adequately supported. Vertebrate bone, however, can be modified during growth by resorption; bone can be preferentially thickened or thinned as required.

Environmental controls have an important influence on adaptation; environment feeds in a range of both external and internal physical, chemical and biological factors, generating an interplay between an evolving organism and its setting. Environmental factors clearly control where and how an organism lives and may induce ecophenotypic variation. Crevice-dwelling bivalves commonly adopt the external shape of their surroundings. These shape changes clearly have functional consequences; however, the modified shapes cannot be transmitted through genetic material to future generations.

The functional or adaptational apex marks the behaviour, function and biomechanics of an organism. The various parts and structures of an organism are thus a fundamental source of information for the ecology and operation of fossil animals and plants. Nevertheless groups of structures can combine to achieve a particular function. Coadaptation emphasizes the total integration of a successful organism.

4.4 GROWTH STRATEGIES

Organisms grow in four main ways. Most shelled animals such as the brachiopods and molluscs grow by accretion, where additional mineralized material is added to the existing exoskeleton from generative zones on the outer part of the mantle which lines the shell. Some animals such as the echinoderms and sponges add new material in the form of calcitic plates or mineralized spicules, respectively, to an existing skeleton. Arthropods grow by ecdysis or moulting, periodically shedding their exoskeletons in favour of a larger, occasionally different-shaped, covering. Vertebrates have adopted growth strategies associated with modification. Bone may grow in all directions; however, it can also be modified by resorption with the redistribution of bone material helping to strengthen some areas of bone under stress. Growth programmes provide a major constraint on the adaptive development of an organism. Adaptation must operate within the framework of genetically controlled construction projects, forming the basis of ontogeny derived from the genome.

4.4.1 Shape description

Detailed description of fossil material is an essential prerequisite to the study of autecology. A range of careful descriptive and illustrative methods, together with codes of systematic practice originating in the mid-1600s evolved and form the basis for modern taxonomy. A large range of necessary terminology is now available for virtually every fossil group together with illustrative aids such as macrophotography, camera-lucida systems, the scanning and transmission electron microscopes together with X-ray techniques. Although adequate data may be derived from high-quality descriptions and illustrations, much more may be extracted by numerical and statistical techniques based on imaging and measuring fossil material (Harper and Owen, 1988; Harper and Ryan, 1990; Jones, 1988). A number of packages, for example PALSTAT (Ryan *et al.*, 1994) and MVSP (Kovach, 1993), offer a range of statistical and graphical techniques, and can analyse imported data matrices from a number of spreadsheets. These packages were developed for PC platforms. Cladistic techniques are available on PAUP (Swofford, 1993) and MacClade (Maddison and Maddison, 1992); the most recent versions of these programs operate on the Macintosh platform.

Computer-generated reconstructions, for example, from serial sections can build a 3D model of an organism; the internal features of brachiopods have been reconstructed with considerable success (Herbert *et al.*, 1995). Moreover, much fossil material, particularly from the world's mountain belts, has suffered tectonic deformation and a range of computer-based techniques has been implemented to remove the effects of tectonic distortion, permitting the accurate description of deformed material. In a study of seven putative species of strained

Cambrian trilobites from the Cambrian rocks of Kashmir, microcomputer subtraction of deformational effects established the existence of only one morphospecies; this species suggested a precise mid-Cambrian age for the strata and an affinity with coeval rocks in China and India (Hughes and Jell, 1992).

Detailed morphological studies generally form the basis of understanding functional and adaptational morphology. The structure and operation of the brachiopod hinge has underpinned a number of functional and evolutionary studies of the phylum as a whole (Jaanusson, 1971). An exhaustive multivariate analysis of the hinge elements of the articulate brachiopods suggested that the two types of teeth, deltidiodont (simple, grow without resorption) and cyrtomatodont (knob-like, grow with resorption), are part of two different types of hinge mechanism (Carlson, 1989). This morphological analysis suggested that the cyrtomatodont teeth, developed by resorption of shell material, evolved across a number of groups as an adaptation for a more advanced and efficient hinging mechanism.

Functional advantage may not always be apparent even from detailed studies of fossil morphology. The verbal, visual and statistical description of the morphology of the bellerophontiform mollusc *Pterotheca* from the lower Silurian rocks of the Pentland Hills, near Edinburgh revealed no functional advantage to the widespread development of asymmetry in a large sample of this algal-grazing gastropod (Fig. 4.3). Rather populations of *Pterotheca* probably used this extreme intraspecific variation to confuse the search images of potential predators such as orthoconic nautiloids (Clarkson *et al.,* 1995).

4.5 INVESTIGATIVE METHODS

Functional morphology can be investigated in a number of ways. Hypotheses arising from these methods, however, should be testable against geological evidence. First, structures can be compared directly with modern, working counterparts. The greatest success is achieved with homologous structures such as the comparison of the wings of the Cretaceous bird *Ichthyornis* with those of a living tern or seagull. In the absence of homologues, analogous comparisons such as comparisons between the wings of the pterosaurs with those of a buzzard or a hang glider are nevertheless still useful. There is, however, a strong element of circular reasoning involved in these techniques. In comparing the wings of fossil and living birds, or even the wings of fossil flying reptiles with those of living birds or aircraft, we have already decided that these fossil structures were also used for flight.

Secondly, one or more functions may be specified for a particular structure. Each function may be tested against the efficiency of a physical or mathematical model for the working structure. This more scientific methodology is termed the paradigm approach and provides a testable hypothesis.

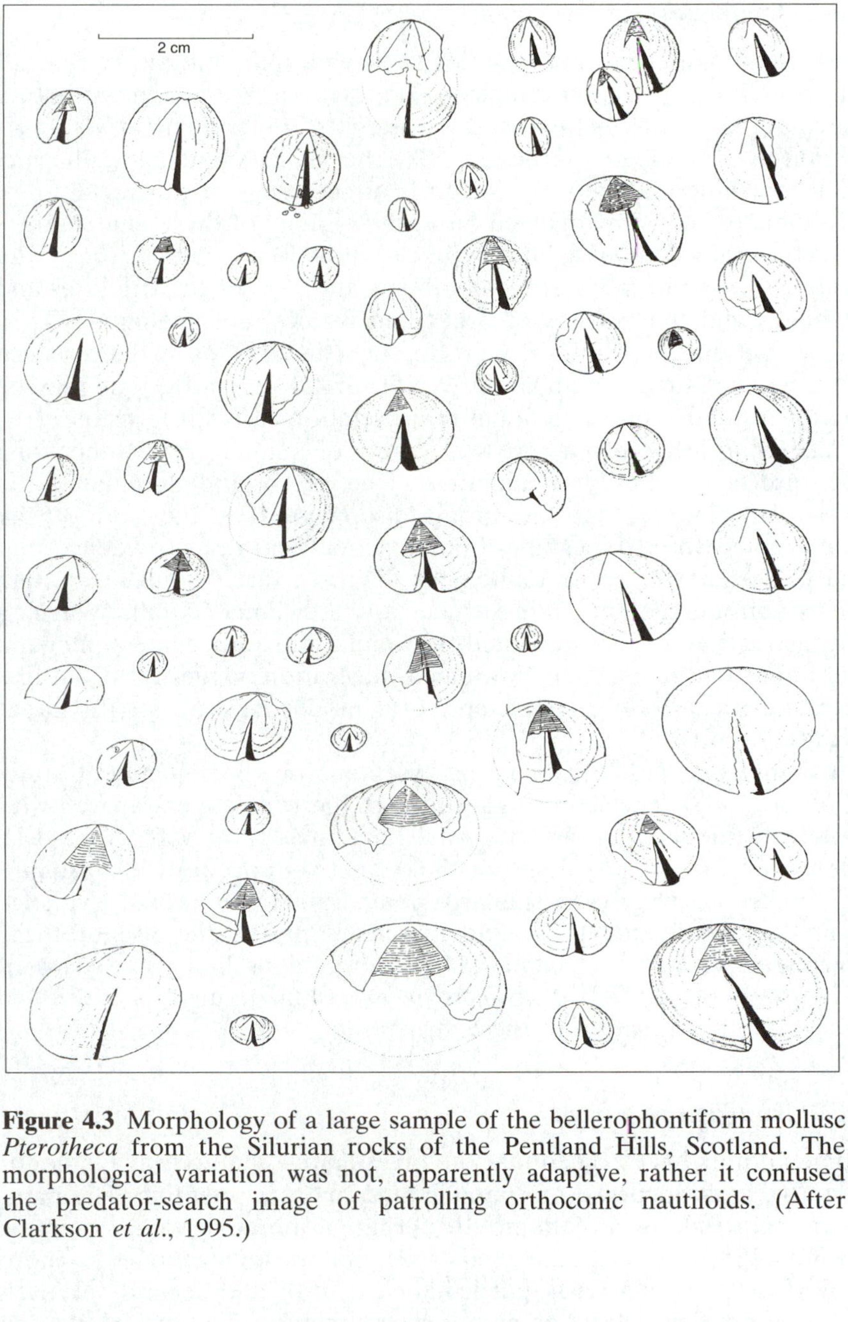

Figure 4.3 Morphology of a large sample of the bellerophontiform mollusc *Pterotheca* from the Silurian rocks of the Pentland Hills, Scotland. The morphological variation was not, apparently adaptive, rather it confused the predator-search image of patrolling orthoconic nautiloids. (After Clarkson *et al.*, 1995.)

Thirdly there is a range of broader experimental techniques where physical models are subjected to simulated environments. These techniques generally monitor the performance of models when subjected to real-life conditions, usually in flume tanks, wind tunnels or other types of environmental chamber.

4.5.1 Analogues and homologues

The use of analogues and homologues is widespread in most branches of palaeontology. The morphology and function of a modern structure is compared with an assumed counterpart in an extinct or fossil organism. Homologous structures, like the shells belonging to all articulate brachiopods, the wings of birds or the lungs of the vertebrates, presumably have only evolved once; the biology of these characters is directly equivalent and allows precise functional comparisons. The limbs of the arthropods and amphibians, the eyes of the trilobites and primates and the tail fins of fishes and whales, are analogous. They evolved at different times from different structures; nevertheless since both pairs of structure apparently fulfil similar functions, they may be analogues and useful functional interpretations are still possible.

The wing has evolved across a variety of animal groups, including the insects, reptiles and mammals. The pterosaurs dominated the Mesozoic airways, but the group disappeared at the end of the Cretaceous. Strictly speaking there is actually no direct evidence that the pterosaur wing was adapted for flight. In fact Collini in the late 1700s considered that *Pterodactylus* was a swimmer, while Wagler's reconstruction in 1830 suggested the animal was rather like a penguin; but some ten years later Newman's illustration suggested a bat-like flier and Richard Owen depicted a dragon-like creature for the Great Exhibition of 1851.

Comparison with analogues is essential for this functional study (Benton, 1988). The shape and outline of the wing compares well with those of modern birds and small man-made aircraft (Fig. 4.4). Moreover, the wing loadings, or the weight supported by each square centimetre of wing, are similarly comparable. Additional evidence from the development of wing muscles supports the use of these structures for powered flight. Many pteranodons had wing spans of 7–8 m whereas that of *Quetzalcoatlus* was about 15 m. It is difficult to assign another function to these structures.

4.5.2 Paradigm approach

Some authors have argued that the investigative methods available for adaptive and functional morphology are far from scientific. The paradigm approach is an attempt to bring a more rigorous scientific methodology to functional studies. It first postulates one or more biological functions for a particular structure, and second for each function an ideal model or paradigm is designed. The model may be developed as a mathematical or physical construct. Nevertheless these models must be testable. So, finally, the models are literally compared 'side by side' with actuality. Although models may be complex, the method is relatively straightforward. For example, the brachiopod commissure can be described in simple mathematical terms (Rudwick, 1964). When open, the gape increases from zero at the hinge to a

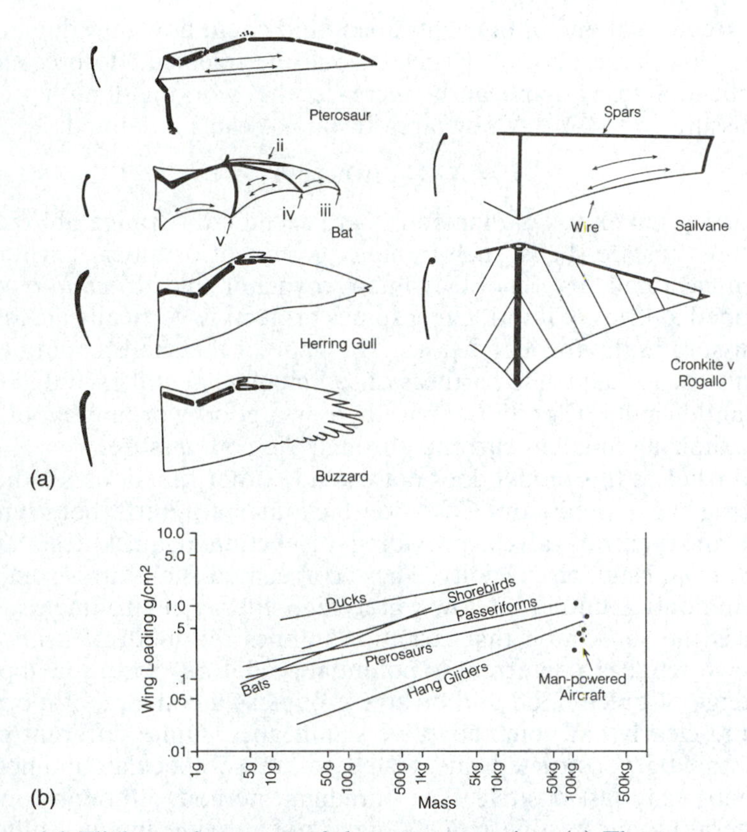

Figure 4.4 Flight performance of the pterosaur wing. (a) The pterosaur wing has an aerodynamic shape when compared with wings of bats, birds and man-made fliers, (b) Pterosaur wing loadings plotted against body mass compared to a range for living birds and man-made fliers. (Redrawn from Benton, 1988.)

maximum (X) at mid-point of the anterior commissure; it decreases again to zero at the other side of the hinge. The area of gape thus equals the area of two triangles of sides $Y/2$ and X, where the length of commissure is Y and the maximum gape is X. The area available for intake (A) is thus:

$$A = X^*Y/2$$

Particles of harmful detritus of diameters up to X can, however, enter through the commissure. More nutrients could presumably be inhaled by increasing the gape between the valves. For example if the gape was increased to $2X$, the area of intake would be correspondingly increased:

$$A' = X^*Y$$

Unfortunately this would also double the maximum diameter of detrital particles with access to the mantle cavity. But if the active length of the commissure was increased keeping the gape constant, a much greater volume of nutrient-laden fluid could flow into the mantle cavity, while large detrital particles would be rejected. Both costation and plication can significantly increase the working length of the commissure to say Y'. A new area in intake can be defined:

$$A'' = X*Y'/2 \text{ where } Y' \gg Y$$

But particles in excess of diameter X are excluded. A range of brachiopods have costate shells, such as many groups of orthides, spiriferides and rhynchonellides. The Devonian rhynchonellid *Uncinulus* even developed a series of interlocking spines projecting vertically across the commissure in the form of a sieve. The gape of the shells could be significantly increased while particles of sediment were still excluded from the mantle cavity. Clearly the paradigm is a good working hypothesis for maximizing inhalent currents through the commissure.

Nevertheless this model does not exclude other functions for the so-called zig-zag commissure. Costation facilitates a tight fit between the dorsal and ventral valves, providing protection against desiccation, predators and sudden turbidity. These corrugated shells are strong and are commonly found in shallow-water, high-energy environments. Less likely, is the suggestion that costation mimics the teeth of fishes and may have acted as a deterrent to potential predators. Costation appears in a range of brachiopod and bivalve groups as a convergent morphotype and clearly had some adaptive significance. Quite different paradigms can be constructed and tested for these speculated functions and many may test positive. The paradigm method, although capable of generating and testing models, need not provide unique solutions in our investigations.

4.5.3 Experimental palaeoautecology

A number of experimental studies have helped confirm or reject some putative life modes of fossil organisms. These studies are usually visual,

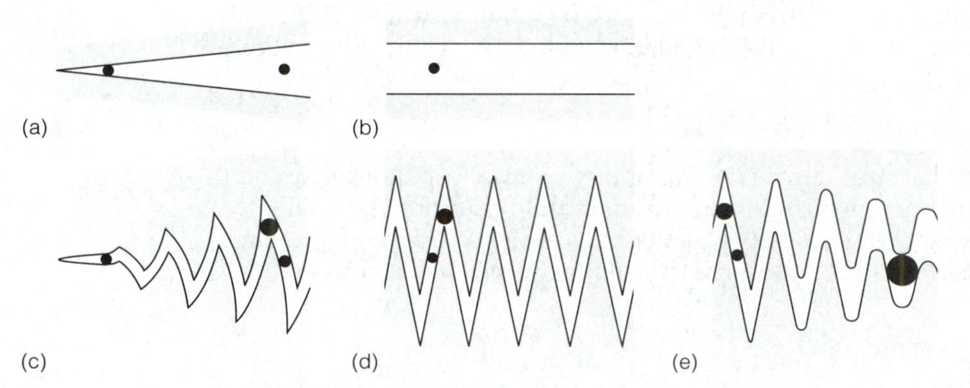

Figure 4.5 Development of costation in brachiopods; a costate commissure increases the area of valve opening between the dorsal and ventral valves while access is restricted for detrital clasts. (After Rudwick, 1964.)

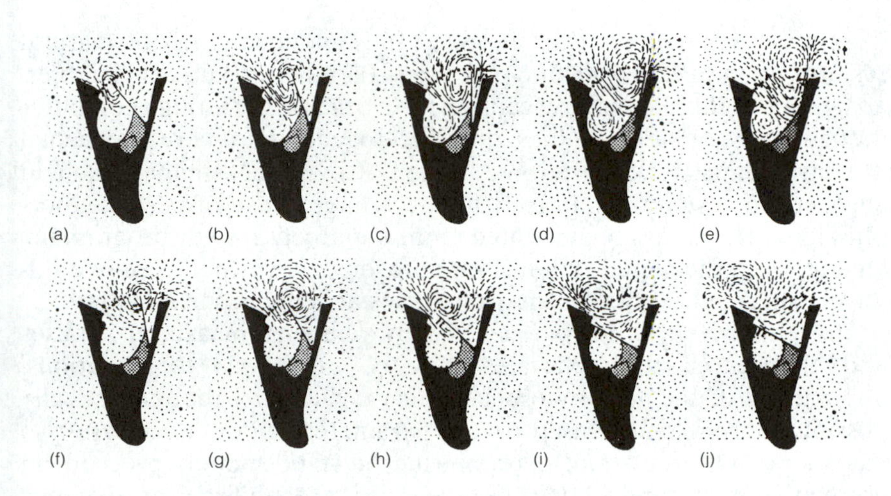

Figure 4.6 Sequential snapshots of the current flow through richthofeniid brachiopod shells as the dorsal valve flaps; nutrient-laden currents are sucked in through the mantle cavity, waste material is flushed out. (After Rudwick, 1961.)

easy to interpret and particularly suited for teaching demonstrations. Ager (1963) and Ager and Wallace (1966) postulated and subsequently modelled under flume conditions the passage of inhalant and exhalent currents, coloured with dye, through the shells of a variety of brachiopods. The majority of experimental runs suggested that inhalant currents entered the anterior commissure of the brachiopod and exited laterally. These flow directions are supported by the position of epifauna on the front of some brachiopod shells and observations on living species that orientate forward into currents. Models of the richthofeniids also simulated feeding currents (Rudwick, 1961), although the paradigm assumed the flow was generated by the flapping of the dorsal valve (Fig. 4.6). More recent studies have suggested that this may be incorrect.

Flow in flumes and wind tunnels can be modelled mathematically, allowing the calculation of drag forces on the skeletons of fossil organisms. The hydrodynamics of swimming ammonoids have been investigated with estimates of drag factors together with comparisons with living *Nautilus*. These studies suggest the fossil ammonoids were relatively poor swimmers compared, at least, with fishes (Chamberlain, 1981).

More sophisticated and mobile models have simulated the life modes of some of the fossils of the mid-Cambrian Burgess Shale; replicas of these giant and bizarre arthropods from the middle Cambrian have been put through their paces in swimming pools, demonstrating the possible locomotion modes of these extinct animals.

4.5.4 Computer simulation

Computer simulations can improve the understanding of natural systems by mimicking organic processes. Parameters and sets of rules, analogous to those in nature, can generate organisms through computer modelling. Simulation is thus opposite to analysis where the actual organisms themselves are investigated. Computer simulations add another dimension to adaptive studies. A large range of possible morphologies can be easily generated from a defined growth programme, analogous to the genetic control on ontogeny. Nevertheless only a relatively small subset of these morphologies actually occur in reality.

The parameters in any model usually relate to measured variates and changes in these variates may be analogous to mutations. Simulations define a morphospace which includes all the theoretically possible morphologies based on the parameters input. There are two main types of simulation: deterministic or static models proceed on the basis of starting conditions, whereas probabilistic or dynamic models are generated by parameters that may be changed, often randomly, as the model develops. To date most morphological simulations have favoured the more simplistic deterministic models.

Two main areas of modelling have generated useful results: simulation of shell shape and the simulation of colonial forms. Most valves of any incrementally grown shelled organism can be modelled as an expanding cone, either straight or coiled (Fig. 4.7); in fact the ontogeny of living *Nautilus* was known to approximate to a logarithmic spiral in the 18th century. The ontogeny of shells may be modelled on the basis of four parameters (Raup, 1966):

1. the shape of the generating curve or axial ratio of the shell's cross-sectional ellipse (S),
2. the rate of whorl expansion per revolution (W),
3. the position of the generating curve with respect to the axis (D); and
4. the whorl translation rate (T).

Shells are generated by rotating the generating curve around a fixed axis, with or without vertical and horizontal translations (Fig. 4.8). For example when $T = 0$, shells lacking a vertical component, such as some bivalves, planispiral ammonites and brachiopods, are simulated, whereas those with a large value of T are typical of high-spired gastropods. Only a small selection of the theoretically possible spectrum of shell shapes occurs in nature. More recent work has applied more complex techniques to simulate ammonite heteromorphs using a different set of parameters (Okamoto, 1988):

1. the enlarging ratio (E);
2. standardized curvature (C); and
3. the standardized torsion (T).

Much more complex shapes can be generated, and assuming the shell substance is homogenous, the shell length:living chamber length

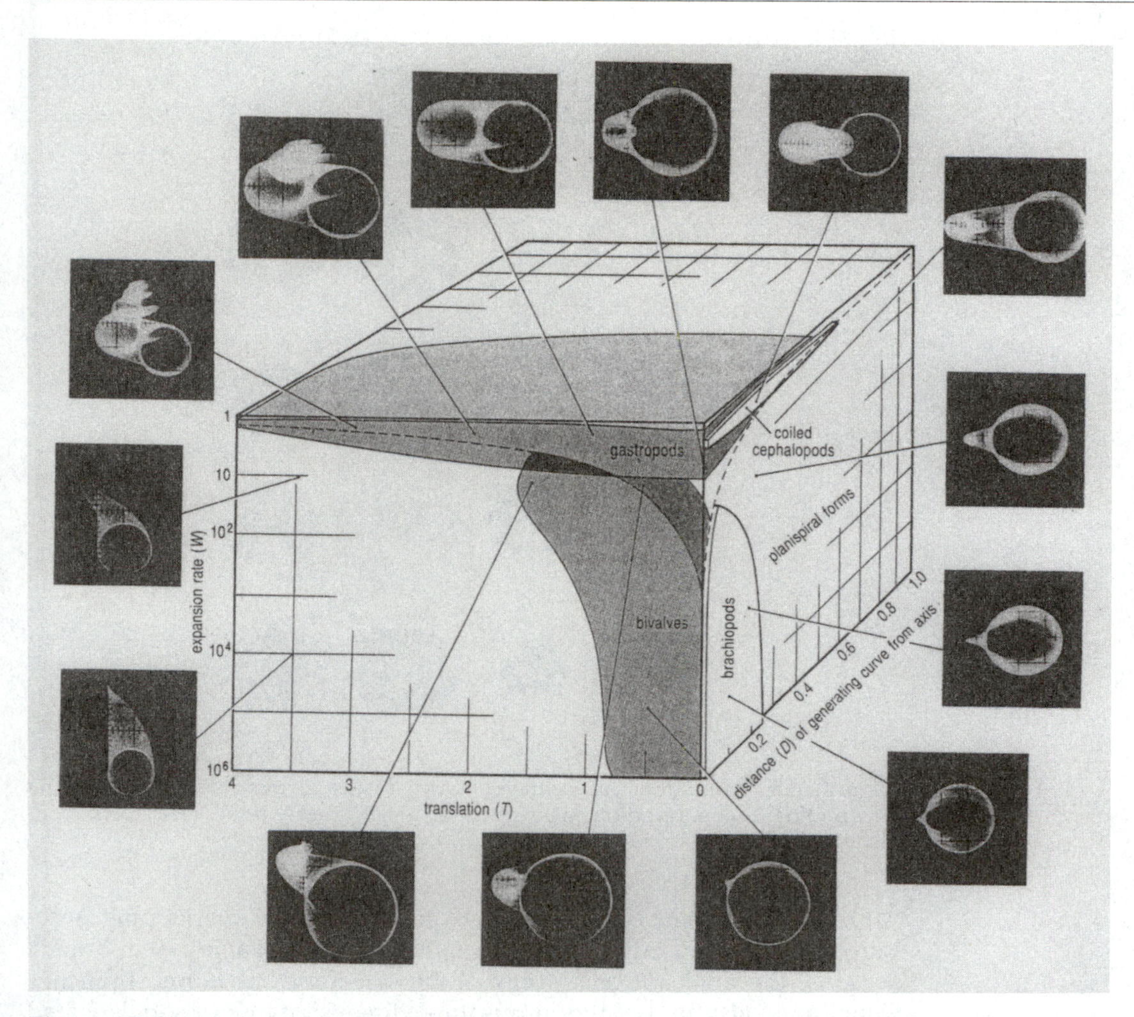

Figure 4.7 Simulation of the shape growth of shelled organisms: the matrix shows the distribution of actual shapes for the bivalves, brachiopods, cephalopods and gastropods generated by the model. (After Raup, 1966.)

is constant throughout ontogeny, the centre of gravity and the floating position of the animal can be restored. Nevertheless, only a relatively small percentage of the theoretically available morphospace has actually been used by fossil and living shelly organisms. Clearly some fields map out functionally and mechanically improbable or disadvantageous morphologies; other fields have yet to be tested. These simulations provide a range of theoretical morphologies suitable for adaptive studies.

Branching models relate mainly to colonial organisms such as bryozoans, corals and graptolites, together with plants (Fig. 4.8). Most models are static, with the mode of branching predetermined by fixed parameters. Static branching models are based on the iterative addition of new branches based on a fixed branch length and thickness together

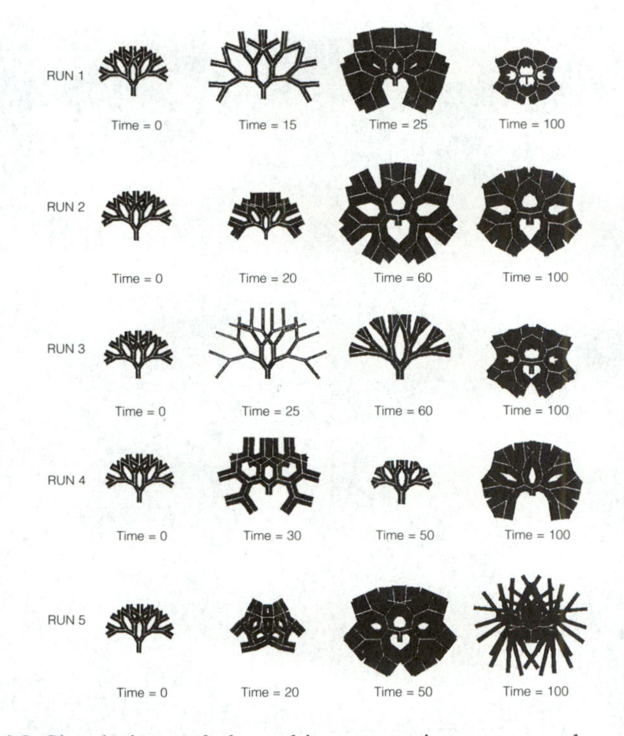

Figure 4.8 Simulation of branching organisms: examples of branching morphologies generated during a series of 'evolutionary runs'. (Redrawn from Swan, 1990a.)

with a defined angle of dichotomy. Commonly dense overlapping structures are formed; these can be modified by changing the branch lengths, the angle of dichotomy or by selecting only some branches which may also be twisted into a third dimension. More sophisticated dynamic models may only develop branches when and where appropriate, usually to occupy empty space; these models can generate some very complex organisms based on a few very simple rules.

An element of evolutionary and adaptive modelling can be introduced into such models (Dawkins, 1986) by generating a hypothetical organism with some mutant descendants; the efficiency of the offspring may be tested against a set of defined conditions and selected either for or against (Swan, 1990a).

4.6 ADAPTATION AND PREADAPTATION

Adaptation, as noted above, reflects how well an organism is fitted to its environment or habitat and how structures or complexes of structures are best suited for particular functions. Neo-Darwinian philosophy suggests that through natural selection these structures are best fitted or adapted to fulfil and sustain these functional requirements.

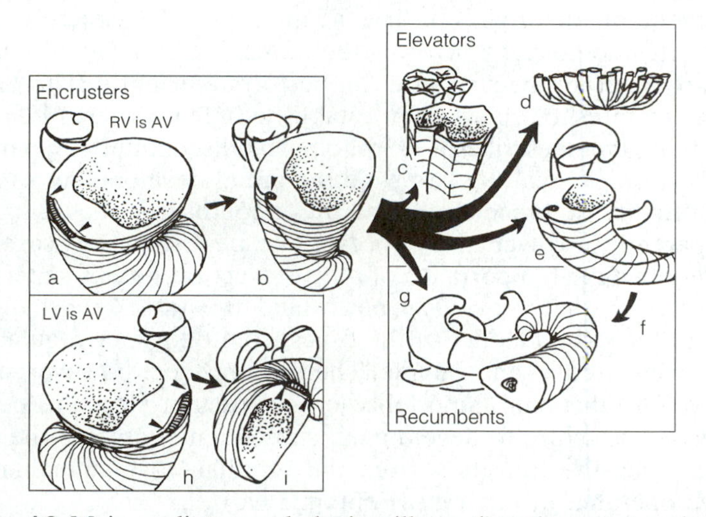

Figure 4.9 Main rudist morphologies illustrating the constructional morphology of the group: Encrusters (right valve (RV) is attached (AV)): (a) *Diceras*, (b) *Valletia*; left valve (LV) is attached (AV): H. *Epidiceras* or *Plesiodiceras*, I. *Toucasia* or *Requiena*), Elevators (c) *Agriopleura*, (d) *Hippurites* or *Hippuritella*, (e) *Pachytraga* or *Caprina*, Recumbents (f) *Immanitas*, (g) *Titanosarcolites*. (Redrawn from Skelton, 1985.)

Some adaptive features, however, may have first operated in a quite different context before their eventual deployment. Preadapted structures, such as the swim-bladder of fishes first used for buoyancy and then respiration, are not uncommon and the concept of preadaptation is an important part of both ecological and evolutionary analysis. The ligament of the aberrant rudist bivalves has been described as a preadapted structure (Skelton, 1985). The rudists display three main types of shell morphology related to particular environmental circumstances (Fig. 4.9): elevators (tall, conical shells) occurred in turbid water where their commissure was kept above the sediment–water interface by rapid upward growth; encrusters (bun-shaped forms encrusted on the substrate) required a stable surface for attachment; whereas recumbents (large, extended shells) were free-lying in shoals on mobile substrates. The earliest rudists, the diceratids, were encrusters attached by their right valves to a hard substrate. Commissural growth was restricted by the external ligament to so called spirogyrate modes. The progressive shortening and indentation of the external ligament disengaged the rudists from spirogyrate growth programmes, permitting uncoiling and the possibility of elevated and recumbent strategies. These latter morphologies explosively diversified in the early Cretaceous during a clear adaptive radiation.

4.6.1 Functional thresholds

The development of functions usually takes place against a background of macroevolutionary change involving the large-scale biological improvement of an organism. The adaptation of a structure for a particular function usually involves the crossing of functional barriers or thresholds before the structure can perform efficiently. Three main categories of threshold have been established (Jaanusson, 1981). The all-or-nothing type describes a morphological discontinuity apparently crossed by a single evolutionary step; the initial development of articulatory structures in brachiopods and the fusion of free cheeks in trilobites apparently involved this process. The peak type envisages the increasing functional importance of a new structure until it supersedes those structures that previously performed the task; the change in function of the swim bladder of the fishes from buoyancy regulator to lung has been cited as an example. The descent type, however, charts the decline in functional importance of a particular structure permitting a second structure to develop and subsequently assume the same functional role; the transition from the reptilian to mammalian jaw may have operated in this way (Benton, 1990)

The transit across functional thresholds provided organisms with a new opportunities and has also formed the basis for many adaptive radiations.

4.6.2 Size as an adaptation

According to Cope's Law, evolving organisms tend towards large size (Fig. 4.10). Some authors have considered the adaptive dimension of size change (Stanley, 1973). There are a number of clear advantages associated with large size: improved ability to both catch prey and escape predators, greater intelligence and reproductive success, decreased mortality with increasing size together with increased heat retention. Giant size can evolve quickly; nevertheless, relatively few taxa have achieved true giant status. There are probably more constraints on large size than there are advantages. Suspension-feeding strategies require massive increases in efficiency to sustain relatively small increases in size in animals such as the brachiopods, bryozoans, many echinoderms and many molluscs. Moreover, shelled organisms require a disproportionately large increase in shell thickness, precipitated from extracted minerals in seawater, to match increase in the body volume of the animal. Arthropods would have to moult many times to achieve large size whereas the load-bearing capacity of vertebrates must be enhanced to cope with the increasing volume and weight associated with large size.

4.6.3 Heterochrony

Macroevolutionary change, requires a mechanism. Changes in the developmental rates of an organism, or **heterochrony**, provides a set of growth strategies that can achieve significant morphological change between parent and daughter populations. By altering the relationships between changing size, shape and time, novel morphologies can

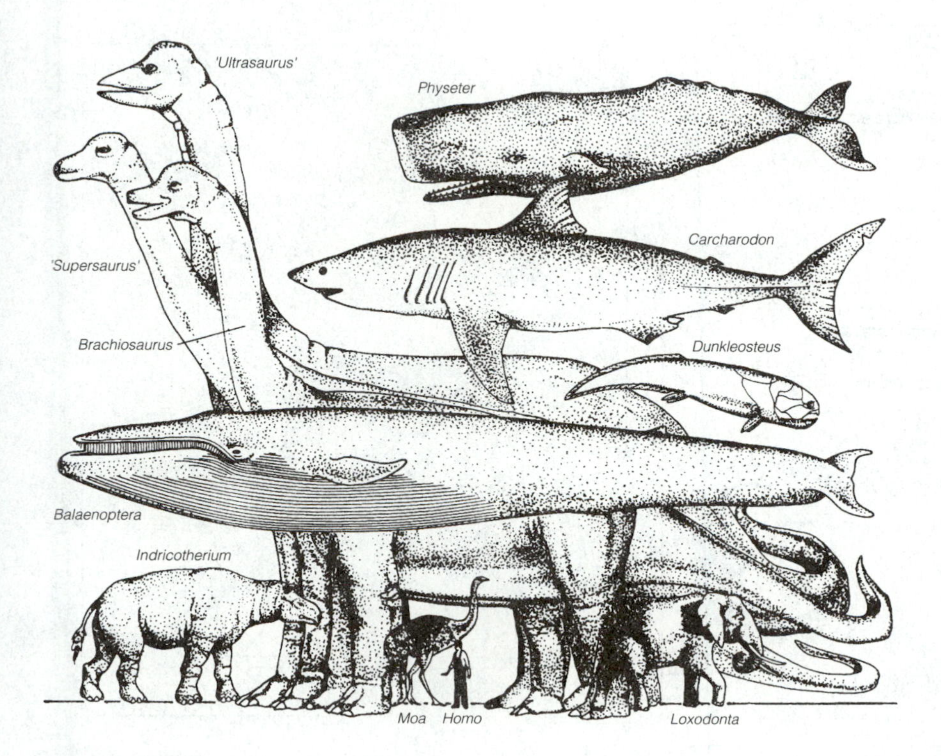

Figure 4.10 Some giant and large animals drawn to scale. (Redrawn from Benton, 1990.)

be generated along evolving lineages. There are two main groups of processes: paedomorphosis and peramorphosis. In **paedomorphosis** ('juvenile development'), sexual maturity occurs early, in the more juvenile stages, whereas with **peramorphosis** ('over development') sexual maturity occurs relatively late in ontogeny. There are a variety of heterochronic mechanisms (McNamara, 1986) and most occur as part of an organism's regulatory system. Heterochronic changes may, nevertheless, be adaptive, environmentally driven responses (Fig. 4.11). The peramorphocline established for the evolution of the echinoid *Protenaster* facilitated feeding from finer-grained sediments; the paedomorphocline described for the brachiopods *Tegulorhynchia* and *Notosaria* followed a regressive track from quiet deep-water to shallow-turbulent marine environments.

4.6.4 Vestigial structures

Functional analysis can be thwarted by vestigial structures. Organisms can possess structures with no apparent function; their necessity is apparently redundant. In humans, the wisdom teeth, ear-lobe muscles and the appendix, although often variably developed, have no

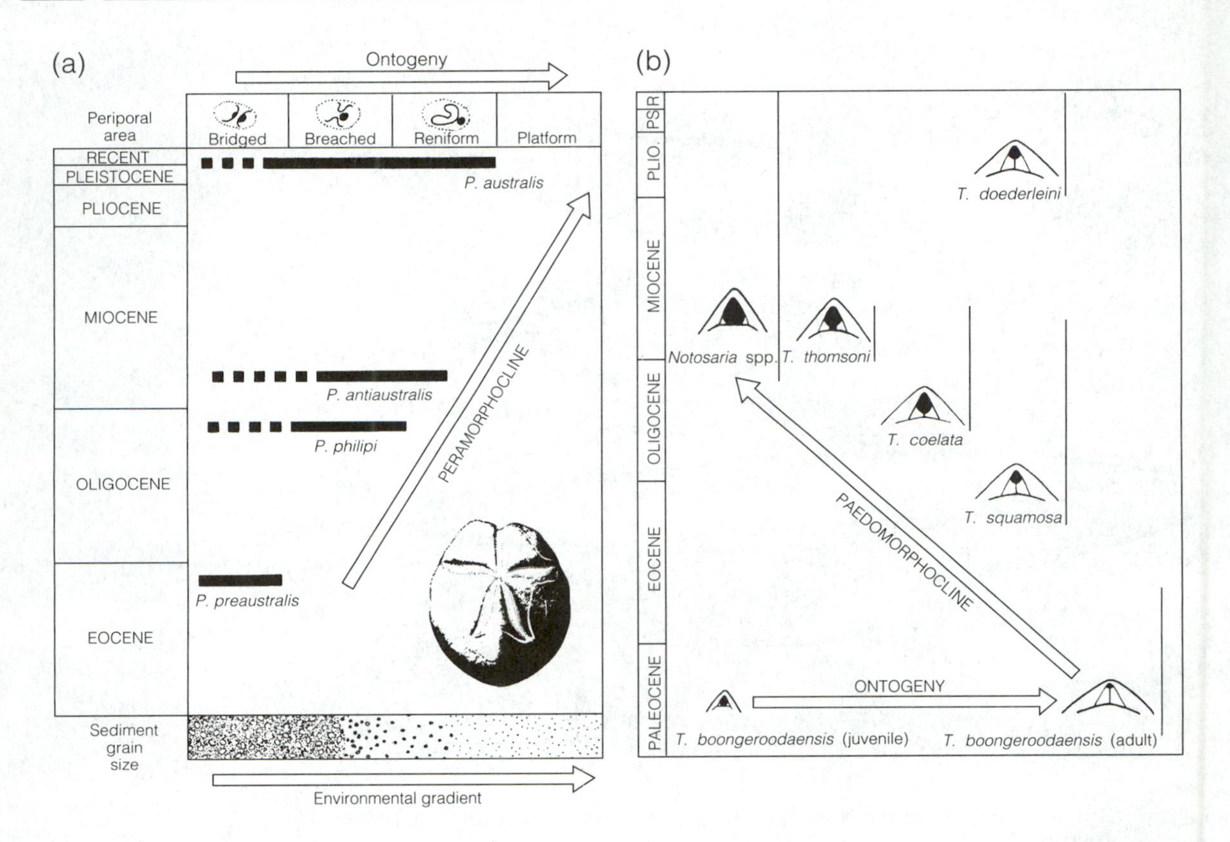

Figure 4.11 Heterochronic change: (a) Peramophic evolution of the periporal area of the Cenozoic echinoid *Protenaster* to facilitate feeding from finer-grained sediments. (b) Paedomorphic evolution of the Cenozoic brachiopods *Tegulorhynchia* and *Notosaria* related to a move from deep-water quiet conditions to shallow-water, high-energy environments. (Redrawn from McNamara, 1990.)

identifiable roles and if removed surgically do not affect bodily functions. Not all structures identified in an organism must be functional and many organisms carry some form of non-functional baggage. Modern whales still possess small, rod-like hind limbs; their ancestors, such as the Eocene whales described from the Middle East, have small hind limbs, which unlike their Palaeocene relatives, could not have supported the animal on land.

4.7 MORPHOLOGY AND ENVIRONMENT

Throughout geological time organisms have taken advantage of new opportunities. The development of adaptations has opened up successive expanses of ecological space. Adaptations are fine-tuned, thus in broad terms there is a relationship between the overall morphology of an organism and its environment. The study of adaptation has not only been fundamental in studies of evolution; approached from the other side, analysis of adaptation has provided a significant amount of data for palaeoenvironmental reconstructions. Adaptation can be studied across taxonomic groups. This autecological method relies heavily on a strong taxonomic base and is followed in many palaeontological texts (Clarkson, 1993). Another approach, followed here, targets adaptations within the context of a number of broad ecological-type zones through the six great evolutionary biotas spanning the Precambrian and Phanerozoic history of the biosphere. There are many possible ways of classifying organisms in ecological terms (see Chapter 1). Organisms may be assigned to positions in a trophic web where there is competition for food and space. Alternatively, organisms can be classified in terms of adaptations for life at the sediment–water interface, within tiered epifaunal and infaunal systems, within the water column or oceanic surface waters or on land and in the air. In this way adaptations are related to the development of environments through time against a taxonomic background.

4.7.1 Pre-Vendian biota

The Pre-Vendian fossil record is dominated by stromatolites, sheets of calcium carbonate associated with cyanobacteria. Carbonate material is trapped and precipitated on the surface of the filamentous bacteria to generate a distinctive laminated structure. Stromatolites are generally rare in Archaean rocks, becoming more common during the late Archaean and early Proterozoic. By the end of the early Proterozoic a range of stromatolite architectures had developed including stratiform, columnar, conical, domal and nodular growth forms; during the mid- and late Proterozoic the columnar growth types diversified. Many groups of stromatolite were much reduced during the early Cambrian as burrowing and grazing metazoans together with the meiofauna and the influx of coarse siliciclastic sediment inhibited and often destroyed stromatolite growth. The stromatolites provide an example of the functional morphology of a pre-Phanerozoic design.

Stromatolites were built by photosynthesizing organisms occupying shallow-water environments within the photic zone. The range of growth forms shows adaptations to a range of environmental conditions controlled by current and wave energy together with sedimentation rates (Fig. 4.12). These simple morphologies show a very direct relationship to environmental conditions, indicative of early modes of adaptation.

4.7.2 Vendian evolutionary biota

The main and most conspicuous elements of the Vendian biota belong to the Ediacara fauna. The fauna is entirely soft-bodied and was probably adapted to relatively low oxygen conditions in a variety of usually nearshore marine environments. The apparently unique morphology and mode of preservation of the Ediacara fauna led to much debate

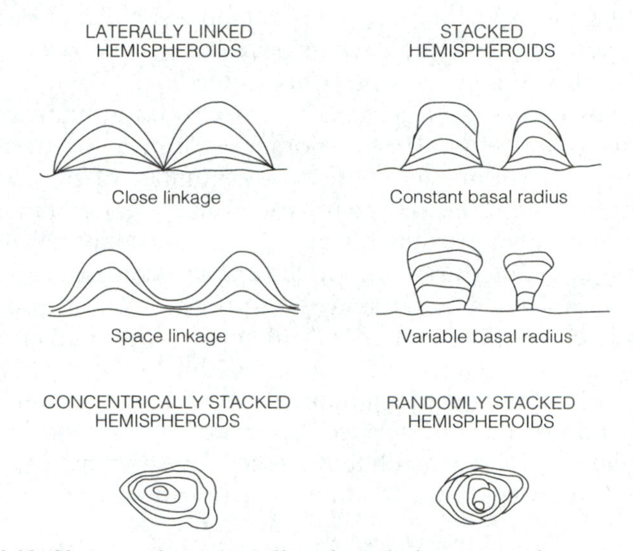

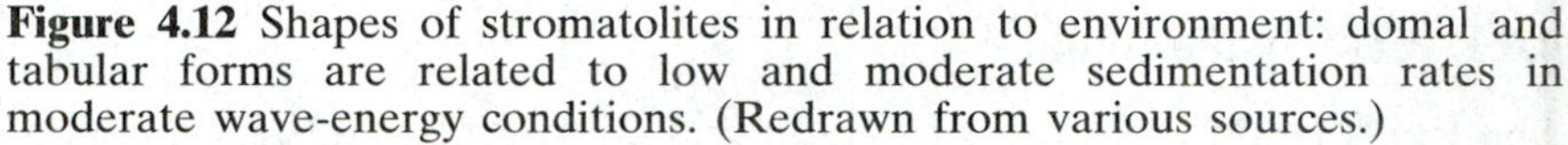

Figure 4.12 Shapes of stromatolites in relation to environment: domal and tabular forms are related to low and moderate sedimentation rates in moderate wave-energy conditions. (Redrawn from various sources.)

about the identity and origins of the assemblage. Are the Ediacarans some of the first true metazoans, or the impressions of an entire ecosystem populated by quite a different type of organism? Seilacher (1989) has reinterpreted the fauna in terms of its constructional and functional morphology. Apart from a distinctive mode of preservation, the fauna shares the following features: quilted pneu (rigid, hollow, balloon-like) structures with sometimes additional struts and supports together with a significant flexibility. If the Ediacara animals are in fact divorced from the true metazoans and indeed may be grouped together as a separate grade of organization – termed by Seilacher and others, the Vendozoa or Vendobionta (Buss and Seilacher, 1994) – certain generalizations about their anatomy and behaviour, some speculative, may be made. Reproduction may have been by spores or gametes, and growth was achieved by both isometric and allometric modes. The skin or integument had to be flexible, although it could crease and fracture. Moreover the skin must have acted as an interface for diffusion processes, whilst providing a water-tight seal to the animal. This stimulating and original view of the fauna, however, remains controversial. A range of adaptive morphologies has been recognized in the fauna (Fig. 4.13).

There is little doubt that the Ediacara biotas dominated the latest Precambrian marine ecosystem, occupying a range of ecological niches and pursuing varied life strategies probably within the photic zone. It is also possible that these flattened animals hosted photosymbiotic algae, maintaining an autotrophic existence in the tranquil 'Garden of Ediacara' (McMenamin, 1986). The ecosystem, however, was domi-

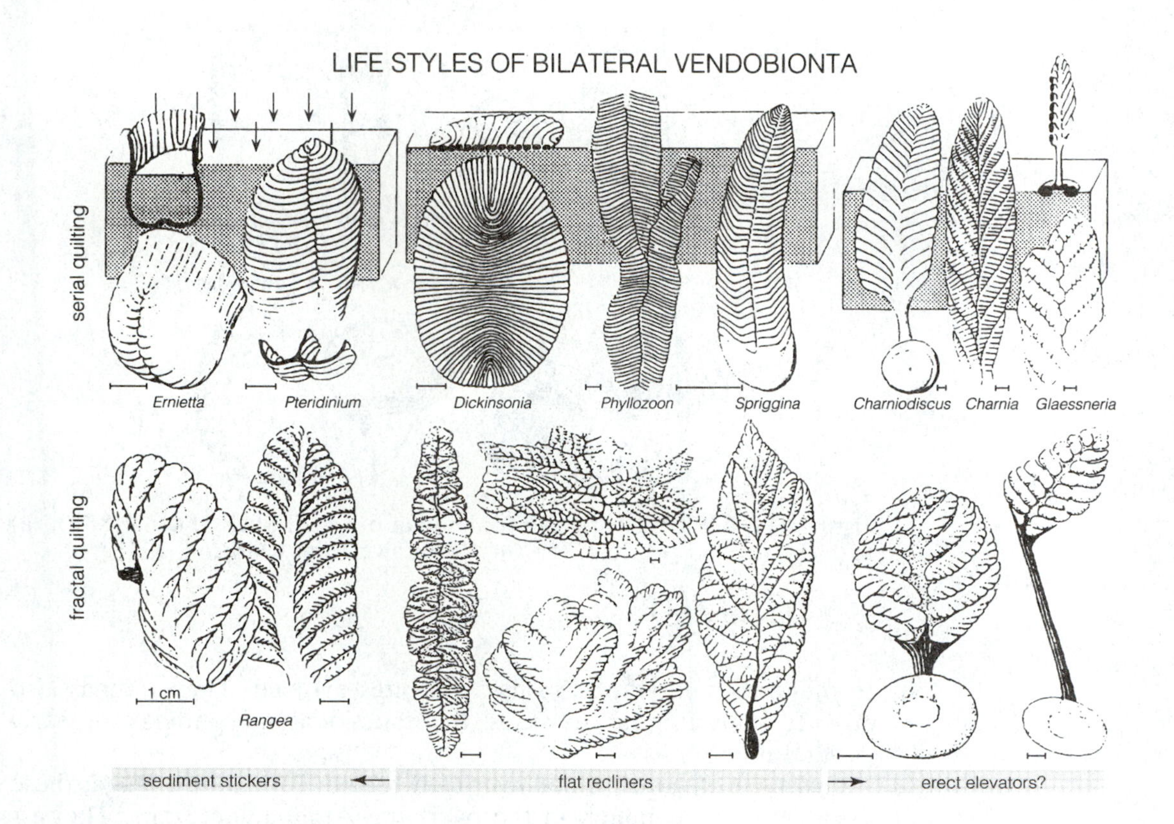

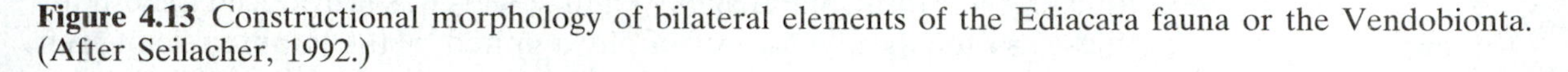

Figure 4.13 Constructional morphology of bilateral elements of the Ediacara fauna or the Vendobionta. (After Seilacher, 1992.)

nated by medusoid pelagic animals and attached, sessile benthos; infaunal animals were sparse; food chains were probably short and the trophic structure was apparently dominated by suspension- and deposit-feeders.

4.7.3 Tommotian evolutionary biota

A distinctive assemblage of small shelly fossils (Fig. 4.14), traditionally labelled the Tommotian fauna, appeared at the Precambrian–Cambrian transition; the assemblage is most extravagantly developed in the lowest Cambrian stage of the Siberian Platform, the Tommotian, which gives its name to the fauna. Much is now known about the stratigraphy and palaeobiogeography of this biota through current interest in the definition of the base of the Cambrian System. Nevertheless the biological affinities of many members of the fauna have yet to be established and, although dominated by minute species, together with small sclerites of larger species, the biota represents the first appearance of diverse skeletal material in the fossil record, some

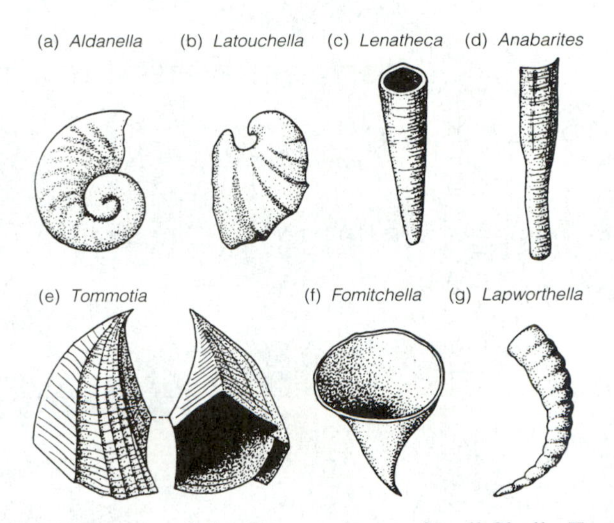

Figure 4.14 A Elements of the Tommotian or Small Shelly Fauna (SSF); (a) X20, (b)–(e) and (g). X15, (f) X40. (Redrawn from Clarkson, 1993.)

10 million years before the first trilobites evolved. These faunas also provide opportunities to study the functional morphology of early skeletal metazoans.

This type of fauna is not restricted to the Tommotian Stage; these fossils are also common in the overlying Atdabanian Stage. The less time-specific term, Small Shelly Fauna (SSF) was introduced to describe these assemblages. A variety of phyla united by the minute size of their skeletal components and a sudden appearance at the base of Cambrian are now included in the Tommotian fauna. It probably dominated the earliest Cambrian ecosystems where many metazoan phyla developed their own distinctive characteristics, initially at a very small scale. Soft-bodied organisms had developed some protection against predation and desiccation and support for organs and muscle systems. Functional interpretations are difficult since many of the elements of the fauna are, in fact, sclerites of larger animals such as the halkieriids; moreover, there is still uncertainty regarding the precise identity and affinities of many cap-shaped shells, such as the helcionelloids, in the fauna. Nevertheless some of the mollusc-like shells have been functionally analysed and suggest that these animals pursued a variety of mobile epifaunal and semi-infaunal life strategies (Peel, 1991); the halkieriids may have constructed U-shaped burrows with their anterior and posterior shells covering the entrance and exit to the burrow system.

4.7.4 Cambrian evolutionary biota

During the first 10 million years of the Cambrian Period the Tommotian fauna was replaced by a more diverse biota of larger metazoans, participating in more complex communities. The Cambrian fauna was dominated by low-level suspension feeders such as the nonarticulate brachiopods and eocrinoids together with monoplacophoran and hyolith molluscs. Two parazoan groups developed colonial strategies: the archaeocyathans and possibly the sponges. Colonies tend to develop iteratively with new iterative units or modules derived by continuous growth from existing units. The colonial or iterative body plan thus contrasts with the unitary or solitary life mode of most organisms. A number of groups show clear trends towards a greater integration of individuals within the colony and in some cases a differentation of functions across the colony.

Trilobites were the most abundant members of the Cambrian benthos; in many assemblages over 90% of the animals were trilobites. Cambrian communities were loosely organized and considerable experimentation and morphological flexibility were features of many groups (Hughes, 1994). Cambrian Lagerstätten such as the Burgess Shale contain a wide range of apparently morphologically disparate organisms.

4.7.4.1 Benthos

The Cambrian benthos had already developed a simple tiering system with two levels of 0–50 mm and 50–100 mm. The lower tier was dominated by archaeocyathans, which formed small reefs together with echinoderms, nonarticulate brachiopods and sponges; the higher level included archaeocyathans, echinoderms (eocrinoids and crinoids) and sponges. Echinoderms were rare but nevertheless included some bizarre forms such as the helicoplacoids, whereas the infauna was generally shallow, burrowing close to the sediment–water interface with the exception of vertical *Skolithos* burrows, which were often deep. Reef frameworks were developed early in the Cambrian by the sponge-like archaeocyathans.

Trilobites dominated the mobile benthos, accounting for over 50% of Cambrian species. Most were probably detritus-feeders crawling across or swimming above the sea-floor, whereas some may have lived in shallow burrows. Both body and trace fossil representatives of the trilobites are common in Cambrian strata. The functional morphology of the olenids targets the already sophisticated designs of the Cambrian trilobita.

4.7.4.2 Nekton and plankton

Relatively little is known of Cambrian nekton. Some molluscs such as the hyoliths may have been pelagic, together with primitive agnathan fishes. The Cambrian plankton was more abundant and diverse than that of the Precambrian oceans. The acritarchs radiated during the Cambrian, radiolarians occupied tropical latitudes, whereas chitinozoans were present in Cambrian plankton but not abundant. Larval

Box 4.1 Growth strategies and function of archaeocyathans

The archaeocyathans were exclusively marine, occurring most commonly in water depths of 20–30 metres on carbonate substrates. The phylum developed an innovative style of growth, during astogeny, based on modular organization (Wood *et al.*, 1992). Such modularity permitted encrusting abilities and the possibilities of secure attachment on a soft substrate; moreover growth to large size was encouraged together with a greater facility for regeneration. The archaeocyathans were thus ideally suited to participate in some of the first reefal structures of the Phanerozoic, in intervals of high turbulence and rates of sedimentation, during the early Cambrian. However, although archaeocyathan reefs were probably not particularly impressive, usually up to 3 m thick and between 10–30 m in diameter, the archaeocyathans were nevertheless the first animals to build biological frameworks and establish this complex type of ecosystem.

In an experimental biomechanical study (Savarese, 1992) three morphotypes (aseptate, porous septate and aporous septate) were constructed and subjected to currents of coloured liquid in a flume. The first morphotype, a theoretical reconstruction, performed badly, with fluid escaping through the intervallum while also leaking through the outer wall. The porous septate forms, however, suffered some slight leakage through the outer wall but no fluid passed through the intervallum. The aporous septate form was most efficient, with no leakage through the outer walls and no flow through the intervallum. Significantly, ontogenetic series of initially porous septate morphotypes develop an aporous condition in later life, perhaps to avoid leakage through the outer wall.

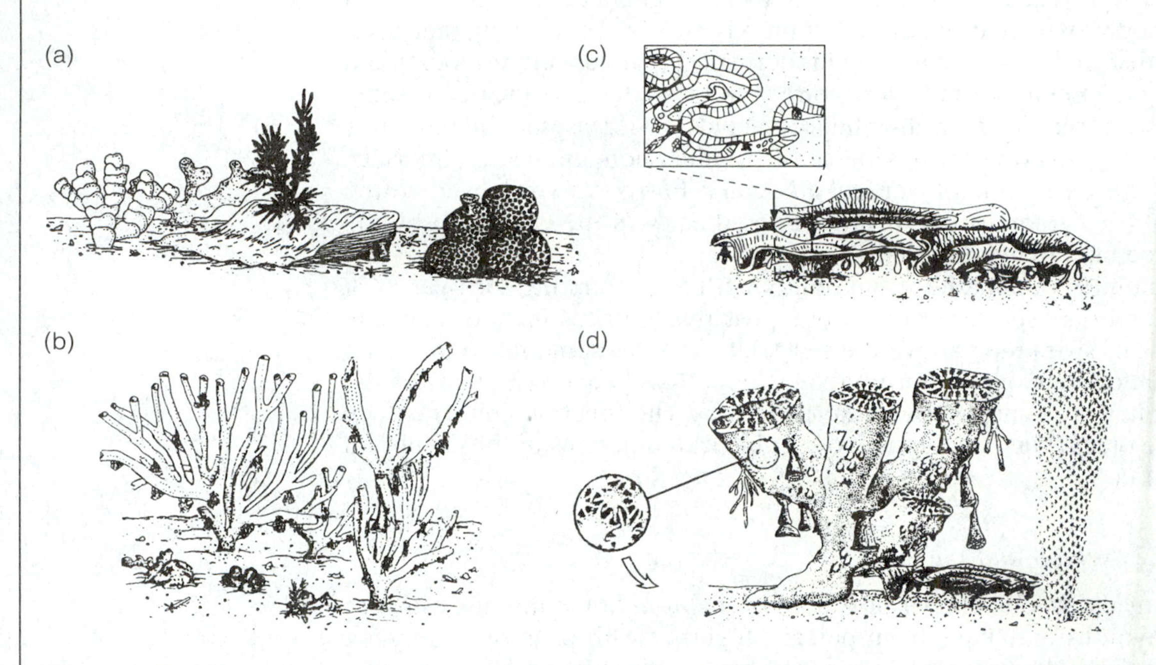

Figure 4.15 Schematic reconstruction of archaeocyathan build-up communities during the Cambrian (after Wood *et al.*, 1992). (a) Calcimicrobe boundstone and bioherms, (b) *Cambrocyathellus*-calcimicrobe bafflestone, (c) *Okulitchicyathus* bindstone, (d) Radiocyath-archaeocyath-cribicyath bioherm.

Box 4.2 Cambrian trilobites from the Alum Shale

The olenid trilobites dominated the Alum shale environments of Baltoscandia during the late Cambrian. The Alum shales were deposited as a dark anoxic mud; the sea-floor was probably quite unpleasant. Although the olenids are fairly typical trilobites, they developed a range of adaptations associated with a benthic life mode (Clarkson and Taylor, 1995). Both *Olenus* and *Parabolina* have broadly similar morphologies, though *Parabolina* is more spiny. But *Leptoplastus*, for example, had very long genal spines that may have cleared its gills above a stinking anoxic seabed. Similarly *Sphaerophthalmus* possessed vertically orientated spines, keeping the gills well clear of the mud. Despite the coherence of the olenid group there were opportunities for quite marked morphological divergence. At a time of experimentation and plasticity in many groups, these divergent morphologies were probably adaptive.

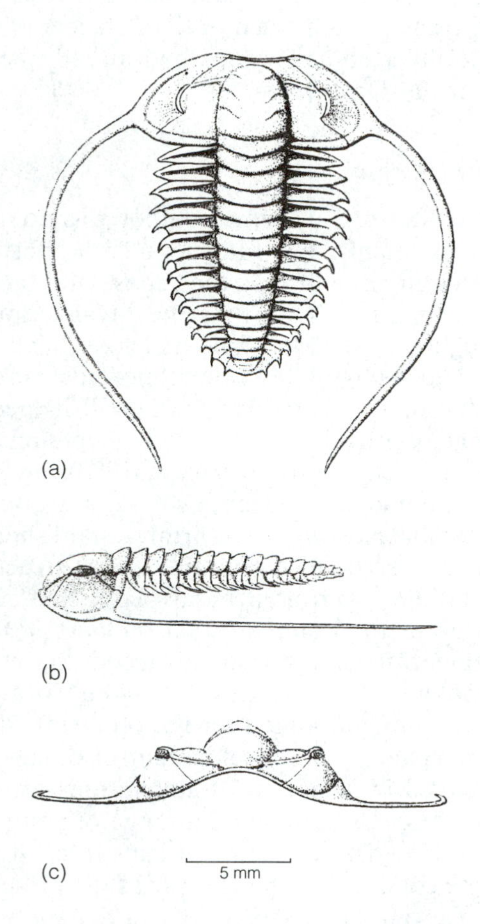

Figure 4.16 Morphology of *Leptoplastus*: Dorsal, lateral and anterior views of an olenid adapted for life on a dark, oxygen-deficient muddy substrate. (Redrawn from Clarkson and Taylor, 1995.)

phases of benthic organisms together with the agnostic trilobites dominated a zooplankton still apparently free of macrophagous predators.

4.7.5 Palaeozoic evolutionary biota

The Palaeozoic evolutionary fauna originated and diversified during an early Ordovician radiation event. Many of the adaptations highlighted in the Palaeozoic marine benthos are associated with soft substrates. Articulate brachiopods, stenolaemate bryozoans, stalked echinoderms (crinoids and blastoids), corals, ostracodes diversified together with graptolites within the water column. Most plankton groups may have been recruited from the benthos while events within the plankton ecosystem were shadowed by changes in benthic systems (Rigby and Milsom, 1996). The vigorous early Ordovician radiation set the agenda for much of the Palaeozoic; the majority of adaptations in the invertebrate groups had already been tried and tested by the end of the Ordovician.

4.7.5.1 Benthos

The low-level Palaeozoic benthos was dominated by fixed suspension-feeders, mainly the brachiopods supported by the bryozoans and microcarnivores such as the rugose and tabulate corals. Molluscs were generally rare, although some bivalve-dominated associations occur throughout the Palaeozoic and may have inhabited specialized habitats. The sponge-like stromatoporoids were responsible for carbonate buildups, particularly during the mid-Palaeozoic, when they developed a range of growth modes including columnar, dendroid, encrusting and hemispherical forms (Kershaw, 1990).

By the mid-Ordovician, two major groups of anthozoans, the rugose and tabulate corals, were firmly established as microphages, although neither ever built substantial reefal structures, lacking many of the adaptations, such as a basal plate, that helped the scleractinians to develop as the dominant reef-builders of the Modern fauna.

Palaeozoic corals also advanced the development of colonial or compound life strategies. Tabulate corals developed only colonial forms, many fine-tuned for life on a soft substrate; the location, orientation, spacing and development of offsets in the corallites have been developed in a tool-kit of adaptations for different substrates.

Brachiopods dominated a range of nearshore to shelf-edge environments, suited to a variety of substrates at a range of water depths in marine conditions. The group was the principal epifauna of the Palaeozoic. Certain basic aspects of the brachiopod morphology, such as the presence or absence of a pedicle, had a major control on the brachiopod life strategy. More subtle features of the animal, for example the size, shell shape and ornament, also had clear adaptive significance.

Crinoids, together with brachiopods, dominated the Palaeozoic sessile benthos. The crinoids developed a range of articulatory

Box 4.3 Life strategies of rugose corals

The scleractinian corals of the Modern Evolutionary Fauna are adapted for life on a hard substrate; during the Palaeozoic, however, solitary rugose corals evolved a range of adaptations and strategies to cope with life on soft substrates (Neuman, 1988). Ambitopic forms were initially attached to a substrate but subsequently lay recumbent on the sea-floor; liberosessile corals, with dilated septae and an incipient basal disc, were fixed for a short interval to sediment grains before a subsequent life in the sediment; fixosessile forms, with fixing grooves and commonly talons, were encrusted or cemented to a hard substrate; rhizosessile corals were probably initially attached to small clasts but were subsequently supported by holdfasts on a soft sea-floor. Vagile corals, such as the mobile *Diaceris* and the small discoidal *Palaeocyclus*, may have crept over the substrate and possibly some may have even joined the pseudoplankton, being attached to floating material.

RUGOSE CORALS	
CATEGORIES OF LIFE STRATEGIES	NATURE OF SUBSTRATE WITH COMMENTS ON CHANGING HABITS THROUGH LIFE
AMBITOPIC	Initially attached for a short period to a small sediment grain, thereafter recumbent on soft bottom. 
LIBEROSESSILE	Initially attached to hard bottom, thereafter recumbent on soft bottom 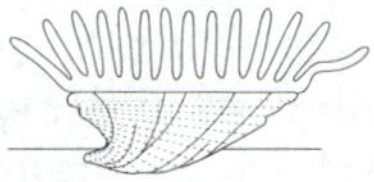
FIXOSESSILE Encrusting Apically cementing	Attached to hard bottom with a basal disc or scar, with or without additional attachment structures during the whole life 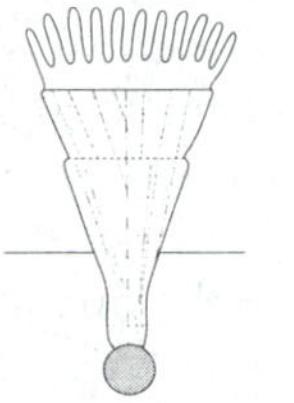
RHIZOSESSILE	Initially attached for a short period to a small sediment grain (?), thereafter supported by holdfasts on soft bottom 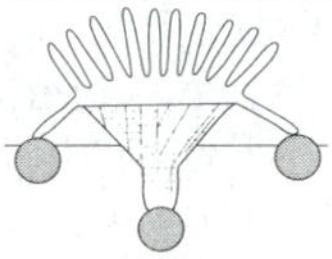
VAGILE Active movement Planktic Pseudoplanktic	Creeping on the substrate with tentacle movement? Larval stage only. Corals attached to floating objects? 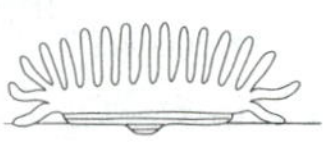

Figure 4.17 Representative life strategies of rugose corals. (Modified and redrawn from Neuman, 1988.)

Box 4.4 Brachiopod life modes

With the exception of the oyster-like brachiopods, such as the richthofeniids, all the main adaptations had appeared during the early Ordovician radiation of the Brachiopoda (Harper and Moran, 1997). Early studies, in the 1920s and 1930s, relating the Palaeozoic shell morphologies of the brachiopods to environments have been confirmed by more modern research (Ager, 1967). A range of strategies and adaptations has been associated with a spectrum of shell morphologies (Bassett, 1984).

Most brachiopods are and probably were fixosessile, attached to substrates by a variety of structures including at least two variable pedicle types (Richardson, 1981): stout plenipedunculate structures, the conventional fleshy stalk, and rhizopedunculate forms where the pedicle is divided into a number of threads or rootlets. Plenipedunculate structures were attached to a variety of hard substrates, whereas rhizopedunculates could root into soft sediment. The majority of pedunculate taxa are and probably were epifaunal, anchored to rocks, boulders, hardgrounds, sediment, skeletal debris and even other living animals as secondary tiers above the sea bed. Nevertheless a number of micromorphic brachiopods such as the living articulate *Gwynia* and the fossil nonarticulate acrotretides probably pursued life strategies within the interstitial fauna.

A variety of brachiopods were cemented to hard substrates by small areas umbonally, marked by a scar or cicatrix such as the nonarticulate *Craniops* and the articulates *Schuchertella* and *Waagenoconcha*, whereas the nonarticulate cranioids *Crania* and *Petrocrania* together with the discinoids *Orbiculoidea* and *Schizocrania* encrusted hard substrates. Fossil associations with nektonic and planktonic organisms such as algae and nautiloids suggest a few taxa may have filled epiplanktonic or pseudoplanktonic niches.

Some Carboniferous productoids, such as *Linoproductus* and *Tenaspinus* continued a fixosessile life style after their pedicles had atrophied. Rows of spines, developed at and near the posterior margins of the shells, clasped supportive crinoid stems. Minute perforations in the umbones of some orthotetoids, so-called koskinoid holes, may have been extensions of the mantle attaching the brachiopod to the substrate by a series of threads analogous to the byssus of living mussels.

Following a short phase of larval attachment many groups of brachiopods atrophied their pedicles and pursued a life lacking peduncular support. These liberosessile forms adopted a range of innovative life strategies: large biconvex brachiopods, such as the articulate pentamerids and the nonarticulate trimerellids, formed cosupportive clusters commonly preserved as thick shell beds; ambitopic brachiopods were initially pedunculate, attached to a variety of substrates, but during ontogeny detached and developed life strategies on soft substrates, lying with their more convex valves downwards in the sediment.

Pseudoinfaunal types were partly submerged in the sediment and many developed a range of spines for anchorage and support. The Permian *Waagenoconcha* developed a battery of spines assisting anchorage and stability while, like snowshoes, these structures prevented the valves sinking too deeply within the sediment (Grant, 1966). The excessively long spines of *Marginifera* probably allowed the productoid to survive inversion, resting on its curved spines and its posterior margin of the dorsal surface; the anterior commissure was probably still free to open above the sediment–water interface.

The spectacular conical gemmellaroiids and richthofeniids both crudely copied the external appearance of solitary corals. Feeding currents were possibly initiated by the

LIFE STYLE	BRACHIOPOD TAXA
ATTACHED BY PEDICLE	
Epifaunal – soft substrate[1] (plenipedunculate)	orthides, rhynchonellides, spiriferides and terebratulides
Epifaunal – hard substrate[2] (rhizopedunculate)	*Chlidonophora* and *Cryptopora*
Cryptic	*Argyrotheca* and *Terebratulina*
Interstitial	acrotretides and *Gwynia*
CEMENTED	*Craniops* and *Schuchertella*
ENCRUSTING[3]	craniids and disciniids
CLASPING SPINES[4]	*Linoproductus* and *Tenaspinus*
MANTLE FIBRES	orthotetoids
UNATTACHED	
Cosupportive[5]	pentamerids and trimerellids
Coral-like[6]	gemmellaroids and richthofeniids
Recumbent	strophomenides
Pseudoinfaunal[7] and inverted[8]	*Waagenoconcha* and *Marginifera*
Free-living[9,10]	*Cyrtia, Chonetes, Neothyrisa* and *Terebratella*
MOBILE	
Infaunal[11]	linguloids
Semi-infaunal[12]	*Camerisma* and *Magadina*

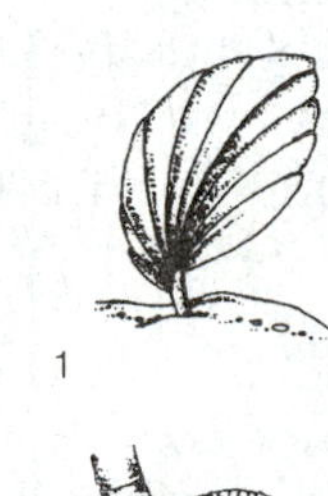

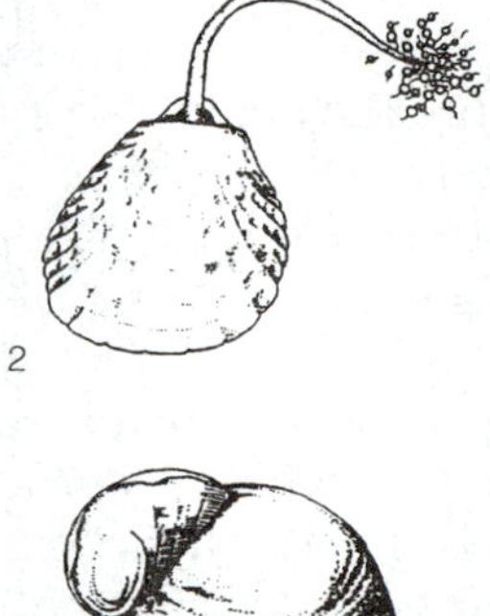

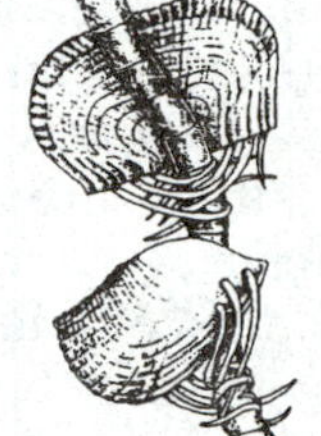

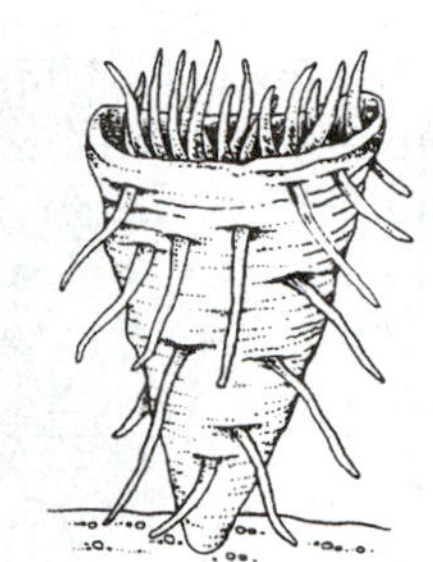

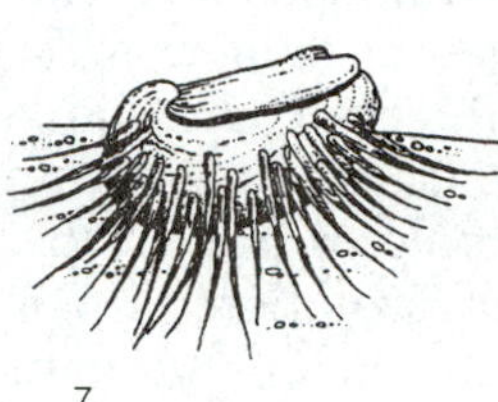

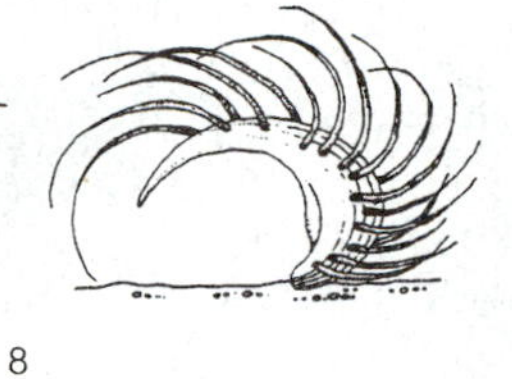

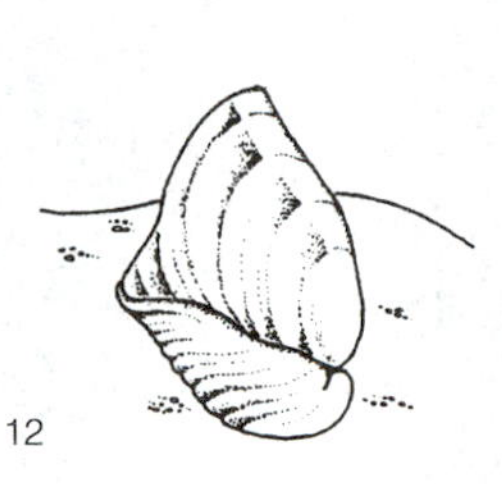

Figure 4.18 Main life strategies of Palaeozoic brachiopods. (After Harper and Moran, 1997.)

vigorous flapping of the cap-like, upper dorsal valves, although these conical forms probably fed by the ciliary pump action of the lophophore. These bizarre animals were anchored in the sediment and supported by a range of ventrally directed spines.

Free-living brachiopods survived at the sediment–water interface by developing large flat shells (*Rafinesquina*), high interareas (*Cyrtia*), extravagant anterior flanges (*Stenoscisma*), posterior spines (*Chonetes*) and rhizoid spines (*Aulostega*) to aid stability and prevent submergence.

Infaunal strategies were adopted early in the history of the phylum. The muscular pedicles of many lingulides probably anchored to or were supported on the substrate, allowing the two valves to construct a U-shaped burrow system. Attached forms such as the majority of orthides, rhynchonellides, spiriferides and terebratulides dominated many Palaeozoic faunas, whereas unattached taxa, such as the strophomenides, diversified during the Ordovician, reaching an acme in the Permian when the spiny *Marginifera* and *Waagenoconcha* together with the coral-like gemmellaroiids and richthofeniids participated in some of the most bizarre and spectacular benthos to inhabit the oceans. These, rather than the fixed and infaunal groups, were targeted by the end Permian extinction event. Perhaps these liberosessile taxa could no longer survive in the bioturbated and bulldozed marine sediments of the post-Palaeozoic sea bed; moreover, most strophomenides may have lacked the armour and mobility to join the Mesozoic arms race. In contrast the linguloids continued, successfully, as part of an evolving infauna.

structures allowing the stem considerable flexibility (Fig. 4.19). Although most crinoids are and were fixed, rheophile organisms, orientating their crowns into the current when feeding, some such as the Recent comatulid *Antedon* are mobile or attached by small roots or cirri, whereas the Jurassic *Saccocoma* may have been either epiplanktonic or part of the free-living benthos.

Evolution, initially, in the crinoids of discoidal holdfasts permitted attachment to hard substrates; however, the development of more versatile root-like holdfasts allowed the group to target soft, muddy substrates commonly located in more offshore environments (Sprinkle and Guensburg, 1995). This transition between these two types of attachment structures may have driven a large-scale diversification of the group and its expansion into deeper-water environments during the early Palaeozoic.

Bryozoans diversified during the Ordovician radiation, building larger and more complex colony types. Initially colonies were low with few zooids; multistorey complexes were developed to occupy available space and evolve efficient feeding strategies.

Only a shallow infauna was well developed in the Palaeozoic fauna, although there were exceptions; it included bivalves, scaphopods, trilobites and crustaceans occupying depths of up to about 100 mm. Nevertheless, by the early Carboniferous, trace fossil data suggests that depths of up to 1 m of sediment were penetrated by bivalves. Bivalves first appeared during the early Cambrian as part of a shallow infauna.

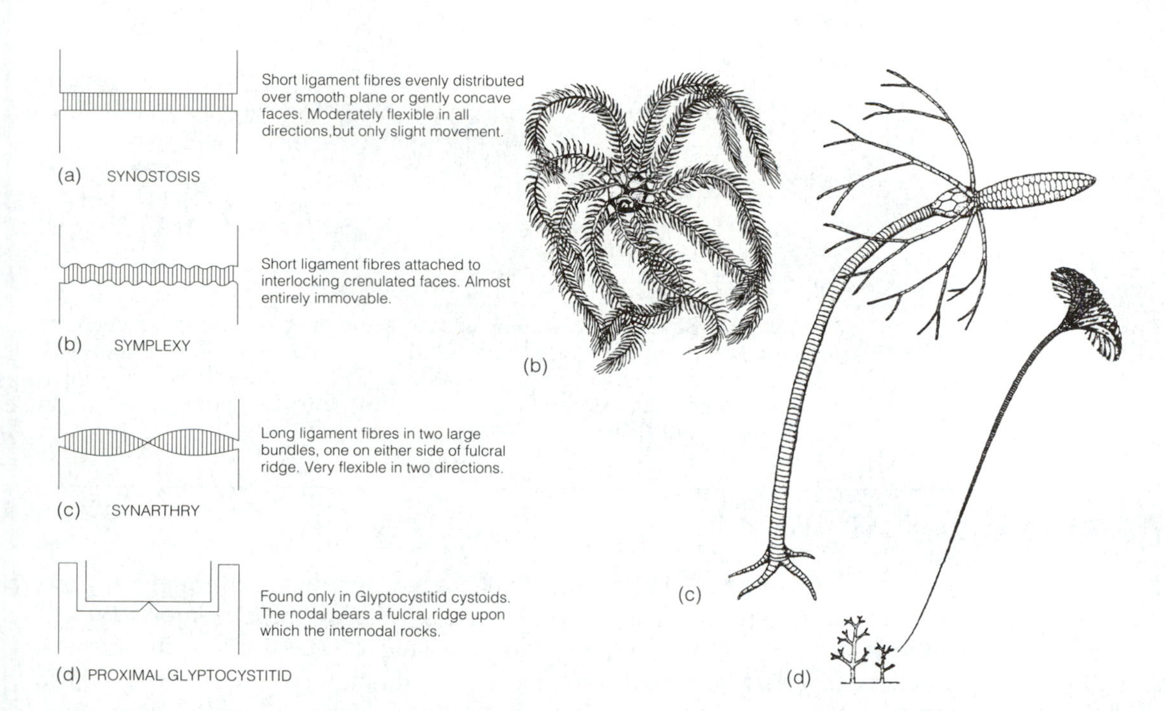

Figure 4.19 (a) Principal types of ligamentary articulations in echinoderm stems (modified and redrawn from Donovan, 1983); (b) Free-living Recent *Antedon*, (c) Ordovician *Dictenocrinus* rooted to substrate (redrawn from Harper *et al*., 1996); (d) Silurian rheophile crinoid attached to a bryozoan. (After Brett, 1984.)

Box 4.5 Function of bryozoan colonies

Although bryozoans can show a marked plasticity in morphology, even within individual taxa, the bryozoan colony through its shape and development of zooids has evolved recognizable adaptations in a variety of environments. Stenolaemates dominated Palaeozoic bryozoan communities. Groups such as the fenestrates formed spectacular cone, mat and vase-shaped structures rising from the seabed, whereas *Archimedes* developed a helical or spiral colony. Functional analysis of the fenestrate colony (Cowen and Rider, 1972) suggested that nutrient and waste-laden currents moved through the highly -integrated bryozoan framework, similar to the passage of fluids through sponges (Fig. 4.20). The polarity of movement was determined by the facing direction of the zooecia. The gross architecture was developed to maximize the feeding capacity of the colony.

Throughout the Phanerozoic the group has demonstrated a remarkable spectrum of adaptive morphologies for life above, at and within the sediment–water interface. Aspects of the Palaeozoic bivalve fauna heralded the intense infaunal radiation of the group during the Mesozoic Marine Revolution, when the group became much more dominant.

Tiered profiles evolved during the Palaeozoic (Ausich and Bottjer, 1982). The intermediate-level benthos (50–200 mm) was dominated by

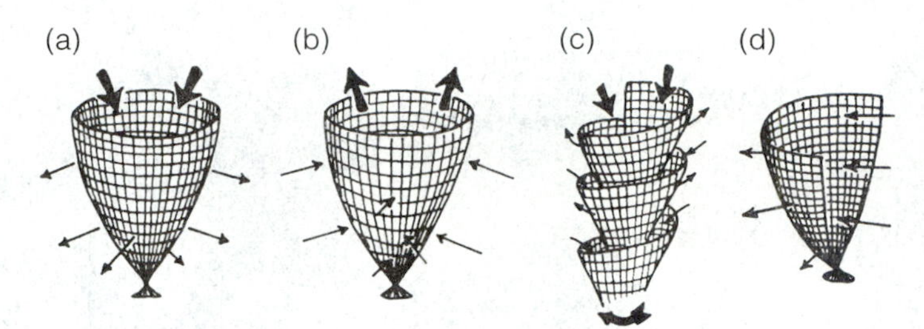

Figure 4.20 Bryozoan colonial feeding strategies, direction of currents indicated with arrows: A. Fenestellid with inward-facing zooecia, B. Fenestellid with outward-facing zooecia, C. Spiral *Archimedes* colony, D. Fan-shaped fenestellid. (Modified and redrawn from Cowen and Rider, 1972 and Clarkson, 1993.)

Box 4.6 Palaeozoic bivalve life modes

Bivalves were not a dominant feature of the Palaeozoic benthos, although several Palaeozoic faunas have abundant bivalves often to the exclusion of the more typical brachiopod associations. These bivalve faunas often consisted of free-burrowing suspension and detritus feeders together with semi-infaunal and epifaunal taxa. The range of life modes was apparently much more restricted than those of the Modern bivalve fauna (see below). Bivalve-dominated palaeocommunities are characterized by epifaunal, infaunal and deep infaunal tiers with commonly 90% detritivores; the deep infauna was relatively uncommon with most taxa restricted to the upper 20 mm of sediment while some bivalves were attached to organic material above the sediment–water interface (Liljedahl, 1985).

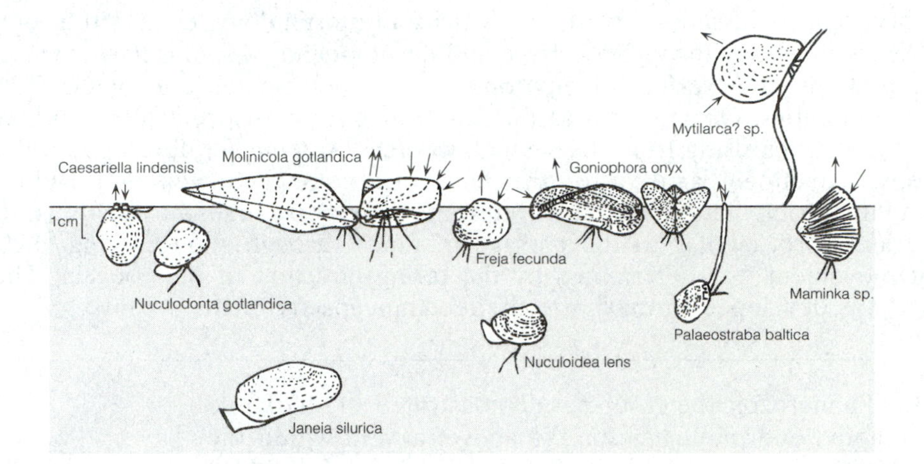

Figure 4.21 The bivalve-dominated palaeocommunity from the Silurian: life positions of the bivalves in the Möllbos fauna, Gotland, Sweden (Liljedahl, 1985); the fauna is dominated by surface and semi-infaunal detritus feeders. (Redrawn from Clarkson, 1993.)

Box 4.7 Trilobite ecomorphs

Many new groups of trilobites appeared during the early Ordovician radiation event. The class shows a number of evolutionary trends, many with adaptive significance. During the subsequent Palaeozoic, coaptative structures, improving enrolment, evolved; eye loss together with increasing spinosity is also a feature of the group. The trilobites, however, also display a set of repeated morphotypes (Fortey and Owens, 1990). A pelagic morphotype with hypertrophied eyes and reduced pleurae appeared suited to life in the surface waters whereas placomorphs, with isopygous and convex exoskeletons, tuberculate sculpture and prominent eyes may be typical of shelf environments, where they grazed and hunted. The illaenimorphs were smooth and were probably associated with carbonate environments such as reefs with the body and pygidium buried in the sediment and the cephalon resting on the surface. Atheloptic taxa have reduced eyes and probably inhabited deeper-water habitats whereas the olenimorphs (see also above) had thin exoskeletons together with a number of other features associated with life in dysaerobic environments. Many of these morphologies were cross-taxonomic, suggesting a morphological convergence with undoubted adaptive significance.

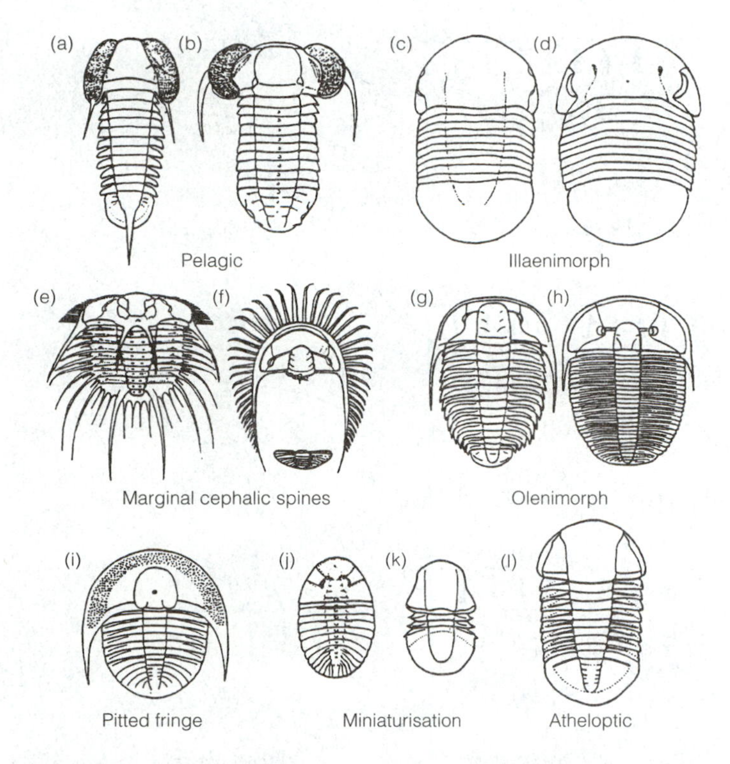

Figure 4.22 Main trilobite morphotypes: (a) *Opipeuter* and (b) *Carolinites* (pelagic), (c) *Illaenus* and (d) *Bumastus* (illaenimorph), (e) *Odontopleura* and (f) *Bowmania* (marginal cephalic spines), (g) *Olenus* and (h) *Aulacopleura* (olenimorph), (i) *Dionide* (pitted fringe), (j) *Schmalenseeia* and (j) *Thoracocare* (miniaturisation - actual size indicated by scale bars), (l) *Illaenopsis* (atheloptic). (Modified and redrawn from Fortey and Owens, 1990.)

Box 4.8 Life modes and radiations of the graptolites

Although there remains some controversy regarding the life modes of many graptolite taxa (Rigby, 1993) most of the evolutionary changes within the graptolites as a whole, from a bushy sessile Cambrian dendroid to a sleek Silurian monograptid, are viewed as adaptational. There are three main trends: a progressive decrease in the number of stipes, for example, from the many-branched dendroid *Rhabdinopora*, through *Dichograptus* with eight, *Tetragraptus* with four, *Didymograptus* with two and *Monograptus* with one. A change in the relative attitudes of the stipes often from pendant, through horizontal and reclined to scandent occurred during the Ordovician. The modification of the shape of the thecae which housed the zooid was carried to extremes during the Silurian. These may have aided the automobility of the graptolite rhabdosome (Kirk, 1969) and certainly maintained their position in the Lower Palaeozoic surface columns and waters (Underwood, 1994) while the group as a whole developed a variety of feeding and harvesting strategies.

During the early Palaeozoic, eustatic changes in sea-level may have driven radiations within the group as various graptolite communities, resident at different levels within the water column with particular adaptations diversified to occupy new ecospace (Fig. 4.23).

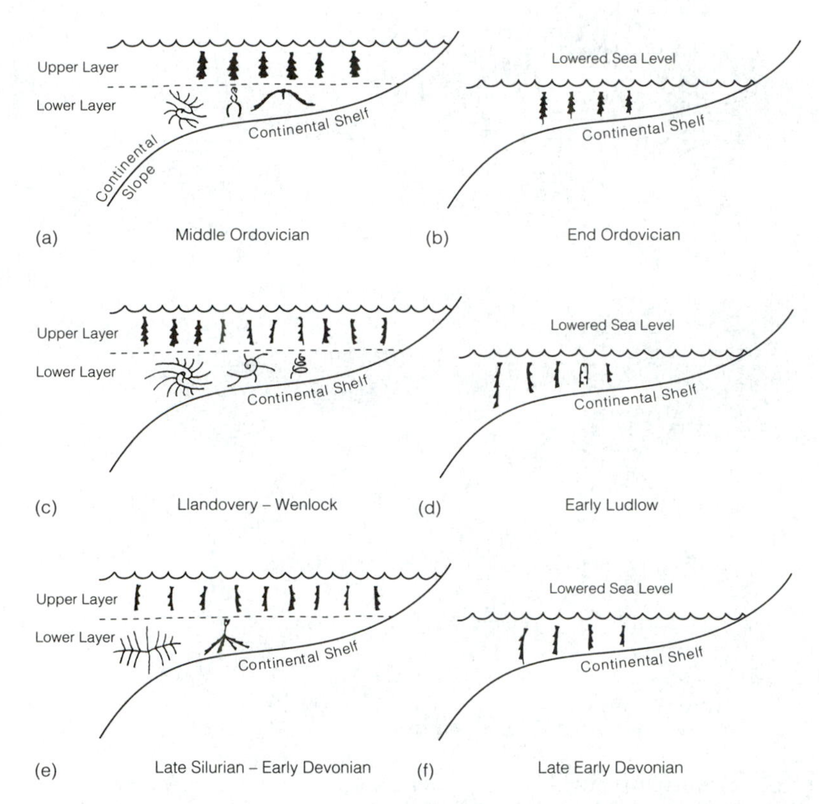

Figure 4.23 Changes in graptolite faunas in relation to automobility, depth zonation and eustatic sea-level changes (after Bates and Kirk, 1984): during transgressive phases (a, c and e) the expanded water column allows for radiations amongst the graptolites through a range of levels; during regressive phases (b, d and f) restricted space is correlated with lower graptolite diversities.

sponges, corals, giant bivalves, giant brachiopods, stalked echinoderms and fixed dendroid graptolites. High-level sessile benthos (200–500 mm) contained mainly crinoids and blastoids. The stalked echinoderms concentrated on improving the efficiency of their feeding techniques with the development of more elaborate calices.

Trilobites continued to dominate the mobile benthos, although bivalves, gastropods and echinoids locally moved across and through the sediment. Cephalopod molluscs such as the orthoconic nautiloids patrolled the benthos, in a role as the main macrophagous predator.

4.7.5.2 Nekton and plankton

Acritarchs, chitinozoans and radiolaria had a dominant place in the microplankton. Graptolites formed a large part of the macroscopic plankton and nekton during the early Palaeozoic together with pelagic and planktonic trilobites; whereas orthoconic nautiloids together with conodont animals were probably major predators above the sediment–water interface to be joined by both agnathan and gnathostome fishes. A few species of small, thin-shelled bivalves and brachiopods may have pursued epiplanktonic strategies attached to floating organic material or even volcanic pumice (see Chapter 8).

4.7.6 Modern evolutionary biota

The Modern evolutionary marine biota radiated after the end Permian extinction event with mobile detritus-feeders both epifaunal and infaunal dominating the seascapes previously occupied by the sessile suspension-feeding benthos. The Mesozoic marine revolution was apparently driven by an arms race (Vermeij, 1987). Thicker shells and the ability to burrow deeper were matched by a variety of improved predatorial skills including the evolution of more advanced jaw mechanisms in groups such as the crustaceans. Moreover specific adaptations such as the evolution of cementation have been related to the appearance of molluscivorous predators (Harper and Skelton, 1993; Skelton *et al.*, 1990).

4.7.6.1 Benthos

Low-level sessile benthos (0–50 mm) held the brachiopods, bryozoans, bivalves, sponges and corals. Intermediate-level sessile benthos included sponges, corals, giant bivalves and soft corals in a second tier (50–200 mm). Whereas the high-level sessile benthos contained crinoids, sponges and soft corals occupying the highest tiers, in excess of 200 mm. The adaptive radiations and offshore migrations of many Mesozoic groups corresponded with developments in the Mesozoic arms race (see Chapter 9).

Post-Palaeozoic brachiopod faunas, together with the crinoids, in general, moved offshore, occupying deeper-water and more cryptic environments in crevices and on submarine cliffs in dark and apparently unpleasant conditions; some developed poisonous tissues. The majority of brachiopods in the Modern fauna are fixed rhynchonellides and terebratulides attached to a wide variety of large and small substrates (Surlyk, 1972); the more robust and globose terebratulides such as *Terebratella* and probably some species of *Tichosina* (Harper *et al.*, 1995) were free to roll on the substrate. Semi-infaunal strategies are less common, however, the living terebratulide *Magadina* can move within surface sediment. Despite the extraordinary longevity of the phylum, the Recent brachiopod fauna has a relatively low diversity, reflecting an apparent evolutionary stasis across the group. Nevertheless some living brachiopods such as *Terebratella* have developed a range of life styles in a variety of habitats (Richardson, 1986); this genus has adapted to a range of substrates but also developed a stable free-living strategy.

The Modern bryozoan fauna is dominated by cheilostome and cyclostome bryozoans. During the early Cretaceous, however, the cyclostome bryozoans markedly declined in abundance while the cheilostomes rapidly diversified (Lidgard *et al.*, 1993). Mechanisms for this transition in dominance are unclear, although a combination of the relative interactions of both groups with competitors and predators together with nutrient sources may have been responsible.

Mobile benthos now included a variety of more familiar invertebrates dominated by molluscs and echinoderms. Bivalves had adapted to a wide range of habitats and pursued a variety of life strategies. Gastropods, however, also have a wide range of trophic modes across a spectrum of marine, freshwater and terrestrial environments, with many gastropods such as *Littorina* and *Nassarius*, crawling over either soft or hard substrates. Echinoids had diverged to develop infaunal strategies together with a mobile existence at the sediment–water interface.

The shallow infauna included bivalves, scaphopods, crustaceans; the deep infauna established during the Carboniferous continued and diversified with bivalves, heart urchins and shrimps dominant.

4.7.6.2 Nekton and plankton

The water columns of the Mesozoic were populated by the fishes and a variety of cephalopod molluscs including the ammonites and the belemnites. Both molluscan groups were adapted for speed and mobility. But by the Cenozoic, following the end Cretaceous extinction event, the fishes dominated neritic faunas. The planet's surface waters continued to supply the base of the food chain with phytoplankton such as coccoliths, diatoms and silicoflagellates together with a zooplankton dominated by foraminifera and radiolarians. Many megafaunal groups such as the amphipod, decapod and isopod crustaceans together with nudibranch molluscs and the annelid and polychaete worms were also probably part of the plankton.

Box 4.9 Modern bivalve life modes

A significant range of innovative life modes was developed during the Mesozoic radiation of the group, associated to a great extent with the Mesozoic arms race. The evolution of thick-shelled, highly sculptured taxa together with the deep infaunal strategy has been directly related to increased predator pressure. The Mesozoic and Cenozoic bivalve faunas display a wide range of morphologies and adaptations far in excess of those of their Palaeozoic counterparts (Stanley, 1970). Many different groups developed convergent morphologies adapted for infaunal life modes (Fig. 4.24). The following life modes have been documented in the group and related to morphological features: shallow burrowing (equivalved with muscles scars of equal sizes and commonly strong ornament), deep burrowing (elongated valves, lacking teeth with permanent gape and long siphons), reclining (different shaped valves, sometimes with spines for anchorage or to prevent sinking), byssal attachment (elongate valves with flat ventral surface; reduction of anterior part of valve and anterior adductor – attached by threads or byssus), cemented (markedly different shaped valves, single large adductor scar and crenulated commissure), boring (elongate cylindrical shells with strong sharp external ornament) and swimming (valves dissimilar in shape and size with very large single adductor muscle and hinge line extended as ears).

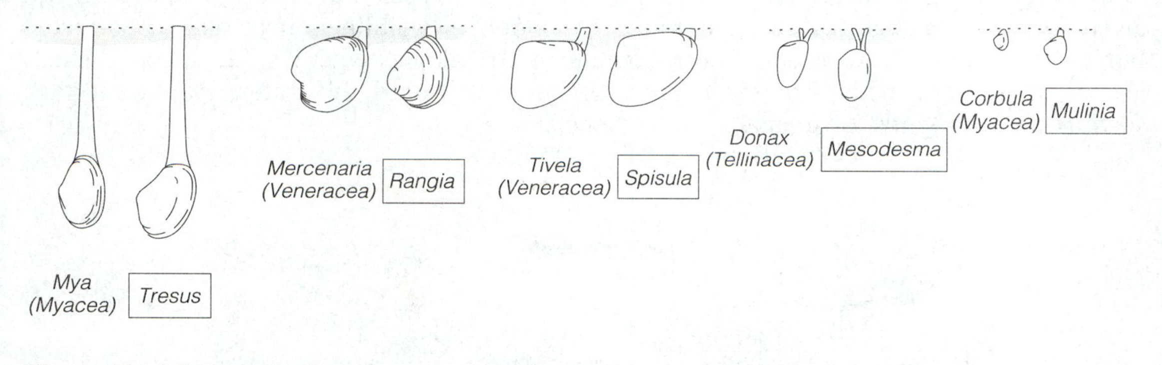

Figure 4.24 Modern bivalve life modes: a range of unrelated genera belonging to different families have converged in form and habit with the mactracean bivalves. (After Stanley, 1970.)

Box 4.10 Gastropod morphology and adaptation

Gastropods developed a wide range of shell morphologies from patellate forms through convolute, trochiform, turbinate, pupiform, turreted, discoidal, biconical, isostrophic, digitate to irregular morphologies (Clarkson, 1993). Nevertheless, with a few exceptions it has proved difficult to relate these diverse morphologies to specific life strategies. An obvious exception is the patellate shell form adopted by limpets for a mainly fixed strategy. Turreted shells, however, possess a marked ratchet or terraced sculpture which can be related to a burrowing life mode. Ratchet sculpture is a combination of continuous ribs with cross-orientation and frictional symmetry; it has a constant height, becoming more densely spaced with age, and a smooth perimeter. The frictional asymmetry and the negative allometry of the sculpture are most significant in a morphological group adapted for burrowing; reports of epifaunal ratcheted taxa have yet to be verified (Signor, 1993). Other groups, such as the strombacean gastropods, also developed a range of adaptive morphologies related to burrowing life modes (Fig. 4.25).

A set of more generalized morphologies has been related to a range of locomotory modes together with the type of substrate and degree of water turbulence (Dodd and Stanton, 1981); there are, however, no general rules and plenty of exceptions. The typical coiled and highly ornamented taxa browsed across the substrate–water interface associated with both low and high turbulence on a range of substrate types; burrowing forms need not always have a ratchet sculpture, a range of smooth streamlined taxa also constructed burrows. Cemented taxa usually have irregular morphologies whereas flat uncoiled shells are commonly attached; both forms are associated with hard substrates. Planktonic taxa such as the pteropods have small, reduced shells.

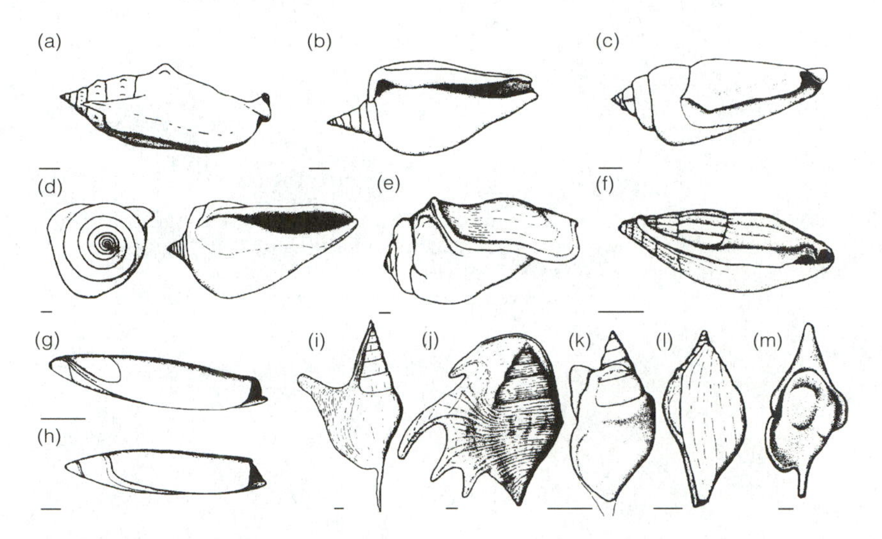

Figure 4.25 Morphology of burrowing strombaceans gastropods: a–c, h. Observed infaunal positions of Recent taxa (a–c. Species of *Strombus*, h. *Terebellum*); d–g. Inferred fossil posture in fossil taxa (d. *Oostrombus*, e. *Thersitea*, f. *Rimella*, g. *Paraseraphs*); i–m. development of posterior canals in the strombids (i. *Hippochrenes*, j. *Chedevillia*, k. *Cyclomolops*, l. *Orthaulax*, m. *Calyptraphorus*). (Redrawn from Savazzi, 1994.)

Box 4.11 Echinoid life strategies

Palaeozoic echinoids were a variable part of the mobile benthos, their regular tests moving about over the substrate as they pursued a detritivore diet. Their role changed dramatically after the end Permian extinction event, as a range of adaptations was rapidly developed to suit infaunal strategies in first the heart urchins and, later, the sand dollars (Kier, 1982). Irregular echinoids evolved rapidly during the early Jurassic with a tool chest of adaptations for an active infaunal life mode: a heart-shaped to ellipsoidal test, the migration of the anus or periproct to a lateral position at the posterior side of the test, the development of a food groove from one of the ambulacral areas and the modification of feet to act as digging paddles and respiratory organs equipped the animal for burrow construction and an infaunal existence. The sand dollars appeared during the early Tertiary; their flattened test was adapted for burrowing and burial in the sediment where accessory tube feet encouraged food along food grooves and helped cover the test with sand; the accentuated and characteristic petals of the group aided respiration, whereas the evolution of low jaw apparatus with horizontal teeth signalled a new mode of infaunal feeding.

The heart urchin *Micraster* has been intensively studied for nearly a century. The adaptations once thought associated with deeper burrowing strategies have been revised. Very few *Micraster* are ever found in a deep burrow, most apparently died on the surface; moreover, many tests have an epifauna attached after death and the burrows in the Chalk, for example, do not fit these taxa. Nevertheless it is still most likely that *Micraster* burrowed. Rather the adaptations were developed for more efficient, rather than deeper burrowing strategies.

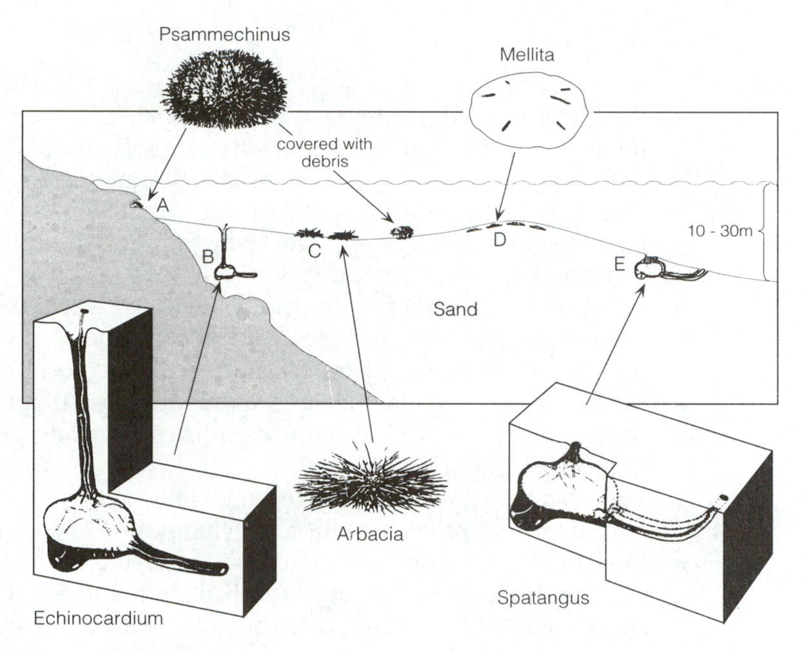

Figure 4.26 Life modes of five echinoids from the Pliocene of Virginia: epifaunal regulars such as *Psammechinus* and *Arbacia* co-occurred with the infaunal sand dollar *Mellita* and the burrowing irregulars *Echinocardium* and *Spatangus*. (After Kier, 1972.)

Box 4.12 Cephalopod life styles

The evolution of the suture patterns of the shelled cephalopods (the nautiloids and ammonoids) has been related to a variety of functions, including development of secure attachments for muscles, buttresses and supports to protect the shells from implosion at depth, or even to aid the removal of liquid from the chambers. Most likely the complex sutures, reflecting the crinkled partitions between chambers strengthened the shell against hydrostatic loading, allowing those with complex patterns to occupy a range of depths within the water column (Hewitt and Westermann, 1987). Alternatively the last-formed septum may have been initially unaragonitized, functioning as a balloon or diaphragm with an ability to alter pressure and rise and fall in the water column like a cartesian diver (Seilacher and La Barbera, 1995).

The cephalopod shell was geared to maintaining a stable position in the water column between bouts of rapid movement both laterally and vertically. In addition the shape and ornament of the ammonoid may be related to the environment. In the Cretaceous of the United States evolute, heavily ornamented forms may have been nektobenthonic together with spiny and nodous taxa. The planulates and serpenticones were probably pelagic with oxycones preferring shallow water (Batt, 1993).

4.8 SUMMARY POINTS

- Adaptation or the fitness of an organism to its environment may be assessed through the phylogenetic, fabricational, functional and environmental constraints on its development.
- The functional morphology of an organism is investigated by homologies and analogies with living animals and plants; paradigms can provide testable models for functions.
- Computer simulations for the growth of shelled and branched organisms generate both theoretical and actual morphotypes; the functional efficiency of morphologies can be tested.
- Many organisms may have preadapted structures; but to make evolutionary progress organisms must cross functional thresholds. Giant size may be adaptive.
- Heterochrony, with both paedomorphic and peramorphic modes, may drive adaptive evolutionary changes.
- During the late Precambrian and earliest Cambrian, prokaryotes and early Metazoans such as the Ediacara and Small Shelly Faunas show a variety of adaptive strategies; colonial growth styles were pursued by the archaeocyathans.
- Palaeozoic benthos, including the brachiopods, corals, stromatoporoids and trilobites, developed a range of adaptive morphologies related to various life modes.

- In the early Palaeozoic water column, graptolites evolved a range of streamlined morphologies and feeding strategies; amongst the invertebrates, orthoconic nautiloids were the primary nektonic and nektobenthic predators.
- During the Mesozoic marine ecosystems were dominated by functional and morphological change associated with the Mesozoic Marine Revolution or arms race; bivalve and gastropod molluscs together with echinoids responded to the challenge with greater armour and deep-burrowing strategies.
- Functional changes and adaptive responses at the population level can trigger major adaptive radiations and large-scale biological change.

4.9 FURTHER READING

Benton, M.J. and Harper, D.A.T. 1997. *Basic palaeontology*. Addison Wesley, Longman, Harlow. 342 pp.

Briggs, D.E.G and Crowther, P.R. (eds) 1990. *Palaeobiology – a synthesis*. Blackwell Scientific Publications, Oxford. 583 pp.

Fretter, V. and Graham, A. 1976. *A functional anatomy of invertebrates*. Academic Press, London.

Gould, S.J. 1989. *Wonderful Life. The Burgess Shale and the nature of history*. Hutchinson Radius, London. 347 pp.

Lane, N. G. 1986. *Life of the past*. Bell and Howell Company, Columbus. 326 pp.

Lehmann, U. 1981. *The ammonites – their life and their world*. Cambridge University Press, Cambridge. 246 pp.

McMenamin, M.A.S. and McMenamin, D.L.S. 1990. *The emergence of animals. The Cambrian breakthough*. Columbia University Press, New York. 217 pp.

McMenamin, M.A.S. and McMenamin, D.L.S. 1994. *Hypersea. Life on land*. Columbia University Press, New York. 343 pp.

Manten, S.M. 1977. *The Arthropoda*. Oxford University Press, Oxford.

Morton, J.E. 1967. *Molluscs*. Hutchinson, London. 244 pp.

Rudwick, M.J.S. 1970. *Living and fossil brachiopods*. Hutchinson, London. 199 pp.

Smith, A. 1984. *Echinoid palaeobiology. Special topics in palaeontology*. Allen Unwin Inc. London. 190 pp.

4.10 REFERENCES

Ager, D.V. 1963. *Principles of paleoecology*. McGraw-Hill, New York. 371 pp.

Ager, D.V. 1967. Brachiopod palaeoecology. *Earth Science Reviews* 3, 157–195.

Ausich, W.I. and Bottjer, D.J. 1982. Tiering in suspension-feeding communities on soft substrata throughout the Phanerozoic. *Science* 216, 173–174.

Bassett, M.G. 1984. Life strategies of Silurian brachiopods. *Special Papers in Palaeontology* 32, 237–263.

Batt, R. 1993. Ammonite morphotypes as indicators of oxygenation in a Cretaceous epicontinental sea. *Lethaia* 26, 49–63.

Benton, M.J. 1988. Recent trends in vertebrate palaeontology. *In:* Harper, D.A.T. (ed.), *William King D.Sc. a palaeontological tribute*, 62–79. Galway University Press, Galway.

Benton, M.J. 1990. *Vertebrate palaeontology*. Chapman & Hall, London. 377 pp.

Buss, L.W. and Seilacher, A. 1994. The phylum Vendobionta: a sister group of the Eumetazoa? *Paleobiology* 20, 1–4.

Carlson, S. 1989. The articulate brachiopod hinge mechanism: morphological and functional variation. *Paleobiology* 15, 364–386.

Chamberlain, J.A. Jr. 1981. Hydromechanical design of fossil cephalopods. *In:* House, M.R. and Senior, J.R. (eds) *The Ammonoidea*, 289–336. Special Volume of the Systematics Association 18. Academic Press, London.

Clarkson, E.N.K. 1993. *Invertebrate paleontology and evolution*. Chapman & Hall, London. 434 pp.

Clarkson, E.N.K. and Taylor, C. 1995. The lost world of the olenid trilobites. *Geology Today* 11, 147–154.

Clarkson, E.N.K., Harper, D.A.T. and Peel, J.S. 1995. The taxonomy and palaeoecology of the mollusc *Pterotheca* from the Ordovician and Silurian of Scotland. *Lethaia* 28, 101–114.

Conway Morris, S. 1978. A new metazoan from the Burgess Shale of British Columbia. *Palaeontology* 20, 623–640.

Cowen, R. and Rider, J. 1972. Functional analysis of fenestellid bryozoan colonies. *Lethaia* 5, 147–164.

Dawkins, R. 1986. *The blind watchmaker*. W.W. Norton, New York. 332 pp.

Dodd, J.R. and Stanton, R.J. Jr. 1981. *Palaeoecology, concepts and applications*. John Wiley and Sons, New York. 559 pp.

Fortey, R.A. and Owens, R.M. 1990. Trilobites. *In:* McNamara, K.J. (ed.) *Evolutionary Trends*, 121–142. Belhaven Press, London.

Gould, S.J. and Vrba, E. 1982. Exaptation – a missing term in the science of form. *Paleobiology* 8, 4–15.

Grant, R.E. 1966. A Permian productoid brachiopod: life history. *Science* 152, 660–662.

Grant, R.E. 1981. Living habits of ancient articulate brachiopods. *In:* Broadhead, T.W. (ed.) *Lophophorates notes for a short course*. University of Tennessee, Department of Geological Sciences, Studies in Geology 5, 127–140.

Harper, D.A.T. and Moran, R. 1997. Brachiopod life styles. *Geology Today* (in press).

Harper, D.A.T. and Owen, A.W. 1988. The vital statistics of fossils. *Teaching Geology* 13, 74–78.
Harper, D.A.T. and Ryan, P.D. 1990. Towards a statistical system for palaeontologists. *Journal of the Geological Society of London* 147, 935–948.
Harper, E.M. and Skelton, P.W. 1993. The Mesozoic Marine Revolution and epifauna; bivalves. *Scripta Geologica Special Issue* 2, 127–153.
Harper, D.A.T., Doyle, E.N. and Donovan, S.K. 1995. Palaeoecology and palaeobathymetry of Pleistocene brachiopods from the Manchioneal Formation of Jamaica. *Proceedings of the Geologists' Association* 106, 219–227.
Herbert, M.J., Jones, C.B. and Tudhope, D.S. 1995. Three-dimensional reconstruction of geoscientific objects from serial sections. *Visual Computer* 11, 343–359.
Hewitt, R.A. and Westermann, G.E.G. 1986–1987. Function of complexly fluted septa in ammonoid shells. I. Mechanical principles and functional models. II. Septal evolution and conclusions. *Neues Jahrbuch für Geologie und Paläontologie Abhandlungen* 172, 47–69 and 174, 135–169.
Hughes, N.C. 1994. Ontogeny, intraspecific variation and systematics of the late Cambrian trilobite *Dikelocephalus*. *Smithsonian Contributions to Paleobiology* 79, 1–89.
Hughes, N.C. and Jell, P.A. 1992. A statistical/computer-graphic technique for assessing variation in tectonically deformed fossils and its application to Cambrian trilobites from Kashmir. *Lethaia* 25, 317–335.
Jaanusson, V. 1971. Evolution of the brachiopod hinge. *Smithsonian Contributions to Paleobiology* 3, 33–46.
Jaanusson, V. 1981. Functional thresholds in evolutionary progress. *Lethaia* 14, 251–260.
Jones, B. 1988. Biostatistics in palaeontology. *Geoscience Canada* 15, 3–22.
Kershaw, S. 1990. Stromatoporoid palaeobiology and taphonomy in a Silurian biostrome in Gotland, Sweden. *Palaeontology* 33, 681–706.
Kier, P.M. 1982. Rapid evolution in echinoids. *Palaeontology* 25, 1–10.
Kirk, N. 1969. Some thoughts on the ecology, mode of life and evolution of the Graptolithina. *Proceedings of the Geological Society of London* 1659, 273–293.
Kovach, W. 1993. MVSP Multivariate Statistical Package *Plus* Version 2.1. Kovach Computing Services, Anglesey.
Lidgard, S., McKinney, F.K. and Taylor, P.D. 1993. Competition, clade replacement, and a history of cyclostome and cheilostome bryozoan diversity. *Paleobiology* 19, 352–371.
Liljedahl, L. 1985. Ecological aspects of a silicified bivalve fauna from the Silurian of Gotland. *Lethaia* 18, 53–66.
McMenamin, M.A.S. 1986. The garden of Ediacara. *Palaios* 1, 178–182.
McNamara, K.J. 1986. A guide to the nomenclature of heterochrony.

Journal of Paleontology 60, 4–13.

Maddison, W.P. and Maddison, D.R. 1992. *MacClade. Analysis of phylogeny and character evolution.* Sinauer Associates Inc, Massachusetts.

Neuman, B. 1988. Some aspects of the life strategies of early Palaeozoic rugose corals. *Lethaia* 21, 97–114.

Okamoto, T. 1988. Changes in life orientation during ontogeny of some heteromorph ammonoids. *Palaeontology* 31, 281–294.

Peel, J. S. 1991. Functional morphology, evolution and systematics of early Palaeozoic univalved molluscs. *Grønlands Geologiske Undersøgelse* 161, 116 pp.

Pickerill, R.K. and Brenchley, P.J. 1991. Benthic macrofossils as paleoenvironmental indicators in marine siliciclastic facies. *Geoscience Canada* 18, 119–138.

Ramsköld, L. and Hou Xianguang 1991. New early Cambrian animal and onychophoran affinities of enigmatic metazoans. *Nature* 351, 225–228.

Raup, D.M. 1966. Geometric analysis of shell coiling: general problems. *Journal of Paleontology* 40, 1178–1190.

Richardson, J.R. 1981. Brachiopods and pedicles. *Paleobiology* 7, 87–95.

Richardson, J.R. 1986. Brachiopods. *Scientific American* 255, 96–102.

Rigby, S. 1993. Graptolite functional morphology: a discussion and critique. *Modern Geology* 17, 271–287.

Rigby, S. and Milsom, C.V. 1996. Benthic origins of zooplankton: an environmentally mediated macroevolutionary effect. *Geology* 24, 52–54.

Rudwick, M.J.S. 1961. The feeding mechanism of the Permian brachiopod *Prorichthofenia*. *Palaeontology* 3, 450–457.

Rudwick, M.J.S. 1964. The function of the zig-zag deflection in brachiopods. *Palaeontology* 7, 135–171.

Ryan, P.D., Harper, D.A.T. and Whalley, J.S. 1994. *PALSTAT package*. Chapman & Hall, London.

Savarese, M. 1992. Functional analysis of archaeocyathan skeletal morphology and its paleobiological implications. *Paleobiology* 18, 464–480.

Scrutton, C.T. 1997. Growth strategies and colonial form in tabulate corals. *Boletin de la Real Sociedad Espanola de Historia Natural Seccion Geologia* (in press).

Seilacher, A. 1989. Vendozoa: organismic construction in the Proterozoic biosphere. *Lethaia* 22, 229–239.

Seilacher, A. and La Barbera, M. 1995. Ammonites as cartesian divers. *Palaios* 10, 493–506.

Signor, P.W. 1993. Rachet riposte: more on gastropod burrowing sculpture. *Lethaia* 26, 275– 386.

Skelton, P.W. 1985. Preadaptation and evolutionary innovation. *In:* Cope, J.C.W. and Skelton, P.W. (eds) *Evolutionary case histories from the fossil record*. Special Papers in Palaeontology 33, 159–173.

Skelton, P.W. 1990. Adaptation. *In:* Briggs, D.E.G and Crowther, P.R. (eds), *Palaeobiology – a synthesis*, 139–146. Blackwell Scientific

Publications, Oxford.

Skelton, P.W., Crame, J.A., Morris, N.J. and Harper, L. 1990. Adaptive divergence and taxonomic radiation in post-Palaeozoic bivalves. *In:* Taylor, P.D. and Larwood, G.P. (eds), *Major Evolutionary Radiations.* Special Volume of the Systematics Association 42, 91–117.

Sprinkle, J. and Guensburg, T.E. 1995. Origin of echinoderms in the Paleozoic evolutionary fauna: the role of substrates. *Palaios* 10, 437–453.

Stanley, S.M. 1970. Relation of shell form to life habits of the Bivalvia. *Geological Society of America Memoir* 125, 1–296.

Stanley, S.M. 1973. An explanation for Cope's Rule. *Evolution* 27, 1–26.

Surlyk, F. 1972. Morphological adaptations and population structures of the Danish Chalk brachiopods (Maastrichtian, Upper Cretaceous). *Det Kongelige Danske Vidensabernes Selskab Biologiske Skrifter* 19, 57 pp.

Swan, A.R.H. 1990a. A computer simulation of evolution by natural selection *Journal of the Geological Society of London* 147, 223–228.

Swan, A.R.H. 1990b. Computer simulations of invertebrate morphology. *In:* Bruton, D.L. and Harper, D.A.T. (eds) Microcomputers in palaeontology. *Contributions from the Palaeontological Museum, University of Oslo* 370, 32–45.

Swofford, D.L. 1993. *PAUP. Phylogenetic analysis using parsimony.* Illinois Natural History Survey, Champaign, Illinois.

Underwood, C.J. 1994. The position of graptolites within Lower Palaeozoic planktic ecosystems. *Lethaia* 26, 198–202.

Vermeij, G.J. 1987. *Evolution and escalation. An ecological history of life.* Princeton University Press, New Jersey.

Vogel, K. 1986. Origin and diversification of brachiopod shells: viewpoints of constructional morphology. *In:* Racheboeuf, P.R. and Emig, C.C. (eds) *Les Brachiopodes Fossiles et Actuels.* Biostratigraphie du Paleozoique 4, 399–408.

Wood, R., Zhuravlev, A.Yu. and Debrenne, F. 1992. Functional biology and ecology of Archaeocyatha. *Palaios* 7, 131–156.

5 Trace fossils

Trace fossils, evidence of the activity of organisms, record the behavioural patterns of fossil animals and plants. They include tracks and trackways, trails, burrows and borings; some studies also include coprolites, faecal pellets and casts. Fossil eggs and bioglyphs made, for example, by empty shells and bones bouncing over or dragged through sediment are not usually described as traces, and agglutinated tests and burrows are considered to be body fossils.

Trace fossils, however, give a unique perspective on the history of life on the planet. They can potentially signal changes in organic activity and faunal diversity through time, partly illuminating the taphonomic shadow created by the very poor representation of soft-bodied organisms in the fossil record, particularly in terrestrial and deep-sea environments. Moreover, even the ontogenies of some traces have been documented in detail and some overall patterns and changes in behavioural evolution can be tracked through the fossil record.

5.1 INTRODUCTION

About three and a half million years ago the volcano Sandiman erupted across the lush Savannah of east Africa. Over a few days several ash falls occurred and following rainfall a whole variety of tracks and trails, generated by many animals from millipedes to elephants, was impressed in the mud. The volcanic products from Sandiman, however, contained a high proportion of carbonatite, which when mixed with water acted essentially as a concrete. Further ash falls covered and preserved these tracks and trails. In addition to the traces of invertebrates and various quadrupeds, the ash preserved some of the most impressive and significant ichnofossils recorded to date (Fig. 5.1). Two hominid trackways extend across the ashfall for almost 30 metres, conclusive evidence that two *Australopithecus* strolled upright, between volcanic eruptions, across this part of the Savannah some 3.5 million years ago.

The spectacular terrestrial trace fossils preserved at Laetoli help illustrate some of their main characteristics. First, most assemblages have a narrow facies range and typically develop in response to particular environmental conditions. Secondly, trace fossils usually cannot

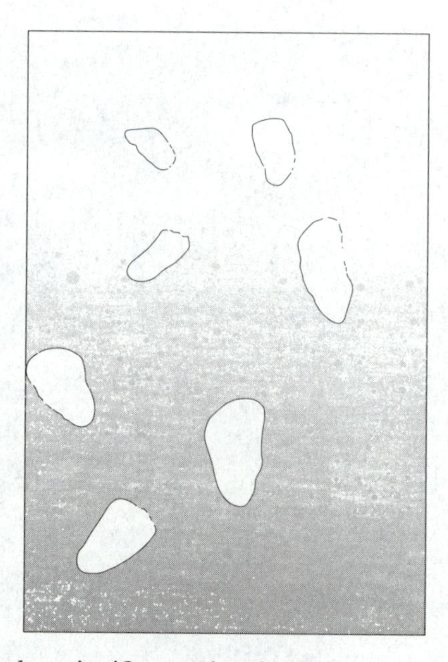

Figure 5.1 Simple but significant: sketch of the outline of the hominid trackways through the concreted volcanic ash at Laetoli, East Africa. (Redrawn from Leakey and Hay, 1979.)

be transported and thus represent evidence of *in situ* communities. However, although we have a good idea of the behaviour of the animals, trace fossils characteristically give us no direct clue as to the identities of their producers. In many cases, however, the nature of the progenitors can be approximated by analogy to structures produced by living organisms; nevertheless, only very rarely is the maker actually preserved with its trace (Fig. 5.2).

Two main palaeoecological approaches have been followed in trace fossil studies. First, the palaeobiology of trace fossils involves the analysis of the behaviour, and more rarely the identity and morphology, of the trace-maker. Secondly, because many trace fossil assemblages (ichnofacies) occur recurrently in particular settings associated with specific water depths and substrates, palaeoenvironmental reconstructions rely heavily on trace fossil data.

5.2 PRESERVATION AND TAXONOMY OF TRACE FOSSILS

The morphology of a particular ichnofossil is often a function of its medium and mode of preservation. Quite different morphologies of the same trace can be generated by preservation in, for example, coarse or fine clastic rocks or at the interface between both lithologies. Stratinomic classifications have been devised, therefore, to describe and identify the relationship of the trace fossil to its casting medium (Fig. 5.3). Both Seilacher (1964) and Martinsson (1965, 1970) have

Figure 5.2 A producer in its trail: *Trentonia shegiriana* (polychaete annelid) from the mid Ordovician Trenton Group, Quebec City; diameter of coin 19 mm. (Photograph courtesy of R.K. Pickerill.)

erected similar classifications which have been reviewed by Ekdale *et al.* (1984), Frey and Pemberton (1985) and Bromley (1996).

Trace fossils are either pre- or post-depositional (Seilacher, 1962). For example Orr's (1995) study of the ichnofauna of the Silurian Aberystwyth Grits recognized pre-depositional components preserved in the sediment and on the soles of turbidites as secondary casts; this background or equilibrium community developed in low-oxygen conditions against background sedimentation. Following an influx of turbidites, post-depositional traces are emplaced with the coarser beds

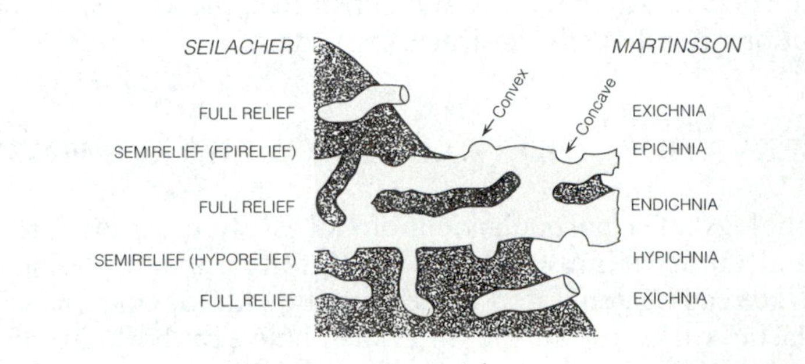

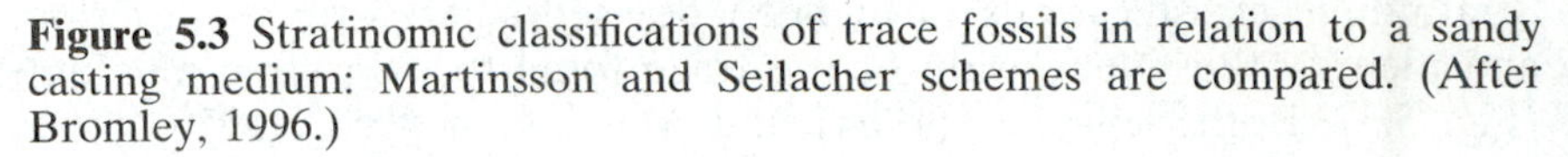

Figure 5.3 Stratinomic classifications of trace fossils in relation to a sandy casting medium: Martinsson and Seilacher schemes are compared. (After Bromley, 1996.)

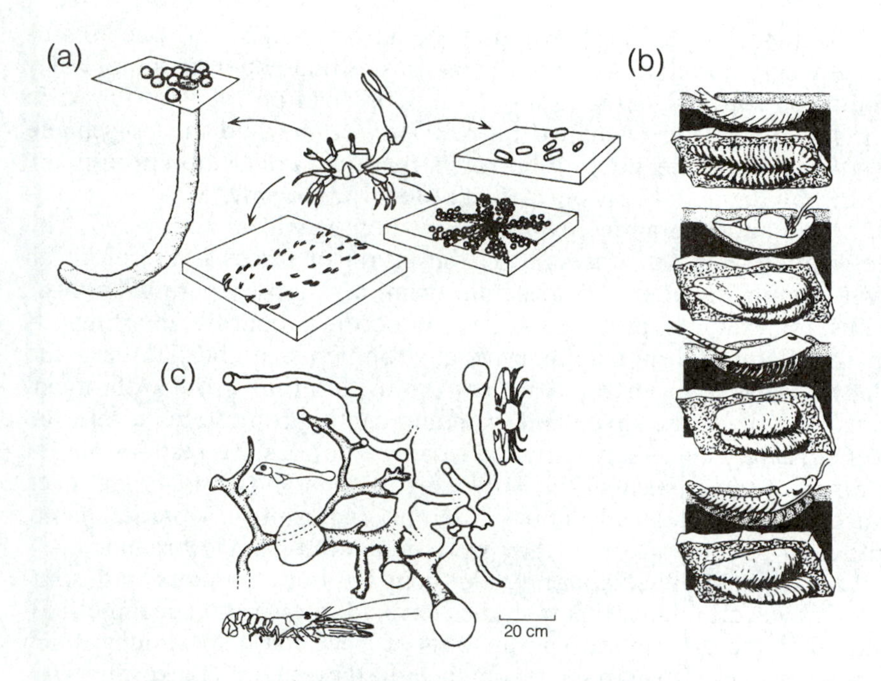

Figure 5.4 (a) Single organisms producing a variety of traces: the fiddler crab, *Uca*, produces a living burrow (*Psilonichnus*), a walking trail (*Diplichnites*), a grazing trace and faecal pellets. b, c. A variety of organisms producing a single trace. (b) From top to bottom, bilobed resting trace, *Rusophycos* produced by a polychaete worm, a nassid snail, a notostracan branchiopod shrimp and a trilobite. (c) Complex system of burrows produced by a fish, a crab and a lobster. (Redrawn from Ekdale *et al.*, 1984.)

related to a freshening of conditions on the seabed. However, as oxygen levels deteriorate and background sedimentation resumes, the seabed is recolonized by the pre-depositional community.

The morphology of most trace fossils is sufficiently characteristic to permit a conventional taxonomic treatment or ichnotaxonomy, based on detailed description of the trace together with photographs, diagrams and, if possible, three-dimensional views. Ichnogenera and ichnospecies are defined, described and illustrated in the same manner as body fossils although to date no higher taxa have been universally accepted.

There are, however, a number of basic assumptions which must be noted in all palaeoichnological studies (Fig. 5.4). First, an individual organism can produce a variety of different traces by assuming different behaviours. Second, an individual may produce different traces through a single behavioural response but in a variety of different substrates. Third, identical structures may be produced by a variety of different organisms pursuing the same behavioural response. And finally, the commensal arrangements of two or more organisms may generate a single structure.

It is important to emphasise that the names assigned to trace fossils usually bear no relationship to the organisms that generate them. Early confusion through, for example, the association of the lugworm *Arenicola* with the U-shaped burrow *Arenicolites* and the polychaete *Nereis* with the meandering burrow structure *Nereites* obscured many of the fundamental concepts of ichnofossil taxonomy. Most examples of both ichnogenera are never generated by *Arenicola* and *Nereis*. The description, nomenclature and classification of traces is therefore an organism-independent exercise in form taxonomy (parataxonomy). Thus, for example, although *Cruziania* occurs in marine envronments in the Cambrian and nonmarine environments in the Triassic, and therefore clearly generated by two quite different groups of organisms, both have the shape and form and can appropriately be referred to *Cruziana*. Conversely early Palaeozoic trilobites are responsible for a range of traces such as *Rusophycus*, *Cruziana* and *Diplichnites*; each trace is morphologically quite different, requiring a separate name, although possibly at times they were generated by one organism.

Like conventional taxonomy, ichnology has both 'lumpers' and 'splitters'; however, relatively few studies have adopted a statistical methodology for the description and analysis of trace fossil specimens which can sometimes also help solve nomenclatural problems (Pickerill, 1994).

5.3 OPERATIONAL CATEGORIES

In practice, ichnofossils are commonly assigned to either traces on bedding planes, generated at or near the sediment–water interface, or traces left within the sediment or an associated substrate. The first category contains tracks, such as the discrete footprints of arthropods and vertebrates, typical of many continental environments, whereas trails are continuous structures left by vagile arthropods, molluscs or worms, dragging their entire body through the sediment. The second group is characterized by burrows, constructed in soft sediment as dwellings or during locomotion, feeding or evasion. Borings are made in hard substrates, commonly, by sedentary organisms and additionally for mineral extraction. Excrement, for example faecal pellets and strings of shallow-marine invertebrates, and the larger coprolites produced by fishes and reptiles, are also usually included as trace fossils. However, eggs and agglutinated burrows are rejected, as are the bioglyph bounce and roll marks of fossil debris such as shells, bones and wood.

5.4 TRACE FOSSILS AND THE BEHAVIOUR OF THE PRODUCER

The association of behavioural patterns and trace fossils was firmly established by Seilacher's (1953) behavioural classification, fine-tuned a number of times as his categories were expanded and modified and new categories were added. Seilacher defined five main groups: resting traces (Ruhespuren), crawling traces (Kriechspuren), browsing traces (Weidespuren), feeding structures (Fressbauten) and finally, dwelling structures (Wohnbauten). His Spurenspektren, or trace spectra, simple piecharts, plotted the relative abundance of these categories across a range of sediments and environments; deep-water flysch was dominated by browsing traces whereas shallow-water epicontinental seas had abundant feeding structures. This work formed the basis for more sophisticated behavioural classifications (Box 5.1, Fig. 5.5) but also anticipated the use of traces in environmental analysis (Fig. 5.6).

5.5 TRACE FOSSILS AND PALAEOENVIRONMENTS

Many trace fossils and trace fossil assemblages are associated with particular sedimentary facies (Fig. 5.6). Behavioural patterns throughout geological time, once established, remained rather constant although the nature of the producers undoubtedly changed. In marine and marginal marine environments most emphasis has been placed on the correlation of trace fossil assemblages to water depth or bathymetry. However, other constraints such as desiccation, disinternment, ambient nutrient and oxygen levels together with generating life modes, for example whether deposit- or suspension-feeding, may have been equally as important under certain circumstances.

5.6 MARINE AND MARGINAL MARINE ICHNOFACIES

Initial studies of fossil marine traces assigned burrows and trails to a variety of worms and seaweeds. However in the late 19th century and the early 20th century, Alfred Nathorst and Rudolf Richter demonstrated the true origins of these structures by comparisons with the traces left by living organisms. Much later in a series of classic and influential papers, Seilacher (for example, 1964, 1967a) established the relationships between recurrent associations of marine trace fossils and changing environmental gradients, mainly water depth. Across an onshore–offshore gradient, from intertidal environments to the deep sea, he described four main associations – the *Skolithos*, *Cruziana*, *Zoophycos* and *Nereites* ichnofacies (Fig. 5.6). In addition, the *Scoyenia* ichnofacies, later defined by Seilacher (1967a) is characteristic of terrestrial redbed sequences whereas the *Glossifungites* ichnofacies occupies firm, often exhumed, substrates such as dewatered muds. With the exception of the last, these ichnofacies developed on and within soft, unconsolidated sediments. The *Trypanites* ichnofacies, however, reflects production in hard substrates, such as hardgrounds and rockgrounds (Frey and Seilacher, 1980); the ichnofacies has been subdivided into the *Gnathichnus* and *Entobia* ichnofacies (Bromley

Box 5.1 Ethological classification of trace fossils

Cubichnia – resting or hiding traces left by stationary arthropods, cnidarians and echinoderms often revealing an impression of the underside of the animal. *Rusophycus, Asteriacites.*

Repichnia – traces associated with directed movement from one station to another. Dinosaur trackways, various linear and curved 'worm' trails. *Cruziana*, *Diplichnites.*

Pascichnia – grazing trails with typical pattern of coiled, meandering or sinuous furrows rarely crossing and usually covering bedding planes. *Helminthoida.*

Fodinichnia – burrows usually inclined or horizontal, often U-shaped, combining stationary/semipermanent dwelling and feeding structures. *Chondrites*, *Dactyliodites.*

Domichnia – residence burrows, usually cylindrical, vertical with horizontal tunnels, often with lining. *Diplocratarion*, *Skolithos* and *Thalassinoides.*

Agrichnia – patterned, deep-water horizontal trail system which may have farmed or trapped food. *Palaeodictyon, Spiroraphe.*

Praedichnia – traces of predation, such as *Oichnus.*

Equilibrichnia – traces that have constantly adjusted to the gradual accretion of sediment.

Fugichnia – escape burrows generated by the rapid upward movement of, for example, echinoderms, molluscs and worms avoiding burial beneath layers of catastrophically emplaced sediment. *Corophioides.*

Aedifichnia – structures constructed and more or less cemented by the architect; mud-dauber wasp nests, sabellarid polychaete 'reefs'. *Chubutolithes.*

Calichnia – structures used exclusively for raising larvae or juveniles; bee cells, scarabeid beetle nests. *Termitichnus.*

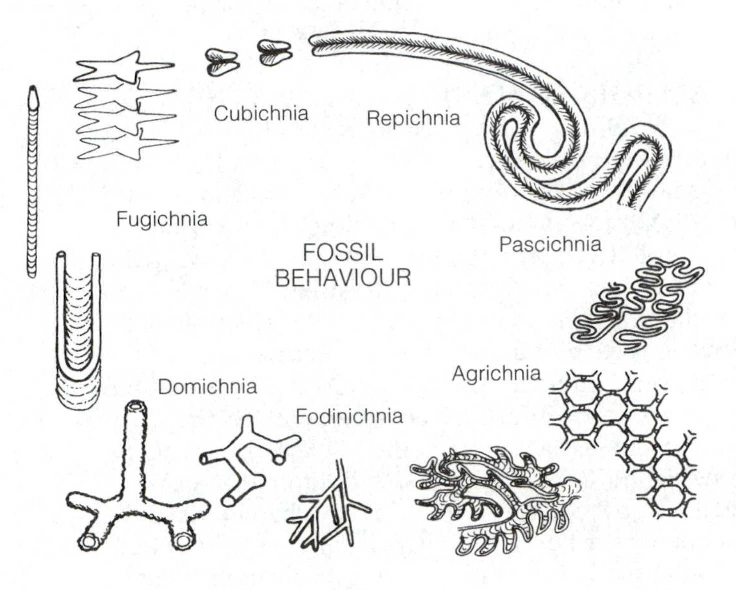

Figure 5.5 Ethological classification of trace fossils: some examples of ichnotaxa clockwise from top right – *Cruziana, Cosmorhaphe, Paleodictyon, Phycosiphon, Chondrites, Thalassinoides, Ophiomorpha, Diplocraterion, Gastrochaenolites, Asteriacites* and *Rusophycus*. (Redrawn from Ekdale *et al*., 1984.)

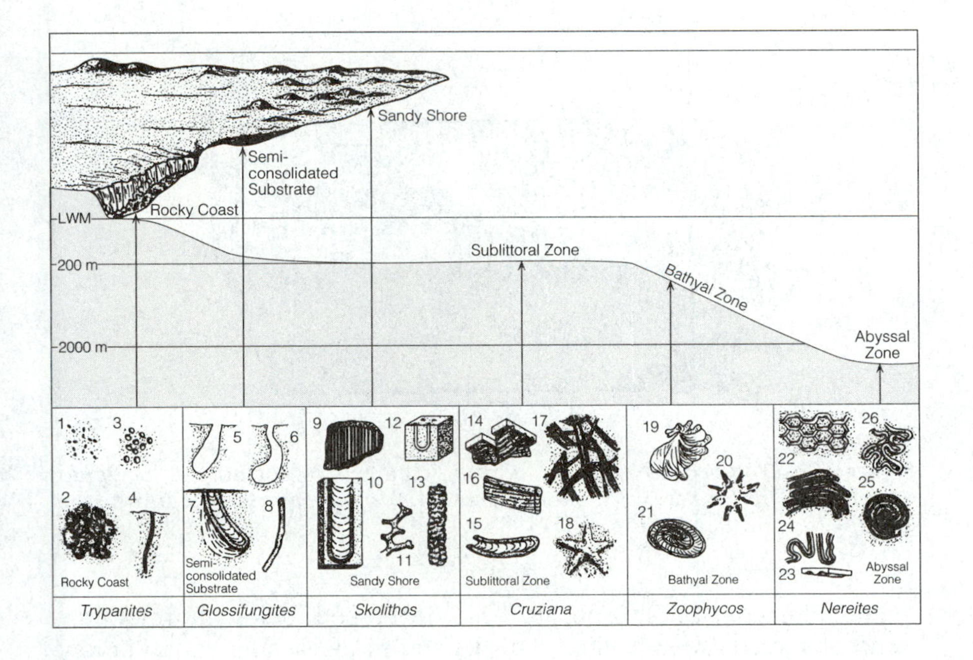

Figure 5.6 Distribution of main recurring ichnofacies across a range of environmental gradients: typical ichnotaxa include – 1. *Caulostrepis*, 2. *Entobia*, 3. Unnamed echinoid borings, 4. *Trypanites*, 5, 6. *Gastrochaenolites* or related ichnogenera, 7. *Diplocraterion*, 8. *Psilonichnus*, 9. *Skolithos*, 10. *Diplocraterion*, 11. *Thalassinoides*, 12. *Arenicolites*, 13. *Ophiomorpha*, 14. *Phycodes*, 15. *Rhizocorallium*, 16. *Teichichnus*, 17. *Crossopodia*, 18. *Asteriacites*, 19. *Zoophycos*, 20. *Lorenzinia*, 21. *Zoophycos*, 22. *Paleodictyon*, 23. *Taphrhelminthopsis*, 24. *Heminthoida*, 25. *Spirorhaphe*, and 26. *Cosmorhaphe*. (Redrawn from Frey, R.W. and Pemberton, S.G. 1984.)

and Asgaard, 1993). The *Teredolites* ichnofacies, however, is developed within wood (Bromley *et al.*, 1984). More recently Frey and Pemberton (1987) added the *Psilonichnus* ichnofacies based on traces associated with sand dunes in both beach and backshore environments. A range of further assemblages have been defined within these nine basic ichnofacies by a number of different authors.

5.6.1 Focus on marine ichnofacies

The *Skolithos* ichnofacies is generally developed in high-energy conditions associated with moving, well-sorted sands, commonly, but not universally, in shallow water, and includes a variety of vertical, uniserial or U-shaped burrows with circular cross-sections. The ichnofacies is usually of low diversity, dominated by abundant burrows of suspension-feeders (Fig. 5.7). Pipe rocks, sandstones permeated by *Skolithos* pipes or tubes, are particularly well developed in the Cambrian rocks of the NW Highlands of Scotland and in Silurian strata in the west of Ireland and southern Norway.

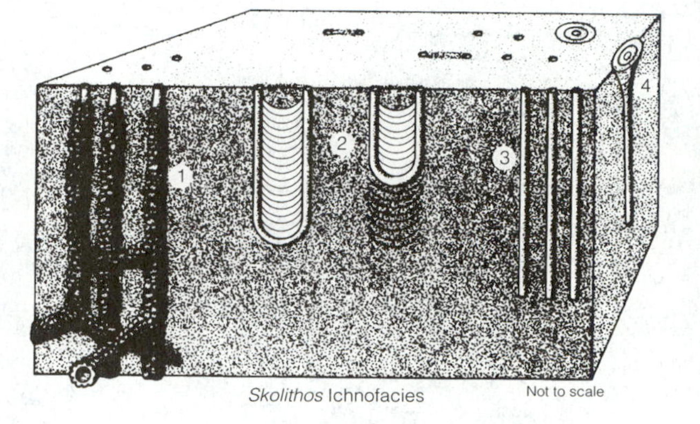

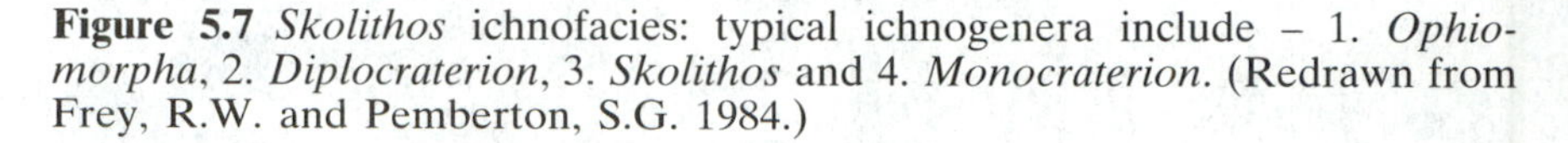

Figure 5.7 *Skolithos* ichnofacies: typical ichnogenera include – 1. *Ophiomorpha*, 2. *Diplocraterion*, 3. *Skolithos* and 4. *Monocraterion*. (Redrawn from Frey, R.W. and Pemberton, S.G. 1984.)

The *Cruziana* ichnofacies is best developed below normal fair-weather wave base in well-sorted silts and sands, accumulating in low-energy environments. This ichnofacies is widely developed in estuarine, lagoonal and shelf environments where conditions are relatively quiet. Mainly crawling traces with some inclined burrows characterise this ichnofacies where both suspension and detritus feeders are joined by mobile carnivores (Fig. 5.8).

Quiet-water and presumably adequate nutrient conditions on the outer shelf and slope favour the *Zoophycos* ichnofacies. The facies is dominated by relatively complex but efficient feeding and grazing trails

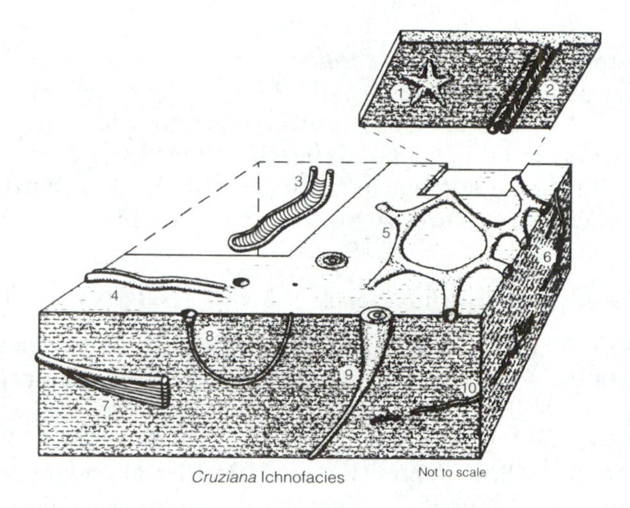

Figure 5.8 *Cruziana* ichnofacies: typical ichnogenera include – 1. *Asteriacites*, 2. *Cruziana*, 3. *Rhizocorallium*, 4. *Aulichnites*, 5. *Thalassinoides*, 6. *Chondrites*, 7. *Teichichnus*, 8. *Arenicolites*, 9. *Rosselia* and 10. *Planolites*. (Redrawn from Frey, R.W. and Pemberton, S.G. 1984.)

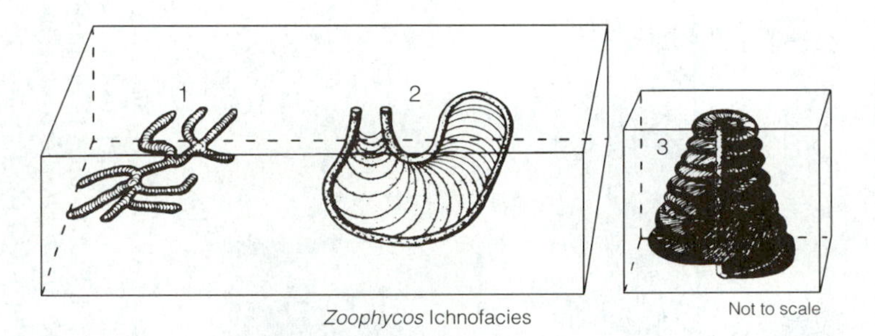

Figure 5.9 *Zoophycos* ichnofacies: typical ichnogenera include – 1. *Phycosiphon*, 2. *Zoophycos* and 3. *Spirophyton*. (Redrawn from Frey, R.W. and Pemberton, S.G., 1984.)

of deposit-feeders with spreiten as sheets or spirals. Although the ichnofacies is typically developed in deep-water settings from the deep shelf to the upper continental slope (Fig. 5.9), *Zoophycos* and some of its associates occupy both shallower and deeper water environments where similar environmental conditions prevailed (for example lagoons). *Zoophycos* although ranging from the Cambrian upwards occurs commonly throughout the Jurassic, Cretaceous and Tertiary of Europe in deep-water settings, generated by a specialized deposit-feeder below the redox boundary in the sediment, preferring low-oxygen conditions.

The *Nereites* ichnofacies is developed on and in quiet but moderately well-oxygenated seabeds, usually at bathyal and abyssal depths. The facies is common in flysch sequences and comprises diverse crawling and grazing trails and networks (agrichnia) which may have trapped or farmed food (Fig. 5.10). The trails and networks are usually

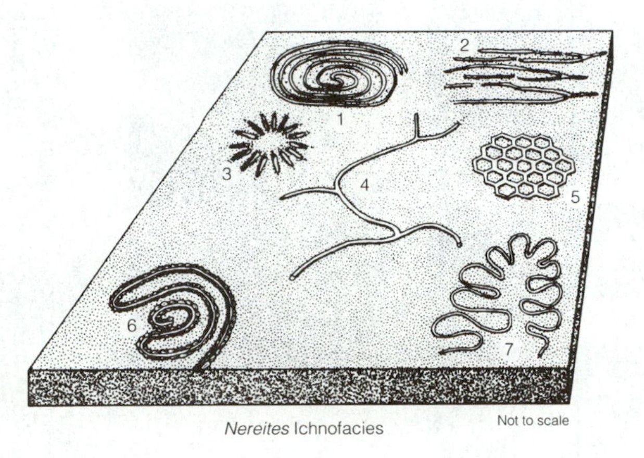

Figure 5.10 Ichnotaxa of the *Nereites* ichnofacies: typical ichnogenera include – 1. *Spirorhaphe*, 2. *Urohelminthoida*, 3. *Lorenzinia*, 4. *Megagrapton*, 5. *Paleodictyon*, 6. *Nereites* and 7. *Cosmorhaphe*. (Redrawn from Frey, R.W. and Pemberton, S.G. 1984.)

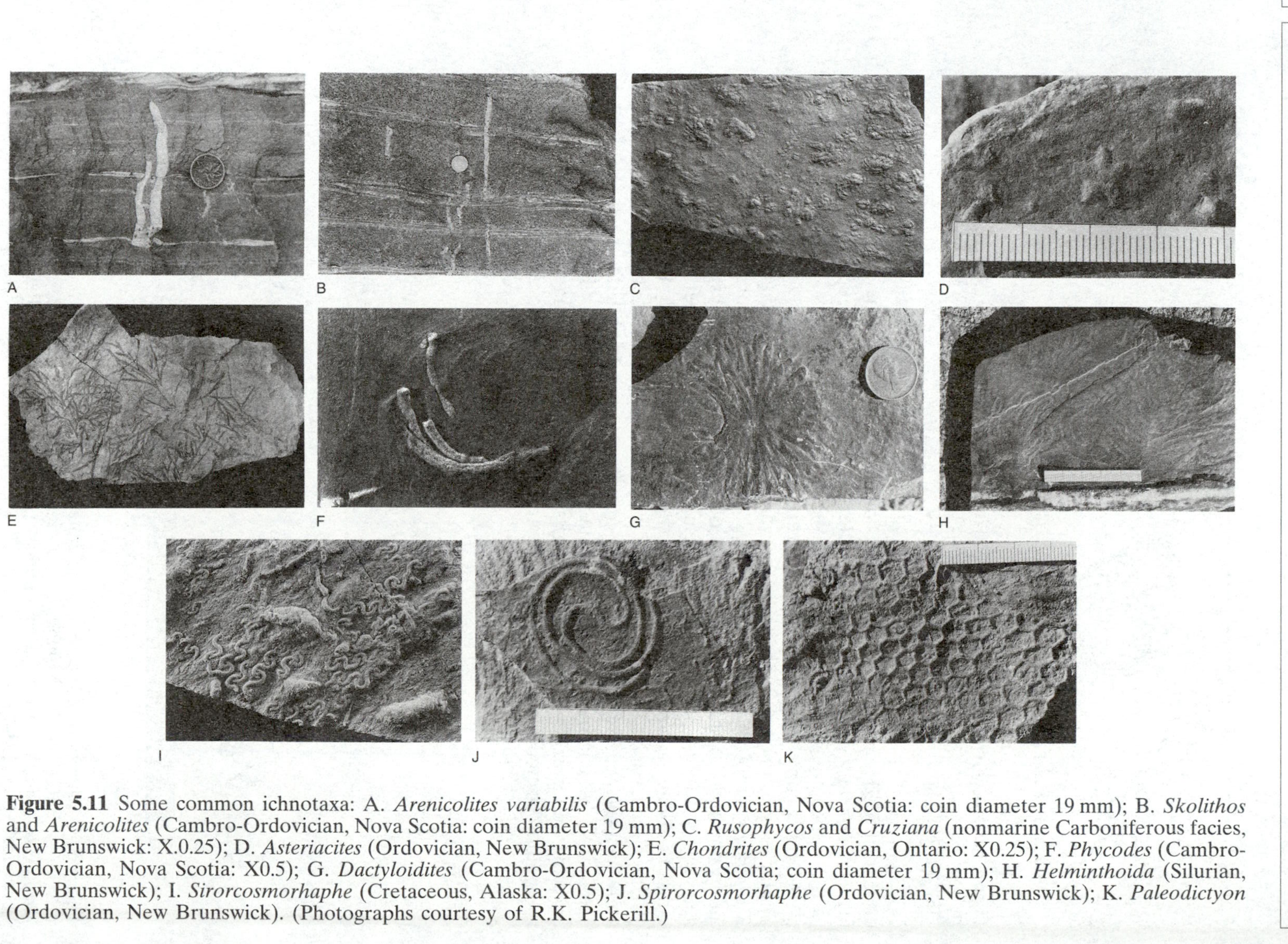

Figure 5.11 Some common ichnotaxa: A. *Arenicolites variabilis* (Cambro-Ordovician, Nova Scotia: coin diameter 19 mm); B. *Skolithos* and *Arenicolites* (Cambro-Ordovician, Nova Scotia: coin diameter 19 mm); C. *Rusophycos* and *Cruziana* (nonmarine Carboniferous facies, New Brunswick: X.0.25); D. *Asteriacites* (Ordovician, New Brunswick); E. *Chondrites* (Ordovician, Ontario: X0.25); F. *Phycodes* (Cambro-Ordovician, Nova Scotia: X0.5); G. *Dactyloidites* (Cambro-Ordovician, Nova Scotia; coin diameter 19 mm); H. *Helminthoida* (Silurian, New Brunswick); I. *Sirorcosmorhaphe* (Cretaceous, Alaska: X0.5); J. *Spirorcosmorhaphe* (Ordovician, New Brunswick); K. *Paleodictyon* (Ordovician, New Brunswick). (Photographs courtesy of R.K. Pickerill.)

mutually exclusive; the *Nereites* and *Paleodictyon* subfacies were, in fact, erected to express their separate distributions. Spectacular examples of the *Nereites* ichnofacies are associated with the late Cretaceous and Tertiary synorogenic flysch deposits of the Alpine belt.

Deep-water trails commonly exhibit a range of behavioural patterns: phobotactic trails never cross previous parts of the same trail or each other, strophotactic trails are generated by a succession of repeated 180° turns, with variable meander lengths, and thigmotactic trails remain in contact with each other, often developed as planar spirals without meanders (see also Box 5.2).

Some authors have extended the set of depth-related marine ichnofacies, established for siliciclastic environments, to carbonates. There are, however, some contrasts between carbonate and siliciclastic ichnofacies. Carbonate facies are rapidly lithified with the common development of hardgrounds and firmgrounds. The lack of sediment colour can prevent the immediate recognition of trace fossils, while many environments such as alluvial, glacial and deep-sea settings are not recorded in carbonate facies. Moreover in terrestrial carbonate facies the traces of plants roots, or rhizomorphs, are widespread.

5.7 BIOLOGICAL ACTIVITY AND ICHNOFABRICS

The behaviour of organisms can significantly alter the initial structure of sedimentary deposits by biological cementation, degradation, reworking, and production of sediments. Bioturbation of sediments occurs at a variety of levels with varieties of penetration and pervasion giving the rocks a distinctive new fabric or ichnofabric. The ichnofabric developed in a sediment thus records considerable ecological information; however, as in the case of the preservation of body fossils the data can be biased by taphonomic factors. For example, deep discrete burrows, carefully constructed, have a better chance of preservation than say random surficial trails. Ichnofabrics must be investigated by sections cut perpendicular to bedding (Droser and Bottjer, 1986); this method is particularly suited to the analysis of core material.

The ichnofabric of a sediment usually preserves infaunal tiers of trace fossils analogous to the epifaunal tiers of benthos above the substrate. Eight tiers have been described from the upper Cretaceous Chalk from Denmark (Fig. 5.13). The shallowest tiers are dominated by trails such as *Planolites* and various shallow burrows, whereas the deepest are occupied by small ichnospecies of *Chondrites*. There is apparently an inverse relationship between the preservation potential of fabrics and the amount of bioturbation in vertical sections. In the same vertical transect bioturbation decreases from 100% at the surface to 0.05% in the deepest tier; whereas the degree of bioturbation preserved varies inversely from 0% at the surface to 100% at depth.

Ichnofabrics have been semi-quantified on a scale of 1–6, from a lack of bioturbation to a sediment homogenized by bioturbation (Fig. 5.14).

Box 5.2 Computer simulation of deep-sea trails

Fossil foraging behaviour has been simulated by computer (Raup and Seilacher, 1969; see also Seilacher, 1977). A simple model can generate meandriform trails on the basis of a series of parameters: a, the turning radius for a 180° turn, b, the mean distance between the developing trail and the existing trail, c, the permissible deviation from this average distance, d, the relative intensities of phobotaxis and thigmotaxis, e, the mean length of meanders not terminated at obstructions and f, the variation in the length of the meanders (Raup and Seilacher, 1969). Simulated trails can be matched with actual foraging patterns. But there is clearly a very wide range of theoretical possibilities, not all practicable. A more sophisticated method involves both goniograms and derivative plots (Hofmann, 1990). Trails may be generated by random numbers. Both simulated and natural traces can then be digitised and the varying orientation may be plotted, as a heading, against the distance along the trail. The changing heading can be converted to a measure of angular change which may be plotted against distance along the trail. The courses of some traces such as *Helminthoidichnites* can be simulated by random number generators whereas segments of *Gordia* show significant directed or nonrandom behaviour.

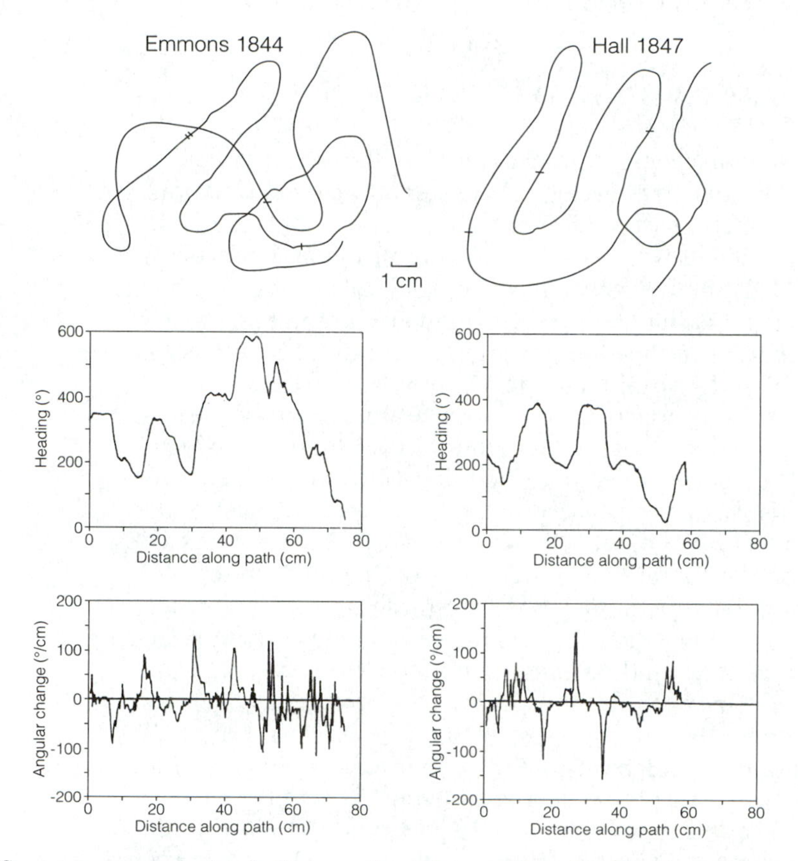

Figure 5.12 Some computer simulations and analyses of trails: Tracings of actual trails (top), converted to goniograms in terms of a headings and distances along the paths (middle), converted to rates of change of headings plotted against distance along path (bottom). (After Hofmann, 1990.)

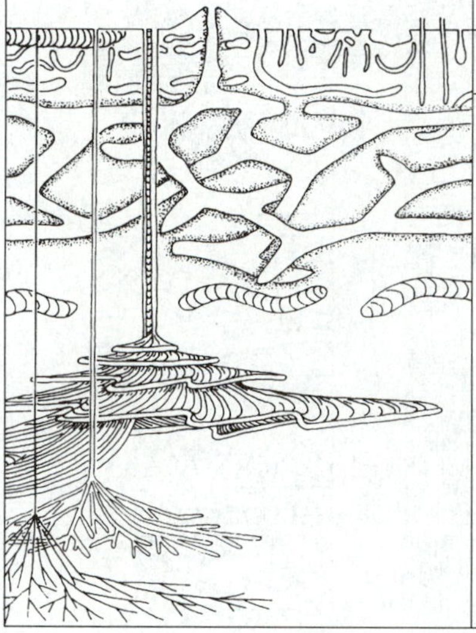

ZONE	TIER	AMOUNT OF BIOTURB'N IN%	% OF WORK OF COM-MUNITY	% OF PRESERVED BIOTURB'N	% OF WORK OF TIER PRESERVED	CHARACTERISTIC TRACE FOSSILS	SUGGESTIVE BODY FOSSILS
MIXED	1	100	70	0	0	NONE	SPATANGOIDS, ASTEROIDS, OPHIUROIDS, GASTROPODS, BIVALVES, SCAPHOPODS, ANNELIDS, ETC., ETC
	2	100	10	0	0	NONE	
	3	100	10	15	5	*Planolites*	
TRANSITION	4	90	5	80	60	*Thalassinoides*	RARE CRUSTACEANS
	5	10	2	2	80	*Taenidium*	NONE
	6	1	2	2	95	*Zoophycos*	NONE
	7	0.5	0.5	1	98	*Chondrites* (large)	NONE
		0.05	0.5	0.5	100	*Chondrites* (small)	NONE

Figure 5.13 Development of a tiered Chalk ichnofabric. (After Bromley, 1996.)

Indices can be standardized in terms of the thickness of rock measured and plotted on a customized histogram or ichnogram (see Fig. 5.15). Standardized data on the degree of bioturbation in strata can be studied through time and across different environments as an aid to changing trends in animal abundance and behaviour in sediments.

The investigation of ichnofabrics also provides windows through which to monitor the development of community succession and distinguish between opportunist and equilibrium ichnotaxa. In shallow-water environments *Skolithos* is a common opportunist, although it may also occur in more mature ichnocoenoses (Droser and Bottjer, 1990). *Chondrites* together with *Zoophycos* commonly form opportunist assemblages in low-oxygen environments. However, as the substrate stabilizes a more complex and mature ichnofabric is developed, with deeper tiers established over intervals of about 10–20 years. The climax ichnocoenosis is dominated by complex burrows constructed over some time by specialized feeders. Intermediate between the climax ichnotaxa and the opportunists are common ichnotaxa such as *Planolites*, *Scolicia* and *Thalassinoides*.

Tiered ichnofaunal systems together with community successions produce cross-cutting relationships. The stratigraphical completeness of succession may be assessed with reference to these profiles (Wetzel and Aigner, 1986). In a typical profile ichnofossils are arranged in tiers usually constructed as sediment is accreted; there may be some erosion, redeposition and casting of trace fossils. In sequences, however, where

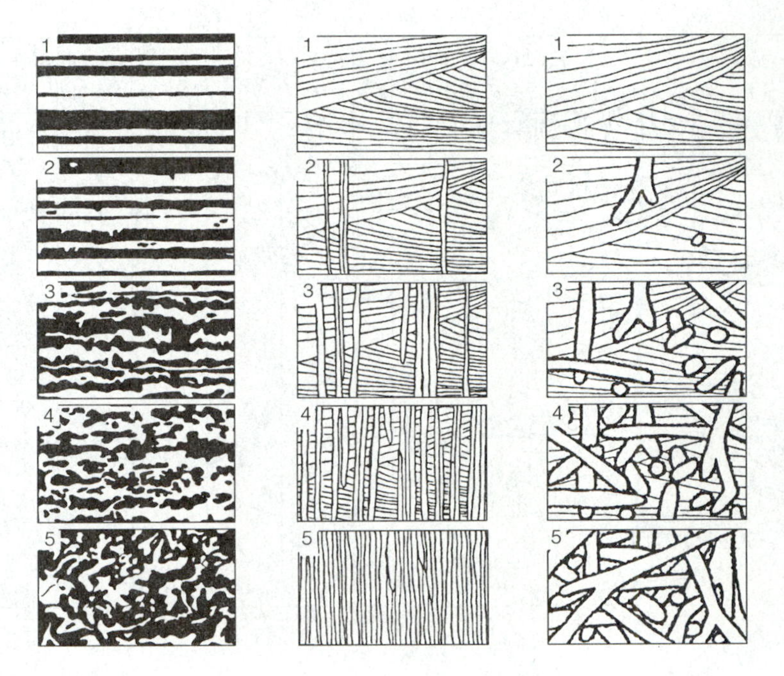

Figure 5.14 Recognition of ichnofabric from 1–5 developed in shelf facies (left), in the high-energy nearshore facies dominated by *Skolithos* (middle) and high-energy nearshore facies dominated by *Ophiomorpha*. (Redrawn from Bottjer and Droser, 1992.)

more intense erosion occurs, the truncation and casting of specific tiers are useful in estimating the depth of erosion in the sequence and the overall completeness of the profile. For example tiers, in the Upper Muschelkalk of southern Germany, with increasing depth are occupied by *Rhizocorallium*, *Planolites*, *Teichichnus* and *Thalassinoides*. Cast and truncated *Thalassinoides* burrows indicate a deep erosional event whereas those of *Rhizocorallium* suggest a less invasive surge (Wetzel and Aigner, 1986).

5.8 FROZEN PROFILES

The sequential development of habitat partitioning in a tiered profile can be reconstructed with reference to the cross-cutting relationships between elements of the ichnofauna, together with, or as an alternative to assessment of their relative depths of penetration. Fortuitously, however, sudden changes in the environment may prompt an evacuation of the profile, leaving the tiers intact. These frozen profiles are clearly an important source of ichnological information, having avoided persistent and pervasive bioturbation. Rapid deoxygenation, for example, can force the migration of faunas from the profile.

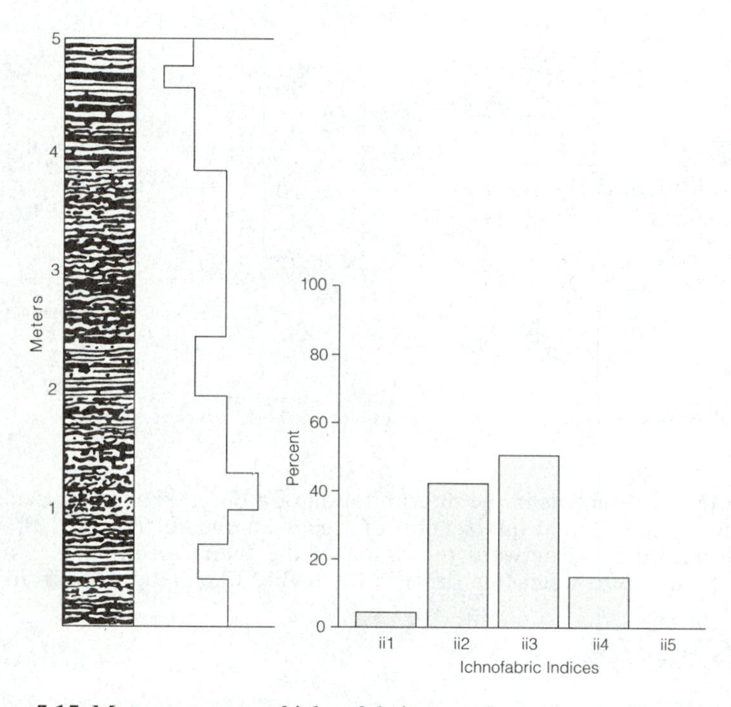

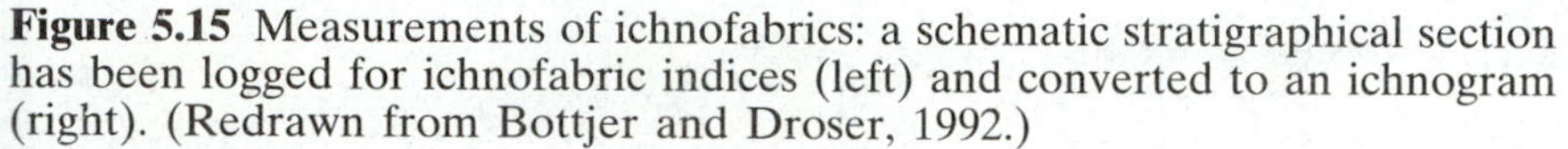

Figure 5.15 Measurements of ichnofabrics: a schematic stratigraphical section has been logged for ichnofabric indices (left) and converted to an ichnogram (right). (Redrawn from Bottjer and Droser, 1992.)

Sedimentary processes, however, may also determine and preserve a frozen profile. Interchannel deposits from middle-fan environments in the Lower Carboniferous of Menorca host two-tiered profiles (Fig. 5.16) with *Nereites* underlain by *Dictyodora* which can be accompanied by *Arthrophycus*. These profiles are preserved by the deposition of an overlying event bed which prompted the relocation of the indigenous infauna (Orr, 1994).

5.9 BIOEROSION

Many organisms living on or within a hard substance modify and even destroy their substrate. This process of bioerosion may be achieved at all scales by a variety of different functions through a variety of different organs. Bioerosion is usually restricted to attacks on hard substrates such as rocks, lithified sediments or skeletal material. Although some borers such as piddocks rasp and grind their way into the substrate, most use some form of chemical solution to construct passages. Unlike traces in soft substrates, borings often show some indications of their makers and thus the fossilization barrier is not so relevant in the description of the boring traces. Organisms living on a hard substrate have been classified as epiliths whereas endoliths live within this

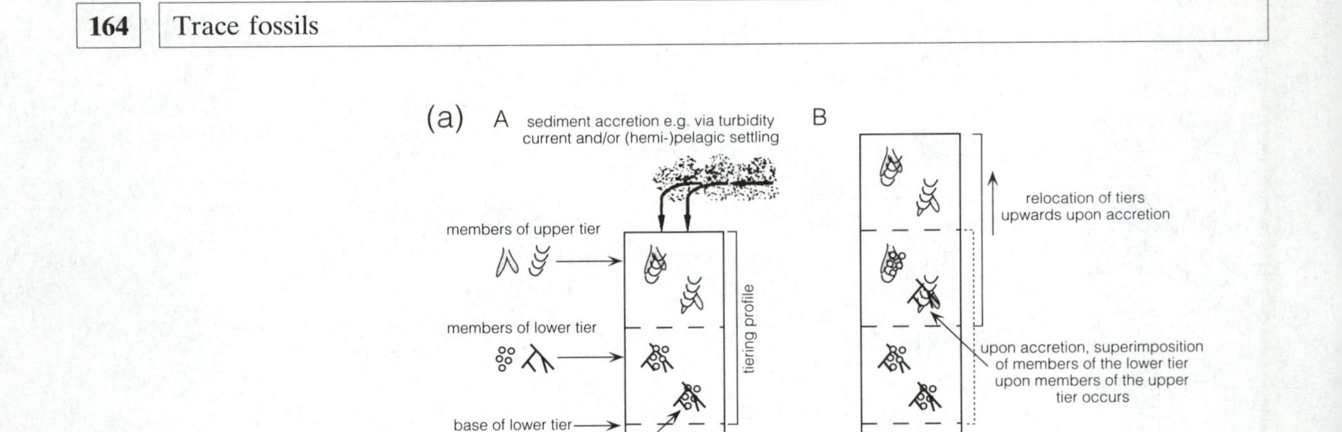

Figure 5.16 (a) Concepts in the interpretation of a tiered profile: A. Each tier is defined by the mutual intersection of traces unique to that tier; B. With sediment accretion the upward relocation of the fauna results in the superposition of the lower faunas on those in the higher tiers. (Redrawn from Orr, 1994.)

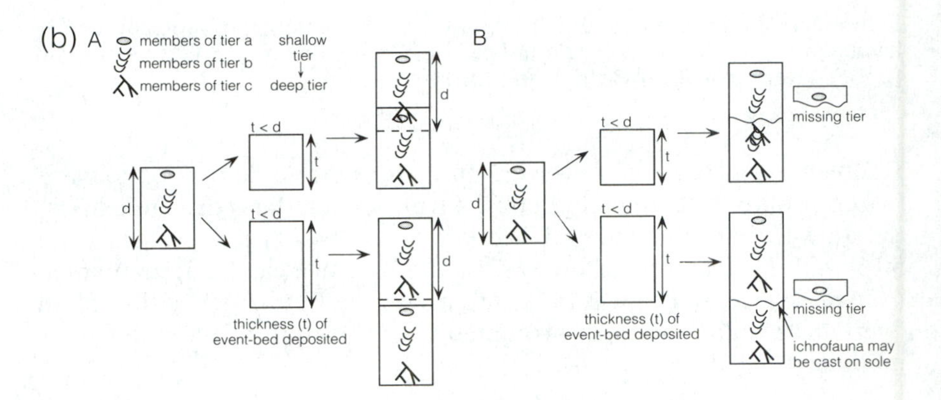

Figure 5.16 (b) Models for the preservation of frozen profiles: A. During passive deposition where there is no erosion of the underlying sediments by the succeeding beds; B. Situation where the event bed has an erosive base with the removal of the upper part of the underlying bed with possible casting of ichnofauna exposed on the sole of the event bed. d = thickness of the tiered profile, t = thickness of event bed; where t<d modification of the frozen profiles occurs with the superposition of deeper on shallower tiers, where t>d the entire frozen profile is preserved. (Redrawn from Orr, 1994.)

substrate. Golubic *et al.* (1981) have refined this crude classification to include passive nestlers in existing cavities or cryptoendoliths, active cavity borers or euendoliths and the chasmoendoliths, occupying minute, natural fissures. To date infaunas dominated by borers are relatively poorly known. Nevertheless these faunas apparently radiated extravagantly during the Mesozoic, thus avoiding the attentions of diversifying marine predators during the era. During the Jurassic, various borings by bivalves and sponges became increasingly important, particularly in reef environments where they form a significant part of hardground communities.

Some of the principal taxa with bioeroding members are listed in Box 5.3. Most of the borings fall into three behavioural or ethological groups: domichnia, fodinichnia and pascichnia, the organisms mainly producing dwellings or feeding and grazing trails. Bromley (1970) has reviewed in detail many of the commoner boring organisms.

Box 5.3 Bioeroders

A huge variety of organisms eat rocks for fun and profit (Bromley, 1992); exquisite detail can often be obtained from SEM investigation of these traces (Palmer and Plewes, 1993).

Porifera – chemical processes, constructing domichnia. *Entobia*.

Polychaeta – chemical and occasional mechanical processes, constructing domichnia. *Trypanites*.

Phoronida – chemical processes, constructing domichnia. *Talpina*.

Bryozoa – chemical processes, constructing domichnia. *Foraripora*.

Brachiopoda – chemical process, constructing domichnia. *Podichnus*.

Gastropoda – chemical and mechanical processes, constructing praedichnia, pascichnia and domichnia. *Oichnus*, *Radulichnus* and *Gastrochaenolites*.

Bivalvia – chemical and mechanical processes, constructing domichnia and equilibrichnia. *Gastrochoenolites* and *Centrichnus*.

Arthropoda – mainly mechanical processes, constructing domichnia. Variety of cavities and tunnels.

Echinoidea – mechanical processes, constructing pascichnia and domichnia. *Gnathichnus* and various cavities and tunnels.

5.10 TERRESTRIAL ICHNOFACIES

Terrestrial ichnofacies, although usually assigned to a single ichnofacies (the *Scoyenia* ichnofacies), contain a spectacular variety of traces generated by a huge range of organisms, and are probably as diverse as their marine counterparts. Invertebrates, particularly the arthropods, together with both the tetrapod and bipedal vertebrates have left a variety of tracks and trails reflecting a diversity of behavioural patterns. Several authors have attempted to define further non-marine ichnofacies (for example, Buatois and Mángano, 1995). The *Termitichnus* ichnofacies is characterized mainly by the dwelling burrows of invertebrates, and vertebrates together with traces left by plants in terrestrial sediments. Termite, bee and beetle nests are typical ichnofossils. The *Mermia* ichnofacies is defined to include grazing trails and other feeding traces usually developed in freshwater aquatic

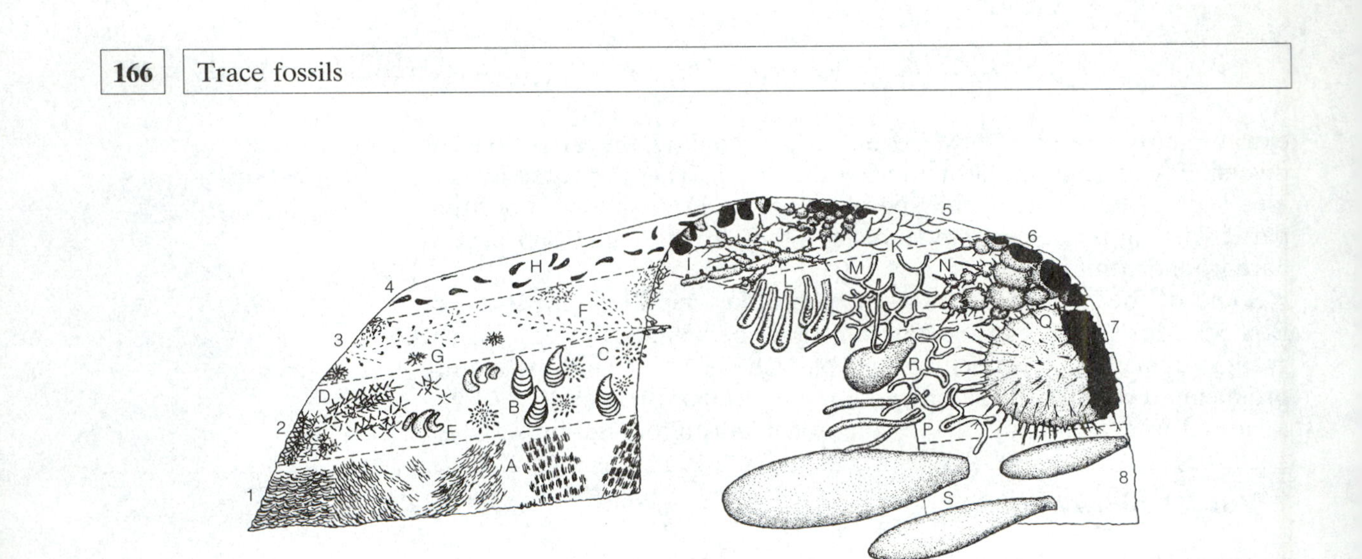

Figure 5.17A Tiering of bioerosion trace fossils arranged in eight tiers, 1–8, characterized by the following ichnotaxa: A. *Radulichnus inopinatus*, B. *Centrichnus eccentricus*, C. *Podichnus centrifugalis*, D. *Gnathichnus pentax*, E. *Renichnus arcuatus*, F. *Spathipora pungens*, G. Algoid microborings, H. *Rogerella lecointrei,* I. *Entobia cateniformis*, J. *Entobia ovula*, K. *Talpina ramosa*, L. *Caulostrepsis taeniola*, M. *Conchtrema canna*, N. *Entobia magna*, O. *Maeandropolydora decipiens*, P. *Trypanites solitarius*, Q. *Entobia gigantia*, R. *Gastrochaenolites lapidicus*, S. *Gastrochaeonolites torpedo*. (After Bromley, 1992.)

Figure 5.17B Some boring ichnotaxa: (a) *Teredolites* in wood, (b) *Oichnus paraboloides* in bivalve shell, (c)*Oichnus simplex* in bivalve shell; all material from the Pliocene Bowden Shell Bed, Jamaica, scales indicated. (Photographs courtesy of R.K. Pickerill.)

environments whereas the *Scoyenia* ichnofacies is confined to transitional terrestrial–aquatic nonmarine environments such as floodplains and ephemeral lakes. But many more terrestrial ichnofacies remain to be recognized and defined (Buatois and Mángano, 1995).

5.10.1 Invertebrate traces

Many invertebrates such as arthropods and molluscs leave tracks and trails in nonmarine environments. In fact about 15 animal phyla are represented in aquatic and terrestrial environments and the majority contain groups with the potential to leave traces in one form or another. Nonmarine arthropods, such as insects, spiders and myriapods have left a variety of terrestrial traces (Donovan, 1994). The traces of myriapods dominate the mid-Palaeozoic record, although by the Permian, ichnofossils attributed to spiders and beetles had joined myriapod-generated ichnofacies.

The environments and ichnology of dune fields are now known in considerable detail and illustrate the diversity of organisms and traces in a relatively inhospitable set of continental environments. A wide range of organisms including ants, beetles, flies, spiders and wasps together with gastropods and toads have constructed burrows in three separate environments from the dry interdune, through the dune itself to the wet interdune.

Some of the oldest, non-controversial nonmarine traces have been recorded from rocks of late Ordovician to early Devonian age, intimately associated with the early stages of the colonization of the land (see Chapter 10).

5.10.2 Vertebrate tracks

Reptilian footprints have been recognized for a long time. By the mid-1820s footprints had been reported from the red Permian sandstones of northeast Scotland by the Reverend Henry Duncan and during the 1830s the Reverend William Buckland was already simulating fossil footprints by marching turtles across piecrust. The following terms are used in the description of vertebrate footprints: manus – front foot of a quadruped; pes – back foot of a quadruped or biped; stride – distance between two prints made by same foot; pace – distance between prints made by two opposite feet; pace angulation – angle made by two sequential paces; and gait – style of lomotion (Fig. 5.18).

Footprints may be the only evidence of vertebrates. For example, few tetrapods and reptiles have been described from the Devonian and the Permo-Trias, respectively, yet many different types of tracks and trackways have been documented in rocks of these ages from many parts of the world. Most Cenozoic tracks can be assigned to particular animals since living representatives may be easily identified. Mesozoic tracks are more difficult to identify and most Palaeozoic tracks, having no modern analogues and few associated pedal bones, defy confident assignment.

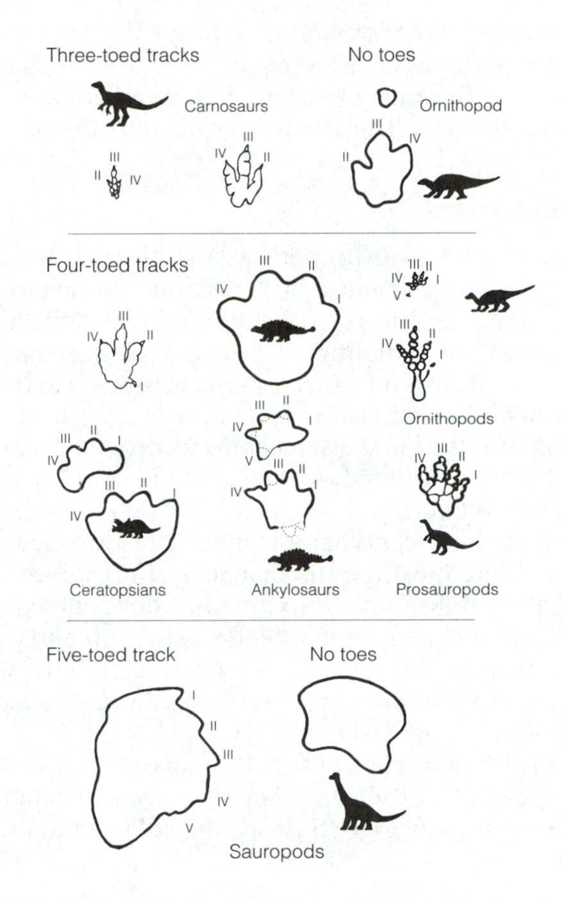

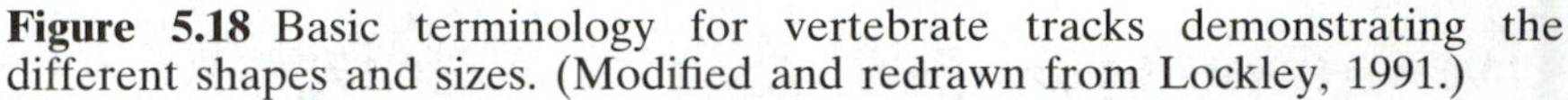
Figure 5.18 Basic terminology for vertebrate tracks demonstrating the different shapes and sizes. (Modified and redrawn from Lockley, 1991.)

The study of dinosaur tracks, the impressions left by the feet of dinosaurs, has now become an important part of palaeoichnology (Fig. 5.19). Description and analysis of these vertebrate footprints have significantly added to our understanding of the individual and social behaviour of the dinosaurs, the environments where they lived and also to the patterns of evolution within the Dinosauria (Lockley, 1991).

Dinosaur tracks and trackways are now known from over a thousand locations worldwide in rocks of Triassic–Cretaceous age. Study of tracks and trackways can establish the speed and direction that dinosaurs were moving at a given time. The mode of locomotion, whether walking, running or swimming, can be determined whereas the data from the trackways may describe the herding behaviour of many species together with their migration patterns. Herding behaviour was established early in the history of the carnivores, brontosaurs and some ornithopods such as the iguanodontids and the hadrosaurs. Community census based on analysis of footprints is possible and the sizes and shapes of tracks and trackways can aid palaeoenvironmental

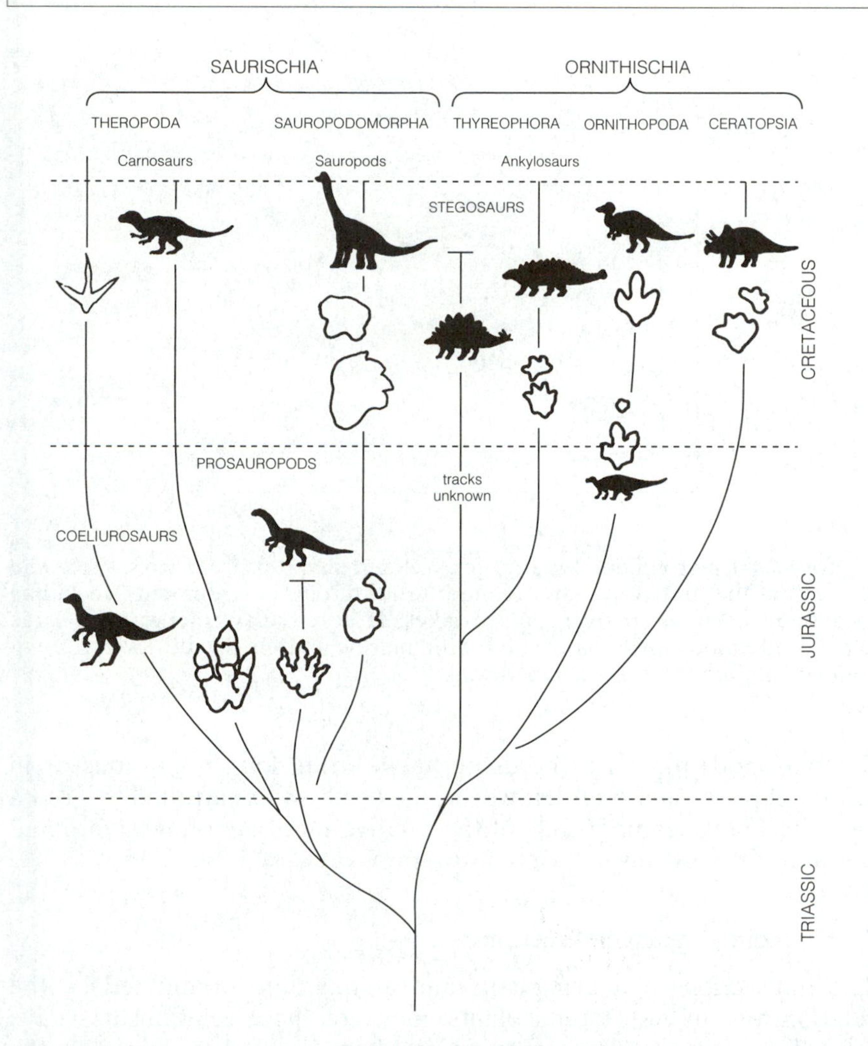

Figure 5.19 Characteristic footprints of the main dinosaur groups related to a phylogeny of the group. (Modified and redrawn from Lockley, 1991.)

reconstructions. Small and large lakes together with river bank and coastal plain settings have been discriminated on the basis of the length, width and shape of dinosaur trackways (Fig. 5.20). Small circular trackways were probably generated in and around small playa lakes, whereas the long trackways stretching up to a thousand kilometres represented marches along coastal plain environments.

To date, the longest dinosaur documented trackways in the world have been described from localities in the Purgatoire River, Colorado. Over 100 different trackways have been recorded including about 1300 individual footprints across 340 metres on a single bedding plane of the upper Jurassic Morrison Formation (Lockley *et al.*, 1986). The tracks are preserved in lake sediments characterized by fluctuating water levels and probably mark the edge of the lake. Giant sauropods, and both bipedal ornithopods and theropods marched west-northwest;

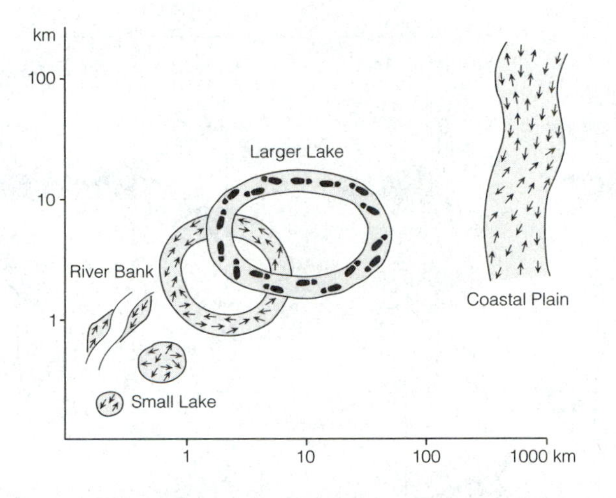

Figure 5.20 Environmental matrix for size and shape of trackways – size and shapes of the trackways are related to different environments including localised trackways in rivers, playa lakes, large circular trackways in large lakes, and ribbon belts associated with marine marginal, and swamp environments. (Replotted from Lockley, 1991.)

the sauropods moved in herds perhaps during long migrations or in search of plants and trees. However, the herds were patrolled by predatory theropods; comparisons of the relative numbers of carnivore and herbivore tracks suggest a predator–prey ratio of 1:30.

5.10.3 Animal–plant interactions

The interactions between plants and animals are dominated by the relationships of insects and plants; many of these relationships have existed for many millions of years and have formed the basis for the co-evolution of many insect–plant couplets (Scott, 1992). A variety of interactions ranging from feeding through shelter, transport, reproduction to transmission of disease are apparent in living ecosystems; some may generate traces preserved in the fossil record. A range of interactions is apparent in modern vegetation (Fig. 5.22). Feeding traces are the most common and recognizable evidence of interactions in fossil material. Nevertheless there is little evidence of widespread herbivory before the Cretaceous, although evidence of simple and continuous marginal feeding dates back to the Carboniferous, as do wood borings (Fig. 5.23).

5.11 EVOLUTION OF TRACE FOSSILS

The facies controls on trace fossils have ensured they are good environmental indicators but not good zone fossils. Nevertheless a few

Box 5.4 Dinosaur trackways

Dinosaurs are responsible for both bipedal and quadrupedal tracks. Tracking dinosaurs (Lockley, 1991) is now a very specialized area of research where a considerable amount of information regarding the behaviour and identitity of dinosaurs can be established from the detailed investigation of tracks and trackways. The Alexander formula has been used to calculate the speeds of moving dinosaurs: speed = $0.25\ g^{0.5} * SL^{1.67} * h^{-1.17}$ (where g = acceleration due to gravity, SL = stride length and h = hip length), which has been simplified to suggest the dinosaur was walking if the step is less than four times the length of the foot and running if the step is greater than that value.

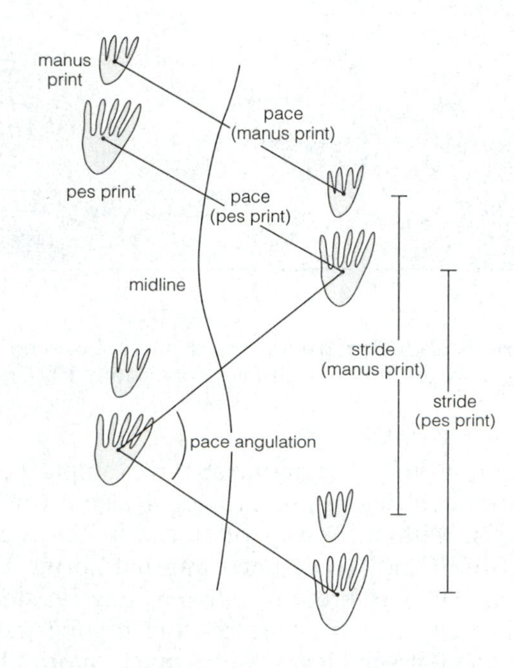

Figure 5.21 Dinosaur track measurements. (Redrawn from Schult and Farlow, 1992.)

horizons are characterized by specific assemblages of some biostratigraphical value. A spectacular example is their use in the definition and correlation of the Precambrian–Cambrian boundary (Landing, 1993). Basal Cambrian rocks are characterized by the first complex feeding burrows and trilobite trails together with the impressions of cnidarians (Crimes, 1987; Crimes *et al.*, 1977). The boundary is placed between the *Harlaniella podolica* trace fossil zone, with unique forms such as *Harlaniella*, *Nenoxites* and *Palaeopascichnus* and the *Phycodes pedum* trace fossil zone with more familiar ichnotaxa such as *Arenicolites*, *Conichnus*, *Monomorphinus*, *Nereites*, *Protopaleodicton* and *Skolithos*. More local correlations have been attempted with

Figure 5.22 A variety of interactions between arthropods and plants in modern vegetation. (Redrawn from Scott, 1992.)

specific groups of ichnotaxa, for example *Cruziana* (Seilacher, 1977) with limited degrees of success. These diverse Cambrian assemblages contrast with the rarity of traces in the Vendian but also with the short-lived and sometimes unusual forms associated with Ediacara faunas. The late Precambrian may have hosted a series of failed evolutionary experiments amongst the mobile metazoans, while the early Cambrian assemblages were much more diverse and subjected to macropredation (Crimes, 1994)

Although the main types of trace fossil assemblages have remained more or less constant throughout geological time, there is some evidence that certain behavioural patterns have evolved (Seilacher, 1967b, 1986). In broad terms some authors consider that Cambrian deep-water ichnofacies are poorly organized, with inefficient, loose feeding trails, focused on concentrated nutrient sources on the seabed and often resembling pencil scribbles. However, by the Ordovician *Nereites*-type trails are more regular and disciplined in form. The Cretaceous and Tertiary deep-water trails have the perfect patterns of *Helminthoida* suggesting organisms were meticulously and efficiently mining a soft substrate with limited nutrient resources.

Three major trends, however, have been described in trace fossil evolution throughout the Phanerozoic. First there is an apparent increase in ichnofaunal diversity with time during the early Palaeozoic;

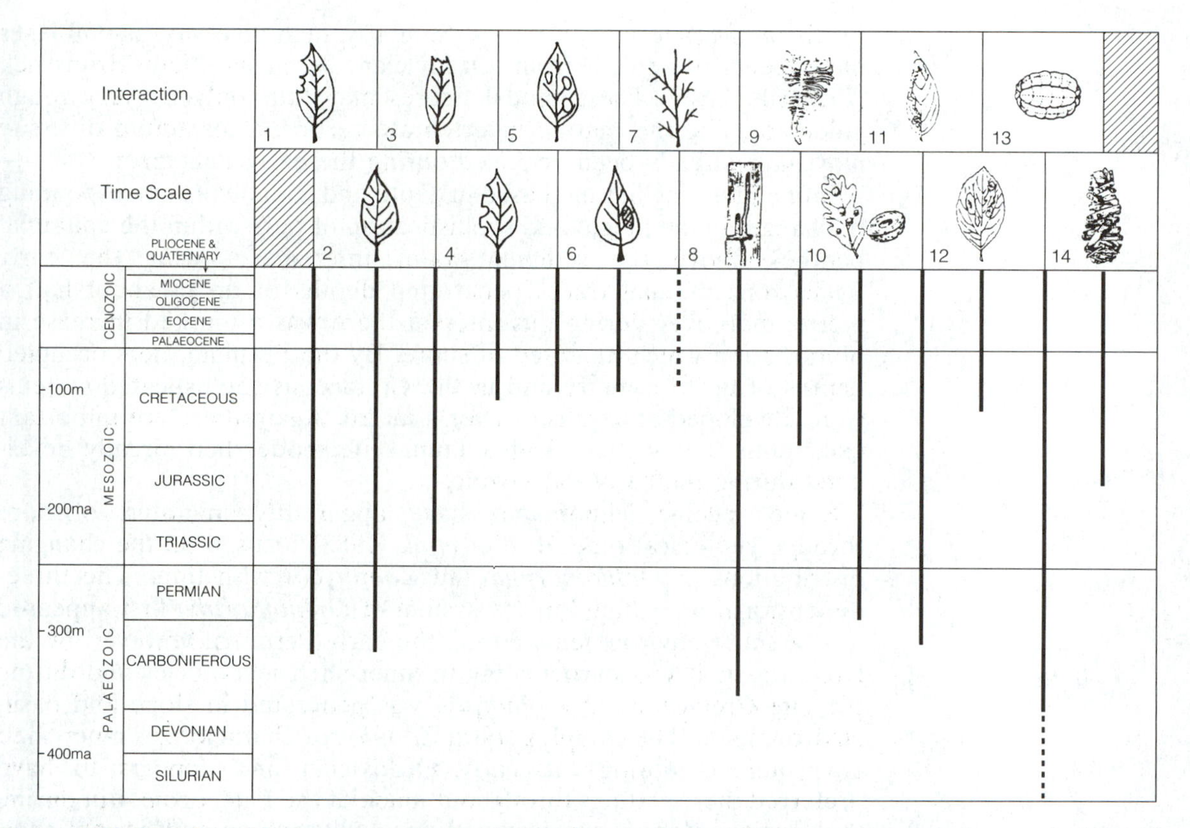

Figure 5.23 Stratigraphical ranges of animal-plant interactions: 1. Leaf showing simple marginal feeding, 2. Leaf showing continuous marginal feeding, 3. Leaf showing terminal feeding, 4. leaf showing interrupted marginal feeding, 5. Leaf showing nonmarginal feeding, 6. Leaf showing surface feeding, 7. Leaf showing skeleton feeding, 8. Wood boring, 9. Bark boring, 10. Leaf galls, 11. Sinuous leaf mines, 12. Leaf blotch mines, 13. Coprolites containing plant material, 14. Caddis-fly case. (Redrawn from Scott, 1992.)

second, there is an increase in the depth of infaunal penetration and the complexity of tiering and third, a move, for some ichnotaxa, from onshore to offshore environments.

Through the early Palaeozoic there is an increase in trace fossil diversity, tracking the changing biodiversity of body fossils as life radiated to occupy new ecological niches and explore new living strategies. A standard shallow-marine ichnofacies in the Cambrian probably contained about 15–20 ichnospecies, a diversity which has remained fairly stable since its establishment. But in deeper-water settings Cambrian *Nereites* ichnofacies are commonly of low diversity, with only about five ichnospecies; this facies is greatly expanded by the Silurian when, commonly, 10 ichnospecies are typical of assemblages of Silurian to Jurassic age; from the Cretaceous to the present day many *Nereites* ichnofacies have a maximum of about 30 ichnospecies. There are, however, many exceptions to this general trend; for example Crimes and Crossley (1991) described 45 ichnospecies from deep-water flysch in the Silurian of Wales. Similarly high-diversity assemblages have been described from Ordovician flysch in New Brunswick (Pickerill, 1980). These modal figures may thus only be very rough guides or, like the marine invertebrate record, a maximum diversity may have already been achieved during the early Palaeozoic.

Infaunal tiering has increased in depth and complexity corresponding to changes in the height and sophistication of tiers within the epifaunal benthos above the sediment–water interface. During the early Palaeozoic infaunal traces penetrated depths of up to about half a metre; moreover during this interval there was a marked increase in bioturbation which migrated offshore. By the Permian, tiers occupied depths of up to a metre and in the Cretaceous sophisticated systems were developed at depths of over a metre. Again there are important exceptions that suggest deep infaunal life modes had already developed during the early Palaeozoic.

Some specific ichnofaunas have apparently migrated offshore through geological time. Bottjer *et al.* (1988) focused on the changing distributions of *Ophiomorpha* and *Zoophycos* with time. The three-dimensional branching burrow system of *Ophiomorpha* first appeared in nearshore environments during the early Permian; however, by the late Jurassic it was participating in inner-shelf ichnofacies and during the late Cretaceous *Ophiomorpha* was generated in slope and basin environments. The complex spiral *Zoophycos* characterizes inner shelf environments during the early Ordovician and appears to have preferred these settings throughout much of the Palaeozoic. But during the Tertiary *Zoophycos* occupied deep-water slope and basinal environments. Many other ichnofossils appear to follow the same general trends. *Palaeodictyon* occurs in shallow-water environments during the early Cambrian but quickly migrated to deep-water settings. Similarly *Scolicia* appeared first in near-shore settings in the Jurassic but had moved to deep-sea fans by the end of the Cretaceous. This dynamic pattern of change is remarkably similar to that reported for macrofossils.

There are, however, alternative models. The movement of deep-water ichnotaxa seaward from their origins in shallow water may signal large-scale migrations or retreats into deeper water, rather than the range expansion of these ichnofacies, while most deep-sea traces had already evolved a high degree of development by the Cambrian with little major change apparent throughout the rest of the Phanerozoic (Crimes and Fedonkin, 1994).

Evolutionary trends are less well known in terrestrial ichnofacies. Nevertheless vertebrate community evolution can be monitored through the evolution of tracks and trackways, in deposits where trace fossils are much more abundant than body fossils (Lockley *et al.*, 1994).

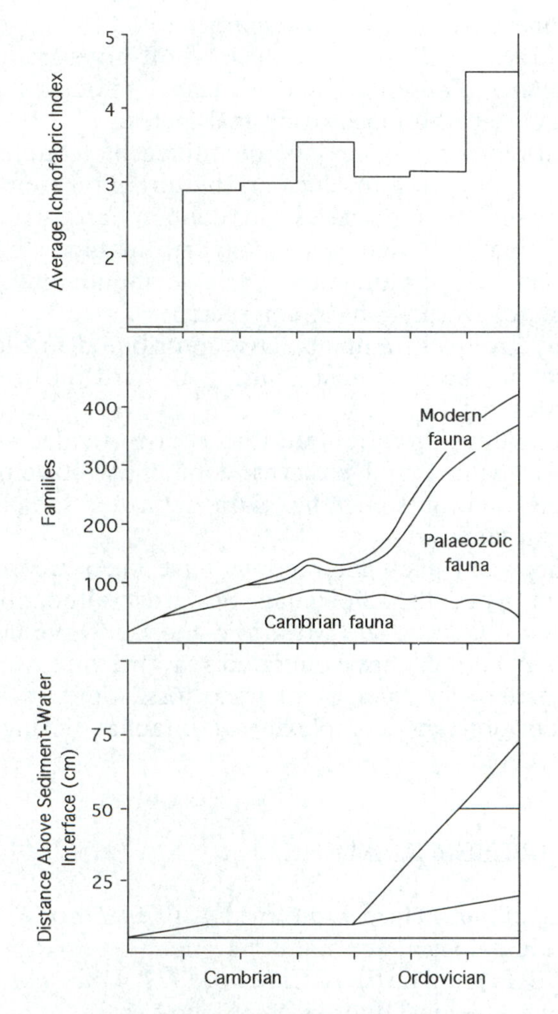

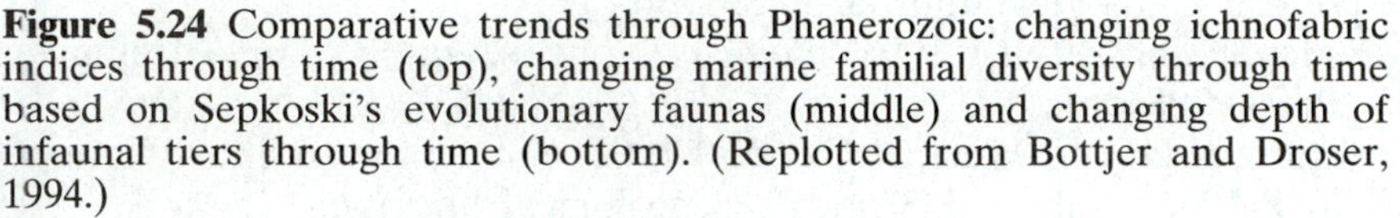

Figure 5.24 Comparative trends through Phanerozoic: changing ichnofabric indices through time (top), changing marine familial diversity through time based on Sepkoski's evolutionary faunas (middle) and changing depth of infaunal tiers through time (bottom). (Replotted from Bottjer and Droser, 1994.)

5.12 SUMMARY POINTS

- Trace fossils are evidence of the activity and behaviour of organisms.
- Ichnotaxa are diagnosed and described on the basis of the morphology of the trace and are named according to Linnaean rules in terms of ichnogenera and ichnospecies. Trace fossils are also classified according to modes of preservation and the behaviour of their producer.
- One organism may be responsible for a variety of morphologically different traces while several organisms may construct similar or identical traces.

- Marine environments are characterized by a spectrum of ichnofacies developed across an onshore–offshore gradient: the *Skolithos*, *Cruziana*, *Zoophycos* and *Nereites* ichnofacies generally range in order from shoreface to basinal facies.
- Total bioturbation may be quantified by ichnofabric indices; these indices may used to elucidate bioturbation trends through time.
- Many sedimentary successions contain tiered profiles of trace fossils reflecting a varied and evolving infauna. Frozen profiles can preserve tiers while their variable taphonomy can aid measurements of stratigraphical completeness.
- Many groups of animals have contributed to bioerosion leaving a range of trace fossils in a variety of hard substrates often in tiered profiles.
- The analysis of vertebrate tracks has provided significant windows on the behavioural patterns, community structures and evolution of the dinosaurs together with a further tool for palaeoenvironmental analysis.
- Animal and plant interactions have been recorded in fossil vegetation from the Devonian onwards, although the majority of evidence is from late Mesozoic and Cenozoic floras.
- Many ichnotaxa have migrated seaward with time, while there was an increase in diversity of trace fossils and an increased depth of penetration and complexity of infaunal tiering during the early Palaeozoic.

5.13 FURTHER READING

Crimes, T.P. and Harper, J.C. (eds). 1970. *Trace Fossils*. Special Issue of the Geological Journal 3. 547 pp.

Crimes, T.P. and Harper, J.C. (eds). 1977. *Trace Fossils 2*. Special Issue of the Geological Journal 9. 351 pp.

Donovan, S.K. (ed.) 1994. *Palaeobiology of trace fossils*. Belhaven Press. 308 pp.

Maples, C.G. and West, R.G. (eds) 1992. *Trace Fossils*. Short Courses in Paleontology 5, 238 pp. Paleontological Society.

5.14 REFERENCES

Bottjer, D.J., Droser, M.L. and Jablonski, D. 1988. Palaeoenvironmental trends in the history of trace fossils. *Nature* 33, 252–255.

Bromley, R.G. 1996. *Trace fossils: biology, taphonomy and applications*. 2nd edition. Chapman & Hall, London. 361 pp.

Bromley, R.G. 1970. Borings as trace fossils and *Entobia cretacea* Portlock, as an example. *Special Issue Geological Journal* 3, 49–90.

Bromley, R.G. 1992. Bioerosion: Eating rocks for fun and profit. *Short Courses in Paleontology* 5, 121–129. Paleontological Society.

Bromley, R.G. and Asgaard, U. 1993. Endolithic community replacement on a Pliocene rocky coast. *Ichnos* 2, 93–116.

Bromley, R.G., Pemberton, S.G. and Rahmani, R.A. 1984. A Cretaceous woodground: the *Teredolites* ichnofacies. *Journal of Paleontology* 58, 488–498.

Buatois, L.A. and Mángano, M.G. 1995. The paleoenvironmental and paleoecological significance of the lacustrine *Mermia* ichnofacies: an archetypical subaqueous nonmarine trace fossil assemblage. *Ichnos* 4, 151–161.

Crimes, T.P. 1987. Trace fossils and correlation of the late Precambrian and early Cambrian strata. *Geological Magazine* 124, 97–119.

Crimes, T.P. 1994. The period of early evolutionary failure and the dawn of early evolutionary success: the record of biotic changes across the Precambrian – Cambrian boundary. In Donovan, S.K. (ed.) *The palaeobiology of trace fossils*. John Wiley and Sons, Chichester, 105–133.

Crimes, T.P. and Fedonkin, M.A. 1994. Evolution and dispersal of deepsea traces. *Palaios* 9, 74–83.

Crimes, T.P., Legg, I., Marcos, A. and Arboleya, M. 1977. ?Late Precambrian – low Lower Cambrian trace fossils from Spain. *Special Issue Geological Journal* 9, 91–138.

Donovan, S.K. 1994. Insects and other arthropods as trace-makers in non-marine environments and palaeoenvironments. In Donovan, S.K. (ed.) *The palaeobiology of trace fossils*. John Wiley and Sons, Chichester, 200–220.

Droser, M.L. and Bottjer, D.J. 1986. Semiquantitative field classification of ichnofabric. *Journal of Sedimentary Petrology* 56, 558–559.

Droser, M.L. and Bottjer, D.J. 1990. Ichnofabric of sandstones deposited in high-energy nearshore environments: measurement and utilization. *Palaios* 4, 598–604.

Ekdale, A.A. 1985. Paleoecology of the marine endobenthos. *Palaeogeography, Palaeoclimatology and Palaeoecology* 50, 63–81.

Ekdale, A.A., Bromley, R.G. and Pemberton, S.G. 1984. *Ichnology: the use of trace fossils in sedimentology and stratigraphy*. Society of Economic Paleontologists and Mineralogists, Short Course 15, 317 pp. Tulsa, Oklahoma.

Frey, R.W. and Pemberton, S.G. 1985. Biogenic structures in outcrop and core. I. Approaches to ichnology. *Bulletin of Canadian Petroleum Geology* 33, 72–115.

Frey, R.W., Pemberton, S.G. and Saunders, T.D.A. 1990. Ichnofacies and bathymetry: a passive relationship. *Journal of Paleontology* 64, 155–158.

Frey, R.W. and Seilacher, A. 1980. Uniformity in marine invertebrate ichnology. *Lethaia* 13, 183–207.

Golubic, S., Friedmann, I. and Schneider, J. 1981. The lithobiontic ecological niche, with special reference to microorganisms. *Journal of Sedimentary Petrology* 51, 475–478.

Hofmann, H.J. 1990. Computer simulation of trace fossils with random patterns, and the use of goniograms. *Ichnos* 1, 15–22.

Landing, E. 1993. Precambrian-Cambrian boundary global stratotype ratified and a new perspective of Cambrian time. *Geology* 22, 179–182.

Lockley, M.G., Houck, K.J. and Prince, N.K. 1986. North America's largest dinosaur trackway site: implications for the Morrison Formation paleoecology. *Bulletin of the Geological Society of America* 97, 1163–1176.

Lockley, M.G. 1991. *Tracking dinosaurs*. Cambridge University Press.

Martinsson, A. 1965. Aspects of a Middle Cambrian thanatotope on Öland. *Geologiska Foreningens i Stockholm Forhandlingar* 87, 181–230.

Martinsson, A. 1970. Toponomy of trace fossils. *Special Issue Geological Journal* 3, 323–330.

McCann, T. 1990. Distribution of Ordovician–Silurian ichnofossil assemblages in Wales – implications for Phanerozoic ichnofaunas. *Lethaia* 23, 243–255.

Orr, P.J. 1994. Trace fossil tiering within event beds and preservation of frozen profiles: an example from the Lower Carboniferous of Menorca. *Palaios* 9, 202–210.

Orr, P.J. 1995. A deep-maxine ichnofaunas assemblage from Llandovery Strata of the Welsh basin, west Wales, UK. *Geological Magazine* 132, 267–285.

Palmer, T. and Plewes, C. 1993. Borings and bioerosion in fossils. *Geology Today* 1993, 138–142.

Pickerill, R.K. 1994. Nomenclature and taxonomy of invertebrate trace fossils. In Donovan, S.K. (ed.) *The palaeobiology of trace fossils*. John Wiley and Sons, Chichester, 3–42.

Raup, D.M. and Seilacher, A. 1969. Fossil foraging behavior: computer simulation. *Science* 166, 994–995.

Seilacher, A. 1953. Studien zur Palichnologie. I. Ueber die Methoden der Palichnologie *Neues Jahrbuch fur Geologie und Palaöntologie* 96, 421–452.

Seilacher, A. 1964. Biogenic sedimentary structures. In Imbrie, J. and Newell, N. (eds), *Approaches to paleoecology*. Wiley, New York, 296–316..

Seilacher, A. 1967a. Bathymetry of trace fossils. *Marine Geology* 5, 413–428.

Seilacher, A. 1967b. Fossil behavior. *Scientific American* 217, 72–80.

Seilacher, A. 1977. Pattern analysis of *Paleodictyon* and related trace fossils. *Special Issue Geological Journal* 9, 289–334.

Seilacher, A. 1986. Evolution of behavior as expressed in marine trace fossils. In Nitechi, M.H. and Kitchell, J.A. (eds), *Evolution of animal behaviour: paleontological and field approaches*. Oxford University Press, Oxford, 62–87.

Wetzell, A. and Aigner, T. 1986. Stratigraphic completeness: Tiered trace fossils provide a measuring stick. *Geology* 14, 234–237.

Fossils as environmental indicators

6

Earlier chapters have shown the various ways in which fossils provide evidence of the environment in which they had lived. This evidence is synthesized in this chapter to show how fossils may help to identify specific environments along a nearshore to deep-basin profile and environments with low oxygen or high or low salinity. Fossils also yield important evidence about the firmness of the substrate and sedimentation rates which can be used in sequence stratigraphy to analyse the development of sedimentary sequences.

6.1 EVIDENCE USED IN ENVIRONMENTAL ANALYSIS

Sedimentary information is the key evidence for environmental interpretation in many circumstances. Observations on composition, grain size, sorting, bed thickness and sedimentary structures allow the discrimination of sedimentary facies which are a reflection of the depositional environment. The associated fossils may, at the very least, help to confirm or refine the interpretation made on sedimentological evidence. However, in some situations palaeontological evidence may be critical. For example, it is commonly difficult to discriminate between packets of cross-stratified sandstones formed in fluvial, tidal or even aeolian environments and fossils may prove to be the vital evidence of a marine or a fresh-water environment. More generally, fossils are particularly important in discriminating between different mudstone facies which yield little sedimentological information. The bathymetry of muddy shelves in particular can be reconstructed mainly from their biofacies and amongst carbonate environments the discrimination between lagoonal micrites and those micrites deposited in deep water may depend on the contained fauna. In addition to the evidence of depositional environment that comes from fossils, they may also yield information about the firmness of the substrate, rates of sedimentation (Chapter 2), and palaeolatitude (Chapter 8). The special characteristics of some carbonates are discussed in a later section. It is important to bear in mind that the most comprehensive environmental analyses are those based on the combined evidence from the physical sedimentology, the faunal content, the ichnofauna and geochemistry, where appropriate.

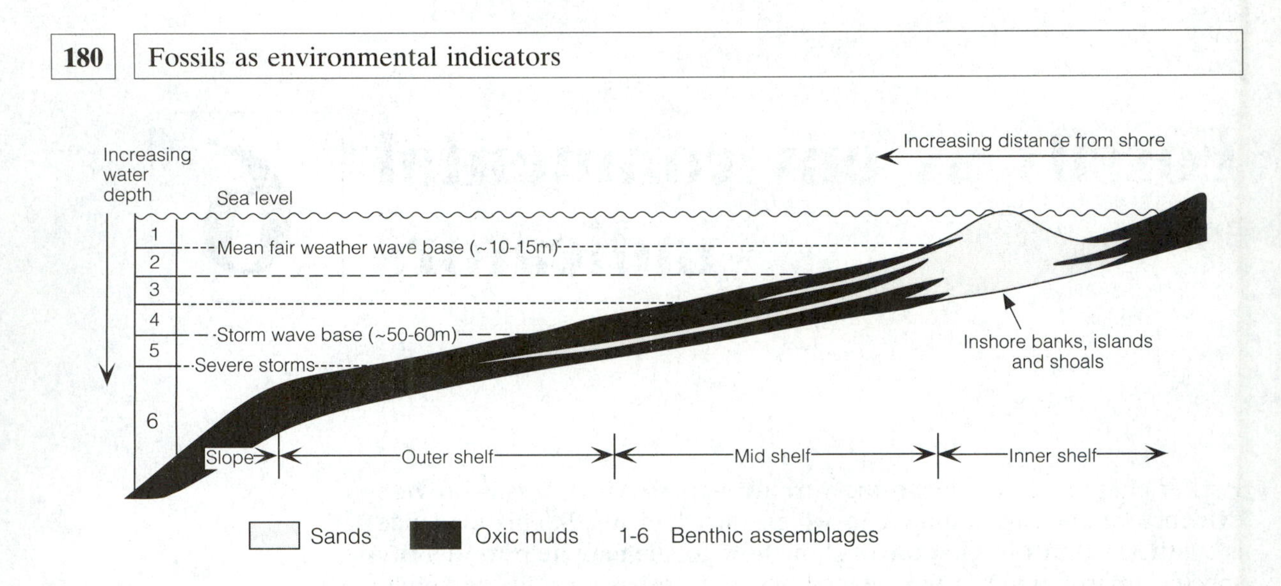

Figure 6.1 Schematic representation of benthic assemblages (BA) 1 to 6, showing their relative position on a shelf and estimates of the depth of the BA2/3 and BA4/5 boundaries (see Section 6.3). (Modified from Pickerill and Brenchley, 1991.)

In the following sections the palaeontological evidence for discriminating between environments is reviewed in the context of onshore to offshore environmental gradients. The initial discussion centres on siliciclastic environments, but can be equally applied to many carbonate environments.

6.2 DISCRIMINATING BEWEEN ENVIRONMENTS ON CLASTIC SHELVES

Nearshore sandy environments can generally be identified by their sedimentary characteristics, but trace fossil associations and their related ichnofabrics may add important supportive evidence (Pollard *et al.*, 1993). In the more offshore muddy environments fossils play an important role in environmental discrimination. It is commonly useful to relate shelf faunas to a series of benthic assemblage zones each of which represents a particular depth-zone on the sea-floor (Fig. 6.1).

6.2.1 Biofacies distribution

The composition of the biota along a profile from the coast to deep marine environments changes with the different environments encountered. These changes are reflected in a broad way by changes in the proportions of the different fossil groups that constitute a number of broadly defined biofacies ranged across shelves and down into deeper water. Thus, House (1975) described marine Devonian faunas of Europe in terms of biofacies regimes comprised of a nearshore regime characteristically with infaunal and epifaunal bivalves and strongly ribbed and robust brachiopods, a shelf regime with coral and stromatoporoid reefs and a variety of gastropods, bivalves, nautiloids, ostracodes, trilobites and brachiopods, and a basinal regime with goniatites, conodonts, tentaculitids and ostracodes amongst others. Communities are much more environmentally specific than the broad biofacies mentioned above and therefore a palaeocommunity is potentially a rather precise palaeo-environmental indicator. However, there are two severe limitations to the usefulness of palaeocommunities in palaeoenvironmental analysis. First, it is necessary to determine the environmental range of a palaeocommunity before it can be used elsewhere as an environmental indicator. Secondly, there is a continuous turnover in species, so the composition of palaeocommunities change with time and their taxonomic composition and environmental ranges have to be repeatedly defined for successive stratigraphic levels. In spite of these problems Ziegler (1965) and Ziegler *et al.* (1968) showed how palaeocommunities in the Lower Silurian of the Anglo-Welsh borderlands, (Fig. 6.2 and see also Chapter 7) could be effectively used to monitor changes in sea-level and to reconstruct relative depth zones across a Silurian shelf. The relative depth distribution of five communities was established from their stratigraphic position in a transgressive sequence above the upper Llandovery unconformity (Fig. 6.2a). The *Lingula* community typically occurs in shallow marine deposits a little above the basal unconformity and the other communities follow in regular sequence in progressively deeper marine sediments. Evolutionary lineages in the brachiopods *Eocoelia, Stricklandia* and pentamerids, together with the presence of some graptolites allow the succession to be divided into several 'time slices' (C1 to C6) that can be identified across the shelf area. The distribution of the five communities across the shelf within any one time-slice allows a reconstruction of the relative bathymetry of the shelf (Fig. 6.2b).

Based on their correlation with particular depth zones the Llandovery communities have been widely used to construct bathymetric curves showing relative sea-level changes through the Llandovery. Accordingly, several deepening and shallowing cycles have been identified in the UK, Norway, Estonia, China and the USA, that appear to be the same age even though they were on different Silurian plates and hence may be regarded as probably being eustatic (Johnson *et al.*, 1991). The success in using the Llandovery communities as global bathymetric indicators appears to derive from their unusually wide geographic range and the relative consistency of community composition through several millions of years. Brachiopod communities in general are useful bathymetric indicators throughout the Palaeozoic but commonly have a more confined distribution than those in the Silurian. Lower Palaeozoic trilobite communities (Fortey, 1975) and graptolitic biofacies have also proved useful in recognizing marine environments of different depths.

The ecological dominance of different faunal elements changed throughout the Phanerozoic, with the times of greatest change being

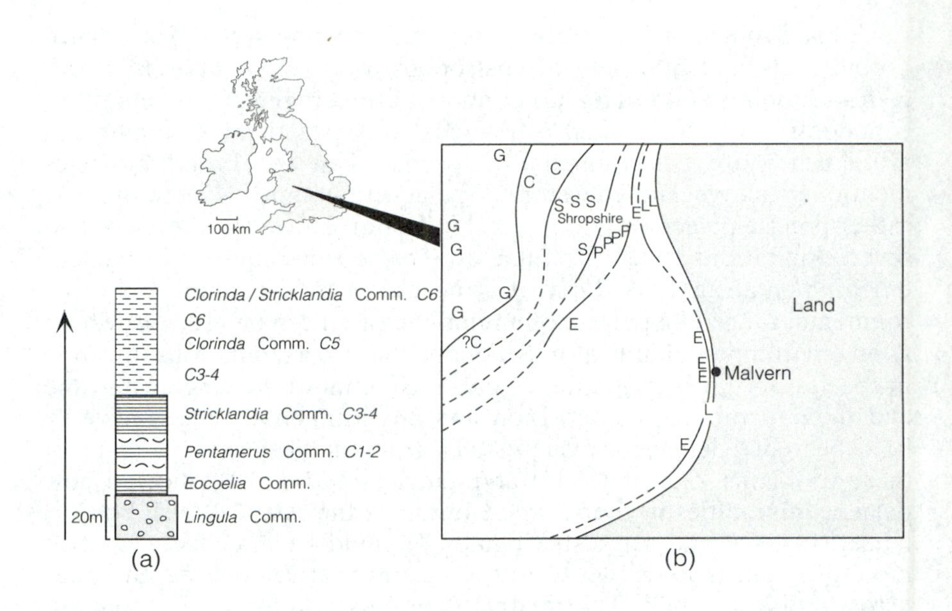

Figure 6.2 (a) A schematic representation of the upper Llandovery succession on the Midland Platform, England. The upper Llandovery is divided into informal units C1 to C6. The sequence of communities, from the *Lingula* community to the *Clorinda* community reflects progressive deepening of the sea with time. (b) A palaeogeographic map for C1 time showing the distribution of communities across the shelf, with the *Lingula* (L) and *Eocoelia* (E), communities in nearshore positions and the *Clorinda* community (C) near the shelf margin (P = *Pentamerus* community, S = *Stricklandia* community and G = graptolitic facies). (Based on Ziegler, 1965.)

those associated with the transitions between the three evolutionary faunas of Sepkoski, described in Chapter 9. Thus the ecological composition and distribution of Cambrian faunas (Fig. 6.3.) is markedly different from those of the Ordovician to Permian (Figs 6.4 and 6.5) and the Mesozoic to Recent are different again in their nature (Fig 6.6) (see McKerrow (1978) for an illustrated guide to Phanerozoic communities). The Cambrian shelly faunas are generally of low diversity and on clastic shelves they are dominated by trilobites, inarticulate brachiopods and hyolithids (Fig. 6.3) and form rather broadly distributed shelly benthic assemblages. The more varied shelly benthic faunas that appeared and diversified during the Ordovician and persisted until the end of the Permian generally formed about five benthic assemblages disposed across the shelves of the time. The communities are generally brachiopod dominated, but include bryozoans, crinoids, corals, molluscs and trilobites. Molluscs are commonly associated with brachiopods in nearshore communities (Fig. 6.4), corals and brozoans may be present in mid-shelf communities and trilobites are present in a range of depths but generally increase in diversity towards the outer shelf (Fig. 6.5).

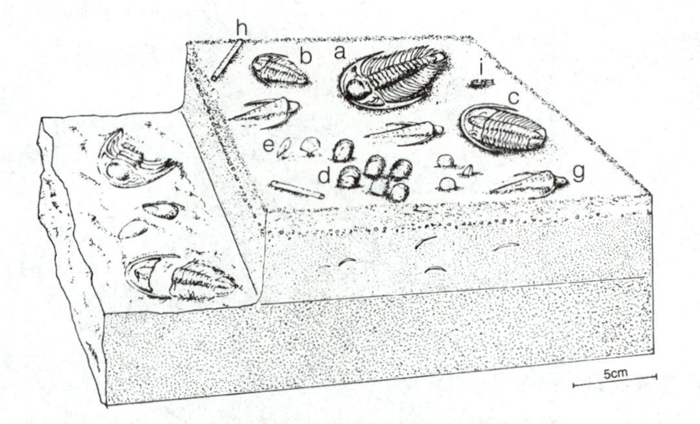

Figure 6.3 A typical low-diversity, shallow marine Cambrian community, a. *Paradoxides* (Trilobita), b. *Bailiella* (Trilobita), c. *Solenopleura* (Trilobita), d. *Lingulella* (phosphatic brachiopod), e. *Billingsella* (billingsellide brachiopod), f. *Micromitra* (phosphatic brachiopod), g. *Hyolithes* (Caliptoptomatida), h. *Hyolithella*, (uncertain affinities), i. agnostid (Trilobita). (From McKerrow, 1978.)

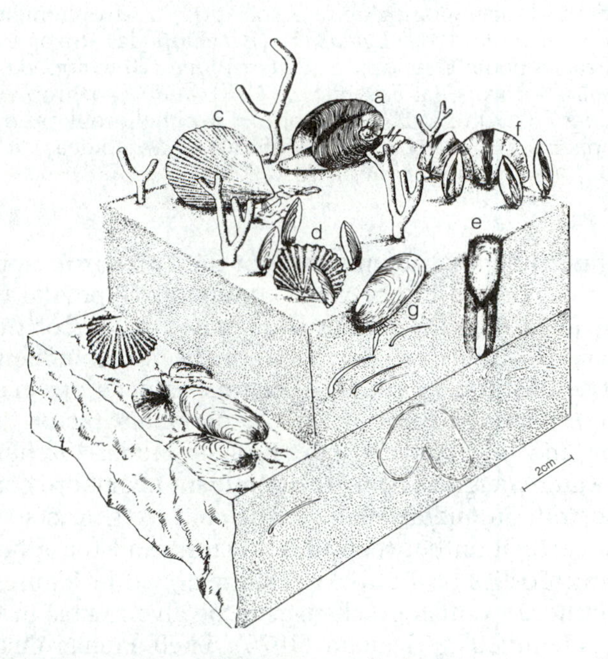

Figure 6.4 A shallow marine (BA2) upper Ordovician community, showing a bivalve-brachiopod-bryozoan association. a. *Plectonotus* (Mollusca: Monoplacophora), b. *Hallopora* (Bryozoa), c. *Ambonychia* (Mollusca: Bivalvia: Pterioidea), d. *Dinorthis* (Brachiopoda: Orthida), e. *Palaeoglossa* (lingulate brachiopod), f. *Harknessella* (Brachiopoda: Orthida), g. *Modiolopsis* (Mollusca: Bivalvia: Modiomorphoida). (From McKerrow, 1978.)

Figure 6.5 An outer shelf (BA4 to BA5) Upper Ordovician community with a high diversity brachiopod-trilobite association. a. strophomenid (Brachiopoda: Strophomenida), b. *Leptaena* (Brachiopoda: Strophomenida), c. *Onniella* (Brachiopoda: Orthida), d. *Platystrophia* (Brachiopoda: Orthida), e. *Brongniartella* (Arthropoda: Trilobita), f. *Calyptaulax* (Arthropoda: Trilobita), g. *Platylichas* (Arthropoda: Trilobita), h. bellerophontid (Mollusca: Monoplacophora), i. orthoceratid (Mollusca: Nautiloidea), j. *Tentaculites* (Criconarida). (From McKerrow, 1978.)

The epifaunal benthic communities of the Palaeozoic appear to provide a simpler relationship between community type and the depth in which they lived than the mollusc-dominated faunas of the Mesozoic and Tertiary (Fig. 6.6), which appear to be more influenced by the nature of the substrate. In a study of benthic associations in the Jurassic, Ager (1965) identified substrate as the dominant factor in brachipod distribution and in a study of late Jurassic faunas Oschmann (1988) identified water energy as being the dominant nearshore control on the faunas, substrate in middle shelf regions and oxygen in some offshore muds. The correlation between these factors, and the lithological and faunal characteristics of the rocks is summarized in Figure 6.7.

Broad onshore to offshore changes in bivalve faunas in the Jurassic have been identified by Hallam (1976). Shelf faunas in the Jurassic form a substantial number of communities and differences between communities associated with substrates that superficially appear similar are a good indication that the environments are, in fact different.

6.2.2 Trace fossils

The palaeoecology of trace fossils has been discussed in Chapter 5. The particular importance of trace fossils in environmental analysis is

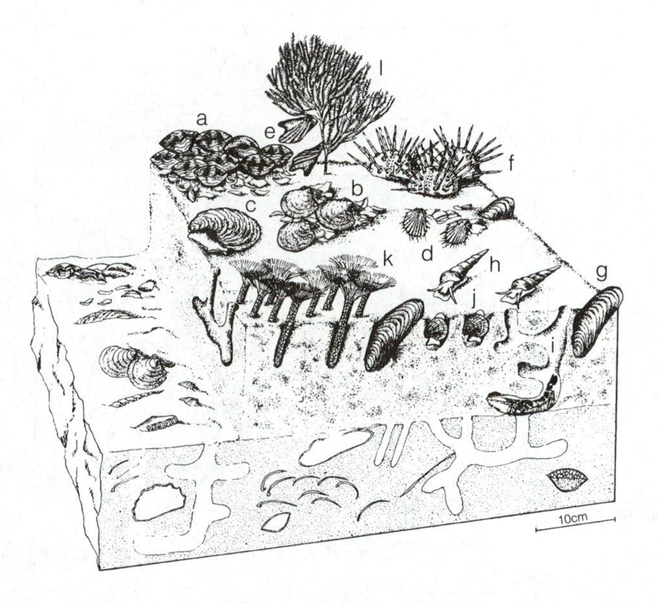

Figure 6.6 A shelly lime mud community from the middle Jurassic, belonging to a sheltered shallow marine environment with a varied infaunal and epifaunal bivalve fauna. a. *Epithyris* (Brachiopoda: Terebratulida), b. *Camptonectes* (Mollusca; Bivalvia: Pterioida), c. *Liostrea* (Mollusca; Bivalvia: Pterioida–oyster), d. *Pseudolimea* (Mollusca; Bivalvia: Pterioida), e. *Costigervillea* (Mollusca; Bivalvia: Pterioida), f. *Acrosalenia* (Echinodermata: Echinozoa), g. *Modiolus* (Mollusca; Bivalvia: Mytiloida), h. *Fibula* (Mollusca: Gastropoda: Archaeogastropoda), i. *Thalassinoides* (trace fossil made by a crustacean), j. *Anisocardia* (Mollusca; Bivalvia: Veneroida), k. terebellid worms (Annellida), l. gorgonian (Cnidaria: Octocorallia). (From McKerrow, 1978.)

that they record behavioural activities that are environmentally controlled and the traces were formed in the environment in which they are found. Furthermore, the types of activity and ichnofacies distribution have remained broadly similar throughout the Phanerozoic. Trace fossil associations are helpful in identifying both marine (Fig. 5.5) and non-marine environments (Fig. 6.8). The scheme of bathymetrically related ichnofacies first proposed by Seilacher (1967) and subsequently modified by later authors, has been criticized in detail but remains one of the most useful guides to relative bathymetry, though the burrowing fabric within the sediment (ichnofabrics) may be equally useful, particularly in the study of cores, where sedimentary structures may be ambiguous and trace fossils are not easily identified. The most pertinent criticism of Seilacher's ichnofacies model is that the behaviour recorded in the trace fossils is influenced by a variety of factors such as turbulence, sedimentation rate and availability of food, which are broadly correlated with depth, but not invarariably so. Consequently, trace fossils can occur far outside their normal ichnofacies if the conditions were right (Chapter 5). In addition to the value of ichnofacies in environmental analysis, the nature

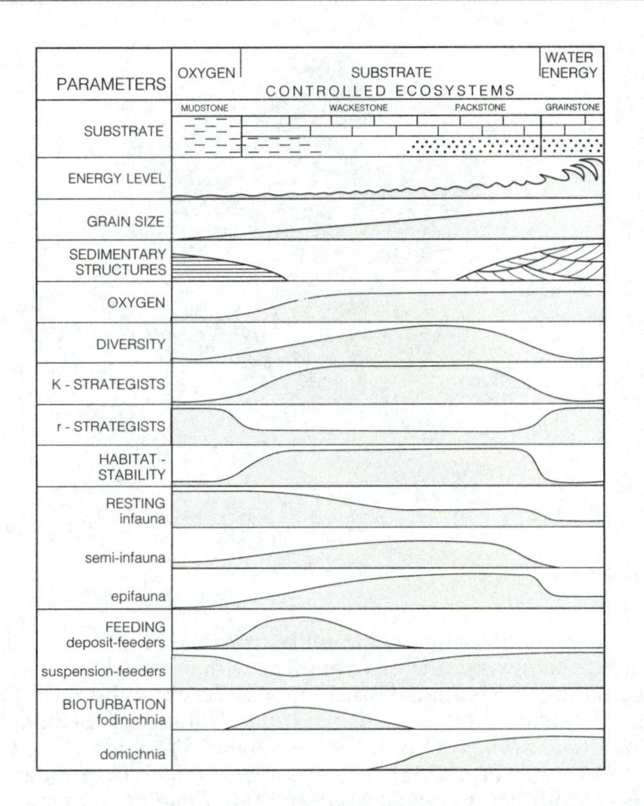

Figure 6.7 Variations in physical and biological aspects of the environment in response to differences in water energy, substrate and oxygen levels. The changes from right to left approximately correlate with nearshore to offshore trends. (From Oschmann, 1988.)

of trace fossil assemblages can be helpful in identifying varying degrees of oxygen starvation (see Chapter 2).

6.2.3 Diversity, biomass and size gradients

The pronounced changes in diversity and biomass along environmental gradients (see Chapter 7) can provide important evidence for environmental analysis. Diversity is generally low in the stressed nearshore environments, particularly in brackish or hypersaline waters and areas with shifting substrates, such as beaches and tidal shoals. Diversity can, however, be substantially higher where shells from different environments have been mixed together, for example in a tidal delta. In clastic settings, the diversity of shelly fossil associations generally increases with increasing environmental stability outwards across the shelf until it reaches a maximum towards the outer shelf and then decreases, possibly as food becomes a limiting factor. In carbonate areas the profile is probably similar but the distribution of shelf faunas appears to be more patchy, particularly where there are reefs, so the diversity pattern is more irregular. Diversity profiles probably change

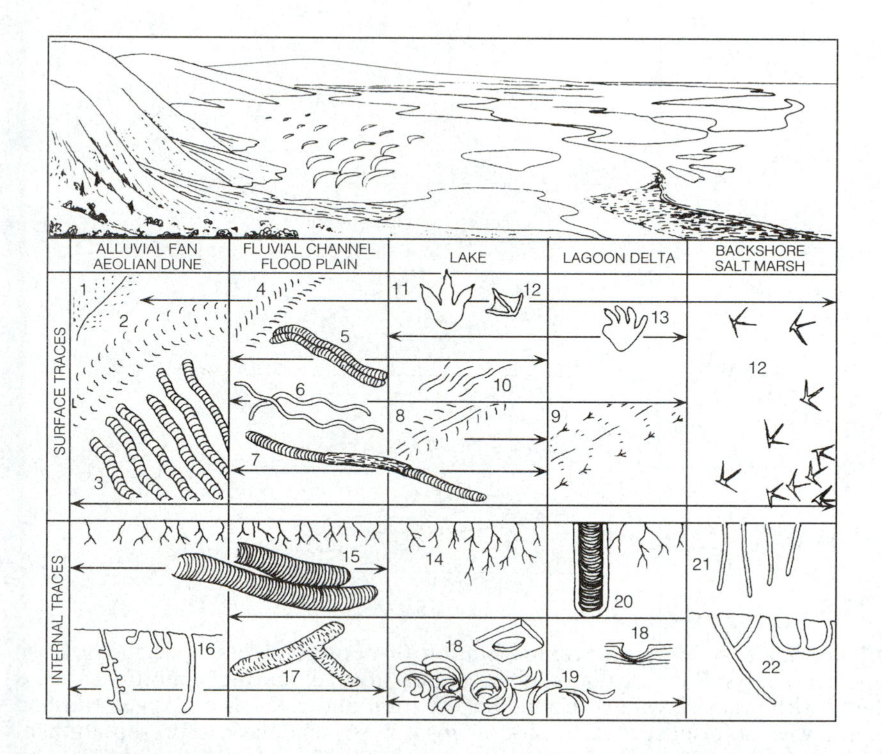

Figure 6.8 Distribution of some trace fossils in non-marine environments. 1. *Paleohelcura* (scorpion trackway): 2. *Mesichnium* (insect trackway): 3. *Entradichnus* (exogenic insect traces): 4. *Acripes* (crustacean trackway): 5. *Cruziana problematica* (branchiopod crustacean burrow): 6. *Cochlichnus*: 7. *Scoyenia gracilis*: 8. *Siskemia* (arthropod trackway): 9. *Kouphichnium* (xiphosuran trackway): 10. *Undichnus* (fish swimming traces): 11. reptile track: 12. bird tracks: 13. amphibian tracks: 14. roots: 15. *Beaconites*: 16. insect burrows: 17. *Spongeliomorpha carlsbergi* (insect burrow): 18. *Lockeia siliquaria*: 19. *Fuersichnus communis*: 20. *Diplocraterion parallelum*: 21. *Skolithos*: 22. *Psilonichnus* and other crab burrows. (From Bromley, 1996, based on Pollard, unpublished.)

considerably with changing biota through the Phanerozoic, but information on this is sparse. A generalized profile of diversity of shelled fossil faunas across a marine shelf with normal marine waters is shown in Figure 6.9a. This would, however, be modified where salinity, oxygen levels or soft substrates reduced the diversity to substantially lower levels.

The abundance and the mass of fossil shells, which are a crude reflection of total biomass, vary greatly according to environmental conditions. In nearshore facies abundance is generally moderate to low, particularly in those sandstones reflecting the mobile substrates of the shoreface and tidal shoals, but it can be high on rocky shores or where opportunistic species have successfully colonized stressed environments, such as brackish-water lagoons, producing beds crowded with a low diversity bivalve, gastropod or ostracode fauna. Abundance can also be particularly high where shells have been transported and

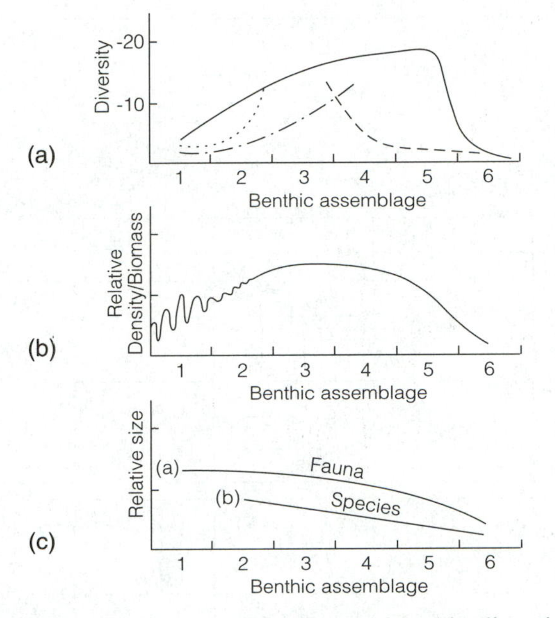

Figure 6.9 a A schematic representation of benthic diversity of shelly fossil species across Phanerozoic shelves under normal marine conditions (solid line); with brackish waters (dotted line); with reduced levels of oxygen (dashed line); with soft substrates (dash-dot line). b. A schematic representation of biomass across Phanerozoic shelves as reflected by abundance and size of shells. c. Schematic representation of (a) the average size of whole faunas and (b) the average size of individual species across a shelf. (Modified from Pickerill and Brenchley, 1991.)

concentrated by nearshore currents to form thick shell beds (see section 6.2.4). In general, abundance increases from inner into mid-shelf mudstone facies and then decreases steadily into deep shelf facies. Most deeper-water facies have low abundances (Fig. 6.9b).

Size of shells is very variable between the members of an assemblage and between assemblages in different environments. Shells of individual species appear to decrease in size with increasing depth. There is little data on variation of size between assemblages, but overall the mean size of brachiopod shells within an assemblage appears to decrease with increasing depth particularly towards the outer shelf and this might be true of bivalves too (Fig. 6.9c).

6.2.4 Taphonomic state

The state of preservation of shells varies according to the turbulence of the environment and rates of sedimentation (see concept of taphofacies, section 3.10 and Fig.3.20). The degree of fragmentation, sorting, abrasion and proportion of articulated shells generally decreases with decreasing turbulence in an offshore direction. However, the pattern is not always simple because these features are also affected by the

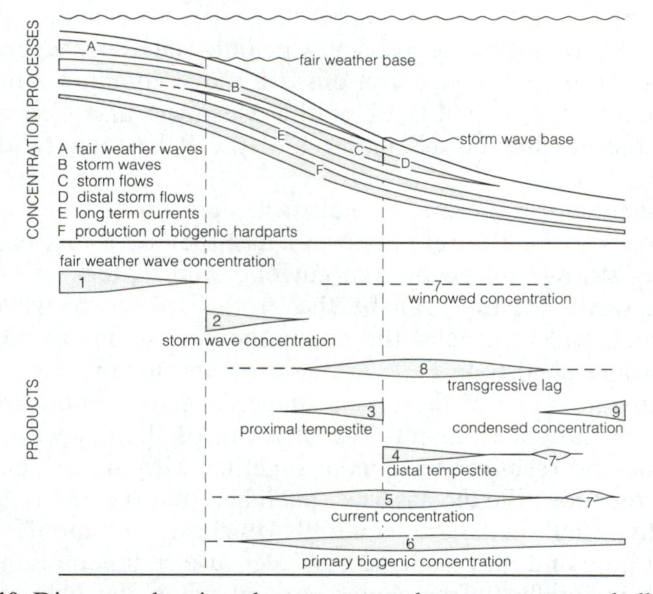

Figure 6.10 Diagram showing the processes that concentrate shells and the types of concentration they produce. Based on Jurassic shelf of Kachchh, India. The extent of some products is somewhat different on other shelves. (Modified from Fürsich and Oschmann, 1993.)

rate of sedimentation, which determines the length of time the shells are subjected to mechanical, chemical and biological erosion (Fig. 3.20). Long-term rates of sedimentation are closely related to subsidence rate, so that low subsidence rates result in low sediment accumulation rates which in turn tend to be associated with a high degree of fragmentation. Hiatal surfaces, at which there has been minimal sedimentation, commonly have fossils highly fragmented by bioerosion even though they may form in relatively low-energy, offshore environments.

6.2.5 Types of shell concentration

The thickness, geometry and and internal fabric of shell concentrations varies according to the processes that produce the shell accumulation (Fig. 6.10). Shell concentrations in nearshore facies are very varied according to the availability of shell material and the hydrodynamics of the depositional area. The planar-laminated or cross-stratified sandstones that reflect the mobile substrates of a beach, upper shoreface, tidal channel or a shoal rarely contain substantial concentrations of shells. Where they do occur, they are likely to consist of shell lenses lying in erosional depressions. Shell concentrations of nearly autochthonous shells are more commonly found in the scour hollows of hummocky cross-stratified sandstones of lower shoreface environments. Thick accumulations of shells transported by nearshore storm-driven currents may occur as cross-stratified or parallel-bedded lenses, or as tabular sheets, in tidal deltas, overwash lobes or in channels within the shoreface. Shells in such deposits may be imbricated. As a general rule, thick nearshore shell concentrations are more common amongst the bivalve-dominated assemblages of the Mesozoic and Cenozoic than amongst the brachiopod assemblages of the Palaeozoic (Kidwell and Brenchley, 1994).

Most shell concentrations in shelf facies are the product of storms. The deposits formed during a storm can either be a mixture of sand carried by storm-induced bottom currents and material resuspended by storm waves, or they can be the product of storm waves alone, which erode and resuspend the autochthonous sediment. On muddy shelves where sandstone beds are rare or absent, the taphonomy of fossils provides some of the best evidence available about the hydrodynamics of the environment. The effect of oscillatory wave currents is to erode and resuspend sediment together with any included shells and then redeposit the bioclasts as a shell bed that is commonly graded. Proximally, shell beds are generally thickest, commonly have an erosional base and may be lenticular. Because sedimentation is rapid, shells are generally unfragmented, except when the exhumed shells had been previously broken (see section 3.10). Shell beds tend to become thinner and more tabular offshore, but in distal shelf environments concentrations may be so thin that they occur as isolated lenses or shell pavements (Fig. 6.11). The facies transition from mudstones with shell concentrations to mudstones with a fauna that is commonly articulated but dispersed, except for life clusters, is a good indicator of the position of effective storm wave base, i.e. the depth at which storm waves can erode and resuspend appreciable amounts of sediment. In addition to mobilizing shells and sand, storm waves may put large volumes of mud into suspension. Mud layers deposited by storm events are generally difficult to detect, especially after the sediments have been bioturbated. However, some events may be recorded by the blanketing of living populations in their life orientation; for example, crinoids are recorded with their cups intact and their stems still in an upright position.

Shell concentrations that were deposited in association with transported sand commonly occur as lenticles in the erosional hollows preserved at the base of a sandstone bed. The sandstone beds and the shell concentrations become thinner distally and the erosional scours are less deep (Fig. 6.11). In some circumstances sand can be transported by storm-driven currents and deposited below storm wave-base to form tabular sandstone beds, but these rarely contain shell concentrations (Fig. 6.11), because no wave erosion was involved.

The palaeontological evidence for different environments along a shore to shelf margin profile is summarized in Table 6.1.

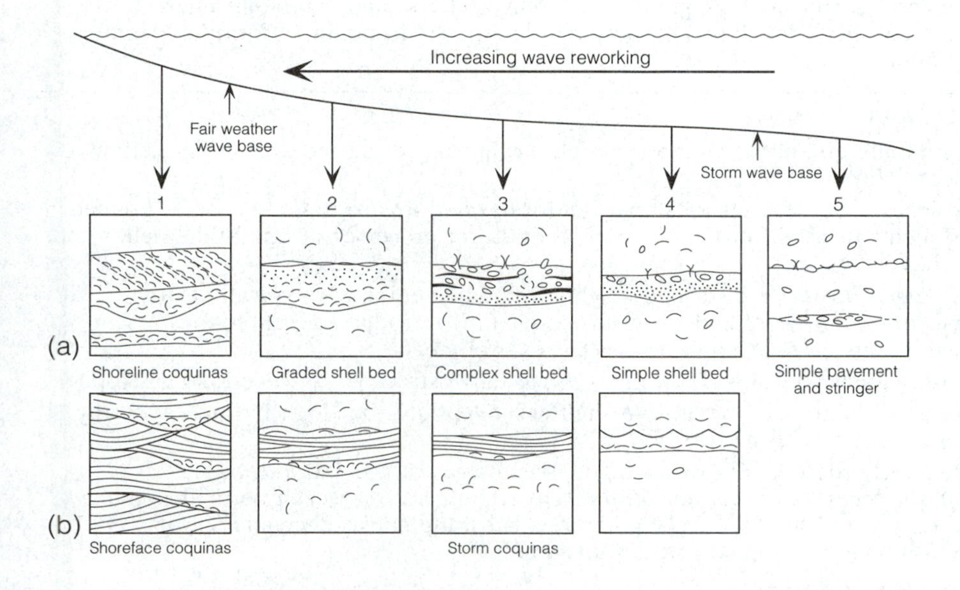

Figure 6.11 Shell accumulations across a shelf. (a) Reworking of shells with little addition of sediment, (b) Shells deposited with sand transported by storm currents (tempestites), but commonly reworked by storm waves. a.1. Cross-stratified current-reworked shell bed, and wave-reworked shells in an erosional hollow. a.2. A graded nearshore shell bed. a.3. a complex shell bed with mud partings, reflecting multiple storm events. a.4. A storm bed with bioclastic material at the base and *in situ* preservation of shells above. Articulated shells occur in the interbedded mudstones. a.5. Shell pavements and stringers reflecting only slight reworking of living assemblages. Most shells in mudstones undisturbed except by bioturbation.
b.1. Shell accumulations in swaley storm-stratified sandstones. b.2. and b.3. Shell concentrations at the base of sandstone tempestite beds with hummocky cross-stratification indicating wave reworking. b.4. Parallel stratified current deposited sandstone bed with shell lag at base. (Modified from Pickerill and Brenchley, 1991.)

6.3 FOSSILS AS BATHYMETRIC INDICATORS FOR MARINE SHELF SEDIMENTS

Although changes in sea-level have important effects on the stratigraphic record and much attention has been paid by geologists to estimating amounts of sea-level change, there are few sound criteria for estimating absolute depth. Potentially the most useful evidence in shelf sequences is that which identifies fair weather wave-base, effective storm wave-base and the base of the photic zone (see Brett *et al.*, 1993, for a discussion of absolute depth). The depth of modern shelf margins ranges from about 200 m to 600 m; 200 m is commonly accepted as a reasonable value for most ancient shelf edges. However, the depth is likely to be varied depending on the tectonics of the shelf, rate of sediment supply and the ability of waves to smooth the surface at storm wave-base.

Table 6.1 Summary of general lithologic, taphonomic, macroinvertebrate and ichnologic characteristics of brackish lagoonal, shoreface, shelf and slope and basin siliciclastic environments. (Modified from Pickerill and Brenchley, 1991.)

BRACKISH LAGOONAL ENVIRONMENTS	
Context	Commonly associated with terrestrial, barrier bar, estuarine, marsh or shallow marine shoreface facies.
Lithology	Typically shales though siltstone, sandstones and less commonly coarser grained sediments may be introduced by tidal or storm processes in flood-tidal deltas and washover fans.
Taphonomy	Dispersed *in situ* or disturbed neighbourhood assemblages, typically as thin shell concentrations. Tidal and storm processes may introduce allochthonous assemblages from shoreface and inner shelf environments.
Taxonomy/Diversity	*In situ* faunas typically of low diversity; usually ostracodes, bivalves and gastropods.
Abundance	Variable, commonly low in thin shell accumulations, but high in thick, typically monospecific, shell beds.
Ichnology	Imperfectly described; low diversity assemblages of the *Psilonichnus, Glossifungites, Cruziana* and *Zoophycos* ichnofacies may each occur depending on energy and salinity levels, grain size, substrate consistency and specific depositional environment within a lagoon.
SHOREFACE ENVIRONMENTS (BA1)	
Context	Nearshore and associated beach facies, distinguished from inner shelf deposits by generally having only rare mudstones interbedded with sandstones. Shoreface facies commonly cap upward-coarsening shallow marine cycles.
Lithology	Typically sandstones with locally developed mudstones. Sandstones exhibit planar- or cross-stratification formed by wave, tidal or wind-forced currents, or hummocky cross-stratification formed by oscillatory or combined flow.
Taphonomy	Allochthonous shell concentrations that may occur as coquinas, on bedding planes or cross-stratified foresets.
Taxonomy/Diversity	Palaeozoic assemblages characterized mainly by brachiopods with more rarely bivalves and gastropods. Mesozoic/Cenozoic assemblages characterized by varied bivalves (including deep burrowers and attached forms) and gastropods. Taxa affected by substrate mobility (e.g. Corals) are rare. *In situ* faunas rare. Diversity, though still generally low, can be elevated by introduction of allochthonous faunas by storm, tidal and wave activity.
Abundance	Generally low in Palaeozoic rocks, but Cenozoic strata typically possess more abundant remains. In more subdued shoreface environments shells may be relatively common in coquinas.
Ichnology	Typically *Skolithos*, more rarely *Psilonichnus* and *Cruziana* ichnofacies depending on substrate mobility and subtle gradients in hydrologic sedimentologic and ecologic parameters. Diversity typically low.
INNER (SHALLOW) SHELF ENVIRONMENTS (BA2)	
Context	Facies subject to frequent storm waves and tidal processes, commonly occurring below shoreface sandstones and above mid-shelf sequences in coarsening-upward cycles.
Lithology	Sandstones dominate, but may occur in equal abundance with interbedded shales. Storm-influenced facies exhibit hummocky cross-stratification; tidal facies may exhibit herringbone cross-stratification, sigmoidal cross-stratification or mud drapes.
Taphonomy	Allochthonous or disturbed neighbourhood assemblages at the base of storm sandstones, typically as coquinas. Tidally influenced sandstones commonly exhibit transported assemblages or shell lags distributed along foresets. Shales possess disarticulated and dispersed shells and bedding plane assemblages. *In situ* faunas rare or absent.

Table 6.1 Continued

Taxonomy/Diversity	Palaeozoic strata dominated by moderately but variably diverse assemblages of brachiopods, particularly orthoids and strophomenoids; Mesozoic/Cenozoic strata typically possess variable infaunal and epifaunal bivalves. Representatives of other benthic macroinvertebrates may occur, sometimes in abundance.
Abundance	Generally moderate, but high in coquinas. Faunas can, however, be generally sparse where a freshwater influence is established, as, for example, in inner shelf environments in front of deltas.
Ichnology	Typically moderate to high diversity assemblages of the *Cruziana* and/or *Skolithos* ichnofacies. Mudstones commonly intensely bioturbated and possess trace fossils typical of the *Cruziana* ichnofacies; associated sandstones contain representatives of the *Skolithos* ichnofacies providing substrate mobility does not result in their destruction.
MIDDLE SHELF ENVIRONMENTS (BA3 and 4)	
Context	Commonly form portions of upward-coarsening sequences, but may also form part of a predominantly mudstone sequence. Below fair weather wave base, above storm wave base.
Lithology	Mudstones, typically intensely bioturbated, and interbedded sandstone tempestites with hummocky cross-stratification are common.
Taphonomy	Mudstones may contain *in situ* clumped or dispersed shells that are still commonly articulated, or as thin stringers or locally reworked disturbed neighbourhood assemblages. Sandstones commonly contain allochthonous lag assemblages or coquinas that are taxonomically similar or dissimilar to those assemblages within the associated mudstones.
Taxonomy/Diversity	Palaeozoic strata possess moderate to high diversity assemblages dominated by brachiopods represented by several orders. Crinoids, bryozoans, corals, trilobites and gastropods also common. Mesozoic/Cenozoic strata dominated by varied infaunal and epifaunal bivalve assemblages with gastropods, echinoderms, bryozoans and less commonly brachiopods also present.
Abundance	Generally high, decreasing with depth.
Ichnology	Similar to inner shelf environments with high diversity assemblages typical of the *Arenicolites* ichnofacies in sandstones, typically as opportunists following storm activity, and intense bioturbation by members of the *Cruziana* ichnofacies in associated mudstones.
OUTER (DEEP) SHELF ENVIRONMENTS (BA5)	
Context	Mudrock-dominated sequences in association with shelf or slope/basin environments. Mainly below effective storm wave base, though extreme storms have disturbed shells.
Lithology	Bioturbated mudstones with uncommon, and where present typically thin, distal siltstone or sandstone tempestites.
Taphonomy	*In situ* dispersed, articulated faunas predominate. Thin stringers and bedding plane assemblages also common; disturbed neighbourhood assemblages may occur. Distal tempestites may contain allochthonous shells, though not commonly.
Taxonomy/Diversity	Palaeozoic strata dominated by brachiopods and trilobites; Mesozoic/Cenozoic strata by bivalves, particularly infaunal species – nuculoids are very common. Other groups also occur but in lower numbers. Pelagic forms (eg. graptolites (Palaeozoic), ammonites, belemnites (Mesozoic) also commonly present. Diversity variable depending on parameters such as nutrients, oxygen levels, etc., but can be high, then typically markedly decreases at the shelf edge.
Abundance	Typically low.

Table 6.1 Continued

Ichnology	Discrete ichnotaxa commonly difficult to recognise as a result of intense bioturbation. Representatives of *Cruziana, Zoophycos* or rarely *Nereites* ichnofacies may predominate depending on specific environmental conditions. Bioturbation intense though diversity commonly low to moderate.
SLOPE AND BASIN ENVIRONMENTS (BA6 AND BELOW)	
Context	Mudrock-dominated sequences with associated sediment gravity flows particularly, but not exclusively, turbidites. Submarine canyons and associated fans commonly occur in association with slope and continental rise sequences. Contourites also common on slope and continental rise environments.
Lithology	Pelagic and hemipelagic mudstones, turbiditic sandstones, and siltstones. Canyons and fans typically possess coarser grained lithofacies.
Taphonomy	Autochthonous faunas, typically dispersed and articulated, in pelagic/hemipelagic mudstones. Allochthonous faunas introduced particularly by turbidity currents.
Taxonomy/Diversity	Dominated by pelagic macroinvertebrates (eg. Palaeozoic – trilobites, graptolites, nautiloids; Mesozoic – ammonites, belemnites). *In situ* forms include Palaeozoic trilobites, hyolithids, etc. and Mesozoic/Cenozoic bivalves, gastropods and crinoids. Other groups may occur but are uncommon. Diversity is low.
Abundance	Very low, but can be higher in sediment gravity flow deposits.
Ichnology	Variable, with representatives of the *Skolithos, Arenicolites* and/or *Glossifungites* ichnofacies commonly occurring in association with well-oxygenated submarine canyons and fans; *Zoophycos* ichnofacies with slopes with restricted circulation and low oxygen levels and the *Nereites* ichnofacies in more stable, classical flysch-like settings in generally quiet, but oxygenated waters.

Fair weather wave-base is estimated to be down to about 10 m and is generally taken to be the base of Benthic Assemblage 2 (BA 2). Sediments above fair weather wave-base show evidence of persistent wave activity, such as stacked wave-ripples and very good sorting. The depth of storm wave-base is variable depending on the fetch of the waves and the exposure of the coastline. Waves can form ripples down to a depth of 200 m, but the maximum depth at which they can erode and concentrate shells is probably about 100 m and may be substantially less on sheltered shelves. Shells in BA 4 are commonly wave-winnowed, but concentrations are much rarer in BA 5, suggesting that the BA 4–5 boundary might be substantially less than 100 m. The depth of the photic zone is about 100 m in clear ocean waters and as little as 20 m in turbid coastal waters. The base of the photic zone in ancient sediments is potentially identified by the disappearance of algae or the maximum depth of reef growth. However, both criteria have their problems. The maximum depth of modern reefs varies from area to area and ranges down to about 110 m, but most reefs have a maximum depth of 60–80 m. Most algae are confined to the photic zone, but some, particularly red algae, may live at greater depths. The dasycladacean green algae are mostly confined to depths of less than 60 m and varied algal floras are in general confined to those depths (Fig. 6.12).

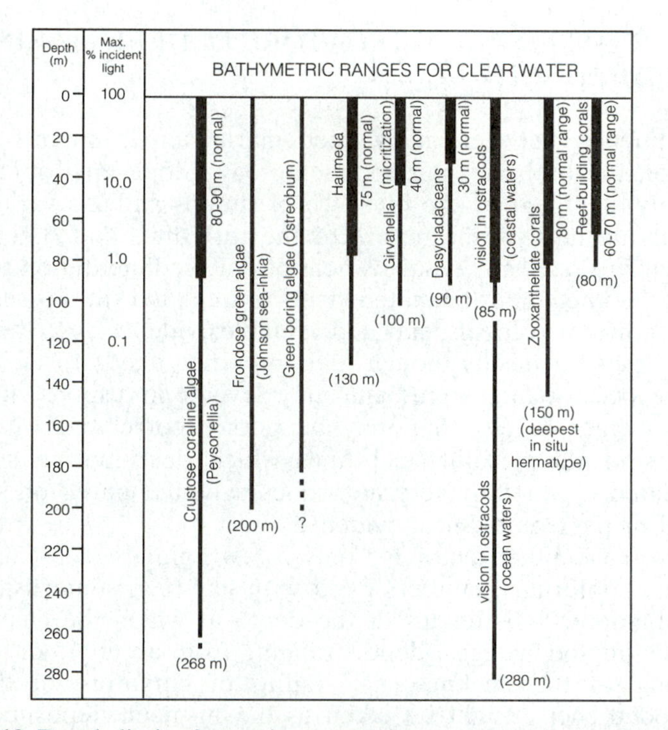

Figure 6.12 Depth limits for various organisms as a guide to estimating the effective depth of the photic zone. (From Brett *et al.*, 1993.)

Photosynthetic algae are relatively common in BA4 and may occur more rarely in BA5. A depth for the BA 4–5 boundary at about 60 m is suggested.

Boring of shells by endolithic algae has also been used to identify sediments above the photic zone (see section 2.4). However, it is difficult to distinguish the minute borings made by algae from those of fungi which are not confined to the photic zone. Until a clear distinction can be made, the value of borings is suspect.

The relationship between benthic assemblages and absolute depth for the Ordovician to Devonian is shown on Figure 6.1 and summarized as follows:

BA 1 beach and upper shoreface	0–5 m
BA 2 shoreface down to fair weather wave-base	5–15 m
BA 3 inner shelf, transition zone with strong storm-waves	15–30 m
BA 4 mid shelf down to effective storm wave-base and base of photic zone and (outer limit of reefs and most algae)	30–60 m
BA 5 outer shelf generally below effective storm wave-base	60–200 m

6.4 ENVIRONMENTAL INDICATORS IN DEEP MARINE SEDIMENTS

The distinction between many deep marine environments is made mainly on the basis of their sedimentology. Slope environments are commonly identified by the presence of slumps and the various parts of submarine fans by the nature of the turbidites and other gravity flows that are present. Areas of hemipelagic sedimentation are identified by the finely interlaminated layers of mud and silt. Fossils, particularly benthic microfossils, may add valuable evidence as to the relative depth of these deposits, though their presence needs to be analysed with care, because faunal gradients may develop in response to factors, such as oxygen levels, that may be poorly correlated with depth. Estimates of the absolute depth at which deep-marine sediments accumulated is generally very hazardous, whether based on sedimentological or palaeontological evidence.

Depth is a limiting factor for only a few animals (see Chapter 2), which have internal chambers filled with gas (e.g. some fish, whales and cephalopods). Estimates of the depth at which the chambers of nautiloids implode suggest depths ranging from about 75 m to 900 m, depending on the thickness and radius of curvature of the shell. Unimploded shells might be taken as having been deposited at less than these depths, unless they filled with water before they sank into deep water.

The depth range of planktonic species can rarely be used as a depth indicator because after death the organisms fall through the water column and may be sedimented at substantial depths. However, it may be possible to identify pelagic assemblages that have a low diversity in shallow facies because they only include surface living forms, but become more diverse with increasing depth as the assemblages include forms from a greater depth range. Depth-related graptolite species have been identified using this method (Fig. 6.13). Shelly benthos has rarely been used as an indicator of depth because faunas are so sparse. However, the behaviour of mobile soft-bodied benthos is reflected in intricate meandering or spiral feeding traces (*Nereites* ichnofacies), which are generally an indicator of relatively deep environments where food is in short supply.

Comparison of fossilized taxa with Recent taxa, whose depth range is known, is the most reliable indicator of depth, particularly using microfossils such as foraminifera (Murray, 1991). However, this application of taxonomic uniformitarianism is only fully reliable in Quaternary deposits and becomes progessively less reliable back through the Cenozoic as the proportion of extant species diminishes. An alternative approach is to record the relative proportions of calcareous benthic foraminifera, planktonic and agglutinating forms in an assemblage as a reflection of depth (Fig. 6.14) (Brasier, 1980; Gibson, 1989). A high proportion of benthic forms is a good indicator of relatively shallow depths, planktonics become increasingly impor-

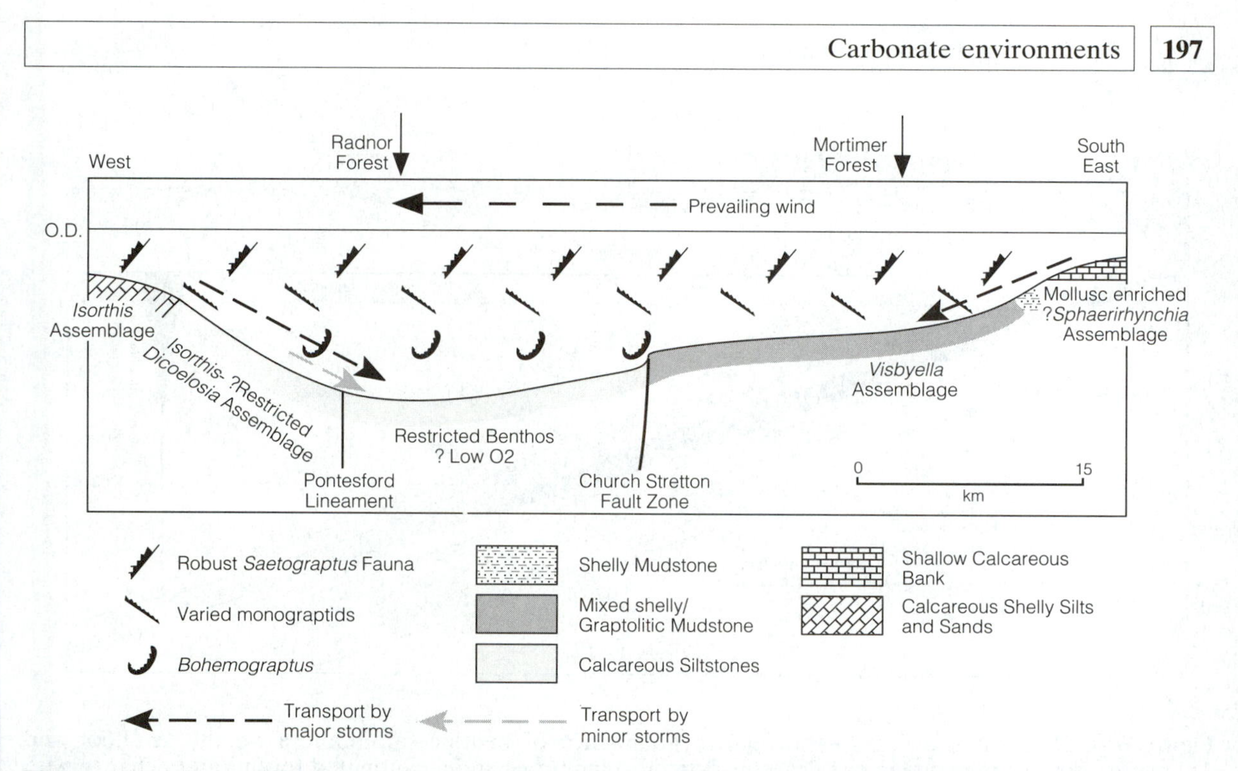

Figure 6.13 Silurian biofacies across the shelf and eastern part of the Welsh Basin. The graptolite faunas are depth related and show how shallow facies potentially have a lower diversity than the deeper facies, which may accumulate faunas from all depths. (From Underwood, 1994.)

tant with greater depths, while a dominance of agglutinated forms is commonly an indicator of depths close to, or beyond the Carbonate Compensation Depth (CCD) at which the carbonate shells are dissolved; however, it can also reflect low oxygen levels. Some caution must be applied when using proportions because the abundance of planktonics can be misleading where they have been subject to considerable post-mortem transport. Modern foraminiferal assemblages also change in their proportions of different morphotypes with depth (Fig. 6.15)(Corliss and Chen, 1988) and apparently also show a more subtle change of shape within individual species (van Morkhoven *et al.*, 1986). Knowledge of the depth distribution of foraminifera is best for the Cretaceous onwards and is generally poor for the Palaeozoic.

6.5 CARBONATE ENVIRONMENTS

Carbonate environments differ substantially from clastic ones because much of the sediment is of biogenic origin and produced in the area in which it is deposited. As a consequence the grain size and sorting of carbonate sediments is commonly less closely related to the hydrodynamics of the depositional area. Biogenic production determines the bioclasts available for transport and sedimentation and so influence the nature of the substrate which in turn may influence the biota that

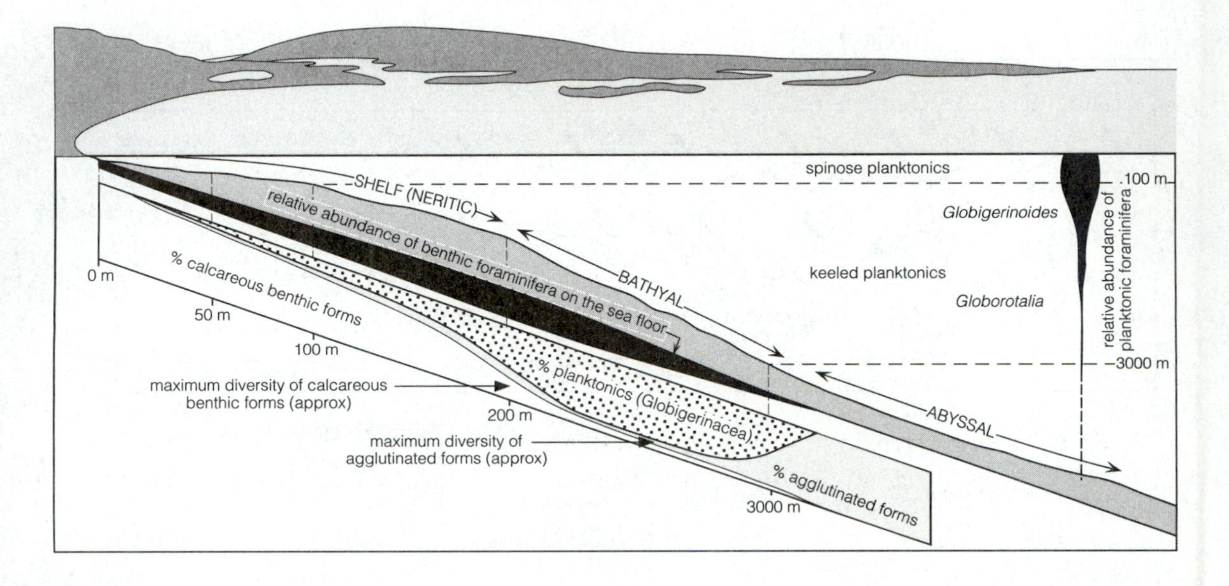

Figure 6.14 Diagram showing 1) the relative abundance of benthic foraminifera on the sea-floor and 2) how the relative percentage of calcareous benthic, planktonic and agglutinated foraminifera change with depth. Note also the change in morphology of planktonics with depth and different peaks of diversity. (Modified from Brasier, 1980.)

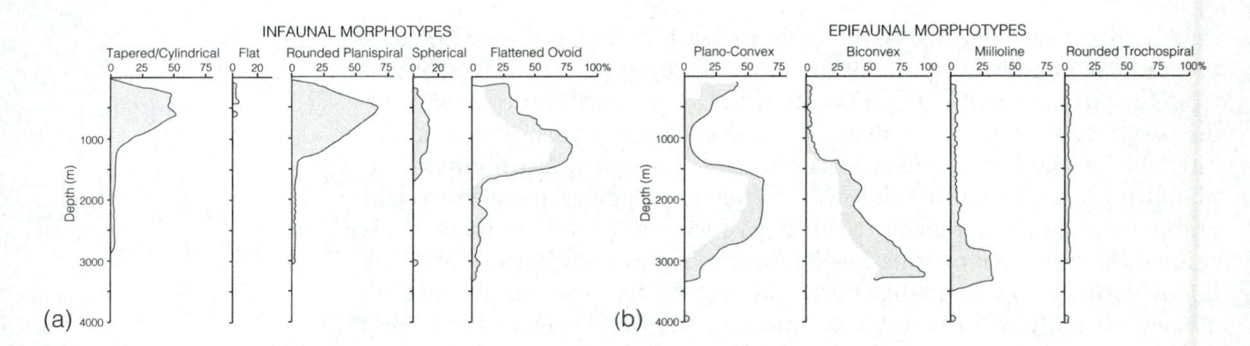

Figure 6.15 Depth distribution of nine different foraminiferal morphotypes with depth, based on samples from the Norwegian Sea. The graphs show the percentage of each morphotype in a series of samples. (a) morphotypes with infaunal microhabitat, (b) morphotypes with epifaunal microhabitat. (Modified from Corliss and Chen, 1988.)

colonizes it. Thus, taphonomic feedback is prevalent in carbonate environments.

Sedimentary grains that are typical of carbonate environments include some that are of non-skeletal origin, e.g. coated grains of various types (ooids, oncoids and grain aggregates such as grapestone) and peloids (faecal pellets, calcareous algae, micritized grains and micrite lithoclasts), but in many facies the majority are skeletal grains.

These may vary from complete shells with a size range from microscopic foraminifera to complete coral colonies more than a metre in diameter. More commonly the bioclasts are fragments of shells that have been created by physical fracture and abrasion or by the boring, crushing and grinding activity of other organisms. Carbonate mud may be formed in many ways. In the Bahamas the main source is the disintegration of the codiacean calcareous green algae, *Halimeda* and *Penicillus*, in oceanic settings other plants, such as the Coccolithophida produce micrite, in lakes algal photosynthesis is a major source and in marginal marine and fresh-water marshes cyanobacteria are the dominant producers. Although the bioclastic composition of a limestone may approximately reflect the nature of the biota inhabiting the area, it is the fabric of the sediment which reflects the prevailing hydrodynamic conditions. The classification of limestones into mudstone, wackestone, packstone, grainstone and boundstone (Dunham, 1962) is aimed to reflect different energy levels of deposition.

Carbonates show bathymetric and latitudinal differences in their levels of production. Most carbonate production is concentrated within the photic zone and particularly in depths of less than 50 m. Consequently most benthic production and a substantial proportion of planktonic production is on or over shelf regions where accretion rates tend to be high. Nevertheless, from the Mesozoic onwards carbonate oozes accumulate over vast areas of the ocean floor, so their total volume is very large.

The composition of biotic communities and the rate of carbonate production varies greatly with latitude. Communities are more diverse and production rates are highest in the tropics and decrease towards high latitudes. The highest rates of production and the site of most reef-growth is on the east side of continents where warm equatorial currents spill on to the continental shelves. Thick limestone sequences have mainly accumulated in tropical latitudes, whereas there has been little carbonate accumulation in polar latitudes. Tropical carbonates are characterized by the presence of oolites, grapestones, assemblages with hermatypic corals and carbonate-producing codiacean green algae (the chlorozoan assemblage of Lees and Buller, 1972) but where there are elevated salinities green algae predominate (the chloralgal assemblage). Large coralline colonies of hermatypic corals, sponges and calcareous algae are common as isolated growths, patches on the seafloor or in reefs. Molluscan shells tend to be large, thick and strongly ornamented. In contrast carbonates in temperate regions and on some cooler west-facing shelves in the tropics commonly have molluscan/foraminiferal bioclasts (foramol assemblages) which can form extensive shell banks and sheets though the thickness of temperate carbonates is usually modest. The biota may include red algae, solitary corals and branching ahermatypic corals, common bryozoans, echinoids, barnacles and vermetiid gastropods.

A substantial proportion of shallow marine carbonates can be broadly related to one of four depositional regimes:

1. carbonate ramps
2. rimmed shelves
3. epeiric platforms
4. isolated platforms.

(See Tucker and Wright, 1990 for a fuller treatment of carbonate platforms.)

6.5.1 Carbonate ramps

Carbonate ramps broadly resemble linear clastic coastlines in that they slope steadily seaward with a diminution of hydrodynamic energy with increasing depth (Fig. 6.16). Waves and tides commonly influence nearshore facies but offshore storms exert a strong influence. Bioclastic and ooidal grainstones predominate in nearshore facies while packstones and wackestones are interbedded with storm-influenced event beds in more offshore facies. Benthic communities form broadly parallel benthic assemblage zones as in clastic environments and give way to mainly planktonic assemblages in the deeper part of a ramp. Most ramps lack well-developed reefs with corals, stromatoporoids and large calcareous algae, but they are the main site for the growth of large carbonate mud mounds which are probably constructed of microbial carbonate that was mainly lithified as successive layers in the build-up were formed, so forming a rigid stucture. (See also Chapter 2, Box 2.1.).

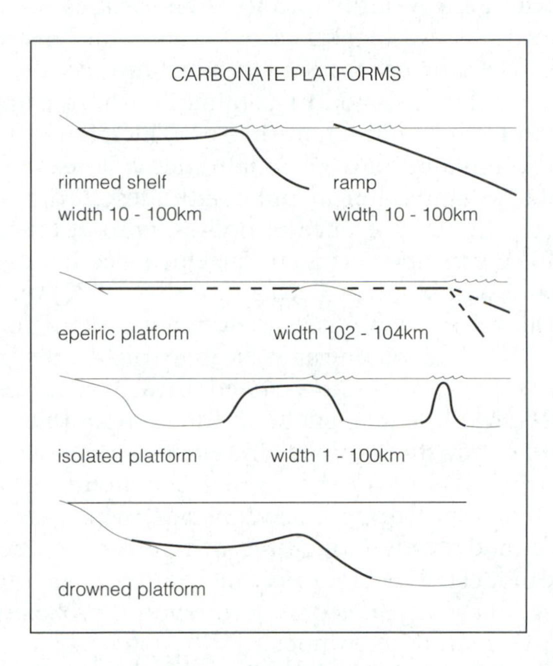

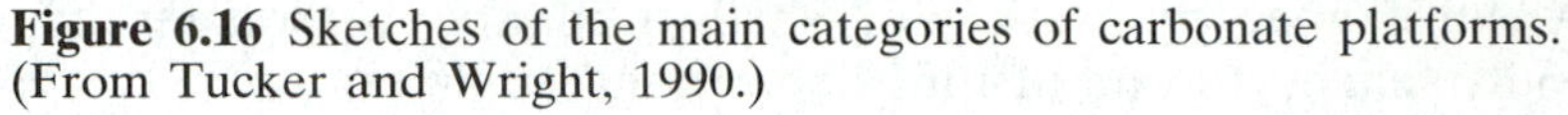
Figure 6.16 Sketches of the main categories of carbonate platforms. (From Tucker and Wright, 1990.)

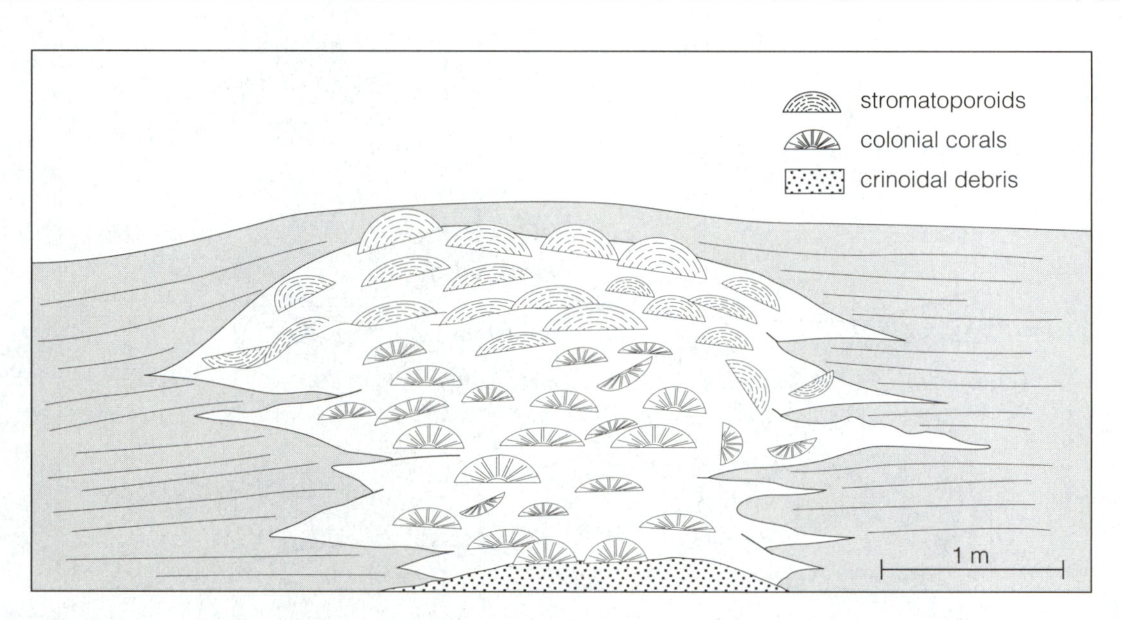

Figure 6.17 Schematic diagram of a late Ordovician to late Silurian patch reef, showing a progressive development from a crinoidal grainstone, through a colonial coral/calcareous algal phase to dominance by stromatoporoids at the crown of the reef. The unshaded area is likely to included fine-grained microbial carbonate, skeletal debris, the shells of organisms that lived in cavities within the reef and coarser sparry cements filling voids within the reef. The microbial carbonate can be the main framework constructor.

6.5.2 Rimmed shelves

Many carbonate shelves are rimmed by a shallow-marine zone characterized by reefs and associated shoals (Fig. 6.16). The rim encloses a shelf lagoon that is protected from open-ocean waves and is a site where fine-grained sediments are deposited. Lagoons that have a restricted access to open marine waters tend towards hypersalinity, whereas those associated with a barrier that is widely breached have normal marine salinities and may have patch reefs in clusters or dispersed on the shelf. The smaller patch reefs are generally composed mainly of core facies (Fig. 6.17) while the larger ones may have bioclastic flank facies dipping away from core. The shelf margin is a high-energy zone, exposed to ocean waves and in some instances to strong tidal currents. Reef facies, oolites and a variety of bioclastic deposits are typical of this zone. The reefs are generally asymmetric with a steeper slope facing the ocean and a gentle slope into the lagoon (Fig. 6.18). In Cenozoic reefs the morphology of corals and calcareous algae varies according to the wave energy impinging on the different zones of the reef (Figs. 6.18 and 6.19). Where wave energy is highest at the reef crest organisms mainly form encrusting sheets, but at somewhat lower intensities colonies form low domes, stubby branches or table-like shapes. On the fore-reef the colonies may have longer, more spreading branches or be plate-like. In the lagoon globular forms predominate. In many fossil reefs microbial carbonates commonly

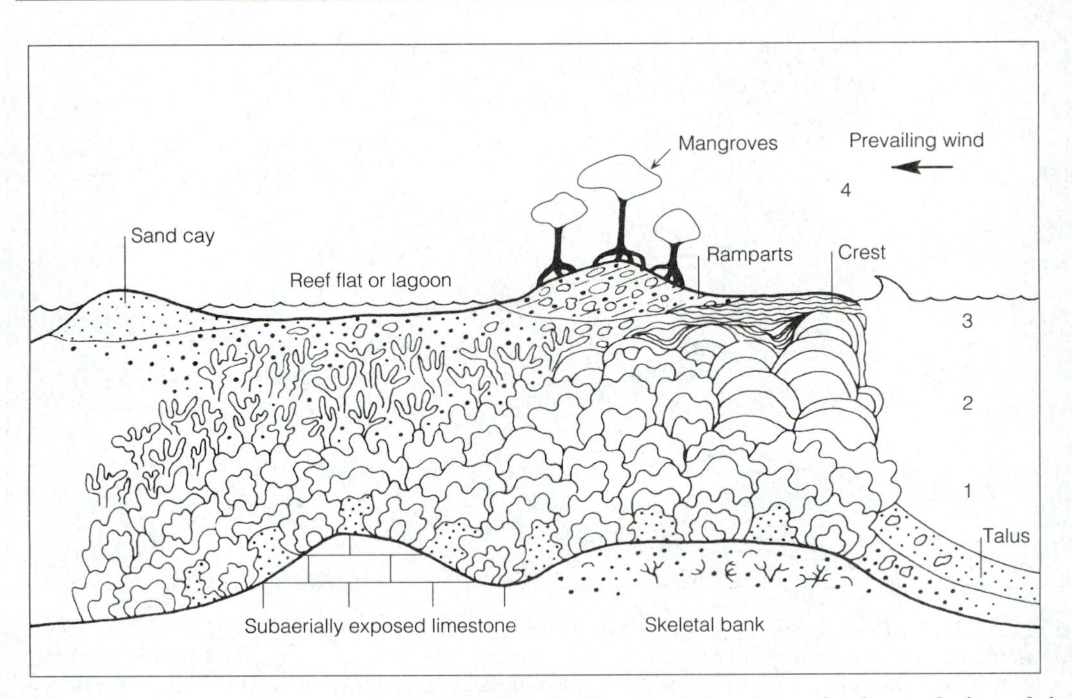

Figure 6.18 Diagram of a modern coral-algal patch reef, showing three phases in the evolution of the reef (levels 1, 2 and 3). (From Tucker and Wright, 1990.)

contribute to the construction of the reef framework and in some build-ups they may the main constituent of the reef mound. The microbial carbonates are seen as micrites, stromatolites, thrombolites and other laminated structures in micrite and fine sparry calcite. The shoreline of a rimmed shelf is more sheltered than that of a ramp and commonly has relatively fine-grained intertidal deposits, though higher energy sediments can occur on more exposed shelves.

6.5.3 Epeiric basins

In the past, shallow seas have covered extensive areas of the continents and thick accumulations of carbonates have accumulated in broad tectonic sags. Facies belts are commonly very wide and lithological units may be traceable over great distances (Fig. 6.16). The model of Irwin (1965) for such ancient epeiric basins has three main facies belts whose distribution was determined largely by wave energy. Where wave base intersected with the sea-floor, shoals of wave-reworked carbonates accumulated to form a relatively narrow belt of grainstone facies. The shoals acted as a barrier which protected the nearshore regions from the effects of waves and separated a quiet-water coastal zone with micritic sediments from a deeper offshore zone, below the effects of frequent waves, also with micritic sediments. The nearshore micrites commonly formed in intertidal and shallow subtidal environments, some of which were hypersaline. Algal-

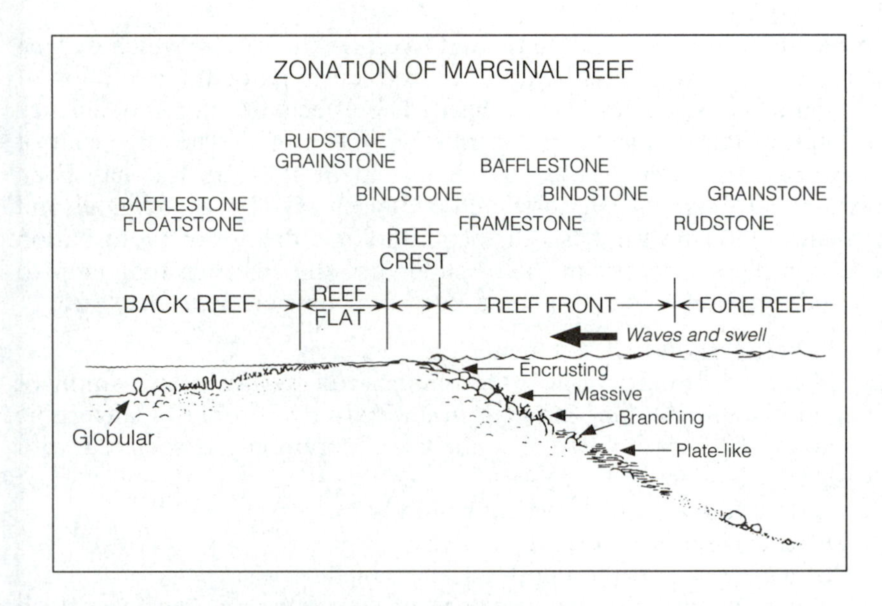

Figure 6.19 Variations in the morphology of coral colonies across a generalized modern reef. (From James, 1984.)

laminated micrites and variably domed stromatólites are common, but the macro-benthic fauna tends to be of low diversity and include ostracodes, gastropods and bryozoans. The grainstone shoals, that formed the barrier to the lagoon, commonly have a varied open marine benthic fauna and locally patch reefs may be present. Seawards of the shoals the sea-floor declined into deeper water and had moderate to deep-water communities in the inner part and mainly pelagic faunas towards the centre of the basin.

6.6 OXYGEN-DEFICIENT ENVIRONMENTS (see also chapter 2, section 2.6)

The development of anoxic or dysoxic bottom waters may occur at a wide range of water depths and may disrupt the onshore–offshore biotic gradients previously described. Anaerobic environments are commonly associated with black shales, but it is a common misconception that black muds have always formed in deep marine environments; more commonly they accumulated on the deeper parts of shelves and in restricted basins. The preservation of organic carbon, which is the primary cause of the dark colour, results when the supply of organic matter exceeds the rate at which it is oxidized. This situation can arise where the rate of organic sedimentation is particularly high or where the level of oxygen in the bottom waters is low (Pedersen and Calvert, 1990). High rates of supply of organic matter are generally related to high surface productivity, while low levels of oxygen in bottom waters are related to either the rate at which oxygen is replenished by wave-stirring or bottom currents or the depletion of oxygen as organic matter is oxidized. The effects of oxygen deficiency on marine faunas has been described in Chapter 2. The spectrum of environments from aerobic to anaerobic in the Jurassic has been divided into six oxygen restricted biofacies (ORB) by Wignall and Hallam (1991) on the basis of their lithology, degree of bioturbation and benthic species richness. In summary, the features that help to identify environments with some degree of oxygen deficiency, are:

1. darker, carbon-rich sediments;
2. decreased bioturbation, often emphasized by the preservation of delicate sedimentary laminae in severely dysaerobic or anaerobic facies. The trace fossil *Chondrites* is commonly associated with dysaerobic facies;
3. benthic faunas, which typically show:
   (a) a decline in species diversity;
   (b) a decrease in the depth and size of burrows;
   (c) an increase in the numbers of opportunistic species, which amongst bivalves tend to be suspension-feeders.

   However, both epifaunal and infaunal forms can be present in dysaerobic sediments, though deep burrowers are absent (Fig. 6.20). Planktonic faunas alone are found in anaerobic facies.

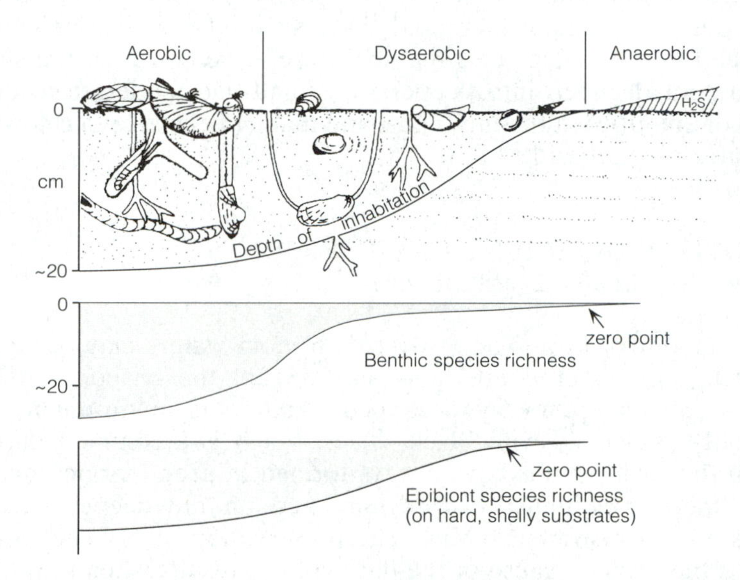

Figure 6.20 Faunal trends under a declining oxygen gradient illustrated by faunas from the British Jurassic. Epifaunal taxa are, from left to right, *Chlamys*, *Gryphaea* with abundant epibionts including serpulids, *Parainoceramus*, small *Gryphaea*, procerithid gastropod. Infaunal taxa are, from left to right, mecochirid crustacean, *Goniomya*, *Solemya* and associated *Chondrites* burrow, *Palaeonucula* and *Protocardia*. (From Wignall, 1993.)

Some care has to be taken in uniquely identifying these changes as being due to reduced oxygen because faunas respond in a similar way to soft substrates (see section 6.8). The most significant differences are the absence of bioturbation in anaerobic environments and the rarity of species attached to hard substrates in most oxygen-deficient environments. The life habits of some of the faunas that are found associated with many dark shale facies have been the subject of much debate. Characteristic of such facies are the 'paper pectens', a varied group of rather flat, thin-shelled bivalves (Wignall, 1994), many of which belong to the superfamilies Aviculopectinacea, Pectinacea and Pteriacea. The bivalves commonly form dense concentrations on bedding planes. The mode of life of different representatives of the 'paper pectens' has been variously interpreted to be epiplanktonic, nektonic and epibenthic, amongst other possibilities. If most of the forms are planktonic they have no bearing on the levels of oxygen on the sea-floor. However, in the opinion of Wignall (1994) a substantial proportion of the bivalves are epibenthic, which implies that the sea-floor was intermittently oxygenated to a degree sufficient for the establishment of a benthic fauna. Such an environment could then be regarded as poikiloaerobic (see Chapter 2, section 2.6).

The sites at which anaerobic and dysaerobic environments develop are quite varied and depend upon the relative rates of oxygen supply and organic sedimentation as described above. There are, however, some types of basins (Figs 6.21 and 6.22) that are particularly prone to anoxia (Demaison and Moore, 1980) and the broader environmental setting in which anaerobic sediments occur can commonly be determined from the facies with which they are associated. Anoxia may occur in relatively deep-water sediments in the oxygen minimum zone. However, over wide areas the levels of oxygen in the OMZ are not sufficiently low to inhibit most marine life. It is mainly under areas of upwelling currents that high surface productivity leads to a rain of organic matter that can exhaust most of the oxygen in the OMZ. Many upwelling areas are adjacent to continental margins and organic carbon can accumulate on the sea-floor where the OMZ abuts against the continental slope or the outer part of the shelf (Fig. 6.21a). Black shales associated with coastal upwelling should be identifiable by:

1. their restriction to low latitudes
2. their deposition in long linear belts parallel with the shelf margin
3. their association with relatively deep offshore sediments that may contain organic cherts, phosphorites and microplankton.

Anaerobic facies are also formed in enclosed basins where there is a high input of nutrients from rivers. In silled basins with a positive water balance, such as the Black Sea, the input of fresh water exceeds evaporation, causing the excess low-density water to form a surface current flowing across the sill into the open ocean. The outflow is countered by a weak, deeper, higher-salinity inflow over the sill (Fig. 6.21b). Within the basin itself there is a permanent halocline between the

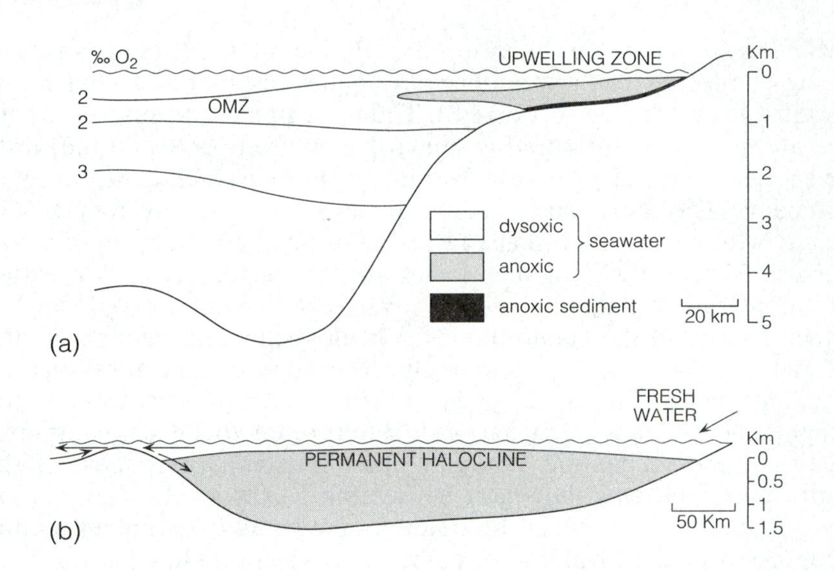

Figure 6.21 Two types of oxygen-deficient regimes. (a) An oxygen minimum zone developed in the ocean and more strongly above an upwelling zone over the outer shelf and upper slope (Peru shelf and trench), (b) A silled basin with a positive water balance caused by fresh-water input exceeding evaporation. Anaerobia results from the relatively high surface productivity and influx of poorly oxygenated ocean water (Black Sea). (Modified from Demaison and Moore, 1980.)

brackish surface water and the more saline waters below. The stable stratification inhibits the circulation of oxygen to the bottom waters and the basin becomes stagnant and potentially anoxic. Silled basins with positive water balance are potentially unstable over geologic time because changes in sea-level or climatic change can alter the flow of water and affect productivity and the oxygenation of bottom waters. Thus anaerobic and aerobic sediments can be intimately interbedded.

A significant number of black shale units in the geologic record are formed within intra-continental basins and are associated with marine trangressions. Many of the basins contain predominantly shallow marine sediments within which the black shales are intercalated. In some instances the black shales lie directly on an unconformity or disconformity and record the first sediments preserved during the transgression. In other cases the black shales form conformably within a shaly interval and are thought to have been formed during a transgression when the rate of sea-level rise was at its maximum (the 'maximum flooding surface' of sequence stratigraphy). Black shales forming in stagnant bottom waters on the initial transgression have been interpreted in terms of a 'puddle model' and those that extend over the lateral margins of the basin were referred to as an 'expanded puddle model' (Fig. 6.22).

Many of the interpretations of black shale facies outlined above have been based, quite reasonably, on uniformitarian principles. However,

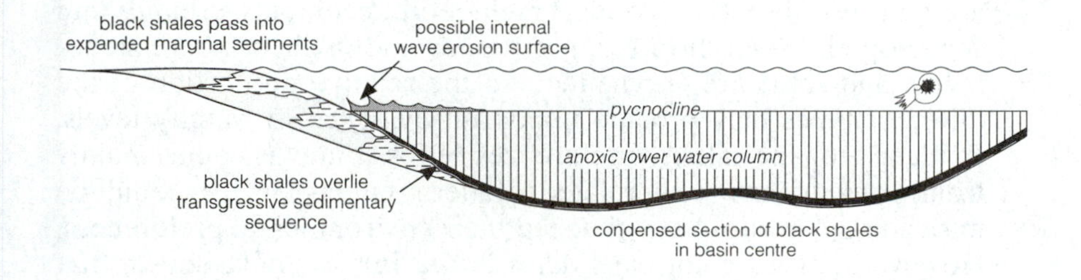

Figure 6.22 The development of anaerobic sediments in relatively shallow intra-cratonic basins: the 'expanding puddle model'. With rising sea-level a pycnocline develops with oxygen deficient waters below it. The area of black shales expands and transgresses across the more marginal shallow-marine facies. (Modified from Wignall, 1994.)

it is not certain that oxygen concentrations in the oceans have been constant throughout the Phanerozoic. The unusually high proportion of black shales in the Cambrian to Devonian interval and the prevalence of the 'graptolitic facies' in the Ordovician and Silurian have been interpreted as reflecting lower oxygenation of the oceans in the early part of the Phanerozoic. Further, 'ocean anoxic events' in the Cretaceous also suggest widespread depletion of bottom-water oxygen during parts of that period.

6.7 ENVIRONMENTS WITH LOW AND HIGH SALINITY (see chapter 2, section 2.8)

Environments with low or high salinity occur mainly in nearshore situations, so the environmental context is likely to be one of the first indications that brackish or hypersaline environments might be present. Brackish environments are likely to be associated with lagoons, estuaries and a variety of deltaic environments, while hypersaline conditions are common in tropical marine lagoons. Sedimentary evidence may be sufficient to identify specific depositional environments, but it provides little evidence of the salinity levels that prevailed, except where evaporites identify hypersaline conditions. Stable isotopes in shell material can provide indications of salinity levels (see section 2.8) but evidence for deviation from normal marine salinities comes mainly from the faunas as summarized below (see also section 2.8. and Fürsich, 1994).

1. Faunal diversity decreases with decreasing salinity down to a level of 5% from where it increases towards the fully fresh-water faunas (Fig. 2.20). Low diversity as a criterion for brackish waters must

be used with caution, because other factors, such as low oxygen and soft bottom substrate, also cause low diversity.

2. Stenohaline taxa are rare in brackish environments. Groups that are typically absent are corals, brachiopods, echinoids, crinoids and cephalopods. Some bivalves, gastropods and ostracodes are stenohaline and some are euryhaline and the relative proportion of the different types may provide important evidence of salinity levels. Evidence for the salinity preferences of fossil faunas comes mainly from comparisons with related modern faunas, which could be misleading if groups have changed their environmental preferences. However, palaeoecological studies in the Jurassic have shown that the taxonomic uniformitarian approach can be effective and that salinity-controlled asssemblages have shown a remarkable degree of evolutionary stability (Hudson, 1963; Fürsich, 1994). As a consequence it is possible to recognize mesohaline, oligohaline, hypersaline and more open marine faunas with some confidence throughout the Cenozoic and Mesozoic (Fig. 6.23). Taxonomic uniformitarianism can only be applied in a limited way to Palaeozoic faunas, but Carboniferous, Coal Measure 'marine bands' also show variations in their faunas that reflect changing levels of salinity (Calver, 1968). The marine bands are shale intervals, commonly only a few metres thick, within delta-top sequences, that show a symmetric cycle from fresh-water fauna to brackish then open marine faunas and then back to fresh-water faunas again (Fig. 6.24). The faunas identify the stages of a marine transgression up to the maximum flooding surface, and then a shallowing phase extending into fresh-water delta-top environments.
3. Microfauna and microflora can be useful guides to abnormal salinities. Fresh-water environments can be identified by distinctive algae, such as charophytes, though these are easily transported. In the Jurassic the planktonic alga, *Botryococcus*, has been shown to be consistently associated with fresh-water mollusc assemblages, though it was also transported in small amounts into marginal marine facies. More marine waters in general are characterized by dinocyst assemblages and rare acritarchs (Andrews and Walton, 1990).
4. In general, individual species decrease in size and shell thickness in the more stressed low salinity environments and ornamentation tends to be less pronounced amongst the faunas.

The above criteria for recognizing brackish water assemblages appear to be sound for Mesozoic to Recent rocks, but it becomes progressively more difficult to apply them back into the Palaeozoic. Low diversity is likely to have been a feature of Lower Palaeozoic brackish water environments, but other diagnostic features are elusive.

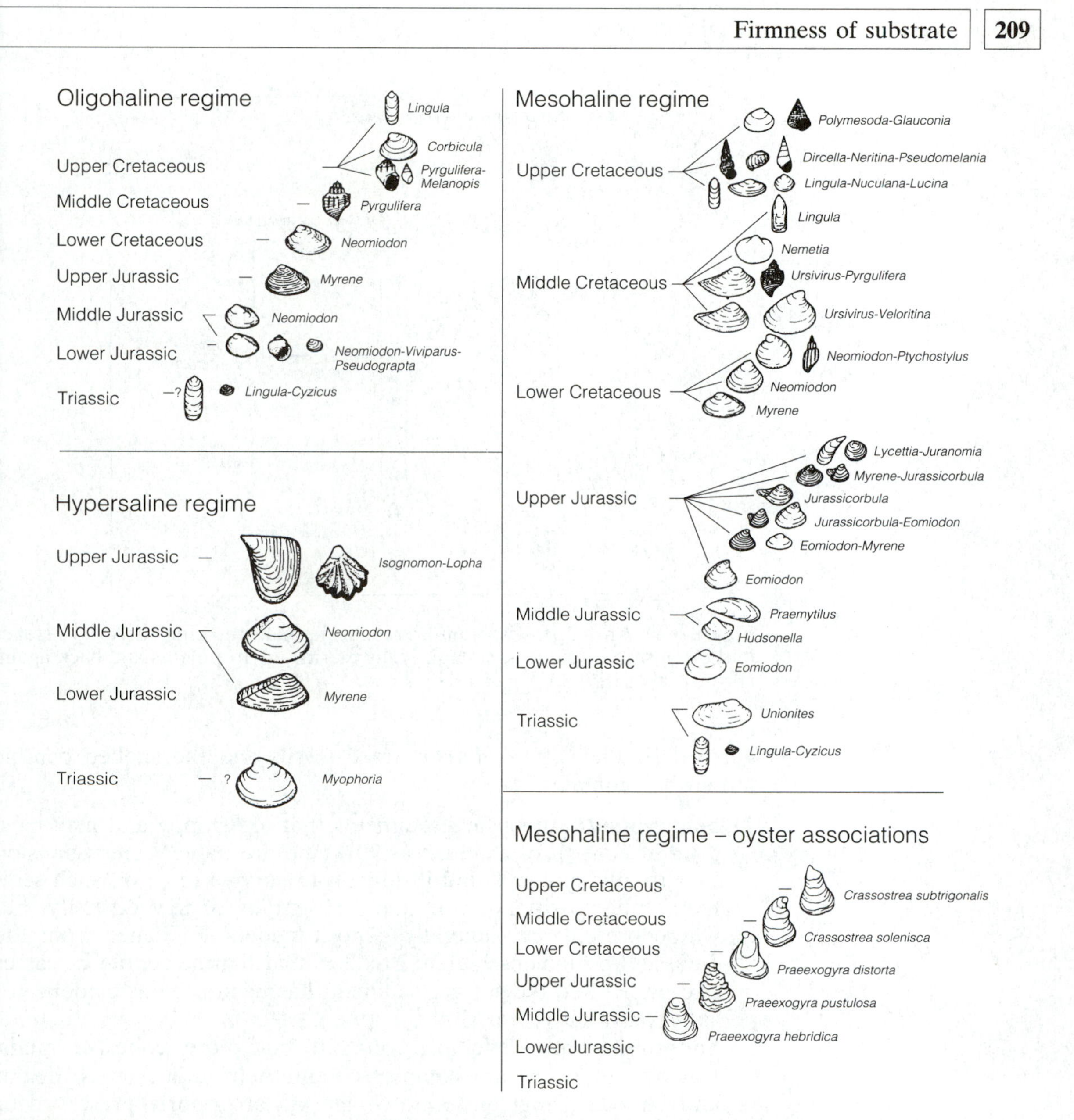

Figure 6.23 Mesozoic faunas related to different salinity regimes. (Modified from Fürsich, 1994.)

6.8 FIRMNESS OF SUBSTRATE

Marine substrates cover a spectrum from watery muds to rock surfaces and each type influences the nature of the biota that colonizes it. Furthermore, the biota commonly changes the nature of the sediment by adding skeletal grains and bioturbation (see McCall and Tevesz, 1982, for an account of these processes). Substrates have been divided into five categories of firmness (Goldring, 1995) and may be distin-

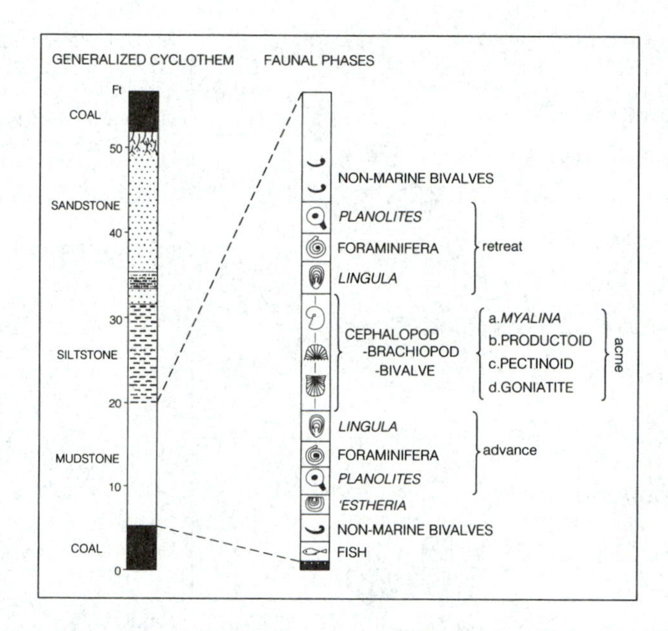

Figure 6.24 An Upper Carboniferous marine band showing successive stages in the transition from freshwater to fully marine environments and back again. (From Calver, 1968.)

guished by the nature of their trace fossils and the shelled benthic fauna they contain:

1. **Soupgrounds** are surface sediments that are sloppy and may have a water content of as much as 90% and are liable to resuspension by currents. Animals find it difficult to survive in or on such sediments and fossils are commonly absent or of low diversity, but where present are generally deposit-feeders. Evidence from the Jurassic Posidonienschiefer (Box 2.2) that marine reptile carcasses had penetrated the surface sediment has been used as evidence of soft, soupy substrates (Martill, 1993).
2. **Softgrounds** are developed on soft but more cohesive muds. Burrows are commonly compressed and their outlines are irregular and blurred. Bioglyphs (scratch marks) are poorly preserved or absent. Benthic faunal diversity tends to be relatively low (Fig. 6.25) and organisms may have adaptations that increase the surface area of the undersurface and so give the shell greater support.
3. **Firmgrounds** are stiff cohesive muds with a substantial supporting strength. Burrows are little compacted and bioglyphs are well preserved. The benthic fauna is relatively diverse and may include surface encrusters.
4. **Loosegrounds** consist of well-sorted silts and sands. If the sediments are thixotropic they may be penetrated by burrowers which may make burrows that collapse behind them or they may make compression burrows where the sediment is forced aside but

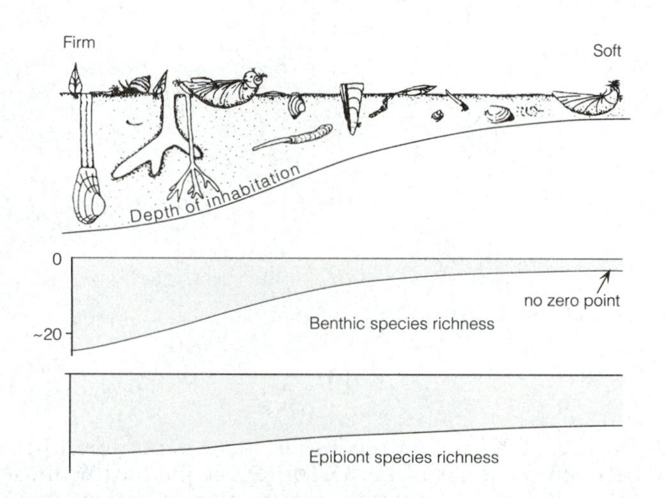

Figure 6.25 Faunal changes in a fine-grained clastic substrate on transition from firm to soft substrates. Epifauna are, from left to right, *Parainoceramus*, serpulid, trochid, *Gryphaea* with diverse epibionts, 'paper pecten', *Gryphaea* with epibionts. Infaunal taxa are, from left to right, *Goniomya*, *Nacaniella*, *Pinna*, *Nucinella*, *Dentalium* and *Palaeonucula*. (From Wignall, 1993.)

maintains a burrow wall. Some burrows may have supported linings, such as in *Ophiomorpha*, which has a wall stabilized by faecal pellets. The substrate is likely to be colonized by a variety of benthic faunas.

5. **Hardgrounds and rockgrounds:** hardgrounds are lithified sediment surfaces that generally form at hiatal horizons, particularly in carbonate environments. They are identified by the presence of organisms encrusting the surface, such as oysters, serpulid worms, sponges corals and crinoid holdfasts and by borings made by bivalves, sponges, bryozoans and endolithic algae (Fig. 2.24). Hardground faunas were sparse until the mid Ordovician when there was rapid adaptive radiation and an increase in diversity. By the Silurian varied biotas had differentiated to occupy the various cavities forming sub-environments within a hard substrate. Rockgrounds have the same substrate properties as hardgrounds but differ in being generally associated with coastal environments where the wave energy of the environment is particularly high and the topography varied.

6.9 SHELL CONCENTRATIONS, SEDIMENTATION RATE AND SEQUENCE STRATIGRAPHY

High rates of deposition in sandstones are most commonly indicated by sedimentary structures such as graded bedding or hummocky cross-stratification, but escape burrows and protrusive spreite in trace fossils (Fig. 6.26) may provide additional evidence. In mudstones vertical

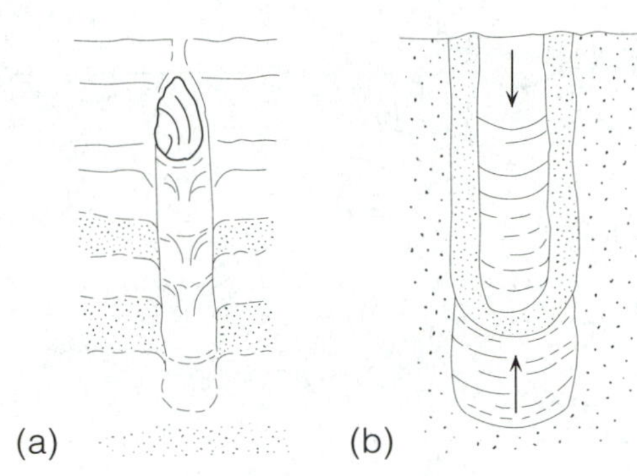

Figure 6.26 (a) Escape burrow in reponse to rapid deposition. (b) A protrusive spreite between the limbs of the U, formed as the burrow moved downwards into the sediment. A retrusive spreite formed below the U as the organism moved upwards in response to accumulating sediment.

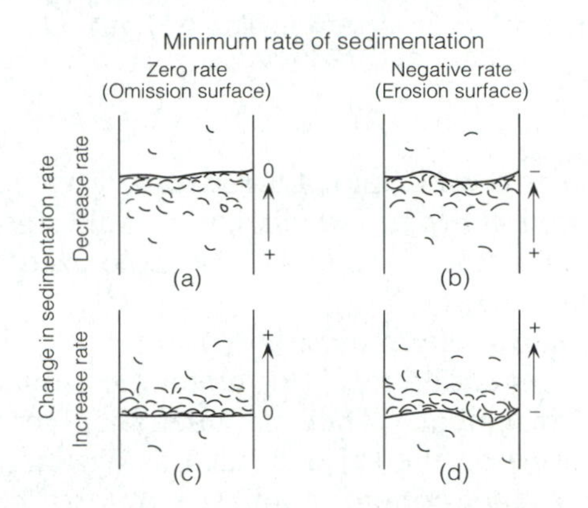

Figure 6.27 Changes in density of shell packing reflecting changes in sediment rate. (a) Showing increase in density upwards as sediment rate decreases. An omission surface forms where there is zero sedimentation. (b) Showing increase of sedimentation upwards. (c) Showing decrease of sedimentation up to an erosion surface. (d) Shows increase of sedimentation above an erosion surface. (From Kidwell, 1985.)

changes in the density at which shells are concentrated in the sediment may reflect changes in sedimentation rate (Fig. 6.27)

Event concentrations, composite concentrations and hiatal concentrations (Chapter 3, section 3.9) reflect different time spans over which the shell concentrations form. An event bed might accumulate in a matter of minutes or hours, while an hiatal bed might form over tens or hundreds of thousands of years. Event beds are commonly present throughout most of a sequence and reflect sporadically repeated episodes of rapid sedimentation. Hiatal horizons reflect long periods of slow accumulation and are important in sequence stratigraphy because they help to identify two key horizons, the flooding surface and the maximum flooding surface, within a sequence (Fig. 6.28). At

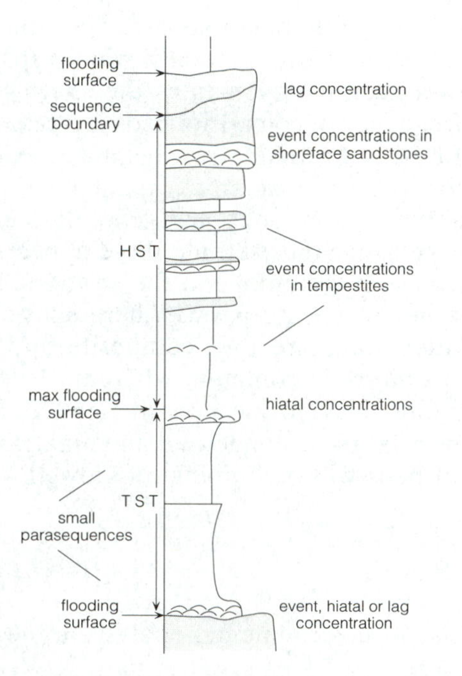

Figure 6.28 Shell concentrations within a sequence. Hiatal concentrations may identify an initial flooding surface or, more commonly a maximum flooding surface. Event beds are typical of the highstand systems tract. Lag deposits may form on the erosional surface at a sequence boundary.

the start of new cycle of deposition, rates of sedimentation are reduced as sea-level starts to rise and commonly floods across deltas and up estuaries where sediment is stored, so reducing the amount that reaches the open sea. As a consequence there is commonly a sharp facies change across the **initial flooding surface** and a rather condensed sequence of transgressive sediments may be formed (the transgressive systems tract); shells may be concentrated on or a little above the flooding surface. The second important level of shell concentrations is at the **maximum flooding surface,** which reflects the period when sea-level is rising most rapidly and therefore coastal retreat is fastest. A large volume of sediment is stored in estuaries and, as a consequence, offshore sedimentation is markedly reduced so that a condensed horizon may result. This is likely to be marked by increased bioturbation, or a hiatal shell concentration or by bone in which phosphate has replaced fossil or formed concretions in the absence of land-derived iron. On initial flooding surfaces and some maximum flooding surfaces where iron was more available, chamosite or berthierine ooids are more likely to be found.

The relative proportion of event, composite and hiatal concentrations in a sequence are an indication of overall sedimentation rates. With high sedimentation rates event beds are rapidly buried and covered by relatively thick accumulations of fair-weather sediment and consequently the event beds are generally widely spaced within the sequence. With lower sedimentation rates the spacing of event beds decreases and the frequency of composite beds increases as the chances that separate event beds becoming amalgamated becomes greater. When sedimentation rates are minimal, hiatal beds result. Because overall sediment-accumulation rates are generally related to subsidence rates and hence to tectonic setting, there appears to be a broad relationship between basin tectonics and the stratigraphic distribution of shell concentrations. In epeiric basins where subsidence rates and sediment accumulation rates are low, composite and hiatal concentrations are likely to be more common, whereas at the other end of the spectrum, in rapidly subsiding basins with high sedimentation rates, fossil concentrations will not be condensed and amalgamated, so more widely spaced event beds will predominate (Kidwell, 1988, 1993).

6.10 SUMMARY POINTS

- The starting point in environmental analysis is the evidence from lithofacies.
- Palaeoecological evidence is particularly helpful in fine-grained sediments and in oxygen-poor and high or low salinity environments.
- Identification of depth-related benthic assemblages on marine shelves is based on the biofacies present, trace fossils, the diversity and abundance of the fauna and its taphonomic state.
- Estimates of the bathymetric depth of benthic assemblages have been related to the base of the photic zone, identified by the disappearance of benthic algae and effective storm wave-base, identified by the disappearance of wave-formed structures. This suggests a depth of about 50–60 m for the depth limit of BA 4. The depth limit of BA 2 appears to be about 10–15 m, based on the depth of fair-weather wave-base.
- The relative abundances of pelagic, benthic calcareous and benthic agglutinated foraminifera provide the best evidence of relative depth in deep water, from the Mesozoic onwards.
- Carbonate environments are partly created by the organisms that inhabit them, so there is a close relationship between fossil biota and lithofacies. Carbonate environments tend to be more patchy than clastic ones and the distribution of benthic assemblages is less easy to discern. The nature of reefs, carbonate mud mounds and intertidal environments are, however, determined to a large degree by the biota they contain.
- Oxygen deficient environments are recognized by their low faunal diversity, the particular types of benthic organisms they contain and the general lack of bioturbation. Care must be taken to distinguish them from environments with a soft substrate.

- Low- and high-salinity environments are recognized by the low diversity of the fauna, the absence of stenohaline taxa, and in Mesozoic and younger rocks, by comparison of the biota with brackish-water modern biotas.
- The firmness of the substrate in an environment can be assessed by the style and preservation of infaunal trace fossils and by adaptations of benthic animals to a soft or hard sediment surface.
- The nature and frequency of different types of shell concentrations provides evidence about rates of deposition and helps to identify key hiatal surfaces in stratigraphic sequences.

6.11 FURTHER READING

Bosence, D.W.J. and Allison, P.A. (eds) 1995. *Marine Palaeoenvironmental Analysis from Fossils*. Geological Society Special Publication, 83. 272 pp.

A collection of papers reviewing different limiting factors in palaeoecology.

Goldring, R. 1991. *Fossils in the Field*. Longman, Harlow. 218 pp.

An account of the palaeontological information to be observed and collected in the field.

McKerrow, W.S. (ed.) 1978. *The Ecology of Fossils*. Duckworth, London. 338 pp.

Well-Presented reconstructions of a representative range of Phanerozoic communities with brief accounts of the contemporary faunas.

Reading, H.G. (ed.) 1986. *Sedimentary Environments and Facies* (2nd edition). Blackwell Scientific Publications, Oxford. 615 pp.

Tucker, E.M. and Wright, V.P. 1990. *Carbonate Sedimentology*. Blackwell Scientific Publications, Oxford. 482 pp.

These two books provide a very thorough background in sedimentary environments and facies.

6.12 REFERENCES

Ager, D.V. 1965. The adaptation of Mesozoic brachiopods to different environments. *Palaeogeography, Palaeoclimatology, Palaeoecology* 1, 143–172.

Andrews, J.E. and Walton, W. 1990. Depositional environments within Middle Jurassic oyster-dominated lagoons: an integrated litho-, bio-, and palynofacies study of the Duntulm Formation (Great Estuarine Group, Inner Hebrides). *Transactions of the Royal Society of Edinburgh: Earth Sciences* 81, 1–22.

Brasier, M.D. 1980. *Microfossils*. Chapman & Hall, London. 193 pp.

Brett, C.E., Boucot, A.J. and Jones, B. 1993. Absolute depth of Silurian benthic assemblages. *Lethaia* 26, 23–40.

Calver, M.A. 1968. Distribution of Westphalian marine faunas in Northern England and adjoining areas. *Proceedings of the Yorkshire Geological Society* 37, 1–72.

Corliss, B.H. and Chen, C. 1988. Morphotype patterns of Norwegian Sea deep-sea benthic foraminifera and ecological implications. *Geology* 16, 716–719.

Demaison, G.J. and Moore, G.T. 1980. Anoxic environments and oil source bed genesis. *American Association of Petroleum Geologists Bulletin 64*, 1179–1209.

Dunham, R.J. 1962. Classification of Carbonate Rocks According to Depositional Texture. *In:* Ham, W.E. (ed.) *Classification of carbonate Rocks*. American Association of Petroleum Geologists, Tulsa, 108–121.

Fortey, R.A. 1975. Early Ordovician trilobite communities. *Fossils and Strata* 4, 331–352.

Fürsich, F.T. 1994. Palaeoecology and evolution of Mesozoic salinity-controlled benthic macroinvertebrate associations. *Lethaia* 26, 327–346.

Gibson, T.G. 1989. Planktonic benthonic foraminiferal ratios: Patterns and Tertiary applicability. *Marine Micropaleontology* 15, 29–52.

Goldring, R. 1995. Organisms and the substrate: response and effect. *In:* Bosence, D.W.J. and Allison, P.A. (eds), *Marine Palaeoenvironmental Analysis from Fossils*. Geological Society Special Publication 83, The Geological Society, Bath.

Hallam, A. 1976. Stratigraphic distribution and ecology of European Jurassic bivalves. *Lethaia* 9, 245–259.

House, M.R. 1975. Faunas and time in the marine Devonian: *Proceedings of the Yorkshire Geological Society* 40, 459–490.

Hudson, J.D. 1963. The ecology and stratigraphic distribution of the invertebrate fauna of the Great Estuarine Series. *Palaeontology* 6, 327–348.

Irwin, M.L. 1965. General theory of epeiric clear water sedimentation. *American Association of Petroleum Geologists Bulletin* 49, 445–459.

Johnson, M.E., Kaljo, D. and Rong, J.-Y. 1991. Silurian eustasy. *Special Papers in Palaeontology* 44, 145–163.

Kidwell, S.M. 1988. Taphonomic comparison of passive and active continental margins: Neogene shell beds of the Atlantic coastal plain and the northern Gulf of California. *Palaeogeography, Palaeoclimatology, Palaeoecology* 63, 201–223.

Kidwell, S.M. 1993. Taphonomic expression of sedimentary hiatus: Field observations on bioclastic concentrations and sequence anatomy in low, moderate and high subsidence settings. *Geologische Rundschau* 82, 189–202.

Kidwell, S.M. and Brenchley, P.J. 1994. Patterns in bioclastic accumulation through the Phanerozoic: Changes in input or in destruction? *Geology* 22, 1139–1143.

Lees, A. and Buller, A.T. 1972. Modern temperate water and warm water shelf carbonate sediments contrasted. *Marine Geology* 13, 1767–1773.

McCall, P.L. and Tevesz, M.J.S. 1982. *Animal-Sediment Relations: The Biogenic Alteration of Sediments*. Plenum Press, New York. 336 pp.

McKerrow, W.S. 1978. *The Ecology of Fossils*. Duckworth, London. 383 pp.

Martill, D.M. 1993. Soupy substrates: a medium for the exceptional preservation of ichthyosaurs of the Posidonia Shale (Lower Jurassic) of Germany. *Kaupia* 2, 77–97.

Murray, J.W. 1991. *Ecology and Palaeoecology of Benthic Foraminifera*. Longman Scientific and Technical, Harlow. 397 pp.

Oschmann, W. 1988. Upper Kimmeridgian and Portlandian marine macrobenthic associations from southern England and northern France. *Facies* 18, 49–82.

Pedersen, T.F. and Calvert, S.E. 1990. Anoxia vs. productivity: What controls the formation of organic-carbon-rich sediments and sedimentary rocks. *American Association of Petroleum Geologists Bulletin* 74, 454–466.

Pollard, J.E., Goldring, R. and Buck, S.G. 1993. Ichnofabrics containing *Ophiomorpha*. Significance in shallow-water facies interpretation. *Journal of the Geological Society* 150, 149–164.

Seilacher, A. 1967. Bathymetry of trace fossils. *Marine Geology* 5, 413–428.

Tucker, E.M. and Wright, V.P. 1990. *Carbonate Sedimentology*. Blackwell Scientific Publications, Oxford. 482 pp.

van Morkhoven, F.P.C.M., Berggren, W.A. and Edwards, A.S. 1986. Cenozoic Cosmopolitan Deep-Water Benthic Foraminifera. *Bulletin des Centres de Recherches Exploration-Production Elf-Aquitaine, Pau*, Memoir 11.

Wignall, P.B. 1994. *Black Shales*. Clarendon Press, Oxford. 127 pp.

Wignall, P.B. and Hallam, A. 1991. Biofacies, stratigraphic distribution and depositional models of British onshore Jurassic black shales. *In:* Tyson, R.V. and Pearson, T.H. (eds) *Modern and Ancient Shelf Anoxia*. Geological Society Special Publication 58, Geological Society, Bath, 291–309.

Ziegler, A.M. 1965. Silurian marine communities and their significance. *Nature* 207, 270–272.

Ziegler, A.M., Cocks, L.R.M. and Bambach, R.K. 1968. The composition and structure of Lower Silurian marine communities. *Lethaia* 1, 1–27.

7 Populations and communities

Populations and communities represent successive levels of ecological organization. A population is composed of individuals of a species that lived together. Differences in the way species utilize energy resources in an ecosystem are reflected in the abundance of individuals in a population, its age structure and its spatial distribution. Communities are an association of species in a particular habitat. Communities are organized according to the way the organisms obtain their food (their trophic structure) and in their competition for space; they show differences in their food chains, the proportions of different feeding types and the adaptations to feeding. The reconstruction of past communities from fossil evidence is described and environmental and evolutionary changes in community structure will be reviewed. Variations in community diversity are shown to be related in general to differences in the depth of marine waters and latitude.

7.1 POPULATION STRUCTURE AND DYNAMICS

The **population** is the fundamental unit in both ecological and taxonomic analysis. The origins, development and long-term change in communities and biotas are initiated at the population level and fluctuations in the abundance of individual species and groups of species, in fact, can trigger large-scale changes in the biosphere as a whole. Ecologists have used so-called life tables and survivorship curves to determine population structure and monitor the dynamics of living populations, and population studies form a major part of many ecological investigations; however, since fossil assemblages are rarely *in situ*, having been modified by some form of sorting, results from similar analyses of fossil material must be treated with considerable caution. The vast majority of fossil assemblages have suffered both taphonomic loss and time averaging (see Chapter 3). Nevertheless, the analysis of size-frequency distributions has impacted on both biological and taphonomic interpretations of the origins of fossil assemblages and fossil concentrates (Boucot, 1953; Johnson, 1960). Two types of plot, mortality and survivorship curves, have proved useful in population analysis. However, it is essential that we understand first the types of assemblage that occur naturally in the fossil record.

7.2 TYPES OF POPULATIONS

Usually fossil populations and communities have suffered a variety of post-mortem modifications. The exceptions are Lagerstätten deposits such as the middle Cambrian Burgess Shale, the lower Devonian Rhynie Chert and the Tertiary Grubbe Messel (Chapter 3). In these, exceptional taphonomic conditions have preserved substantial parts of the life assemblages replete with soft parts and soft-bodied organisms. The life assemblage or **biocoenosis** is thus rare in the fossil record; the death assemblage or **taphocoenosis** is much more common. Both types of assemblage have been described through the eyes of the inhabitants of the Roman city of Pompeii at the time of the eruption of Vesuvius (Ager, 1963). The living population at the time of the eruption on 24 August AD 79 would have included all manner of human inhabitants together with both domestic and wild animals, pets and vegetation. The biocoenosis recorded a normal day in the life of Pompeii and the human population would have included a typical demographic range for Roman populations of the time. A second assemblage, the **catastrophic** assemblage, was entombed by volcanic ash during the eruption. The more mobile elements of the population, such as the youngsters, probably escaped; nevertheless this population, a thanatocoenosis, is fairly close to the biocoenosis. A third assemblage is very different. The Necropolis assemblage, a taphocoenosis, is found in the graveyard of Pompeii and is biased towards infants and the elderly.

Ager's catastrophic assemblage has also been described as a **census** population, preserving an entire living unit more or less *in situ* (Hallam, 1972). Census populations tend to characterize very narrow stratigraphical intervals such as on a single bedding plane and consist of relatively few species. Normal populations, on the other hand, are time-averaged, accumulating over a number of years or even decades. The Pompeii graveyard accumulated a concentration of human bones over decades.

Population studies of fossil organisms are most reliable when applied to census assemblages, which are commonly identified by their size-frequency distribution (Craig and Hallam, 1963; Hallam, 1967) together with a range of other independent lines of evidence.

7.2.1 Size-frequency analyses

The analysis of **size-frequency curves** and **histograms** forms the starting point for most palaeoecological investigations at the population level. Class intervals are chosen for size groups and a frequency table is constructed for all the size intervals. Each class can ideally be equated with an absolute age unit, although in many fossil populations an arbitrary set of age units is more applicable. The data may be plotted as size-frequency histograms, polygons or cumulative frequency polygons (see Box 7.1). Three main types of frequency histograms have been recognized. The right- or positively-skewed curves (Fig. 7.1a) suggest high infant mortality and subsequent relatively lower mortality; this situation is typical of most invertebrate populations where there is a high initial spatfall and high mortality as the larval shells struggle to establish themselves, exposed to a variety of physical stresses and predators. In many living populations and the majority of fossil occurrences the smallest larval shells are rarely preserved, generating a positive displacement of the curve along the X- or size axis. The vertical extension of the distribution as a peak or kurtosis reflects the absolute numbers of specimens in the modal size range.

A normal, Gaussian or bell-shaped curve (Fig. 7.1b) suggests high mortality in the mid-late life group of a population. Under natural conditions this scenario rarely occurs. Normal distributions are more likely to be generated by mechanical sorting, chemical solution and compaction, or with selective predation; the smaller factions of the population may be selectively removed to positively displace the modal value of the size and by implication the age of the population. Normal distributions have, additionally, been recognized in populations exhibiting so-called steady state growth. Rapid intial growth pushes young growth stages into the main part of the distribution where growth slows; this equilibrium distribution, observed in living decapod populations, has also been inferred for assemblages of Ordovician trilobites lacking abundant young moult phases (Sheldon, 1988).

Left- or negatively-skewed distributions (Fig. 7.1c) indicate high senile mortality preceded by relatively low rates of mortality; human populations in most of western Europe, supported by advanced health systems and care for the elderly, exhibit left-skewed distributions. Constant mortality rates, however, are signalled by curves with a slight positive skew and a low-moderate kurtosis.

A number of other curves have been described from both living and fossil populations. Multimodal distributions (Fig. 7.1d) may reflect seasonal spawning patterns in a persistent population with periodic recruitment of fresh cohorts attaching themselves to the overall population structure. The broad developing profile of the population can be assessed by constructing a further size frequency polygon (see Box 7.1) based on the joining peak of each cohort to form a smoothed curve.

Multimodal-peaked distributions (Fig. 7.1e) may also be produced by arthropod ecdysis or moulting. Fossils arthropods such as the microscopic ostracodes and the calcareous-shelled trilobites developed through a series of moult stages; each peak relates to one of a sequence of moult phases. Usually, however, there is a decrease in frequency with increasing size or age, as less individuals survive a life of periodic moulting. A curve joining the peaks thus generally shows a downward trend.

In summary, size-frequency analysis can help differentiate between life, death and transported assemblages, detect seasonal mortality and spawning patterns and investigate the frequency of ecdysis in arthropod populations. This type of analysis, however, can form the basis for much more informative studies of population dynamics when

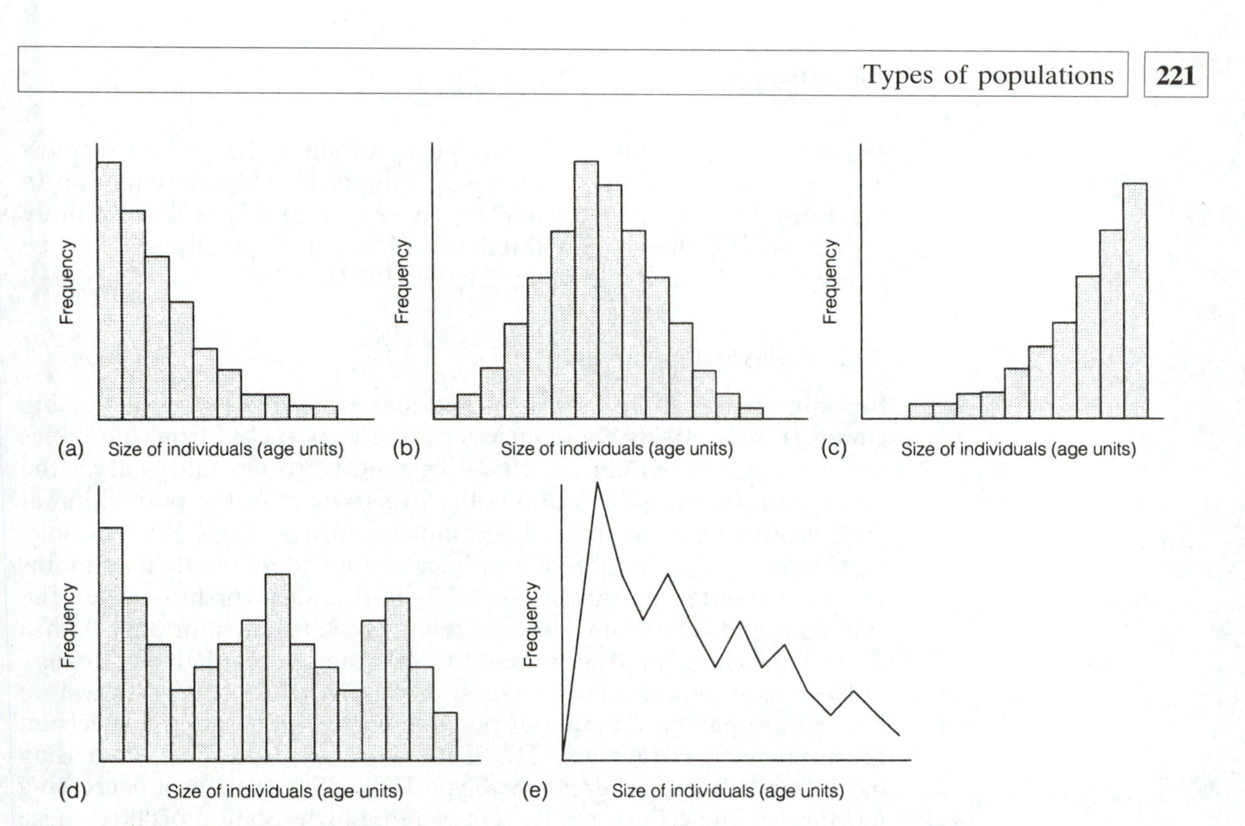

Figure 7.1 (a) Schematic size-frequency histogram; right (positively) skewed, typical of many invertebrate populations with high infant mortality (original). (b) Schematic size-frequency histogram; Normal (Gaussian) distribution, typical of steady-state populations or transported assemblages (original). (c) Schematic size-frequency histogram; left (negatively) skewed, typical of high senile mortality (original). (d) Schematic size-frequency histogram; multi-modal distribution, typical of populations with seasonal spawning patterns (original). (e) Schematic size-frequency histogram; multi-modal distribution with decreasing peak-height, typical of populations growing by ecdysis (original).

the age together with the growth and mortality of rates of fossil populations can be established.

7.2.2 Age of fossil specimens

It is difficult to assess the absolute age of most fossil specimens, except for organisms with annual growth rings, such as trees, corals and a number of other invertebrates with marked seasonal growth patterns. The life span of a number of arthropod taxa, such as ostracodes, has been calculated with reference to the number of moult phases or instars represented in a population. The ages of fossil hominids have been estimated on the basis of the eruption of teeth, whereas tooth wear has provided estimates of the age of some other vertebrate groups, such as cave bears.

For most fossil populations a relationship between size and age must be assumed. Growth can be modelled by a linear equation such as

D (size) = S (constant) * [T (time) + 1], or since the growth of most invertebrates decelerates with age, a simple logarithmic equation in the form D (size) = S (constant) * ln [T (time) + 1] is usually more applicable (Levington and Bambach 1970); alternatively size can be plotted on a logarithmic scale (Thayer 1977).

7.2.3 Survivorship curves

Mortality patterns are best displayed and examined with **survivorship curves** (Cadée, 1988); these curves can be constructed from life tables of fossil organisms (Kurtén, 1954). In contrast to mortality curves the survivorship curve plots the number of survivors in the population at each growth stage or defined age unit or growth stage. For example, the frequency distribution of mortality among adult woolly mammoths can be converted as shown (Fig. 7.2) into a survivorship curve. The raw data and histogram indicate relatively constant mortality with a slight increase of mortality rates towards the end of their life span.

The three representative types of survivorship curves shown in Fig. 7.3 are part of a family of possible curves, illustrating a spectrum of survivorship strategies. The Type I curve depicts an increasing mortality with age, whereas the Type III curve simulates a decreasing mortality with age, essentially representing high infant mortality; these two curves are bisected by the Type II curve tracking constant mortality through the ontogeny of the population. Palaeoenvironmental assumptions have been drawn from the shapes of survivorship curves. Brachiopod populations associated with soft substrates from Ordovician, Silurian and upper Carboniferous strata (Richards and Bambach, 1975) together with faunas in the Danish Chalk (Surlyk, 1974) show high infant mortality, with Type III curves, probably due to difficulties with settlement and function on a soft seabed. Populations associated with hard substrates generate Type I curves, generally indicative of more favourable conditions throughout ontogeny. A Type II curve is recorded for some large terebratulids that may have grown rapidly while attached prior to a recumbent phase with the commissure above the sediment–water interface (Surlyk, 1972). Similar patterns have been described for the terebratulide brachiopod *Tichosina* from the Pleistocene of Jamaica (Harper *et al.*, 1995).

The construction and analysis of survivorship curves can establish the maximum age of a population, its growth and mortality rates; these parameters and their fluctuations between comparable populations are often of considerable environmental significance. Nevertheless, analyses of survivorship curves do assume a direct relationship between size and age together with a constant population structure (Thayer, 1977).

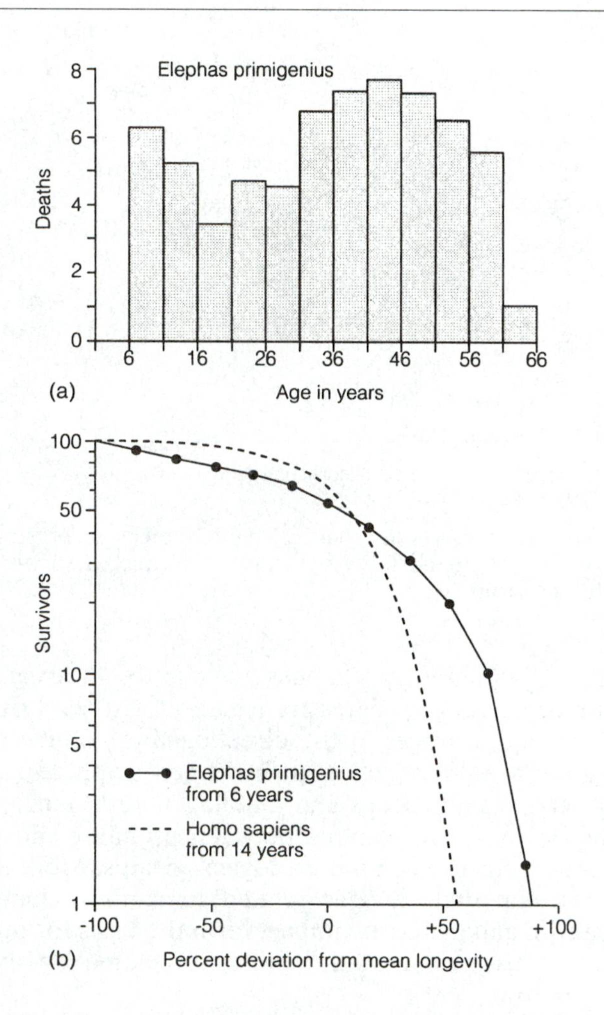

Figure 7.2 Mortality and survivorship curves for adult mammoths (adapted from Kurtén, 1954): (a) Individual age at death for 67 mammoths based on wear of dentition converted to a mortality frequency distribution. (b) Survivorship curves for adult mammoths and adult male humans for comparison.

7.3 VARIATION IN POPULATIONS

The vast majority of populations not only show some morphological variation controlled by ontogenetic, genetic and phenotypic factors (see Chapter 5), but also large variations in **population size**. For example, lemmings swarm over parts of Arctic North America during spectacular 3–4 year growth cycles. At the local level, marine benthic populations may fluctuate due to a variety of factors such as unsuitable substrates, overcrowding and overproduction of spat; moreover the migration of populations into less suitable environments or niches may be responsible for population crashes. There are many more widespread physical, chemical and biological changes that may affect the

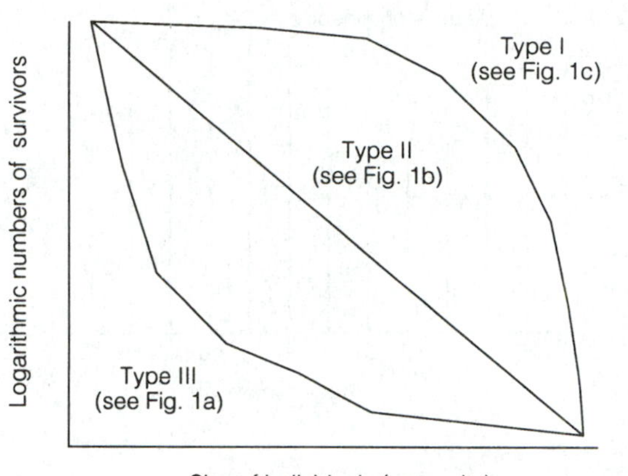

Figure 7.3 Schematic survivorship curves; type I tracks increasing mortality with age, type II constant mortality with age and type III shows decreasing mortality with age (original).

development of populations: changes in salinity in basins, due to increased surface runoff or progressive isolation, sea-level rise and fall, temperature change, changes in the direction and volume of oceanic currents together with production of hydrogen sulphide and natural and artificial oil spills and seeps and dinoflagellate blooms. Moreover biological interactions, for example between predator and prey, may causes population crashes in both ecological groups. More global and long-term environmental changes related to climate change, earthquakes, volcanism and meteorite impact form the basis for mass extinctions and the investigation of the causes of extinction events (see Chapters 9 and 10).

7.4 SPATIAL DISTRIBUTIONS

Populations generally occur as random, regular or clumped distributions. Randomly distributed individuals in a population are located independently from all other members of the population. This type of distribution suggests there is no overall biological or environmental control on the origin and development of the population; this situation is rare in both living and fossil assemblages where environments are rarely uniform and the individual members of a population do exhibit some mutual interactions. Regular distributions are more common, particularly in nonmarine environments where space is developed between individuals by competition or by efficient exploitation of resources. Clumped distributions are much more common in both marine and nonmarine environments.

Box 7.1 Population analysis of a sample of the fossil brachiopod *Dielasma* from the Permian of northeast England

The smooth terebratulide brachiopod *Dielasma* is very common in dolomites and limestones associated with Permian reef deposits in the north of England. Do the samples approximate to living populations and is there a distinctive pattern of growth? Sagittal length was measured on large samples of specimens from different parts of the reef complex (Hollingworth and Pettigrew, 1988). Figure 7.4 shows the population structure relating to a sample from one of the *in situ* nests of *Dielasma* within the reef complex. The size-frequency histograms, polygon, cumulative frequency polygon and survivorship curves all indicate high juvenile mortality, typical of an *in situ* invertebrate population.

The survivorship curve approximates to the Type III pattern, high infant mortality and decreasing mortality with increasing age.

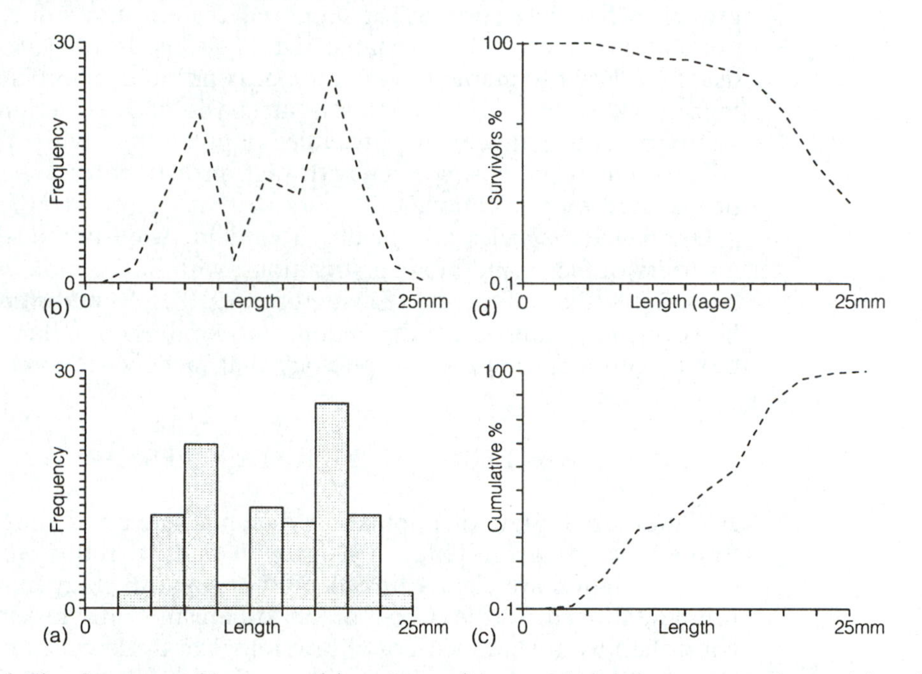

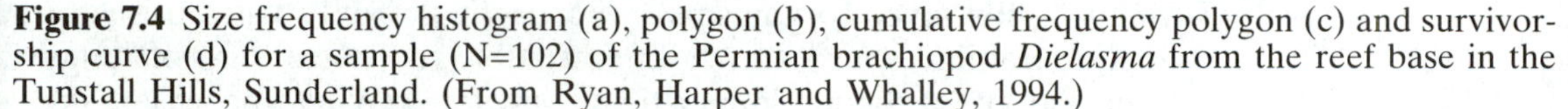

Figure 7.4 Size frequency histogram (a), polygon (b), cumulative frequency polygon (c) and survivorship curve (d) for a sample (N=102) of the Permian brachiopod *Dielasma* from the reef base in the Tunstall Hills, Sunderland. (From Ryan, Harper and Whalley, 1994.)

7.5 OPPORTUNIST AND EQUILIBRIUM SPECIES

There are, of course, correlations between life style, habitat and the life history of an organism displayed in size-frequency histograms and survivorship curves. Two end members of a spectrum of life histories have been identified, on the basis of population structure, and its relationship to the stability of habitats. Species that matured early and produced small but numerous offspring and died young have been labelled **'*r* strategists'**; their high rate of intrinsic increase developed in density-independent conditions. **'*K* strategists'** are long-lived species with low reproductive rates evolving in density-dependent environments. Populations have been described and modelled by the logistic equation which relates the rate of change in size of a population to its carrying capacity and the natural increase in size of the population:

$$dN/dt = rN[(K-N)/K]$$

K = carrying capacity of the population or upper limit of population size, N = actual population size, r = intrinsic rate of increase, t = unit of time.

There are thus two extreme conditions. When N approaches K the rate of population growth will tend towards zero. Thus population growth will slowly approach a supposed equilibrium value where the population will establish a stable size. This mode of development is usually related to stable environmental conditions where species with high K values are selected; this is the basis of K-selection or the K strategy. Alternatively in unstable populations, with presumably adverse environments, species with high growth rates are prominent, propagated by r-selection.

Opportunist species are usually abundant, widespread, dominating a variety of facies and biotic associations, with initial rapid growth and a variable morphology (Levington, 1970). **Equilibrium species**, however, are more facies dependent, moderately abundant in diverse biotas with a specialized morphology that lacks variation.

7.6 LIFE STRATEGIES AND TRADE-OFFS

An alternative view of population dynamics involves the concept of **'trade-offs'** (Stearns, 1992). The growth and reproductive strategies of a population are clearly linked. Rather than invoking external environmental factors, the track of an organism's life history may be controlled by a whole variety of physiological trade-offs. For example, many short-lived organisms, with right-skewed distributions trade considerable resources into reproduction at the expense of longevity. On the other hand, longer-lived species, having more left-skewed distributions, put more energy into physiological processes, such as repair of damaged organs, associated with a longer life span, devoting correspondingly less energy to reproduction. These actions can have long-term evolutionary consequences.

7.7 LONG-TERM SURVIVORSHIP PATTERNS

The study techniques of population survivorship have some application to longer-term patterns of biological change in higher taxa (Van Valen,

1973). For these analyses large data sets are necessary together with tight taxonomy and a complete fossil record. Some of the problems of collecting sufficient data may be overcome by targeting specific cohorts and documenting their decline with time. **Cohort analysis**, based on the timing of origination and extinction of over 17 000 taxa through the Phanerozoic, suggests an average species duration of about 11 million years (Raup, 1978). Cohorts for each of the geological periods from the Cambrian to Tertiary were monitored through geological time. The ten cohorts were plotted as survivorship curves. The generic survivorship curves for the pre-Jurassic cohorts were uniformly concave; these curves were modified by the end Permian and end Cretaceous extinction events.

7.8 COMMUNITY STRUCTURE

Communities represent the next level up from species populations in the ecological hierarchy. They are characterized by a repetitive association of species and a similarity in species dominance between different samples. Species within a community are associated partly because they have similar tolerances of physical aspects of the environment such as temperature, salinity or oxygen and partly because they interact with one another through a food chain or through complementary niche requirements, e.g. one gains shelter from another. Communities are the biotic part of an **ecosystem,** which embraces the community and its environment. The degree of organization and interaction within communities has been a matter of prolonged debate. There is little doubt that complex ecosystems such as tropical rain forests or reefs have many interactions amongst their component species. The development of a reef is influenced by temperature, salinity, turbidity and nutrient supply, but there is a complex food web within a reef and dominant reef constructors create many sub-environments which are the home of other animals with specific niches (Fig. 7.5).

Animals such as barnacles and prawns have become intimately associated with particular species of coral and a multitude of fish species have developed territorial and feeding behaviours associated with particular parts of the reef. Each level of the food chain is represented by many species drawn from many major taxa; suspension-feeders may include corals, crinoids, molluscs, bryozoans and brachiopods, while carnivores include, amongst others, gastropods, starfish and a host of fishes. In contrast to complex communities found in reefs, some communities that occupy uniform parts of the sea-floor form a sparse cover (Fig. 7.6) and may have short and rather generalized food chains with few biotic interactions.

The distribution of marine communities is broadly related to the marine landscape of the sea-floor, which includes its topography, depth and type of substrate. Communities living on the relatively flat muddy

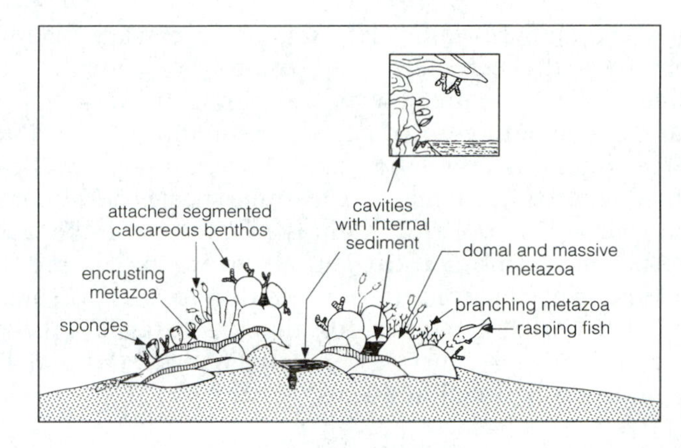

Figure 7.5 A reef community with high density and diversity. There are many interactions between species as they compete for space and resources and fit within the complex food web. (From James, 1984.)

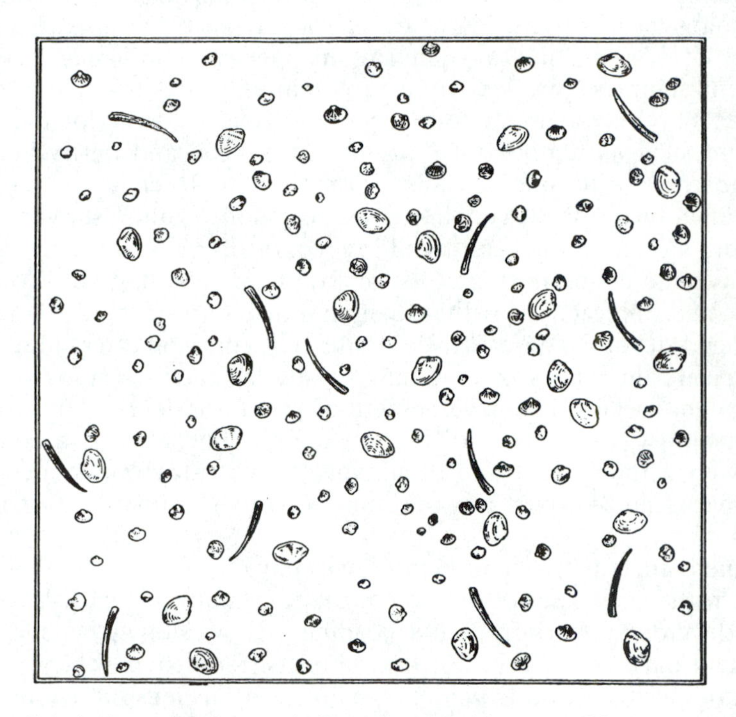

Figure 7.6 A low density and diversity level-bottom *Macoma incongrua* community in the northwest Pacific. The species are *Macoma incongrua*, *Cardium hungerfordi* and *Dentalium octangulatum*. (From Thorson, 1957.)

or sandy shelves were characterized as 'level bottom communities' by Petersen, who, in a series of classic papers in the early part of this century described the distribution of marine communities in the Baltic Sea (see Fig.7.7 and Thorson,1957 for a summary and discussion). Each

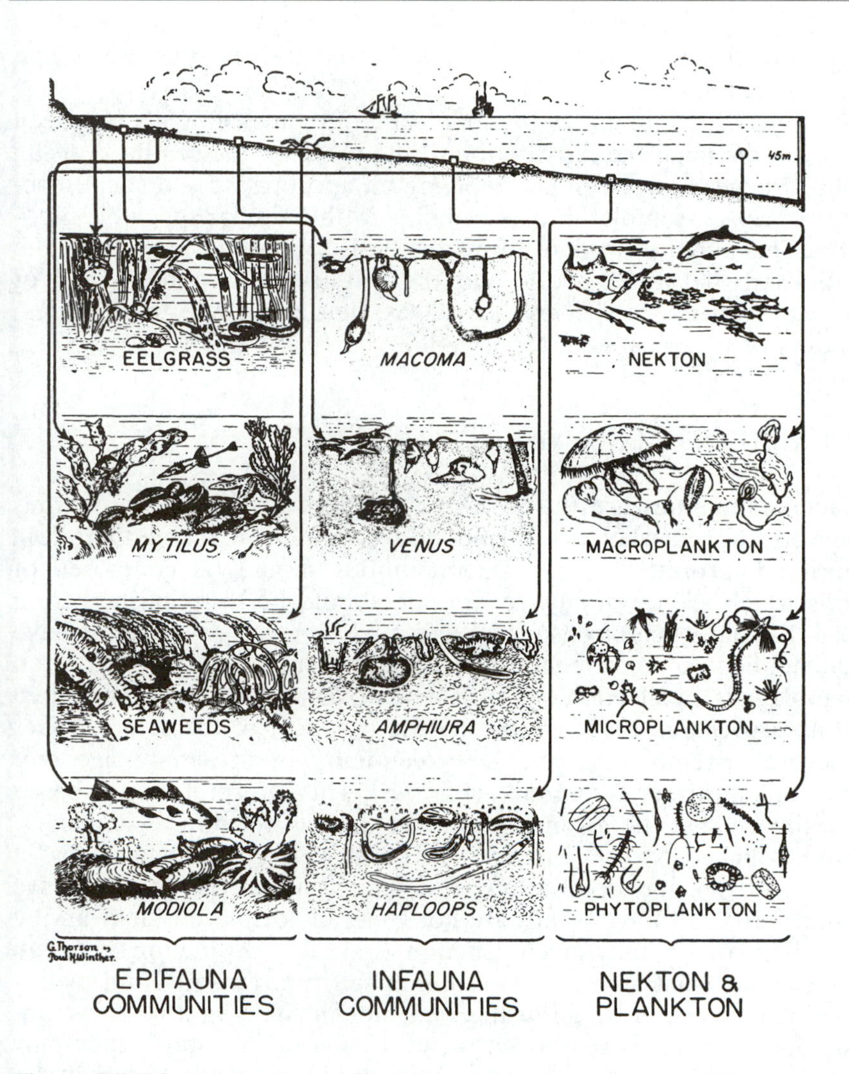

Figure 7.7 Diagrammatic profile with representative communities across the Danish Sound from the coast of Scotland towards Kullen. (From Thorson, 1957.)

of these temperate communities, dominated by bivalves, echinoids and polychaete worms commonly cover large areas of the sea-floor and intergrade with one another, but nevertheless show a clear zonation with changing depth. In contrast, amongst the heterogeneous environments of rocky coastlines, communities tend to be localized and to include sub-communities where there are local differences in the submarine terrain. For example, a study of a rocky-shore terebratulide community in the Bay of Fundy showed three sub-communities, one related to sheltered cavities, a second to exposed rock faces and a third to the upper surfaces of rocks (Noble *et al.*, 1976). Reefs are another example of communities associated with a heterogeneous environment, but it is largely the reefs themselves that create the varied landscape.

Analysis of communities in modern marine environments has been based on mapping the distribution of the biota on the sea-floor, using grabs, dredges and cores and applying an appropriate statistical technique to cluster samples into associations with a similar species composition. This initial procedure forms the basis for further analysis of the relationship between species and the physical environment and of interactions between species themselves, that contribute to an understanding of the ecosystem as a whole.

7.9 PALAEOCOMMUNITIES

Palaeocommunities are the fossilized residues of living communities after the processes of decay and destruction have taken their toll. Samples interpreted as palaeocommunities should be composed of fossils which are essentially *in situ* and should be distinguished from those that have been transported and mixed. A variety of terms have been applied to fossil communities and their component parts; an **assemblage** is a collection of fossils made from a single horizon or bed and an **association** is recognized by the presence of the same species in several assemblages. A **palaeocommunity** is an assemblage that represents a former community (Fig. 7.8). This interpretation is based on evidence that the assemblages are either *in-situ* 'life assemblages' or are believed to be nearly *in-situ* because they show no signs of prolonged transport or of a long residence time on the sea-floor (see section 3.10). Although shells are not generally displaced far from the habitat in which they lived (section 3.3), the distinction between ecological associations and those that arise from transport and hydrodynamic mixing can be difficult. Even though shells in a fossil assemblage have not commonly been mixed by lateral transport, they may have experienced vertical mixing. A fossil accumulation is commonly the result of winnowing and concentration of shells from several levels within the sediment so the assemblage does not represent a community living at a single time but is the 'time-averaged' product of mixing a succession of communities (Fürsich and Aberhan, 1990; Kidwell and Bosence, 1991). In such cases the palaeocommunity commonly represents a composite sample of successive variations of the same community, but it could represent a mixture of two or more communities which successively replaced each other in the same location.

Communities are generally characterized by both their species composition and the relative abundance of individuals of the different species present. Both aspects can present problems; faunas living in the same habitat commonly exhibit both a substantial variability in their species content and in the abundance of different species (**equitability**) in local assemblages, which is likely to be reflected in a variability amongst fossil assemblages belonging to the same community.

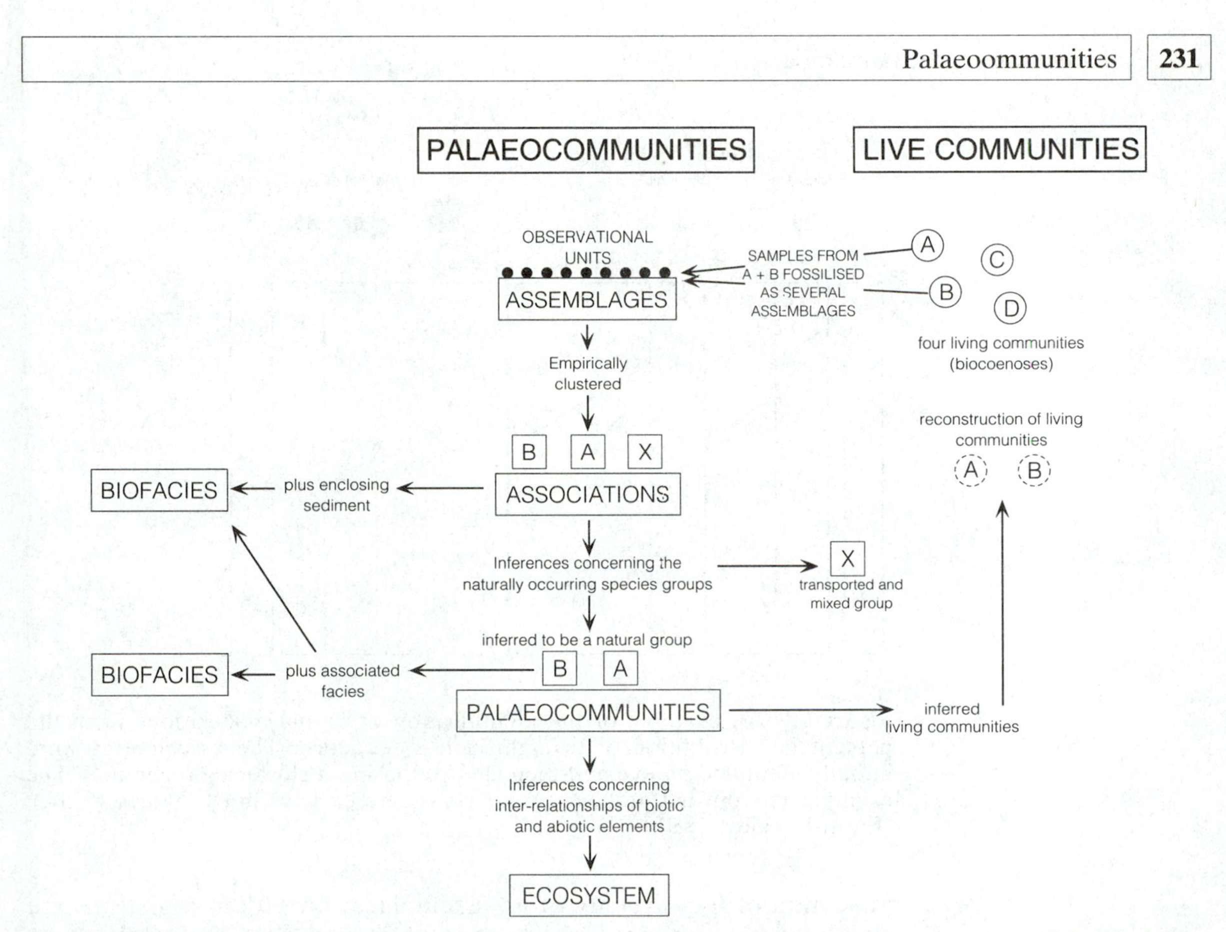

Figure 7.8 Relationship between live communities, palaeocommunities and ecosystems. (Modified from Pickerill and Brenchley, 1975.)

These problems have been addressed in a variety of ways, some qualitative and some quantitative. In their classic paper on Silurian communities, Ziegler *et al.* (1968) took a mainly qualitative approach and defined their communities by selecting a type example of each community, which they described in terms of its species composition and relative abundance. A semi-quantitative approach has been taken by several workers who have collected samples at closely spaced intervals through a vertical sequence and have plotted the species-abundance data graphically so that it reveals levels at which there are marked changes in the species composition or species-abundance (Fig. 7.9). The vertical sequence of communities so defined has been likened to those identified from the environmental transects in modern marine environments, except that the environments in one case are stacked vertically and in the other they distributed horizontally. In geological sequences where there are commonly stratigraphic hiatuses, many of which are cryptic, the change from one faunal association to the next is abrupt, so aiding the discrimination of palaeocommunities.

The faunal list has formed the basis for many assemblage studies. Although this simple method reveals the main components and is an

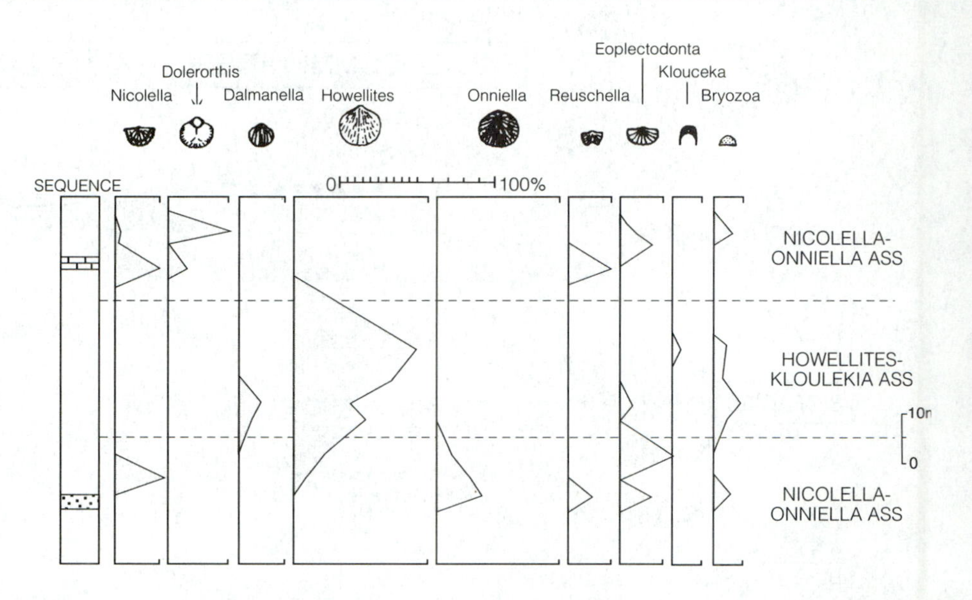

Figure 7.9 An example of the identification of faunal associations from the percentage distribution of taxa through a sequence. The associations were initially identified by eye and then checked, using a clustering technique. The example is from a Caradoc (Ordovician) sequence in north Wales (U.K.). (From Lockley, 1980.)

indication of the diversity of an assemblage, rare fossil organisms are clearly over-emphasized whereas abundant fossils are ranked the same as those merely present. Many lists were based on a fairly random scan of the fauna or flora, with the more obvious components, commonly large with distinctive ornaments, having a much greater chance of discovery than their more subtle associates. Modern palaeoecological studies rely heavily on more scientific sampling methods based largely on techniques developed by plant ecologists. Rigorous sampling methods using line transects, quadrats, bedding plane counts and bulk samples have been adapted for palaeoecological studies; although each method has its own advantages and disadvantages, they all generate controlled data sets amenable to scientific analysis.

7.10 NUMERICAL ANALYSIS OF COMMUNITY DATA

Numerical analysis of fossil assemblages provides a scientific approach to community analysis. First, however, counting conventions are a critical part of any palaeocommunity analysis to allow accurate definition, description and communication of the precise composition of an association. Fossil associations are rarely complete and in place – most of the material has been disarticulated, broken and often fragmented. There is no absolute solution to the problem posed by the separation and fragmentation of shells; however, clear description of the counting convention followed usually allows accurate comparisons between similar associations and communities. A standard procedure must be adopted to record the abundance of multicomponent fossils and fossil colonies which disaggregate after death. Moreover, amongst arthropods, groups such as the trilobites moult and each individual may leave up to ten exoskeletons during its ontogeny (Jaanusson, 1984). Counting procedures for dealing with disarticulated and broken fossil material have been summarized by Goldring (1991).

In most communities several species tend to dominate and there is crude inverse relationship between size and abundance in both living and fossil benthic faunas. In order of decreasing size, the megafauna, meiofauna and microfauna are progressively more abundant (Jaanusson, 1979). Some authors have, however, suggested that the use of biomass rather than individual counts is a better estimate of the relative contributions of organisms to community ecosystems, because a very abundant microfauna may contribute relatively little to the total community biomass. For operational reasons, most palaeoecological studies are focussed on the megafauna.

Palaeoecological information can be displayed in a variety of ways. Raw data on the abundance of each organism may be converted to relative abundance or relative frequency data and plotted as a bar chart to provide a quick visual census. These charts can form the basis for the definition, description and mutual comparisons of palaeocommunities. **Diversity, dominance** and **evenness indices** can be rapidly calculated from these data (see below). Microcomputer packages, such as PALSTAT (Ryan *et al.*, 1994), calculate a range of indices and changing values can be plotted through spatial and temporal gradients. For example, amongst the bivalve-dominated communities of the middle Jurassic of England, ten bivalve-dominated biofacies were recognized in the Lower Oxford Clay of the Midlands on the basis of the relative abundance and diversity of the macrofauna (Duff, 1975); trophic groups were established and plotted on a series of ternary plots. A fauna dominated by infaunal deposit-feeders and high-level suspension feeders characterized these bituminous shales of the mid-Jurassic. In another study, nearly 20 bivalve-dominated associations were described from the Corallian of England and Normandy based on a similar strategy (Fürsich, 1977).

Bar charts can illustrate faunal changes across environmental gradients. The absolute abundance of over 15 late Llandovery organisms was tracked across putative depth gradients in the Anglo-Welsh borderlands (Cocks and McKerrow, 1984). This graphical method delimited a series of intergrading brachiopod-dominated palaeocommunities (Fig. 7.10).

Since each bar chart records a distribution, two or more distributions can be compared statistically using a nonparametric tests such as Kolmogorov-Smirnov or more commonly χ^2 tests. For example, three main brachiopod-dominated associations were recognized in the upper Carboniferous of Illinois, related to water depth and substrate, by comparisons with simple χ^2 tests (Johnson, 1962).

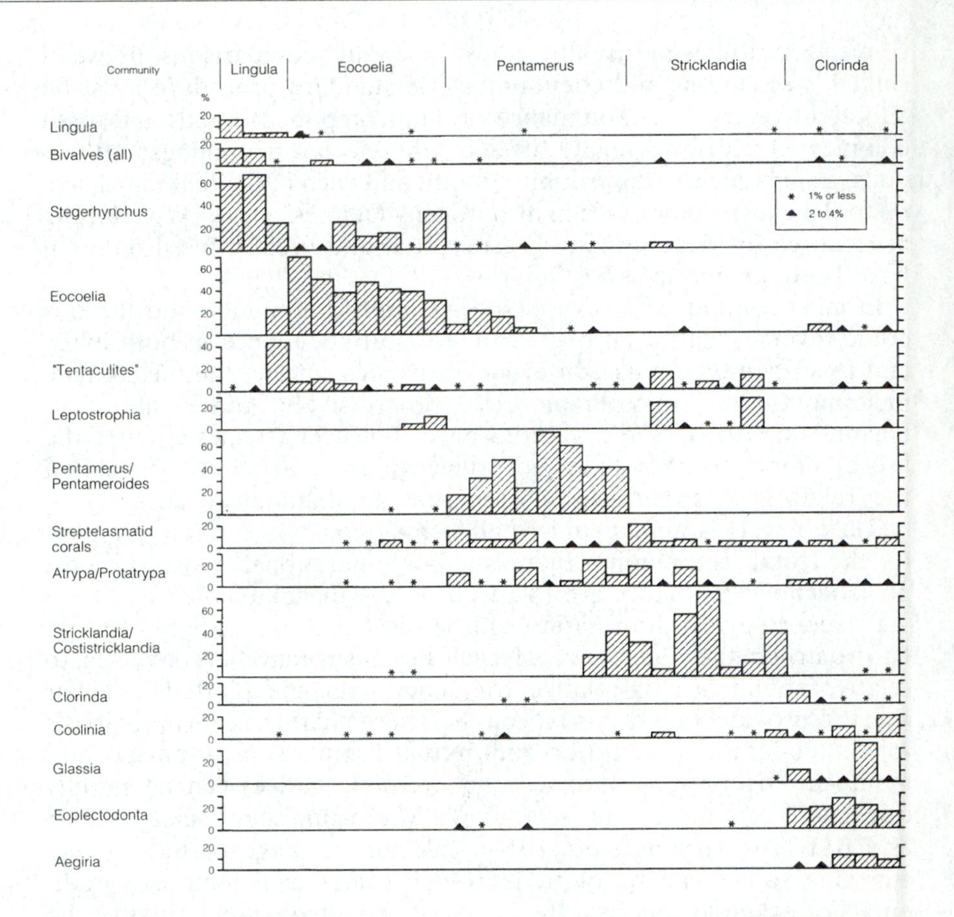

Figure 7.10 Lower Silurian marine communities from the Anglo-Welsh area; the sequential or seriative changes in faunas across the gradient of palaeocommunities indicates the changes of abundance of taxa. (Modified from Cocks and McKerrow, 1984.)

Although raw data, bar charts and diversity and other indices are sufficient for the majority of palaeocommunity investigations, there is a range of more sophisticated techniques based on multivariate statistical analyses. A number of microcomputer packages, for example, MVSP (Kovach, 1986–90) and PALSTAT (Ryan *et al.*, 1994), are built around a spreadsheet for data entry and contain several multivariate programs with graphics. Data may be imported from other spreadsheets and the results exported to a number of drawing and word-processing packages.

There are three main groups of multivariate techniques. The distribution of fossil organisms across sampled sites can be investigated statistically by, first, ordination techniques such as **principal component** and **correspondence analyses**, secondly, gradient or ordering algorithms such as **seriation** and thirdly through a variety of **cluster analyses** based on various distance and similarity coefficients and a number of

Box 7.2 Ecological indices

A range of statistics is available to describe the main aspects of fossil communities. Although the number of species collected from an assemblage provides a rough guide to the diversity of the association, obviously in most cases the larger the sample, the higher the diversity. Diversity measures are usually standardized against the sample size whereas dominance measures, based on relative abundance, have high values for communities with a few common elements, and low values where species are more or less evenly represented; measures of evenness are thus usually the inverse of dominance.

$$\text{Margalef diversity} = S\text{-}1/\log N$$
$$\text{Dominance} = \Sigma(ni/N)^2$$
$$\text{Evenness} = 1/\Sigma(pi)^2$$

S = number of species, N = number of specimens, ni = number of ith species, pi = relative frequency of ith species.

5 m
0
GAP?
GAP?

Horizons	*Pseudolingula*	*Schizocrania*	*Hesperorthis*	*Glyptorthis*	*Dalmanella*	*Salopia*	*Horderleyella*	*Triplesia*	*Oxoplecia*	*Sowerbyella*	*Macrocoelia*	*Rostricellula*	TOTAL BRACHIOPODA	Gastropoda	Nuculid Bivalve	Orthocone	*Basilicus*	Marroithinid	*Tallinella* sp.	Ramose Bryozoa	Prasoporid	Fenestellid	Pelmatozoans	*Hyalostelia*	Margalef Diversity	Dominance	Evenness
CD10			1							7	1		9				(2)			1			(1)		4.9	0.3	3.0
CD9			2		8					36	2		48				(1)			5			(3)		3.4	0.4	2.3
CD8			3		15		1	1		25	4	5	54				(1)			4	1		(8)		5.5	0.2	4.7
CD7			2	1	4					12	1		20							2	1		(3)		4.9	0.3	3.8
CD6	3	3	1		285		4	10		190	10	16	522			1	(330)			18	40		(37)	(11)	4.9	0.3	3.3
CD5					43	6	2			422	13	1	487	1			(3)		(9)	3	4		(1)	(3)	4.9	0.8	1.3
CD4			1		42	1				115	15	1	175		1		1		(10)	3	3	1			5.0	0.6	1.8
CD3					46					288	2	1	337				1	1	1	4	2				3.2	1.0	1.1
CD2					24	4			1	225	6		260				1		1	2			(1)		3.4	0.9	1.2
CD1					44	2	2			306	6		360								1				2.0	0.9	1.0

Figure 7.11 Cuod Duon Quarry, central Wales; changes in faunal abundance, diversity, dominance and evenness through ten levels in the Llanvirn rocks of the quarry. Marked changes are related to changes in facies particularly above unconformities. (Modified and redrawn from Ryan *et al.*, 1994.)

Detailed analysis of these palaeocommunity parameters has permitted recognition of a number of specific community types. Pioneer communities are dominated by one or two very abundant, opportunistic species in contrast to equilibrium communities where there is a relatively high diversity of taxa more or less equally present.

linkage methods. The mathematical basis for these methods is described by Davis (1986) with a number of multidisciplinary case histories. The starting point for most analyses is a simple two-dimensional matrix usually recording the abundance of each organism in a series of samples or sites.

Two modes of analysis based on a distance or similarity matrix are possible. The starting point for **Q-mode analysis** is a matrix of coefficients calculated for each pair of samples; on the basis of these coefficients similar samples or sites are grouped together based on the mutual occurrence of taxa. **R-mode analysis,** on the other hand, operates on the relative co-occurrence of taxa distributed across the investigated samples; groups of genera with a high probability of mutual occurrence are thus grouped together. Commonly both modes are used in palaeoecological analysis. Nevertheless despite the sophistication of the methodologies available, the most important part of any study is the careful collection and accurate description of data. These techniques provide methods of organizing data and a scientific framework where hypotheses may be tested.

Ordination techniques involve a reduction of the so-called 'measurement space'. New axes made up of groups of variates are calculated; in palaeoecological studies the new axes or eigenvectors may be combinations of taxa against which samples may be plotted (Q-mode) or combinations of samples against which taxa may be plotted (R-mode). In either analysis clusters of points may be identified in structured data; with reference to the loadings on the axes the combinations of variates generating the clusters can be easily identified. Unlike cluster analysis the degree of overlap and gradation between clouds can be visualized and assessed.

Principal component analysis (PCA) is an eigen technique which operates on a correlation or variance-covariance matrix (Davis, 1986). The first eigenvector is always orientated in the direction of maximum variation in the sample; the second and subsequent eigenvectors are perpendicular to the first, holding decreasing amounts of variation. Usually only the first few eigenvectors containing most of the sample variation are used in these analyses.

Correspondence analysis (CA) is another eigen technique. Rather than interrogate a matrix of distance or similarity coefficients, CA operates on a matrix of conditional probabilities. The method is more useful when applied to presence/absence data and has been used in the analysis of both living and fossil plant assemblages.

Markov chain analysis provides a more focussed method of investigating transitions and gradients. The probabilities of particular transitions can be defined and entered into a Markov matrix. From this, the most probable transition track can be drawn and commonly related to specific environmental set of factors. The upper Ordovician Martinsberg Formation of southwest Virginia contains rich brachiopod-dominated faunas. Markov analysis suggested probable transitions through the sequence from palaeocommunities dominated, in sequence, by *Lingula*

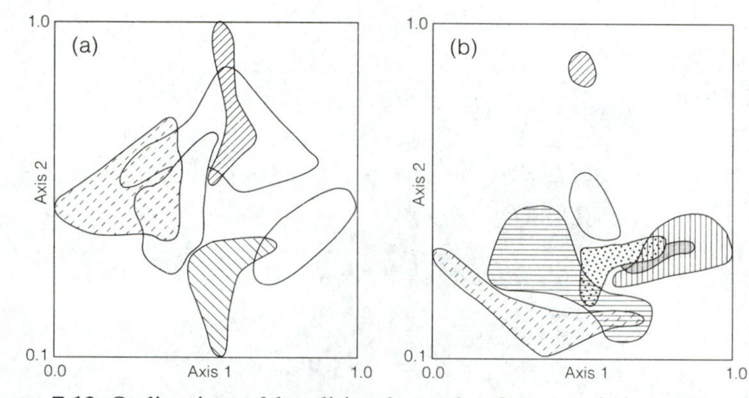

Figure 7.12 Ordination of localities from the Cattawa (A) and Narrows (B) sections based the relative occurrence of fossil taxa; the localities are grouped into palaeocommunities indicated by ornament (modified from Springer *et al.*, 1985). Many of the palaeocommunities overlap but gradients are obvious: for example on eigenvector (axis) 2 palaeocommunities with *Lingula* (solid diagonal – top right to bottom left) have large scores whereas palaeocommunities with *Onniella* (solid diagonal – top left to bottom right: Cattawa) and *Sowerbyella* (diagonal dash: Narrows) have low scores. This may be interpreted as a depth-related gradient – decreasing depth correlated with an increasing score on this axis.

– bivalves – *Rafinesquina* and *Onniella* (Springer and Bambach, 1985); the faunal changes may reflect an onshore–offshore gradient where a variety of factors, such as water depth and distance from shore, controlled these transitions. Nevertheless there was a considerable overlap between the palaeocommunities.

Seriation reorders two-dimensional data matrices, of binary (presence or absence) data, to display gradients; the method aims to concentrate the presence of taxa along the diagonal of the matrix. Most methods first calculate the mean position of taxa in each row and then sort the rows accordingly; secondly the mean of each column is calculated and the columns are then sorted. The algorithm proceeds iteratively until an optimal solution is reached. Although the technique was first used in archaeology and subsequently applied to biostratigraphical problems, it is useful for any data set with gradational properties. Within the mid-Devonian Hamilton Group of New York State, a regressive interval is marked by the Delphi Station Member (Fig. 7.13) (Brower and Kile, 1988) ; through the sequence there is a progressive increase in sediment grain size and the amount of organic matter. Seriation of the raw data, comprising taxa and samples from levels in the Delphi Station Member, generated a clear pattern; the seriated matrix demonstrated a gradation from basal units dominated by nonarticulate brachiopods such as *Craniops* and *Lingula* to the higher units with strophomenide, atrypide and spiriferide brachiopods together with corals and bivalves. These faunal changes were correlated with decreasing water depth.

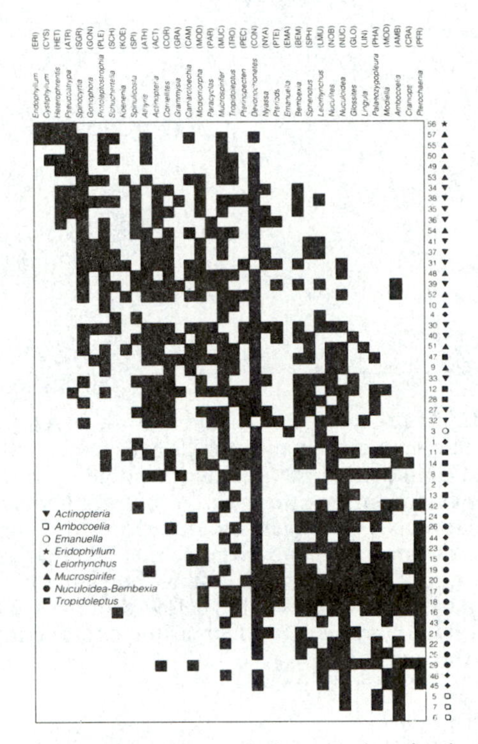

Figure 7.13 Seriation of presence and absence data in the Devonian Hamilton Group; in broad terms the taxa are organized into a depth gradient with increasing depth from the top left to the bottom right. The seriation is far from perfect because many different factors control the distribution of these marine organisms. (Modified and redrawn from Brower and Kile, 1988.)

Cluster analysis is the most commonly applied method to palaeoecological problems, although it can be the most unstable and least well understood; results must be treated with caution. For example a detailed study of the upper Silurian associations from the Anglo-Welsh borderlands suggested the most robust results were achieved using cosine q and an unweighted pair group method of linkage (UPGMA); the use of specific coefficients and linkage methods together with the input order of the data could all affect the composition and topology of the final dendrogram (Lespérance, 1990). Chaotic behaviour may be reflected in dendrograms; repeated runs varying the order of data input can, of course, identify and minimize this problem. Nevertheless there have been many palaeoecological data sets organized by cluster analysis. For example, it has been applied to the abundant and diverse shelly and graptolite faunas that occur in early Silurian successions in the Oslo Region, Norway, that were deposited in a range of deepening environments during a period of global warming. Q-mode cluster analysis of 55 samples, yielding about 15,000 specimens, based on the Jaccard Coefficient and using average linkage, defined seven main recurrent associations (Baarli, 1987). These associations, related to

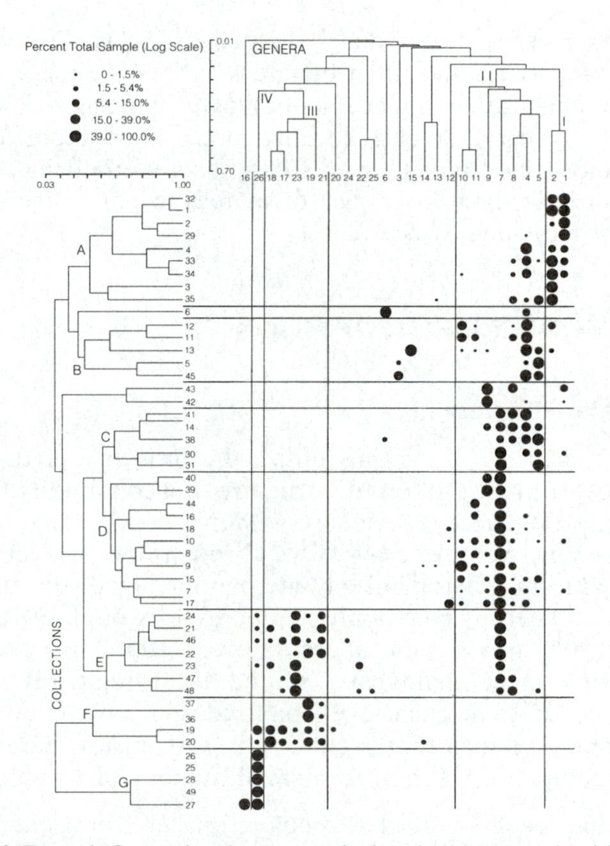

Figure 7.14 R and Q mode cluster analysis of data set of mid Ordovician brachiopods from eastern North America. The R and Q mode dendrograms have been plotted against the original data, modified to show the relative abundance of taxa, to indicate the generic and locality groups behind the clusters. Seven clusters of field collections are recognized along the Y axis whereas four recurrent generic groupings (I – *Rostricellula* and *Doleroides*, II – diverse assemblages dominated by *Strophomena* and *Sowerbyella*, III – dominated by *Paucicrura* and *Eoplectodonta* and IV – mainly inarticulates) are present along the X axis. (Redrawn from Patzkowsky, 1995.)

lithofacies, were dominated by eurytopic species, many surviving the end Ordovician extinction event as possible generalists.

The most informative and visual results have been obtained by combining both modes of analysis superimposed on the ordered data matrix. Ludvigsen and Westrop (1983) have illustrated the use of both Q- and R-mode cluster analysis on an extensive database of latest Cambrian–earliest Ordovician trilobites from the northern part of North America. The Q-mode dendrogram of collections is plotted up the Y-axis whereas the R-mode dendrogram of genera is plotted along the X-axis. If the ordered original data matrix is now plotted in the main part of the graph to correspond with the Q- and R-mode clusters, the basis for the twin dendrograms can be assessed at a glance. On the basis of these analyses the intersections of the Q- and R- mode analyses on an ordered data matrix helped define 11 discrete trilobite associations partly related to carbonate facies.

A more sophisticated enhancement involves flagging the sample size and species diversity for each locality along the sample axis of the Q- and R-mode crossplot (Jones, 1988). This method has the added advantage of indicating the impact of sample size and diversity on the definition of individual clusters.

7.11 COMMUNITY ORGANIZATION

7.11.1 Trophic structure

The manner in which organisms utilize the food resources within an ecosystem determines the **trophic structure** of a community. Resources are consumed at successive levels within a food chain or a more complex feeding web. Energy flows through the system from the primary producers (phytoplankton and benthic algae) through a chain of consumers. There is commonly an energy loss of 20–30%, rising to as much as 90% in some instances, between successive levels in the chain, so that most chains are limited to a relatively few links. Organisms compete at each level for food and have evolved special structures and activities for its collection. Organisms, based on their mode of feeding, are referred to a small number of **trophic groups:**

1. **Suspension-feeders** collect particulate matter from seawater
2. **deposit-feeders** collect particulate matter from sediment
3. **browsers** eat leaf-like vegetation
4. **grazers** rasp algae from the surface of the substrate
5. **carnivores** consume live prey
6. **scavengers** consume the larger particles of dead organisms
7. **parasites** consume the fluids or tissues of another organism over a period of time (Table 7.1).

Members of a trophic group may occupy a variety of life sites and collect food from different levels within the environment (Table 7.1). In marine ecosystems, life sites include surface-living plankton, plankton and nekton living within the water column, epifaunal benthos and infaunal benthos (Fig. 7.15).

Food is generally collected from the level at which the animal lives, but some infaunal animals, such as deep infaunal bivalves with long siphons, may collect their food from above the sediment surface. Amongst each trophic group there are differences in the nature of the food consumed and many variations in the structures designed to collect that food. Suspension feeders may select particles of a particular size or may consume dissolved or organic colloids. They may collect food by creating ciliary currents, or by using mucoid nets or setae; they may increase the collecting surface by developing an extensive network of branches, as in crinoids, or they may increase the length

Table 7.1 Trophic groups

Trophic group	*General definition*	*Life site*	*Location of collection*	*Food resources*
SUSPENSION-FEEDERS	Remove food from suspension in the water mass without need to subdue or dismember particles	EPIFAUNAL	high water mass	Swimming and floating organisms, dissolved and colloidal organic molecules, some organic detritus
		INFAUNAL	low	As above, with additional resuspended detritus
DEPOSIT-FEEDERS	Remove food from sediment either selectively or non-selectively. Without need to subdue or dismember particles	EPIFAUNAL	sediment-water interface	Particulate organic detritus, living and dead smaller members of benthic flora and fauna and organic rich grains
		INFAUNAL	in shallow sediment shallow deep	as above, but excluding living plants
GRAZERS	Acquire food by scraping plant material from environmental surfaces	EPIFAUNAL	sed-water interface	benthic flora
BROWSERS	Chew or rasp larger plants	EPIFAUNAL	sed-water interface	benthic flora
CARNIVORES	Capture live prey	EPIFAUNAL	sed-water interface	benthic epifaunal meio- and macro fauna
		NEKTO-BENTHIC	sed-water interface	as above
		INFAUNAL	in sediment	benthic infaunal meio- and macro fauna
SCAVENGERS	Eat larger particles of dead organisms	EPIFAUNAL	sed-water interface	dead, partially decayed organisms
		INFAUNAL	in sediment	
PARASITES	Fluids or tissues of host provide nutrition	SAME AS HOST	same as host	mostly fluids and soft tissues

or surface area of gills, as in bivalves, or the lophophore, as in brachiopods (Copper, 1990). The different morphologies developed amongst suspension-feeding organisms reflect to an important degree the specific way in which they extract food from the food chain.

7.11.1.1 Food chains

Food chains vary greatly in their length and the dominance of participating trophic groups. In grazing-browsing food chains the primary producers are dominantly benthic algal mats, seaweeds and angiosperms (sea grasses) since the early Cretaceous. These are consumed by grazers and browsers, particularly gastropods, other molluscs and herbivorous fishes, which in turn are consumed by predators,

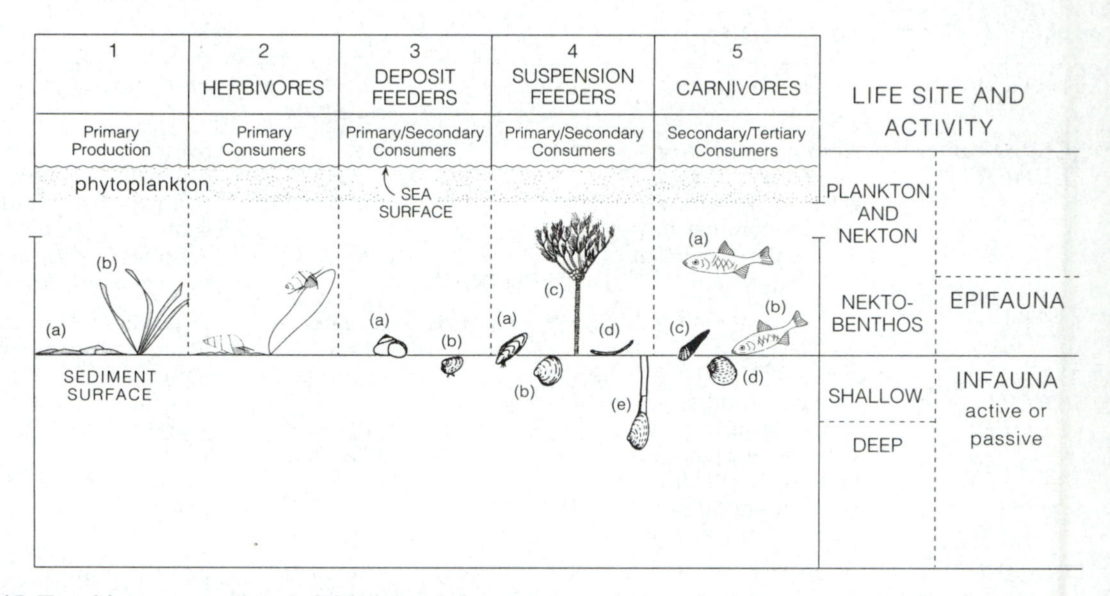

Figure 7.15 Trophic groups, the activities of their members and their different life sites (lettered): 1. Primary producers: phytoplankton in surface waters (a) cyanobacteria, (b) benthic algae. 2. Herbivores: browzing and grazing gastropods. 3. Deposit feeders. (a) deposit feeding gastropod (b) Shallow infaunal bivalve. 4. Suspension feeders (a) semi infaunal byssally attached bivalve (b) shallow infaunal bivalve (c) crinoid (d) epifaunal bivalve (e) deep infaunal bivalve. 5. Carnivores (a) nektonic fish (b) nekto-benthonic fish (c) epifaunal gastropod (d) infaunal gastropod.

commonly fishes (Fig.7.16) (Copper, 1988). Grazing-browsing food chains are particularly common in nearshore environments, particularly those on rocky shores or on intertidal mudflats.

In suspension-feeding chains (Fig. 7.17), the primary producers are phytoplankton, which are consumed by zooplankton and then this mixture of phytoplankton and zooplankton plus organic detritus is consumed by a variety of suspension-feeders (brachiopods, bivalves, bryozoans, sponges, corals and crinoids) which in turn may be

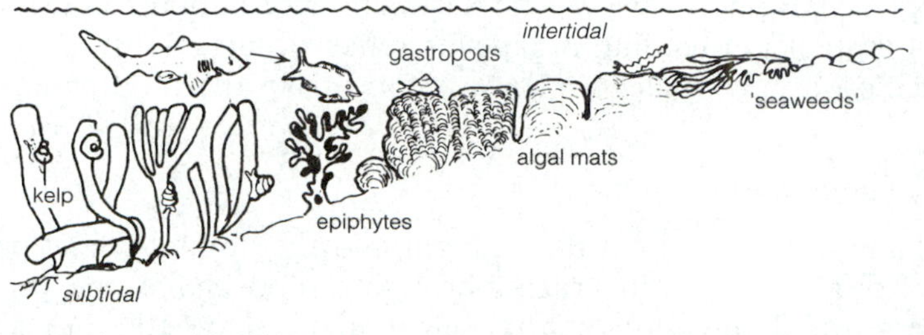

Figure 7.16 Reconstruction of a community with a grazing-browsing food chain. (From Copper, 1988.)

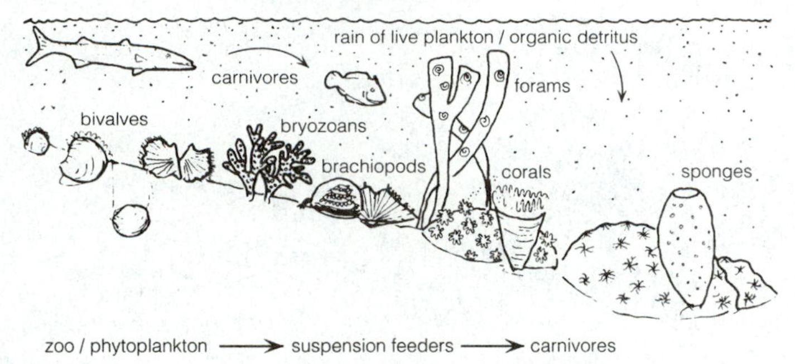

Figure 7.17 Reconstruction of a community with a suspension feeding food chain, showing a variety of suspension feeders that collect food in different ways (bivalves with a mucous trap or setae, brachiopods and bryozoans with lophophores, foraminifera with cilia, corals with tentacles and flagellate sponges). (From Copper, 1988b.)

consumed by predators. Diverse shelled suspension-feeding communities were particularly characteristic of Palaeozoic shelf environments. Detritus-feeding food chains (Fig. 7.18) are common where a large amount of organic detritus accumulates on the sea-floor, most commonly in muddy environments such as tidal flats and lakes. They are dominated by deposit-feeders (polychaete worms, bivalves with labial palps, gastropods, starfish and trilobites). These in turn are consumed by predators. One particularly specialized food chain is the chemotrophic chain (Fig. 7.19) which establishes itself around deep marine vents or smokers. The volcanically derived exhalations provide a source of hydrogen sulphide which can be metabolized by specialist bacteria. These are consumed or live symbiotically with a range of invertebrates that cluster round the vents and may grow to more than a metre in size.

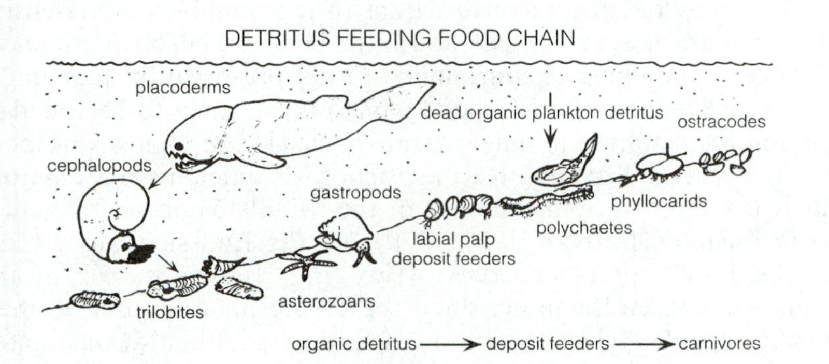

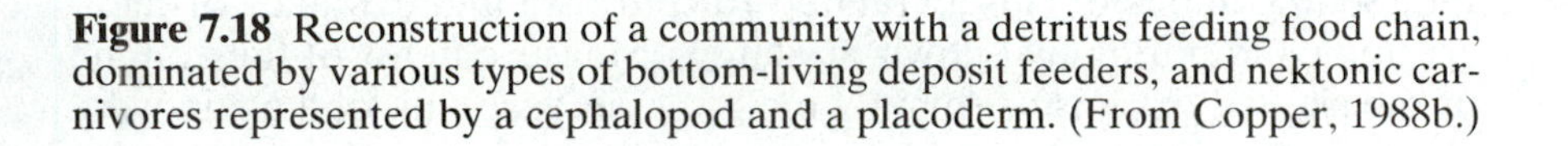

Figure 7.18 Reconstruction of a community with a detritus feeding food chain, dominated by various types of bottom-living deposit feeders, and nektonic carnivores represented by a cephalopod and a placoderm. (From Copper, 1988b.)

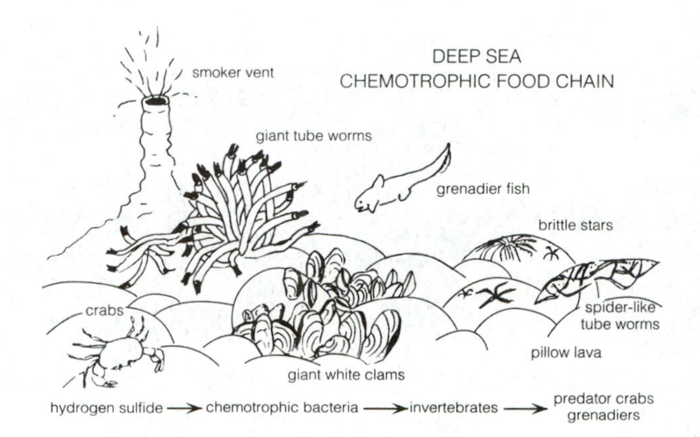

Figure 7.19 A chemotrophic food chain, centred on a deep sea 'smoker' venting hydrogen sulphide which is consumed by chemotrophic bacteria. These are the primary producers for a complex food chain which includes some large and specialized invertebrates. (From Copper, 1988b.)

The length of a food chain can vary greatly, particularly in the length of the predator chain, which can include several links from the zooplankton through a chain of predatory fish or may be very short, for example, when large balleen whales at the top of the food chain subsist on zooplankton low in the food chain. In some communities parts of the food chain may be wholly missing – for example, primary producers are rare or absent below the photic zone.

Studies of trophic structure in palaeocommunities are beset by the many problems posed by the incomplete preservation of communities. In most fossil assemblages the soft-bodied component of the fauna is missing so the reconstruction of the food chain is incomplete. Furthermore, it is commonly difficult to establish which animals ate which prey, particularly as carnivorous members high in the food chain tend to be few in number and not necessarily preserved near their prey. As a consequence, reconstructions of reasonably comprehensive food chains are few in number and have been based on particularly favourable taphonomic circumstances. Good preservation that might allow the reconstruction of a food chain is most likely to be found in stagnation Lagerstätten in quiet marine or lacustrine deposits or more rarely in obrution deposits. Reconstruction of quiet marine environments is exemplified by the studies of the Middle Miocene Korytnica Clays in Poland (Hoffman, 1977, 1979) and the Jurassic Oxford Clay, in England (Martill *et al.*, 1994) (Box 7.3). In his studies of the Korytnica Clays, Hoffman classified the organisms according to their trophic group, life style and position of feeding and then reconstructed the likely food chains and trophic webs on the basis of knowledge of the feeding habits of closely related Recent taxa and evidence of shell-crushing and boring by known predators. Measurements of the volume of the shelled organisms (mainly molluscs) in each link of a chain were used to estimate the biovolume of living material involved and hence the relative importance of the energy pathway of that link.

Box 7.3

The Peterborough Member of the Oxford Clay Formation is composed of dark grey fissile shales with interbedded thin shell concentrations. The shales have a high carbon content of 4–10% and may lack bioturbation or may be pervasively bioturbated; they are believed to have accumulated in a dysaerobic environment in which the muds formed a soft, sometimes soupy substrate. The shell beds were probably the product of large storms.

The Member has yielded a large and varied fauna including a generally low-diversity benthos, but varied nekton, including ammonites, belemnites and other dibranchiates, fish and reptiles. The living biota appears to have been broadly divided between the bottom-living epifaunal benthos, the active nekton in mid-depths and the near-surface plankton and nekton (Fig. 7.20), but the degree of interaction between these levels is a matter of uncertainty.

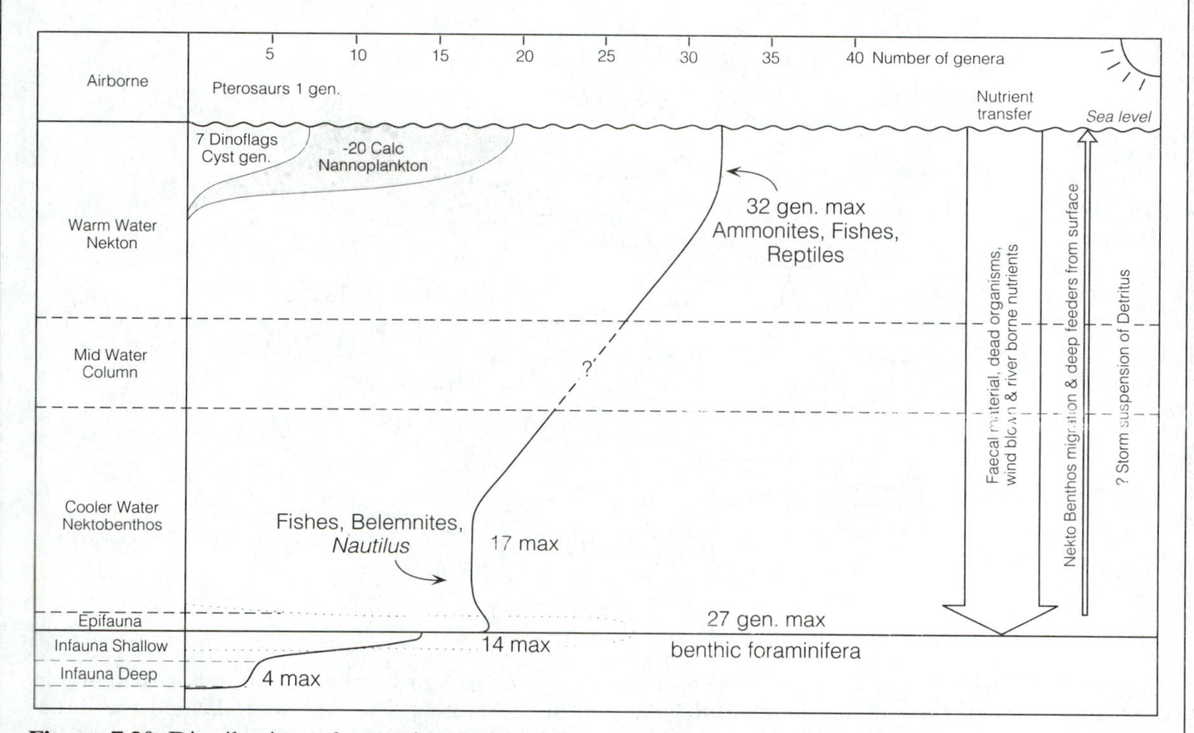

Figure 7.20 Distribution of organisms within the water column during deposition of Peterborough Member (Jurassic). A schematic profile shows the changing number of genera with depth (scale along the top). The maximum number of genera actually recorded in each water/sediment depth realm is indicated. The depth to the sea floor is not accurately known. (From Martill, Taylor and Duff, 1994.)

The fauna has been used to reconstruct the likely food web for a period of about 2 my, rather than at one instant in time. Some of the most interesting trophic reconstructions come from the predator–prey relationships amongst nekton. These have been based largely upon evidence of gut contents or distinctive bite marks, jaw and dental morphologies, general morphological comparisons with modern forms and close association of fossil finds (Fig. 7.21). In many instances it is possible to identify a general type of prey rather than a specific one; for example the giant nektobenthic shark (*Asterocanthus*) has powerful duraphagous teeth capable of crushing the largest bivalves and the icthyosaurs in general have long narrow jaws and a streamlined design adapted to being pursuit-predators that probably fed on fish. The study as a whole gives a qualitative assessment of carnivorous food chains in mid-Jurassic seas.

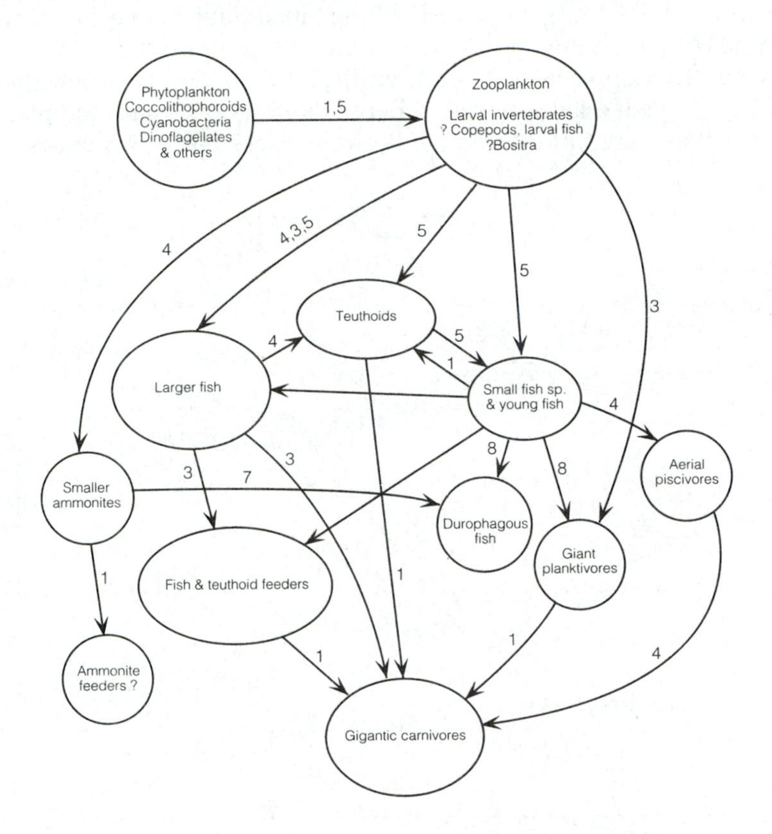

Figure 7.21 Trophic relationships for genera, or groups of genera forming the benthos or living in the lower part of the water column of the Peterborough Member sea. Arrows indicated the direction of the food uptake in the food chain. The numbers on the arrows refer to the evidence for the link, approximately in an order from (1) strong evidence to (8) weak. (1) Preserved stomach contents, distinctive bite marks; (2) close association of fossil finds, e.g. shark teeth near reptile skeleton; (3) functional morphological analysis, e.g. dental morphology; (4) fossil evidence as in (1) above, but from other deposits; (5) by comparison with similar living forms; (6) geochemical data; (7) sedimentological or preservational information: (8) supposed relationships. (From Martill, Taylor and Duff, 1994.)

7.11.1.2 *Predator prey relationships*

Because it is rarely possible to reconstruct even substantial parts of ancient food chains, many studies have been restricted to parts of a chain. Some of these have concentrated on the general impact of predation amongst faunas of a particular period, whilst others have focussed on specific predator–prey relationships. Studies of predation in marine environments in the Mesozoic suggest that there was a radiation amongst predators that included the diversification of large marine reptiles, such as placodonts, nothosaurs, plesiosaurs, ichthyosaurs, mososaurs and marine crocodiles (the Mesozoic marine revolution of Vermeij, 1977). The escalalation of predatory behaviour continued with the diversification of large fishes and sharks in the late Cretaceous and marine mammals in the Cenozoic. The increase in predation amongst shell crushing (durophagous) carnivores is reflected in the increased incidence of damaged shells amongst their molluscan prey (Fig. 7.22a, b). Predatory behaviour also expanded amongst the invertebrates, particularly in the late Cretaceous when the ability of carnivorous gastropods to envelope prey, prise open shells or drill them (Fig. 7.22c), became widespread (Taylor, 1981). The proliferation of predators apparently caused a response amongst the prey who evolved such defensive devices as thicker shells, protective spines and shapes resistant to crushing. Vermeij (1987) sees the predator–prey relationship, the escalation of predation and the response of the prey, as being one of the principle driving forces in the evolution of ecological structure (see Chapter 9, section 9.2.9).

Detailed evidence of predator–prey relationships has come from studies of two groups of carnivorous gastropods, the Muricidae which live epifaunally and the Naticidae which pursue their prey infaunally (see Kabat, 1990 for a review). Members of both groups envelope their prey before drilling a hole in the shell and consuming the contents. Naticid drill-holes are countersunk (bevelled), whereas muricids drill simple cylindrical holes (Fig. 7.22c). Thus it is possible to identify both groups of predators and their prey. Analyses of drilled shells in death assemblages in modern environments have suggested that it is a good measure of predation intensity in the living community, though some discrepancies could arise from the tendency of drilled shells to disintegrate more easily than entire shells and so introduce a taphonomic bias. From studies of naticid predation, Kitchell (1982) derived a model predicting that the predator optimizes its net energy gain by prey selection based on the probability of successful predation, prey availability, its energetic value and the energy costs of handling the prey; subsequent studies have generally supported this model. Naticids have been shown to be selective in the species that they attack, they are selective in the siting of the borehole and they generally reject large individuals that would be difficult to handle. The behaviour is stereotyped in the sense that it does not adapt to particular circumstances. This has been shown by an experiment in which the shell of the bivalve *Mercenaria*

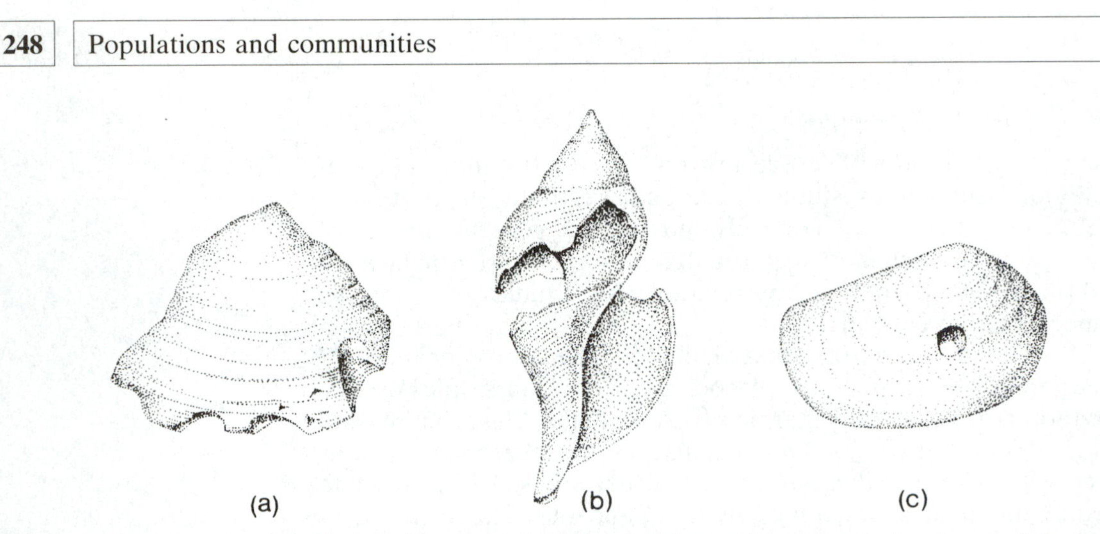

Figure 7.22 Predation of shelled molluscs. (a) Permian bivalve with crescentic breakage of the outer lip, probably the result of fish predation, X1. (b) Recent gastropod shell exhibiting peeling damage inflicted by the predatory crab *Calappa,* X1. (c) Bivalve with naticid (gastropod) boring, X1. (Modified from Brett, 1990.)

was artificially thinned at some points, but this did not affect the site chosen for attack by the gastropod *Polinices* (Boggs *et al.*, 1984). Stereotypic predatory behaviour amongst naticids appears to have evolved at least as early as the Miocene (Kelley, 1988), but the prey of naticids do not appear to have evolved protective devices, apart from a general increase in size and thickness of shell.

7.11.1.3 Symbiosis, commensalism and parasitism

Other interractions between two species of a community that do not involve a predator to prey relationship include **symbiosis** in which both partners benefit from the relationship, **commensalism**, in which only one partner benefits but the effect on the other is neutral, and **parasitism**, in which one species adversely affects the other by gaining nutritional long-term benefit. Two types of symbiosis can be recognized: one in which the relationship is essential for survival of both organisms (this is referred to as **obligatory symbiosis, or mutualism**) and a second type in which the relationship is beneficial but not essential to survival (**non-obligatory symbiosis**). Similarly commensalism can be obligatory or non-obligatory. Mutualism is well illustrated by the relationship between hermatypic (reef-building) corals and the zooxanthellae that live in the cells or tissues of the host. Zooxanthellae are dinoflagellates that gain from the symbiotic relationship by obtaining the essential nutrients, nitrogen and phosphorus, from the host, and from the protection from predation that the host offers. Corals gain from the zooxanthellae because they absorb carbon dioxide during photosynthesis, which facilitates the deposition of calcium carbonate and they synthesize and release carbon compounds that provide energy and also promote calcification. This symbiotic relationship that facilitates the deposition of calcium carbonate is the key to the massive structure of many reef-building corals. The fossil record of such symbiotic relationships is extremely patchy and generally insufficient to detect any evolutionary patterns. Nevertheless, the symbiotic relationship between hermatypic scleractinian corals and the zooxanthellae appears to have been a stable one which persisted through Mesozoic to the Recent. By analogy the massive structure of some Palaeozoic reef-building corals might indicate an early, but polyphyletic origin to the symbiotic relationship. Commensal relationships are widespread in the biological world, where one organism commonly gets a firm anchorage, protection or surplus food without harming the other. Ager (1961) has recorded a range of examples of fossil bryozoans, *Spirorbis* or corals that found a firm substrate by attaching themselves to living brachiopods and in some instances gained further advantage by positioning themselves on the fold on the brachial valve, close to the path of the inhalent or exhalent currents (Fig. 7.23b). In only rare instances did the infestation of the shell appear to have harmed the host. Other recorded examples of commensal attachment include *Spirorbis* on Carboniferous bivalves and productoid brachiopods which embrace the stems of crinoids with their spines. The exploitation of excess organic matter rejected by another organism also appears to be common. For example, *Urechnus*, a type of echiuroid worm that lives in a U-shaped burrow and is an untidy feeder, has an association with small fishes, annelids, crabs and a bivalve which live off the rejected detritus (Ager, 1963). The relationship between some fossil platyceratid gastropods and crinoids has also been seen as a commensal feeding relationship. The fossil gastropods are found firmly attached over the crinoid anus and the close conformity of the gastropod's apertural margin (Fig. 7.23a) with the morphology of the crinoid has been taken as an indication that the gastropod was living off waste products excreted by the crinoid (coprophagy). However, whether the relationship was always benign has been called into question by the

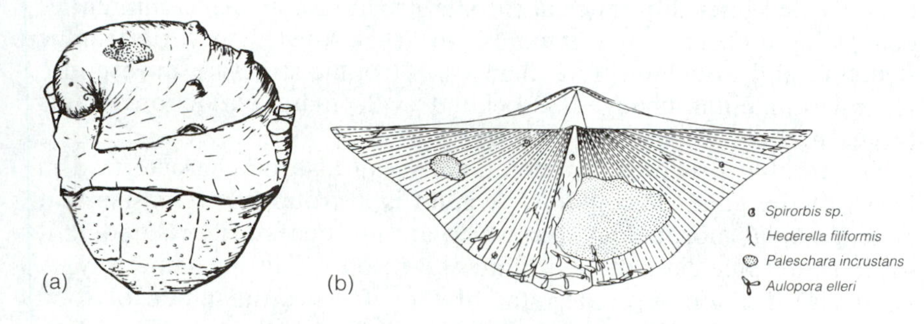

Figure 7.23 (a) Commensalism between the gastropod *Platyceras* and a Devonian crinoid. The gastropod covered the anal aperture of the crinoid and so fed without necessarily harming the crinoid. (b) A hypothetical specimen of *Spinocyrtia iowensis* with epifauna, particularly located on the fold of the brachial valve, close to the inhalent or exhalent area. (From Ager, 1963.)

discovery that some platyceratids attached themselves to the crinoid tegmen away from the anal tube and drilled through the tegmen (Baumiller, 1990). Evidence of attachment scars and multiple drilling suggests that the drilling was not predatory and aimed to kill the host. However, the drill hole may have allowed the gastropod access to richer food supplies in the gut of the crinoid and therefore have been exploitive rather than benign.

Most parasites are soft-bodied, so are rarely preserved. Fossil evidence of parasites ranges from their preservation as fossils to traces of infestation. Cases of preserved parasites include copepods in the gill chambers of Lower Cretaceous phosphatized fishes and nematodes as parasites on beetles in Eocene lignites of east Germany. Traces of parasitism include a range of holes, cysts and swellings in the shells of various organisms, but particularly in echinoderms whose stereome has the ability to respond to penetration or irritation. Many of the deformities have been ascribed to parasitic gastropods, barnacles or misostomid annelids, with variable degrees of confidence. A more extensive record of parasitism would help to answer such questions as, 'How does parasitism arise; is it a development of a symbiotic relationship or does it develop from a formerly independent species?', and 'What happens to the parasite if the host becomes extinct?' (Conway Morris, 1990).

7.11.1.4 Guilds

The problems involved in establishing food chains has prompted an alternative approach to trophic analysis, which is to utilize the preserved shelly fauna to make comparisons between the relative proportions of trophic groups in different communities. Thus palaeocommunities have been characterized in terms of the proportion of fossilized suspension-feeders, deposit-feeders, carnivores that they contain, the levels at which they feed and whether or not they are mobile. This more general, but more accessible information has been used to infer difference in trophic structure between communities occupying different environments, to track the history of trophic structure and to demonstrate changes in trophic structure in response to environmental changes associated with mass extinction events (Hansen *et al.*, 1993).

The trophic structure of a community can also be characterized in terms of the guilds it contains. A **guild** is a group of organisms that have a similar morphology, but which are not necessarily related, that exploit the same class of environmental resources in the same way. A guild classification takes account of the life site, the mode of food collection and morphological adaptations of the organism (Fig. 7.24). In a study of bivalve-dominated Mesozoic faunas Aberhan (1994) has shown how, within the suspension-feeding communities, the constituent guilds differ between environments and may also change with time within a single environment.

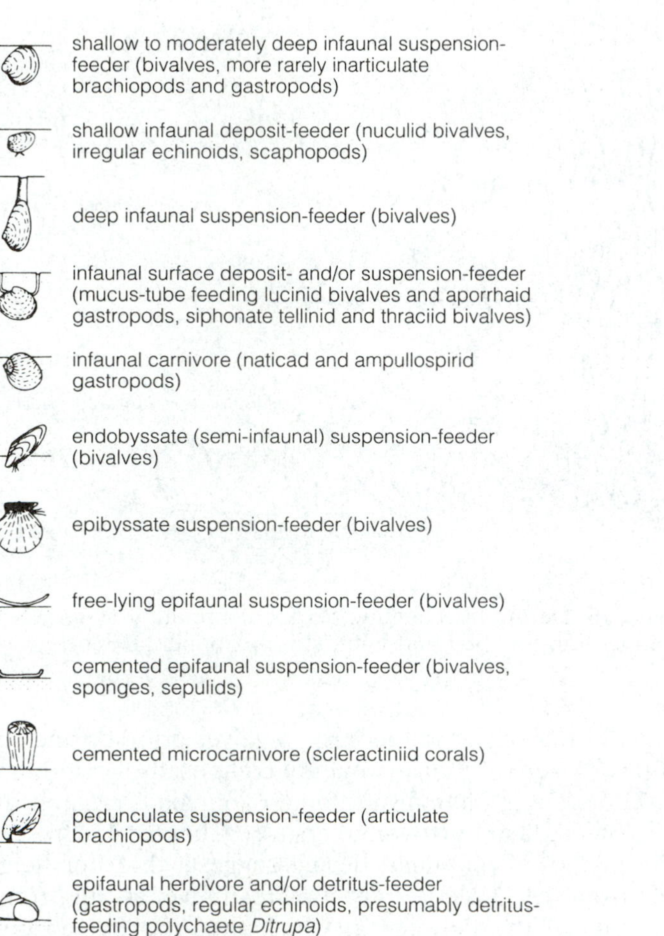

Figure 7.24 The variety of guilds recognized in the study of Mesozoic benthic communities, by Aberhan (1994).

7.11.2 Tiering

Stratification of organisms according to their height is a common feature of many communities. It is particularly prominent amongst rainforests where the different layers represent intricately linked parts of the ecosystem. Layering in marine communities according to their height above the substrate or their depth within the substrate has been termed **tiering.** Competition for suspended food amongst suspension feeders and for light amongst benthic algae leads organisms to optimize their feeding opportunities and to grow to different levels above the sea-floor. In many marine communities there is a layer of low-level suspension-feeders (<5 cm high) which includes algae, brachiopods, bivalves, bryozoa and solitary corals. A second level (5–25 cm high)

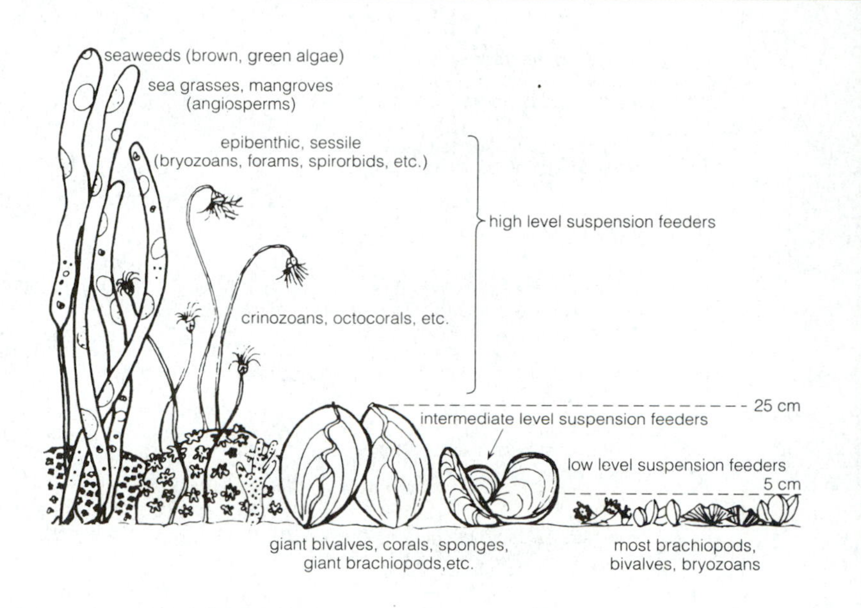

Figure 7.25 Tiering in a benthic marine community as an adaptive response to competition for food and light. (From Cooper, 1988b.)

includes corals, algae, some large bivalves crinoids and sponges and a high-level layer (>25 cm) is mainly composed of crinoids and sponges (Fig. 7.25). Tiering amongst infaunal animals is created by a spectrum of shallow to deep burrow-makers (see Chapter 4). Present knowledge of the history of epifaunal tiering suggests that tier height increased slowly from the Ordovician to a plateau level lasting from the Silurian to the end of the Permian at which time the Palaeozoic community structure was destroyed by the Permo-Triassic extinction. A high tier was subsequently re-established amongst Mesozoic communities, but strangely, high-level suspension-feeders have been rare from the late Mesozoic onwards.

7.11.3 Coevolution

Within the integrated structure of many communities the evolution of one part of the system leads to evolutionary response amongst other members of the ecosystem. For example, as predators improve their techniques of predation the prey improve their defences, so initiating an arms race in which attacker and attacked are interlinked (Vermeij, 1987). The linked evolution of pairs of species has been the subject of intense evolutionary study and has been generally referred to as coevolution. **Coevolution,** in its strict sense, implies that evolutionary developments in a pair of species are related such that changes in each species influence the heritable characters of the other. There are many records of detailed interactions between plants and insects, whereby, for example, the morphology of an insect is adapted to feeding from

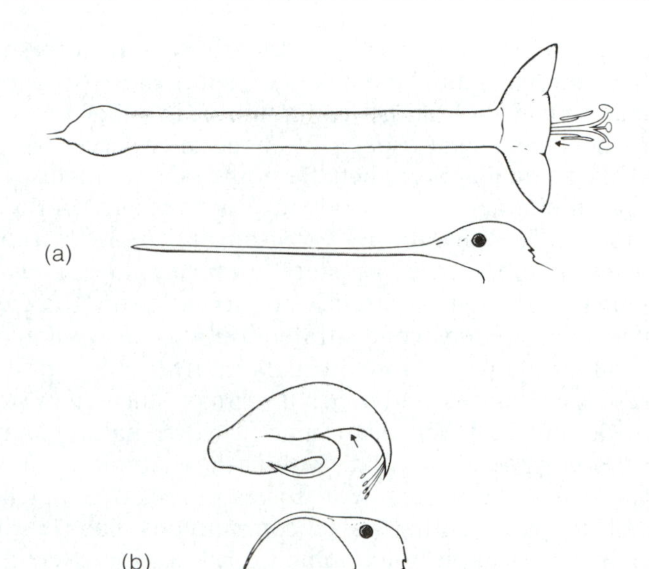

Figure 7.26 Two examples of coadaptation between the bills of specialized hummingbirds and their flowers. The arrows indicate the point of entry of the bills. (a) The swordbill (*Ensifera ensifera*) and the passion vine (*Passiflora mixta*). (b) The sicklebill (*Eutoxeres aquipa*) and *Heliconia* sp. Hummingbirds generally feed while hovering at the entrance to the flower, but sicklebills have to perch on a lower flower and tilt their heads back while feeding; the flowers are exactly spaced in the inflorescence to make this possible. (Modified from Snow, 1981.)

a specific shape of flower, which is consequently successfully pollinated. Similarly, the beak shapes of species of hummingbird are precisely adapted to feeding from particular species of flowers (Fig. 7.26). Whether such examples are strictly examples of coevolution is debatable, because it is not demonstrable whether each species has exerted an influence on the evolution of the other or whether one adapts to independent evolutionary changes in the other. Strict coevolution is difficult to demonstrate in the fossil record, but there are many examples of symbiosis and many predator–prey relationships which suggest that the prey has developed protective structures or behaviour in response to the pressure of predation (Vermeij, 1987 Harper and Skelton, 1993).

7.11.4 Community succession

Ecological succession is the orderly process of community change through time within an unchanging environment. It should be distinguished from ecological replacement, in which faunas succeed one another as a response to changes in the environment. Typically, community succession begins with a pioneer stage and progresses through successively mature stages to a climax stage. The entire sequence of stages is termed a **sere** and the successive stages are **seral**

stages. Community succession is most clearly seen in terrestrial plant communities, though some elements of succession probably occur amongst many plant and animal communities in other environments. One of the most significant aspects of community succession is that it is generally directional and predictable in the development of a particular type of community. Each seral stage is transient and appears to make the habitat less favourable for itself and more favourable for the organisms of the succeeding stage, until equilibrium is attained with the climax community. Studies of succession in Recent marine communities have mainly concentrated on the succession of algae which generally start with a low diversity pioneer stage, pass through more diverse seral stages and reach a climax stage in which a few species are dominant. Local disturbances of the habitat may restart the colonization process so that, within the habitat as a whole, a community can be at different seral stages in localized patches.

Studies of marine benthic palaeocommunities have yielded few wholly convincing examples of community succession except amongst reefs (Copper, 1988). Some fossiliferous beds showing an upward change in fauna have been described in terms of community succession (Walker and Alberstadt, 1975) but it is commonly difficult to determine whether changes of this kind reflect ecological succession or 'taphonomic feedback' whereby organisms colonized the shells of a pioneer population that was already dead and had merely created a suitable substrate.

Reefs clearly show a vertical change in their fauna and algal flora. In many cases this change has been interpreted as ecological replacement in response to falling sea-level which promoted colonization by a progressively shallower-water biota. However, there are many instances of reefs showing faunal change with constant sea-level so there are good grounds for recognizing true succession. In their study of succession in Ordovician reefs, Alberstadt *et al.* (1974) recognized a pioneer stage, a stabilizing stage, a diversification stage and a climax stage (Fig. 7.27).

Copper (1988), on the other hand, sees the initial colonization and the stabilization of the substrate as being all part of the pioneer stage and diversification as being part of the process by which a climax stage is reached. Pioneer species are commonly opportunists, generalists, they show *r* selection, have high fecundity and rapid growth rates and tend to be eurytopic and cosmopolitan. Typical pioneer forms are crinoids, bryozoa, algae and loosely branching corals (Fig. 7.28). During the pioneer stage the community has a relatively low percentage of cover to the substrate, diversity is low, food chains are simple and there is a limited variety of species-to-species interactions.

In contrast, climax species are commonly specialists with narrow niches, they show *K* selection, have low fecundity and low growth rates but long life histories and high biomass. The reef at the climax stage creates varied environments with associated high diversity, boring is common and tiering well developed (Copper, 1988) (Fig. 7.29).

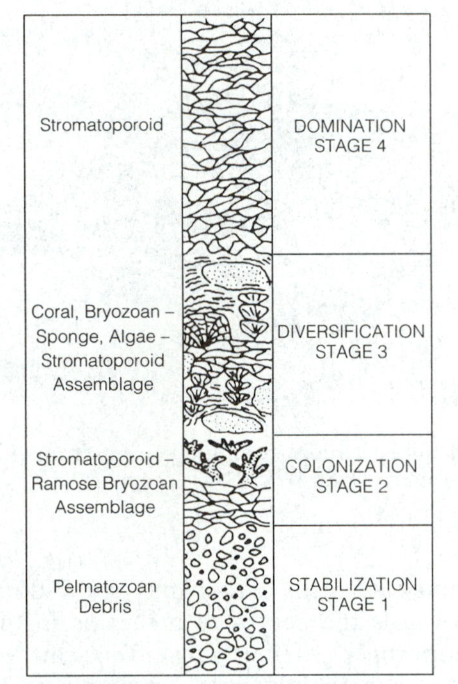

Figure 7.27 Ecological succession in a middle Ordovician reef in the Carters Limestone at Crown Point, Tennessee. (From Alberstadt, Walker and Zurawski, 1974.)

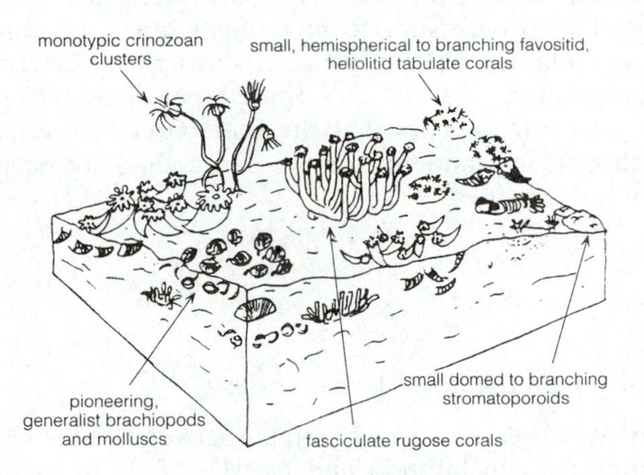

Figure 7.28 An idealized pioneer reef community from the Siluro-Devonian. See text for details. (Modified from Copper, 1988a.)

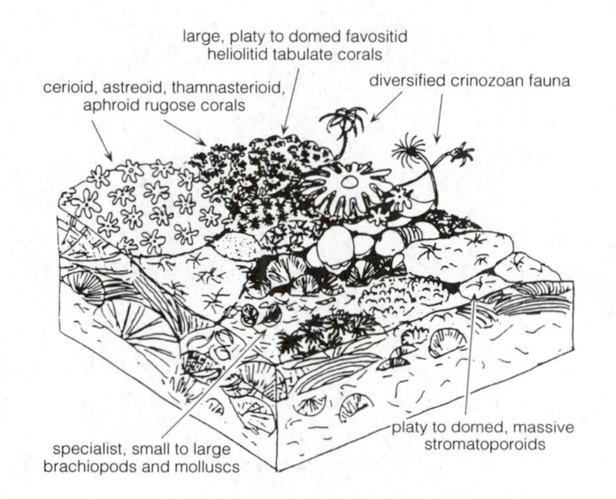

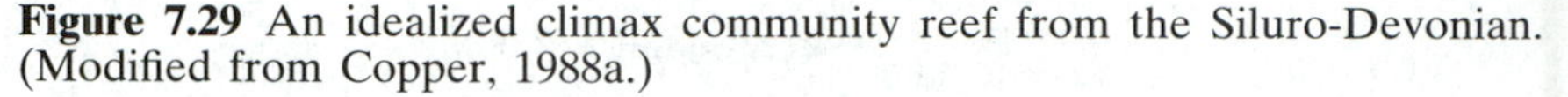
Figure 7.29 An idealized climax community reef from the Siluro-Devonian. (Modified from Copper, 1988a.)

Although the climax stage of a reef may show dominance of a few species locally towards the centre of a reef as in Silurian reefs with massive stromatoporoids at their crest, this can give a misleading impression of a low climax diversity. Commonly on the flanks of such reefs there are varied sub-environments with a wealth of organisms (Fig. 7.29). As a reflection of this Segars and Liddell (1988) recorded 32 different species living in microhabitats created by stromatoporoids.

One aspect of succession that is not clearly seen in reefs is predictability. The progression from pioneer stage to climax stage is far from predictable and can proceed through a variety of intermediate faunas even within the same reef belt (Copper and Grawbarger, 1978). It appears that the reef biota of one stage does not exert such a strong influence on the succeeding colonization as it does in plant communities.

7.12 SPECIES DIVERSITY

7.12.1 Diversity patterns on different scales

Species diversity varies between habitats, between areas of different size, with differences in latitude and between faunal provinces (see Chapter 8). It has also varied through geological time (see Chapter 9). Some of the patterns of diversity variation are reasonably well established, for some the causes are reasonably clear, but for others the pattern remains perplexing (Rozenzweig, 1995).

Diversity within communities is mainly related to the extent to which niches are partitioned. The more that organisms specialize in their life-requirements, the narrower is their niche and the more species will pack into a community. In general, diverse communities are likely to contain more guilds and more species per guild.

Diversity also correlates well with the size of area being considered (the species-area effect) and this relationship appears to hold for an area as small as a flower head to a biogeographic province, though the controlling factors may be different. Where this relationship has been studied on islands there appears to be a maximum diversity for the area (the equilibrium species diversity) that is determined by a negative feedback whereby low diversity allows diversity to rise, but high diversity tends to promote a fall in diversity. The rate of diversity increase is determined by the rate of immigration of species while the extinction rate is increased by raising the number of species available for extinction and by constricting the range of species, so making them more extinction-prone. The increase in diversity with enlarged area appears to arise because larger areas are likely to have a larger range of habitats and the increased area allows larger species-populations, which are less prone to extinction. Species-area effects, similar to those on islands, appear to operate on mainland areas and within the oceans, except that these areas have a long-established biota, so that new species are more likely to be added by speciation within communities than by immigration from outside.

On a global scale, diversity changes with latitude; it is highest in tropical regions and declines polewards. This pattern could be another example of a species-area effect, because the area between lines of latitude decreases from the equator to higher latitudes. Tropical diversity should be particularly high because the tropical zones are not only the largest, but the north and south tropics abut and so double the area of the tropics as a whole. However, other factors may be important and the causes of the latitudinal changes in diversity remains controversial; it could relate to such factors as productivity, stability of resources and spatial heterogeneity, though none of these seems to satisfactorily explain all the observations relating to latitudinal trends.

7.12.2 Diversity trends in marine habitats

Submarine 'landscapes' are most varied near coastlines where rocky shores, sandy bays, lagoons and estuaries create a host of habitats. Habitat diversity decreases across marine shelves as bottom topography and substrate becomes more uniform. Bathyal and abyssal environments, such as the abyssal plains can be extensive and uniform, but the continuing exploration of the deep sea-floor is showing it to be more varied than expected. Most early studies of the deep sea-floor, starting with those of the marine exploration vessel *Challenger* in the 19th century, suggested that deep-sea faunal density and diversity were low, probably as a consequence of very low food resources. The survey of Sanders and Hessler (1969) totally revised our knowledge of diversity in the oceans. With the aid of a more efficient dredge and a newly designed benthic sled to collect the epifauna a much larger number of

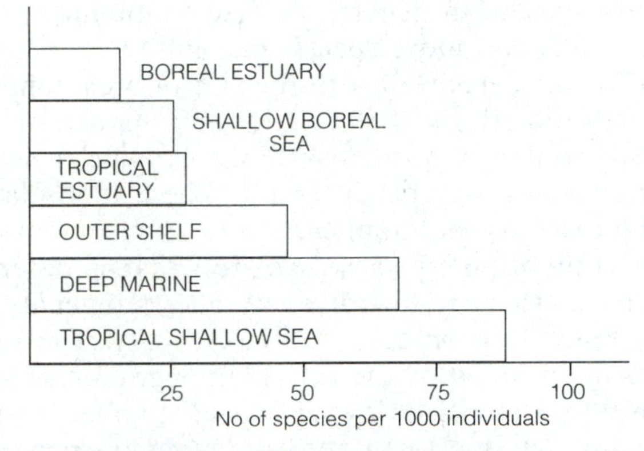

Figure 7.30 Diversity in the different environments within the oceans. Unshaded bars show a progressive diversity increase from shallow to deep water in boreal environments. The shaded bars show higher diversity in tropical environments. (Based on data in Sanders and Hessler, 1969.)

specimens were collected and these proved to belong to a remarkably high number of species, particularly of bivalves and polychaete worms. Comparisons were made with other habitats which had a similar mud substrate and it was shown that, for an onshore–offshore transect in the boreal zone, diversity increased with depth from estuarine environments to the outer shelf, slope and abyssal depths (Fig. 7.30). Latitudinal comparisons of diversity between boreal and tropical estuaries and shallow shelves showed much higher values in the tropics, which was less unexpected. The Sanders and Hessler study showed that for a single habitat – a mud substrate – diversity increased with depth.

This relationship has been partially confirmed by later studies, with the important exception that maximum diversity appears to occur at depths of about 2000 m, below which depth it declines (Rex, 1981) (Fig. 7.31). However, when whole depth zones have been sampled by collecting from a full range of habitats, the pattern appears to be different. Nearshore diversity still appears to be low but diversity increases to a peak near the outer shelf and then decreases towards deeper water, so that the diversity curve has a unimodal peak at shallower depths (Rosenzweig, 1995).

What determines the diversity of communities? The species–area relationship which is relevant to the biota of regions is unlikely to hold good for communities because the high level of diversity within a large region partly arises from summing the diversity of many habitats, whereas a community is confined to a single habitat. High diversity in a community reflects the ability of species to divide the available resources between them and occupy narrow specialized niches. The opportunity to thrive as a specialist appears to be influenced by the surrounding environment. Under adverse environmental conditions, such as low oxygen or extremes of salinity, only a few species adapt to the stress involved and low-diversity euryoxic or euryhaline commu-

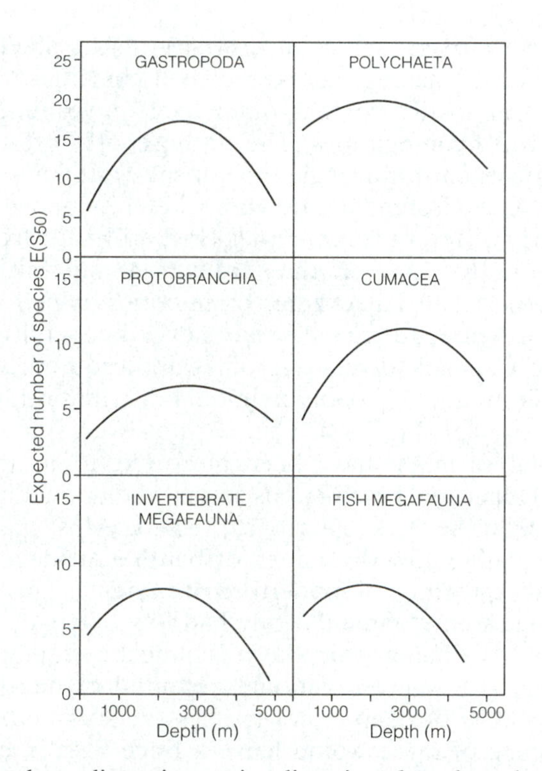

Figure 7.31 Depth gradients in species diversity of modern deep sea benthos and fish megafauna, based on samples dredged with a variety of gear from the western North Atlantic south of New England. (Modified from Rex, 1981.)

nities result. However, diversity varies widely in seawater that has normal levels of oxygen and salinity, so it is clear that these factors alone are not determining diversity levels in general. Though there appears to be no single physical aspect of the environment which shows an overall correlation with diversity, there does seem to be a correlation between environmental stability (constancy of physical factors) and diversity. Where environmental conditions fluctuate widely species are generalists, diversity is low and biotas are commonly *r* selected. Where physical conditions are constant, species specialize in gathering resources, have narrow niches, diversity is high and species are commonly *K* selected (section 7.5). This stability–diversity relationship is consistent with the low diversities in unstable nearshore regions and the high diversities in stable deeper marine environments. However, it would not explain the decrease in diversity below depths of 2000 m, so it appears likely that other factors might also influence diversity. The decrease in diversity towards bathyal depths correlates with a decline in nutritional resources to low levels, so that food availability might become a factor in limiting the total number of species that could be sustained in these deep-water environments.

Studies of diversity in benthic communities of the Ordovician and Silurian generally show a progressive increase in diversity across shelves and a sharp reduction in diversity at the shelf margin (Fig. 7.32). The diversity increase across the shelf coincides with a decrease in abundance of fossils, so that outer shelf environments have few fossils but diverse communities. The onshore–offshore diversity trend is probably consistent for a single type of substrate, but might be more complex where environments are more heterogeneous and different substrates carry different communities (Pickerill and Brenchley, 1979). Similar patterns to those in the Ordovician and Silurian probably persisted amongst late Palaeozoic communities, which have a similar community structure, though this has never been fully documented. Mesozoic and Cenozoic diversity profiles are also poorly documented, though it is likely that a broadly unimodal distribution like that in the Palaeozoic prevailed (Fig.7.32). There is, however, some evidence that Mesozoic bivalves might have been more eurytopic and less depth-related than Palaeozoic brachiopods so that the unimodal depth–diversity curve might be less clearly expressed. Most palaeontological records show rather low diversities of benthos in deep marine environments, unlike records of modern environments. This may be partly because deep sea environments truly had low diversity faunas, particularly in the early Phanerozoic, but it is almost certainly partly related to the rarity and low preservational potential of many deep-marine species: up to 80% of deep–water species are soft–bodied and only a small percentage of these would leave a trace fossil record.

7.12.3 Measuring diversity in palaeontological samples

The simplest measure of diversity is a count of the number of taxa in an assemblage. As, however, larger collections are made the diversity will usually increase, rapidly at first until a plateau is reached where, despite further collecting very few new taxa will appear. This programme is the basis of the rarefaction curve, which can provide a quick and usually reliable measure of the diversity of a sample. Initial attempts to construct rarefaction curves involved the iterative sampling of an assemblage, in blocks of between 10 and 100, plotting increasing sample size along the x-axis and the increasing total number of species along the y-axis. An alternative method involves converting an ordered bar chart of taxa-abundance frequencies into rarefaction curves (Fig. 7.33). These methodologies provide a quick and robust way of estimating, graphically, taxa diversity and also the minimum sample size at which a near-maximum diversity occurs, but it has been pointed out that they can be misleading when samples are taxonomically dissimilar, accumulated by different processes from different habitats and the curve is extrapolated beyond the maximum sample size (Tipper, 1979). However, under controlled conditions diversities across two or more samples may be compared.

A range of other measures of diversity are based on standardizing the number of taxa against the sample size. For example, Margalef Diversity (S-1/log N) is based on the number of taxa (S) divided by

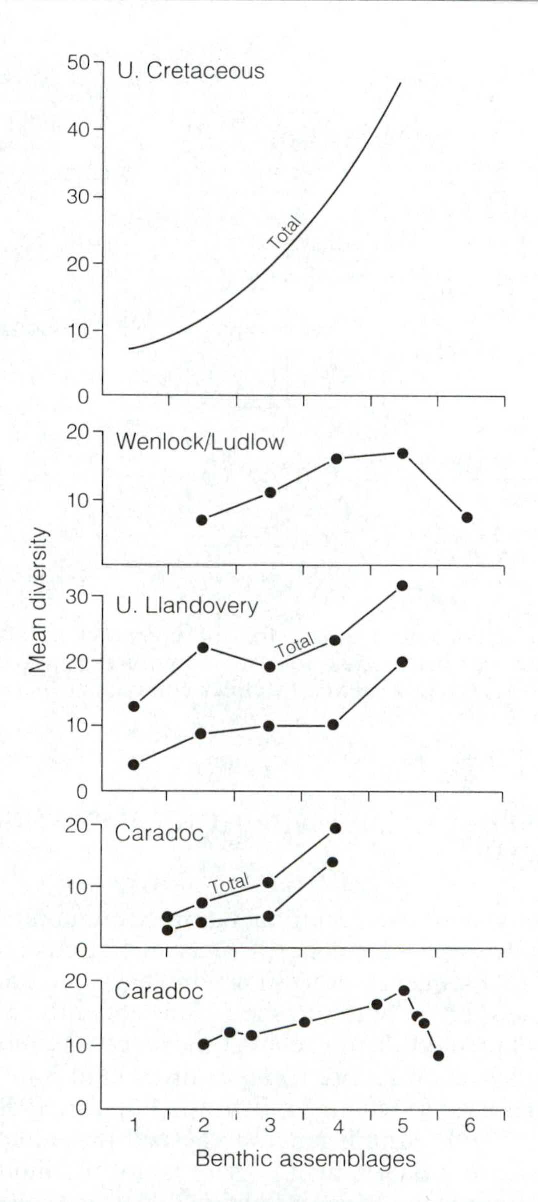

Figure 7.32 Profiles of mean diversities of brachiopods and of total fauna in assemblages from different benthic assemblage zones in the Lower Palaeozoic and Cretaceous, based on five different studies. (From Pickerill and Brenchley, 1991.)

a modification of the sample size (N), accounting for the exponential relationship between diversity and absolute sample size. The Fisher diversity index (a) is derived from the Poisson distribution, which assumes more taxa will have a small number of representatives than those with large numbers. The Fisher index appears twice on the right-hand side of the following equation, $S = a \ln (1 + N/a)$, and thus must be solved by iteration.

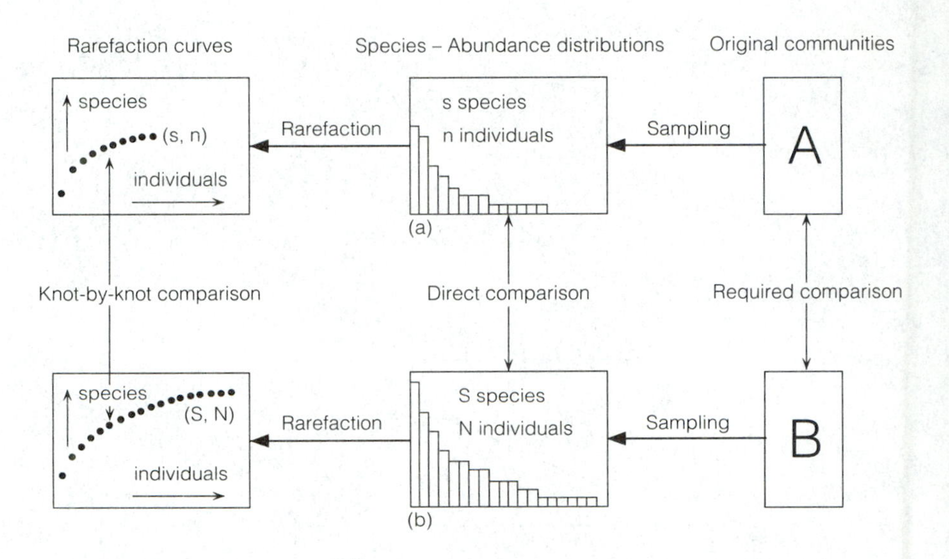

Figure 7.33 An operational framework for the construction of rarefaction curves. The abundance distribution for two communities (A and B) are converted into two rarefaction curves to effect comparison between A and B. (Redrawn from Tipper, 1979.)

7.13 ENVIRONMENTAL DISTRIBUTION OF PHANEROZOIC COMMUNITIES

In his classic study of Lower Silurian marine communities, Ziegler (1965) showed how Petersen's concept of level bottom communities could be applied to a sequence of fossil assemblages that had occupied depth-related zones on a Silurian shelf. Subsequently many other studies have documented depth-related palaeocommunity distributions, but others have shown more complex distributions influenced by other physical factors, particularly substrate (Fürsich, 1976; Pickerill and Brenchley, 1979). It is now generally agreed that although many communities are depth-related, depth *per se* is not the limiting factor, but is related to a particular range of physical factors such as temperature, turbulence, nutrients (section 2.11). Nevertheless communities that occupy specific depth zones commonly resemble one another, even though they may have different taxa and belong to different areas of the globe (the parallel communities of Thorson, 1957) or come from different stratigraphic levels. The constituent species, genera or even higher categories may be different, but commonly belong to a similar range of guilds which generates a similar appearance to the communities. Communities that resemble one another in the range and morphology of the constituent organisms and occupy a similar habitat have been termed **congruent communities.** They are well exemplified by the Ordovician and Devonian shallow marine communities described by

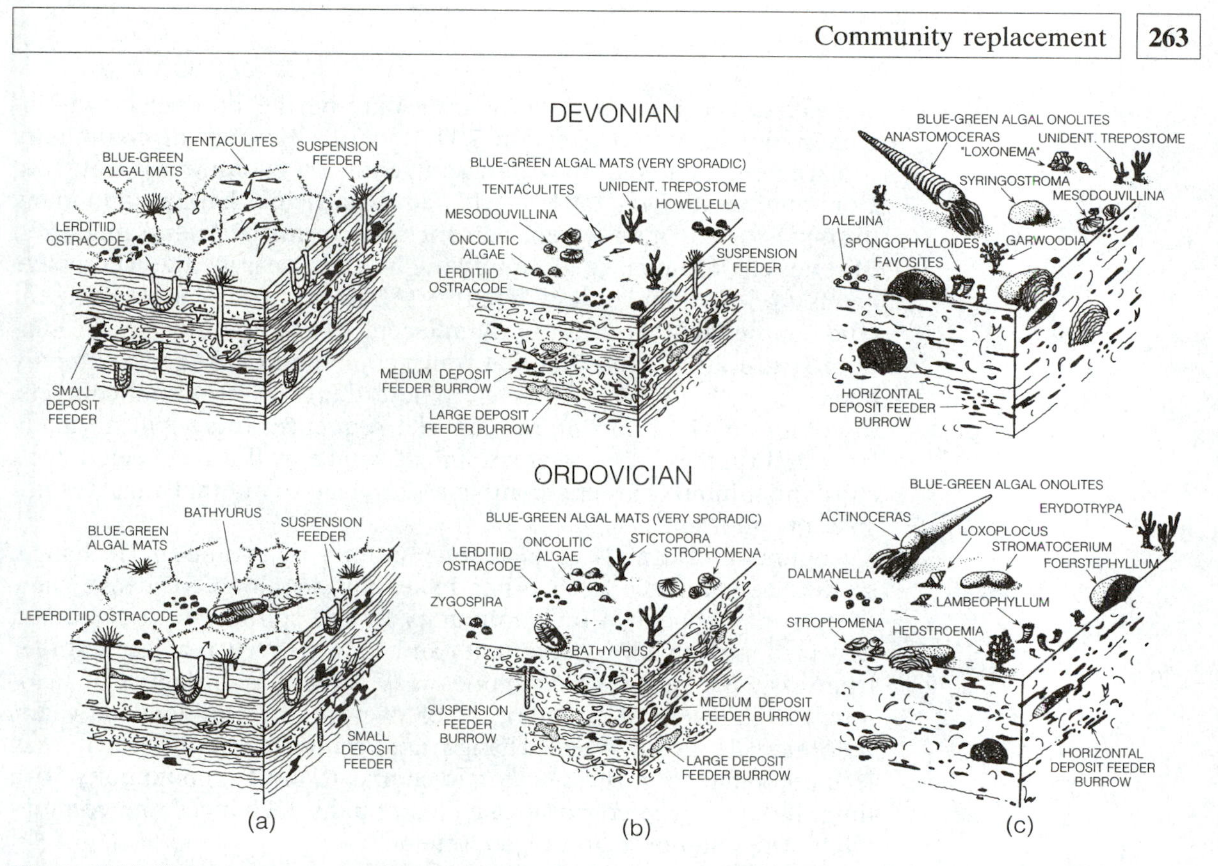

Figure 7.34 Congruent communities from the Ordovician Black River Formation and the Devonian Manlius Formation; (a) high intertidal communities, (b) low intertidal communities, (c) subtidal communities. (Modified from Walker and Laporte, 1970.)

Walker and Laporte (1970) (Fig. 7.34). Similarly, in Mesozoic communities bivalves and gastropods with a similar mode of life commonly maintain similar morphologies through time even though they may belong to different taxonomic groups (Aberhan, 1994).

The term **benthic assemblage** is applied to a group of communities that lived in the same position relative to the shoreline, throughout a geographic region or biogeographic province. The term was first applied to benthic faunas in the Silurian and Devonian (Boucot, 1975) which generally formed about five benthic assemblages distributed according to depth in five benthic assemblage zones, on the marine shelves of that time. The recognition of benthic assemblages provides a more systematic comparison between the organization of communities occupying the same depth zone, but of a different age, so allowing their evolution through time to be tracked.

7.14 COMMUNITY REPLACEMENT

Communities may succeed one another through a geological sequence on a variety of scales, from the bed scale to the formation scale. Faunal changes at the bed or bioherm scale may be the product of within-community succession (section 7.11.4) or the product of community replacement that may have environmental or evolutionary controls. Communities are generally closely tied to particular habitats and many examples of community replacement arise from the lateral migration of one environment over another. In these cases, the successive communities should follow Walther's Law for facies and represent communities that were formerly adjacent to one another on the seafloor. The stratigraphical level at which communities change commonly coincides with a change in facies. Where there are erosional contacts, disconformities or unconformities in a sequence, there will normally be an abrupt change in environment which will be reflected in a substantial jump from one benthic assemblage to another that was not spatially adjacent.

Communities can also replace one another by evolutionary change; species may replace each other by evolution along a lineage, new species may appear in the community by the splitting of a clade and new taxa may invade a community and competitively displace another from a similar niche. Communities may show a high degree of taxonomic stability at the generic level over several millions of years. Alternatively evolutionary change and ecological replacement may lead to a change in the taxonomic composition of a community with time, though if the constituent guilds remain unchanged the communities will commonly appear congruent.

Many palaeontologists have recognized that, in broad terms, similar environments may have been inhabited by the same types of organisms throughout geological time. Clearly the existence of recurrent faunal and floral associations within the fossil record strongly emphasizes the relationships between biotas and physical and chemical habitats.

Recurrence has been explained in a variety of ways (Miller, 1993): first, palaeocommunities may in fact track palaeoevironments through time; secondly, palaeocommunities are essentially stable within larger ecological structures such as ecologic-evolutionary units; thirdly, ecological interactions may fix the components of long-term palaeocommunities; and finally, recurrence may be the manifestation of a long-lived ecological system. Alternatively a uniformity of taphonomic processes through time have conspired to preserve sets of essentially similar assemblages. Nevertheless these types of analyses suggest a fundamental relationship between community types and ambient environments with an important time dimension.

Large changes in the nature of benthic assemblages coincide with periods of extinction. The mass extinctions in the geological record divide the history of community ecology into a succession of **ecologic-evolutionary units** within each of which the benthic assemblages maintain a general congruence but which are reconstituted and restructured following the episodes of mass extinction (Boucot, 1983). In an extension of the concept of ecological-evolutionary units, Sheehan (1993) distinguished nine long periods of community stability between mass extinction and several short periods, which followed mass extinctions when communities were being reorganized and complex ecosystems were being re-established. In addition to the mass extinctions there have been extinction events of lesser magnitude, some of which disrupted ecosystems on a regional scale and others which had global effects but were insufficiently large to seriously affect global ecology (Chapter 9).

7.15 SUMMARY POINTS

- The population is the fundamental unit in palaeoecology; fossil population have been modified by taphonomic loss and time-averaging; changes at the population level can trigger large-scale biological change.
- The structure and dynamics of fossil populations are analysed by size-frequency histograms, polygons and survivorship curves; populations can vary in size, structure and stability.
- There are various types of populations including equilibrium and opportunist populations; *r*- and *K*-strategist taxa together with ecological trade-offs are important parts of population evolution.
- Fossil associations and assemblages can be interpreted as palaeocommunities, defined on the presence and relative abundance or equitability of fossil taxa.
- Palaeocommunities can be described in terms of diversity, dominance and evenness; abundance data can be analysed by multivariate techniques such as principal component and cluster analyses, together with seriation.
- Trophic structures may include a variety of trophic groups occupying different life levels and sites within the community; food chains and predator–prey relationships are important parts of trophic structures.
- Palaeocommunities are the loci for various interactions and relationships including symbiosis, commensalism and parasitism; some relationships form the basis for coevolution between organisms.
- Communities are tiered, both above and below the substrate, and contain guilds of similar, but unrelated, organisms utilizing the same resources and occupying similar ecological niches.
- Patterns of diversity are many and varied, controlled by many factors, in a spectrum of both marine and nonmarine environments; diversity indices and rarefaction curves signal the diversity and diversity changes in palaeocommunities.
- Recognition of congruent and recurrent communities has helped define long-term ecological change. Large-scale community changes have been described in terms of ecologic-evolutionary units, partitioned and reset by extinction events.

7.16 FURTHER READING

Ager, D.V. 1963. *Principles of Paleoecology*. McGraw-Hill, New York. 391 pp.

Boucot, A.J. 1975. *Evolution and Extinction Rate Controls*. Elsevier, Amsterdam. 427 pp.

Boucot, A.J. 1981. *Principles of Benthic Marine Ecology*. Academic Press, New York. 463 pp.

Dodd, J.R. and Stanton, R.J. Jr. 1981. *Paleoecology, Concepts and Applications*. John Wiley & Sons, New York. 559 pp.

Goldring, R. 1991. *Fossils in the Field*. Longman, Harlow. 218 pp.

Rosenzweig, M.L. 1995. *Species Diversity in Space and Time*. Cambridge University Press, Cambridge. 436 pp.

Scott, R.W. and West, R.R. (eds) 1976. *Structure and Classification of Paleocommunities*. Stroudsburg. 291 pp.

Stearns, S.S. 1992. *The Evolution of Life Histories*. Oxford University Press, Oxford.

7.17 REFERENCES

Aberhan, M. 1994. Guild-structure and evolution of Mesozoic benthic shelf communities. *Palaios* 9, 516–545.

Ager, D.V. 1961. The epifauna of a Devonian spiriferid. *Quarterly Journal of the Geological Society, London* 117, 1–10.

Ager, D.V. 1963. *Principles of Paleoecology*. McGraw-Hill, New York. 391 pp.

Alberstadt, L.P., Walker, K.R. and Zurawski, R.P. 1974. Patch reefs in the Carters Limestone (Middle Ordovician) in Tennessee, and vertical zonation in Ordovician reefs. *Geological Society of America Bulletin* 85, 1171–1182.

Baarli, B.G. 1987. Benthic faunal associations in the Lower Silurian Solvik Formation of the Oslo-Asker Districts, Norway, *Lethaia* 20, 75–90.

Baumiller, T.K. 1990. Non-predatory drilling of Mississipian crinoids by platyceratid gastropods. *Palaeontology* 33, 743–748.

Boggs, C.H., Rice, J.A., Kitchell, J.A. and Post, W.M. 1984. Predation at a snail's pace: What's time to a gastropod? *Oecologia* 62, 13–17.

Boucot, A.J. 1953. Life and death assemblages among fossils. *American Journal of Science* 251, 25–40.

Boucot, A.J. 1975. *Evolution and Extinction Rate Controls*. Elsevier, Amsterdam. 427 pp.

Boucot, A.J. 1983. Does evolution take place in an ecological vacuum? *Journal of Paleontology* 57, 1–30.

Brower, J.C. and Kile, K.M. 1988. Seriation of an original data matrix as applied to palaeoecology. *Lethaia* 21, 79–93.

Cadée, G.C. 1988. The use of size-frequency distribution in palaeoecology. *Lethaia* 21, 289–290.

Cocks, L.R.M. and McKerrow, W.S. 1984. Review of the distribution of the commoner animals in Lower Silurian marine benthic communities. *Palaeontology* 27, 663–670.

Conway Morris, S. 1990. Parasitism. *In:* Briggs, D.E.G. and Crowther, P.R. (eds), *Palaeobiology: a synthesis*. Blackwell Scientific Publications, Oxford, 376–381.

Copper, P. 1988. Ecological succession in Phanerozoic reef ecolosystems: Is it real? *Palaios* 3, 136–152.

Copper, P. 1990. Paleoecology: paleosystems, paleocommunities. *Geoscience Canada* 15, 199–208.

Copper, P. and Grawbarger, D. 1978. Paleoecological succession leading to a Late Ordovician biostrome on Manitoulin Island, Ontario. *Canadian Journal of Earth Sciences* 15, 1987–2005.

Craig, G.Y. and Hallam, A. 1963. The interpretation of size-frequency distributions in molluscan death assemblages. *Palaeontology* 10, 25–42.

Davis, J.C. 1986. *Statistics and Data analysis in Geology*. John Wiley & Sons, New York. 646 pp.

Duff, K.L, 1975. Palaeoecology of a bituminous shale – the Lower Oxford Clay of southern England. *Palaeontology* 4, 653–659.

Fürsich, F.T. 1976. Corallian (Upper Jurassic) marine benthic associations from England and Normandy *Palaeontology* 20, 337–385,

Fürsich, F.T. and Aberhan, M. 1990. Significance of time-averaging for palaeocommunity analysis. *Lethaia* 23, 143–152.

Goldring, R. 1991. *Fossils in the Field*. Longman, Harlow. 218 pp.

Hallam, A. 1967. The interpretation of size-frequency distributions in molluscan death assemblages. *Palaeontology* 10, 25–42.

Hallam, A. 1972. Models involving population dynamics. *In:* Schopf, T.J.M. (eds), *Models in Paleobiology*. Freeman, Cooper, San Francisco, 62–80.

Hansen, T.A., Farrell, B.R. and Upshaw, B.I. 1993. The first 2 million years after the Cretaceous-Tertiary boundary in east Texas: rate and paleoecology of the molluscan recovery. *Paleobiology* 19, 251–265.

Harper, D.A.T., Doyle, E.N. and Donovan, S.K. 1995. Paleoecology and palaeobathymetry of the early Pleistocene brachiopods of the Manchioneal Formation, Jamaica. *Proceedings of the Geologists' Association*. 106, 219–227.

Harper, E.M. and Skelton, P.W. 1993. The Mesozoic marine revolution and epifaunal bivalves. *Scripta Geologica, Special Issue* 2, 127–153.

Hoffman, A. 1977. Synecology of macrobenthic assemblages of the Korytnica Clays (Middle Miocene; Holy Cross Mountains, central Poland). *Acta Geologica Polonica* 27, 227–280.

Hoffman, A. 1979. A consideration upon macrobenthic associations of the Korytnica Clays (Middle Miocene; Holy Cross Mountains, Poland). *Acta Geologica Polonica* 29, 345–352.

Hollingworth, N. and Pettigrew, T. 1988. *Zechstein Reef Fossils and their Palaeoecology*. Palaeontological Association. 75 pp.

Jaanusson, V. 1979. Ecology and faunal dynamics. *In:* Jaanusson, V., Laufeld, S. and Skogland, R. (eds) Lower Wenlock faunal and floral dynamics – Vattenfallet section, Gotland. *Sveriges Geologiska Undersøkning* C762, 253–294.

Jaanusson, V. 1984. Ordovician benthic macrofaunal associations. *In:* Bruton, D.L. (ed.) *Aspects of the Ordovician System.* Palaeontological Contributions from the University of Oslo 295. Universitetsforlaget, Oslo, 127–139.

Johnson, R.G. 1960. Models and methods for analysis of the modes of formation of fossil assemblages: *Bulletin of the Geological Society of America* 71, 1075–1086.

Johnson, R.G. 1962. Interspecific associations in Pennsylvanian fossil assemblages. *Journal of Geology* 104, 32–55.

Jones, B. 1988. Biostatistics in paleobiology. *Geoscience Canada* 15, 3–22.

Kabat, A.R. 1990. Predatory ecology of naticid gastropods with a review of shell boring predation. *Malacologia* 32, 155–193.

Kelley, P.H. 1988. Predation by Miocene gastropods of the Chesapeake group: Stereotyped and predictable. *Palaios* 3, 436–448.

Kidwell, S.M. and Bosence, D.W.J. 1991. Taphonomy and time-averaging of marine shelly faunas. *In*: Allison, P.A. and Briggs, D.E.G. (eds) *Taphonomy: Releasing the Data Locked in the Fossil Record.* Plenum Press, New York, 115–209.

Kitchell, J.A. 1982. Coevolution in a predator-prey system. *Proceedings of the North American Paleontological Convention* 2, 301–305.

Kovach, W. 1993. *MVSP. Multivariate Statistical Package.* Abermagwr, Wales.

Kurtén, B. 1954. Population dynamics – a new method in paleontology. *Journal of Paleontology* 28, 286–292.

Lespérance, P.J. 1990. Cluster analysis of previously described communities from the Ludlow (Silurian) of the Welsh Borderland. *Palaeontology.*

Levington, J.S. 1970. The ecological significance of opportunistic species. *Lethaia* 3, 69–78.

Levington, J.S. and Bambach, R.K. 1970. Some ecological aspects of bivalve mortality patterns. *American Journal of Science* 268, 97–112.

Ludvigsen, R. and Westrop, S.R. 1983 Trilobite biofacies of the Cambrian-Ordovician boundary interval in North America. *Alcheringa* 7, 301–319.

Martill, D.M., Taylor, M.A. and Duff, K.L. 1994. The trophic structure of the biota of the Peterborough Member, Oxford Clay Formation (Jurassic), UK. *Journal of the Geological Society of London* 151, 175–194.

Miller. W., III 1993. Models of recurrent fossil assemblages. *Lethaia* 26, 182–183.

Noble, J.P.A., Logan, A. and Webb, G.R. 1976. The Recent *Terebratulina* Community in the rocky subtidal zone of the Bay of Fundy, Canada. *Lethaia* 9, 1–17.

Pickerill, R.K. and Brenchley, P.J. 1979. Caradoc marine communities of the south Berwyn Hills, North Wales. *Palaeontology* 22, 229–264.

Raup, D.M. 1978. Cohort analysis of generic survivorship. *Paleobiology* 4, 1–15.

Rex, M.A. 1981. Community structure in deep-sea benthos. *Annual Review of Ecological Systematics* 12, 331–353.

Richards, R.P. and Bambach, R.K. 1975. Population dynamics of some Paleozoic brachiopods and their palaeoecological significance. *Journal of Paleontology* 49, 775–798.

Rosenzweig, M.L. 1995. *Species diversity in space and time*. Cambridge University Press, Cambridge, 436 pp.

Ryan, P.D., Harper, D.A.T. and Whalley, J.S. 1994. *PALSTAT*. Chapman & Hall, London, 74 pp.

Sanders, H.L. and Hessler, R.L. 1969. Ecology of the deep-sea benthos. *Science* 28, 1419–1424.

Scott, R.W. and West, R.R. 1976. *Structure and Classification of Paleocommunities*. Stroudsburg, 291 pp.

Segars, M.T. and Liddell, W.T. 1988. Microhabitat analyses of Silurian stromatoporoids as substrata for epibionts. *Palaios* 3, 391–403.

Sheehan, P.M. 1993. Patterns of synecology during the Phanerozoic. *In:* Dudley, E.C. (ed.) *The Unity of Evolutionary Biology* vol 1, Discorides Press, Oregon. 103–118.

Sheldon, P.R. 1988. Trilobite size-frequency distributions, recognition of instars and phyletic size changes. *Lethaia* 21, 293–306.

Springer, D.A. and Bambach, R.K. 1985. Gradient versus cluster analysis of fossil assemblages – a comparison from the Ordovician of southwestern Virginia. *Lethaia* 18, 181–198.

Stearns, S.S. 1992. *The Evolution of Life Histories*. Oxford University Press, Oxford.

Surlyk, F. 1972. Morphological adaptations and population structures of the Danish Chalk brachiopods (Maastrictian, Upper Cretaceous). *Det Kongelike Danske Videnskapelige Selskab Biiologiske Skrifter* 19, 1–57.

Surlyk, F. 1974. Life habit, feeding mechanism and population structure of the Cretaceous brachiopod genus *Aemula*. *Palaeogeography, Palaeoclimatology, Palaeoecology* 15, 185–203.

Taylor, J.D. 1981. The evolution of predators in the late Cretaceous and their ecological significance. *In:* Forey, P.L. (ed.) *The Evolving Biosphere*. British Museum of Natural History and Cambridge University Press, Cambridge, 229–240.

Thayer, C.W. 1977. Recruitment, growth and mortality of a living articulate brachiopod, with implications for the interpretation of survivorship curves. *Paleobiology* 3, 98–109.

Thorson, G. 1957. Bottom Communities (Sublittoral or shallow Shelf). *In:* Hedgpeth, J. (ed.) *Treatise on Marine Ecology and Paleoecology*. Memoir of the Geological Society of America, Washington 67, 461–534.

Tipper, J.C. 1979. Rarefaction and rarefiction: the use and abuse of a method in paleoecology. *Paleobiology* 6, 423–434.

Van Valen, L. 1973. A new evolutionary law. *Evolutionary Theory* 1, 1–30.

Vermeij, G.J. 1977. The Mesozoic marine revolution; evidence from snails, predators and grazers. *Paleobiology* 3, 245–258.

Vermeij, G.J. 1987. *Evolution and Escalation*. Princeton University Press, Princeton, New Jersey, 527 pp.

Walker, K.R. and Alberstadt, L.P. 1975. Ecological succession as an aspect of structure in fossil communities. *Paleobiology* 1, 82–93.

Walker, K.R. and Laporte, L.F. 1970. Congruent fossil communities from Ordovician and Devonian carbonates of New York. *Journal of Paleontology* 44, 928–944.

Ziegler, A.M. 1965. Silurian marine communities and their environmental significance. *Nature* 207, 270–272.

Ziegler, A.M., Cocks, L.R.M. and Bambach, R.K. 1968. The composition and structure of Lower Silurian marine communities. *Lethaia* 1, 1–27.

Palaeobiogeography

All living species have a geographical range. These ranges may vary from a single pond to an entire continent; in the case of our own species, *Homo sapiens*' range is virtually cosmopolitan whereas the kangaroo, for example, is restricted to Australia. Studies of modern biogeography have clarified the distribution of many taxa of all ranks, establishing provinces and the types of barriers between them. Both Charles Darwin and Alfred Wallace recognized the reality of provinciality in their respective studies of the Galapagos and East Indies biotas in the mid-19th century. Palaeobiogeography has permitted the recognition of ancient oceans and island complexes, together with a range of land areas from micro to supercontinents. In addition, many types of morphology are directly related to geography as are often the size of animals and plants and the diversities of their communities. Changing palaeogeography through time may also have been influential in the evolution of life at a spectrum of levels from speciation within individual populations to the development of major evolutionary biotas.

8.1 INTRODUCTION

The palaeobiogeography of ancient organisms was obvious to many scientists for over two centuries, but for those opposed to continental drift there were no feasible mechanisms to explain many of the past, apparently disjunct, distributional patterns of fossil faunas and floras. For example 19th-century biologists and geologists recognized that the Lower Palaeozoic faunas of Scotland were most similar to those of North America, whereas faunas of similar age from England and Wales had a greater similarity with those of Europe and Scandinavia. Although the existence of a proto-Atlantic had been speculated for over a century these fossils, together with a range of geological, geophysical and palaeontological evidence, provided Tuzo Wilson (1966) with the critical evidence for a much older early Palaeozoic ocean system dividing Europe from North America, cutting across the line of the present North Atlantic ocean.

Much more obvious was the fit of West Africa into east-facing South America (Fig. 8.1); these two pieces of a jigsaw have probably been

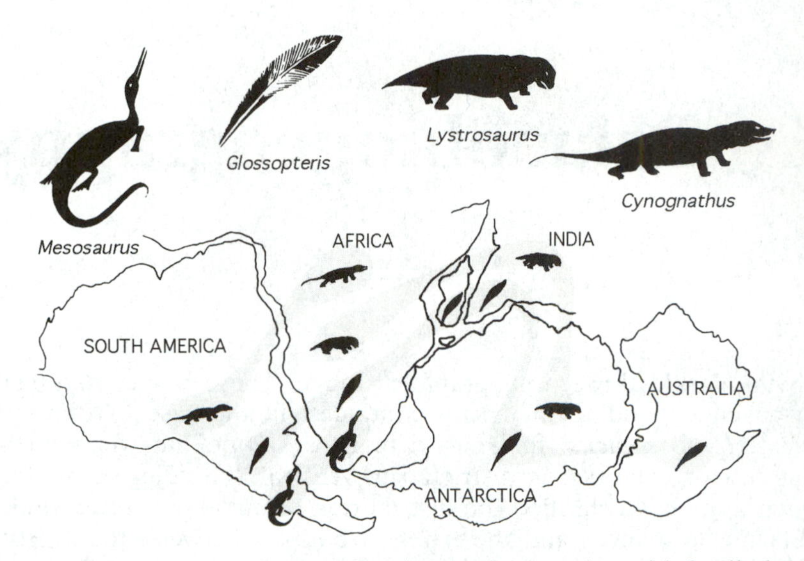

Figure 8.1 The fit of Africa and South America together with fossil plant and vertebrate data. (Adapted from Smith, 1990.)

the subject of comment since the first maps of the two continents were published. Moreover these parts of Africa and South America were all characterized by the late Palaeozoic *Glossopteris* flora and the *Mesosaurus* reptile fauna, together with the early Mesozoic reptiles *Cynognathus* and *Lystrosaurus*. Although various landbridges were invoked to explain these disjunct distributions, the simplest solution, suggested by Wegener and others, involved the divergence of Africa and South America during the development of the southern Atlantic ocean by the process of Continental Drift in the Mesozoic.

Biogeography has long been recognized as a constraint on the evolution of biotas. The eccentric Baron Nopsca (1923), for example, in his detailed synopsis of the late Cretaceous reptiles of Hungary, painted a vivid picture of life on and around the ancient islands of Transylvania. Nopsca argued that these islands and archipelagos, generated during the Alpine orogeny, formed a locus for distinctive, isolated dinosaur faunas during the late Senonian; however, continuing emergence peaked during the Eocene, providing a land bridge for the interchange of mammal faunas between Europe and Asia. Changing palaeogeography clearly had an important effect on the biogeography and migration patterns during the Mesozoic and Tertiary in the Carpathian region and the evolution of its faunas.

8.2 MODERN BIOGEOGRAPHY

Today the surface of the Earth is divided into six main biogeographical regions (Fig. 8.2) based on maps first devised in the last century

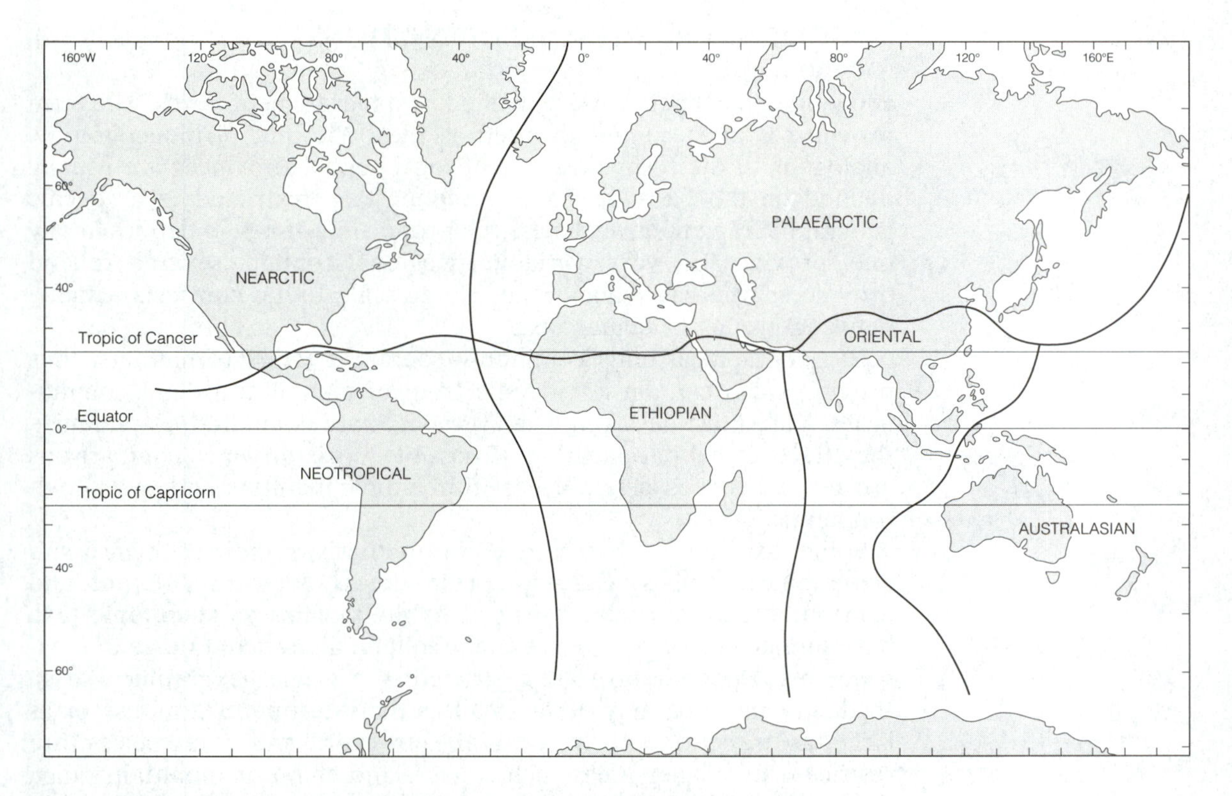

Figure 8.2 Modern biogeography. (Adapted from various sources.)

by Philip Sclater and Alfred Russel Wallace. These regions were initially defined on the basis of characteristic terrestrial biotas including bird and mammal faunas and later flowering plants. These divisions have since been tested by a wide range of other faunal and floral data. Nevertheless biogeographical boundaries may be far from precise. For example much of Wallace's research concerned the bird and mammal faunas of the East Indies, where he defined Wallace's Line, the boundary between the Oriental and Australasian provinces. Subsequent work by Weber on the mammal and molluscan faunas of islands such as Timor established a boundary between the same two provinces farther south. Weber's line highlights the difficulties of biogeographical analysis in island ecosystems. Moreover biogeographical boundaries clearly need not follow continental boundaries.

Two models have dominated biogeographical theories. The majority of biogeographical studies assume dispersal models where most organisms originate from a point and sequentially enlarge their geographical ranges, spreading across continents or ocean basins. In contrast, vicariance biogeography assumes the fragmentation of a biogeographic unit, such as a province.

8.3 SOME CONCEPTS AND DEFINITIONS

Biogeographical patterns are defined by a hierarchy of terms. The term **province** is most commonly used as the basic unit of biogeography, analogous to the formation in lithostratigraphy. Provinces are usually defined on the basis of shared **endemic** taxa restricted to a specific geographic region; nevertheless the proportion of key endemics in any one province is very variable. Regions contain several related provinces, whereas **realms** reflect large-scale biotic contrasts, usually between groups of higher taxa.

Biogeographical units commonly contain a range of habitats, thus provinces can contain a wide spectrum of animal and plant communities. Care must be taken when provinces are compared, to minimize the effects of habitat partition. If possible, cross-province comparisons are made between assemblages from similar habitats and their environments.

Some taxa can be characterized on the basis of their ecological and geographical ranges. **Eurytopic** taxa are ecologically tolerant and distributed across a wide range of habitats, whereas **stenotopic** taxa have limited ecological tolerance and inhabit a restricted range of environments. **Endemic** taxa are restricted to a single geographical area. In theory the area may be as small as the intestine of a mouse or as large as a continent. In general, however, endemic species are restricted to smaller areas such as an island chain or mountain range than, say, endemic families which occur across an entire continent or ocean. In addition so-called young endemics characterize relatively new geological environments; new endemics including snails have colonized the volcanic islands of the South Atlantic. These centres of speciations may in the future source much wider-spread taxa. On the other hand, old or relict endemics are localized remnants of once widespread taxa now essentially occupying a refugium. Fossil refugia may have held and protected so-called Lazarus taxa during extinction events (section 9.3.2). **Pandemic** taxa have more widespread distributions, whereas organisms with global or cosmopolitan distributions are relatively rare. Some taxa have **disjunct distributions**, inhabiting widely separated areas; a variety of explanations are possible for these discontinuous ranges.

A fixed biogeographical area, in theory at least, will only provide habitats and space for a finite number of taxa (section 9.3.3.1). **Hypsographic curves** have been used to determine first the absolute sea-level in a particular region, and secondly the available shelf area for colonization. This type of analysis has been extended to predict species–area ratios for a variety of habitats across the entire globe (see Chapter 9). Using hypsographic models the effects, for example, of sea-level rise are spectacular; relatively small rises in sea level can significantly increase the shelf areas available for colonization. The species–area effect has featured in a number of studies of extinction and radiation events.

8.4 CONTROLS ON BIOGEOGRAPHY

The geographical variation of environmental parameters, commonly referred to as ecological or geographical biogeography, provides the main control on the distribution of living and fossil organisms, although past distributions were influenced also by the geological and tectonic history of the region. George Gaylord Simpson, based on his knowledge of mammals, described three main types of barrier to faunal and floral migration: corridors were restricted passages but were open at all times, filters allowed selected access, while sweepstake routes were only occasionally open and access was nearly random. In addition different types of barrier are especially appropriate, respectively, to continental and marine organisms. In oceanic environments the distribution patterns of marine faunas and floras are governed by a number of factors such as the expanse of oceans, ocean currents and salinity, together with ocean ridges and islands. Nevertheless, cross-latitudinal temperature gradients exert the most effective control on marine biogeography. Over 70% of Earth's surface is covered by oceans, extending from surface waters to depths of 11 kilometres. Four main climatic zones are recognized: cold, cold-temperate, warm-tropical and tropical. This simple latitudinal pattern is interrupted by continental plates and oceanic ridges, together with hydrothermal vents. In general terms endemicity decreases with depth to a certain level, but contrary to most models, assemblages in deep ocean basins are highly endemic.

The thermocline provides a significant vertical barrier for marine organisms but varies in depth according to seabed topography, latitude, seasons and water currents. In many regions water temperature decreases rapidly from the surface to depths of about 100–1000 m; below this threshold the temperature drop, with increasing depth, is much less. Above the thermocline, the thermosphere contains relatively warm but seasonally variable water mass whereas below, the psychosphere holds cold, thermally stable water masses. In low and mid latitudes the thermocline is commonly coincident with the interface between shelf and oceanic waters and partitions shallow and deep water faunas; at high latitudes the definition and effects of the thermocline are less clear. Moreover the development of deep, dense warm but high-salinity bodies of water provides a further barrier to marine biotas. In continental environments, barriers are provided by rivers, canyons, inland seas, mountain ranges and even rain forests together with climatic zones.

Many species or groups of species have disjunct rather than continuous geographical distributions. Disjunct distribution may be created by a range of tectonic processes such as continental rifting and seafloor spreading, the offset of provinces along an active transform system or by the tectonic suturing of two quite different biotic regions (Smith, 1990). Moreover the variable dispersal, establishment and survival of biotas can themselves lead to a discontinuous patterns of distribution.

A number of disjunct fossil distributions have recently been related to the sporadic development through time of chemosynthetic environments in cold seeps (section 7.11.1.1). Modern seep-related faunas, dominated by molluscs and worms, are now relatively well known; but similar chemosynthetic environments in the past have different faunas suggesting these biotas may have evolved with time (Campbell and Bottjer, 1995a). A range of Palaeozoic faunas dominated by brachiopods may also have been related to chemosynthetic environments (Campbell and Bottjer, 1995b).

The distributional patterns of fossil organisms have been described and investigated according to two main paradigms, dispersal and vicariance models.

8.4.1 Dispersal biogeography

The dispersal model assumes that organisms originate and initially populate a centre from where subsequent outward dispersion or migration occurs to expand their geographic ranges. The model arose mainly from Alfred Wallace's studies of the animals of the East Indies and particularly suited 19th-century biologists and geologists when both continents and oceans were assigned fixed positions on the globe. Organisms thus extended their geographical ranges by a variety of dispersal mechanisms involving mobile larvae and spores or the mobile adult organisms themselves. Most organisms have the ability to move and disperse; even fixed or sessile animals and plants have mobile larval stages which can travel considerable distances, aided by agencies such as oceanic currents and wind.

Dispersal mechanisms form the basis for the **founder principle** and **allopatric speciation**. Migrants invading new geographic areas take with them only part of the entire genetic pool of the parent population. If the migrant population is now isolated from its parent gene pool, evolution will probably proceed differently from that in the parent population and both morphological and ecological shifts are likely. **Sympatric speciation** is analogous to allopatric speciation but the barriers are usually ecological, involving the partition of habitats and even behavioural patterns. Usually the compositions of arriving populations are related to their relative distances from the source areas; problems of survival associated with the small size of immigrant populations must be overcome and rapid expansion of both population and community sizes is essential.

Initial dispersal may be followed by a period of intense adaptive radiation when new opportunities in an ecological vacuum are exploited. Islands such as the Galápagos, Juan Fernandez and Hawaiian complexes together with the Micronesian islands all have provided opportunities for the adaptive radiation of animals and vegetation (Carlquist, 1974).

Box 8.1 The role of larvae

In contrast to the direct development programmes of the vertebrates, the majority of immature marine invertebrates are represented by a larval stage. There are a number of types of larvae. **Lecithotrophic** larvae do not require sustenance, living either in surface waters as a pelagic phase or near the seabed as a demersal phase. The majority of modern larvae, however, do feed and float in the surface water as **planktotrophic** forms. About 70% of all marine benthos have planktotrophic larvae. These larvae are abundant and widely dispersed, but many are wiped out. Pelagic lecithotrophic larvae are most common on high-latitude shelves, whereas the demersal phase also inhabits deep-sea environments. In the Boreal regions most larvae have insufficient life spans for large-scale dispersal. Planktotrophic larvae are most abundant in the shallow-water environments of the tropics where many species have a teleplanic larval stage of six months to a year during which they are capable of long-distance dispersal. Most larvae can be related to their mature counterparts; nevertheless it is possible that larvae during their development can be transferred, horizontally, between otherwise unrelated higher-level taxonomic groups, casting some doubt on their use in phylogenetic reconstructions (Williamson, 1992).

The larval development of species, however, exercises a considerable control on their adult ecology and evolution but also on the biogeographical distribution of animals and plants (Fig. 8.3). For example, planktotrophic larvae have a very high capacity for dispersal and thus are widespread geographically; however, their species are generally long ranging with both low speciation and extinction rates (Jablonski and Lutz, 1983). In contrast nonplanktotrophic gastropods, for example from the Tertiary of the Gulf Plain, have shorter geological and narrower geographical ranges (Hansen, 1980).

Two important mechanisms have helped dispersal throughout geological time and may go some way towards explaining many disjunct distributions, island-hopping and rafting (see Donovan, 1992 for summary). The oceans, today, are punctuated by islands. Commonly these are relatively short-lived volcanic seamounts associated with spreading ridges or hot spots and more enduring archipelagos and microcontinents consisting of continental material which rifted from adjacent cratons but are now free to move across the ocean as suspect terranes. Past oceans almost certainly had similar configurations. Oceanic islands are useful stop-overs during migrations; mobile organisms and even the larvae of benthic organisms may migrate between islands during the gradual and step-wise expansion of the geographical range of animal and plant taxa. For example, the taxonomy and range of species of the giant rudist bivalve *Torreites* are now very well documented (Skelton and Wright, 1987). It has a disjunct range, reported from only the Caribbean and Oman and the United Arab Emirates. Migration from a centre in the Caribbean may have occurred, during the late Cretaceous, via a series of oceanic seamounts along the Tethys Ocean.

Many organisms live attached to plankton and other floating objects in the oceanic surface waters. This specialized life style has a long

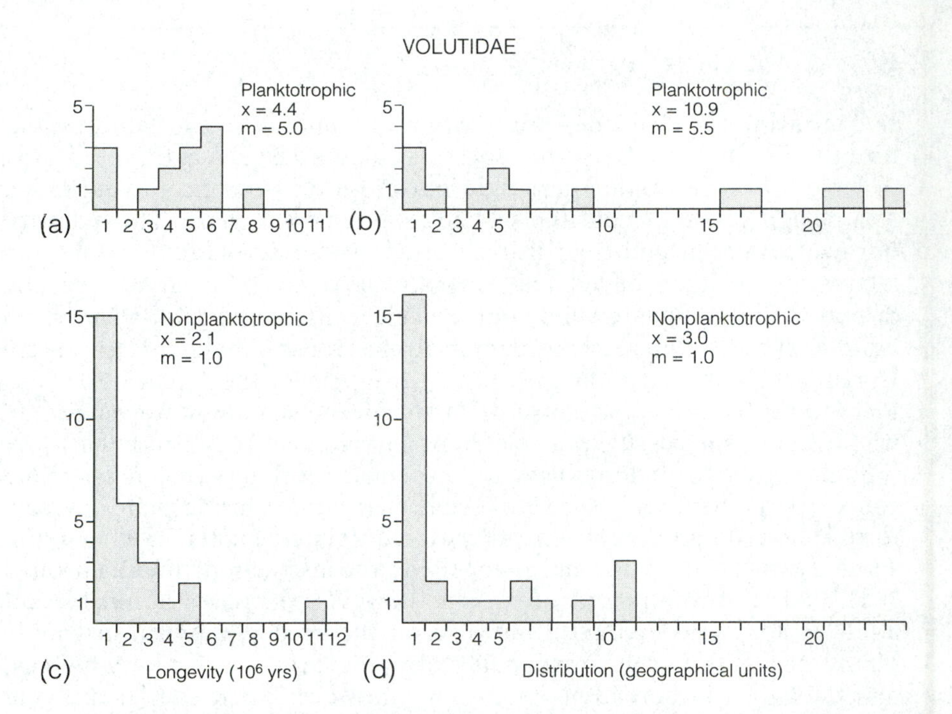

Figure 8.3 Distribution and longevity of late Cretaceous gastropods from the Gulf and Atlantic plains: the planktotrophic larvae of volutid gastropods generally have a greater longevity and geographical distribution than their nonplanktotrophic counterparts. (From Hansen, 1980.)

history (Wignall and Simms, 1990) and clearly drives widespread dispersal. Many different organisms can participate in these communities; Wignall and Simms (1990, text fig. 2) illustrated five taxa from an upper Jurassic log, including bivalves, brachiopods and a dominant serpulid worm, whereas Donovan (1989, fig. 2) illustrated a goose barnacle attached to a floating serpulid from Recent sediments on the Caribbean island of Tobago. Floating material will also provide attachment for a variety of larvae. Admittedly floating logs are unlikely to have punctuated the surface waters of the older Palaeozoic oceans. Nevertheless modern pumice has been recovered from the Pacific Ocean with an epifauna; the Ordovician, for example, was a period of considerable volcanicity, which may have provided such rafts for a variety of attached organisms and possibly their larvae.

Dispersal models favour phenetic methods of investigation such as cluster analysis which group together areas with similar biotas; these methods are particularly suited for the analysis of data based on an assumption of dispersion. Modern dispersal models have now been adapted and modified to account for plate tectonics.

The origin and direction of taxa dispersal has been explained by three main models (Crame, 1993). Darwin and Wallace, for example, accepted that high latitude biotas were generated by the poleward

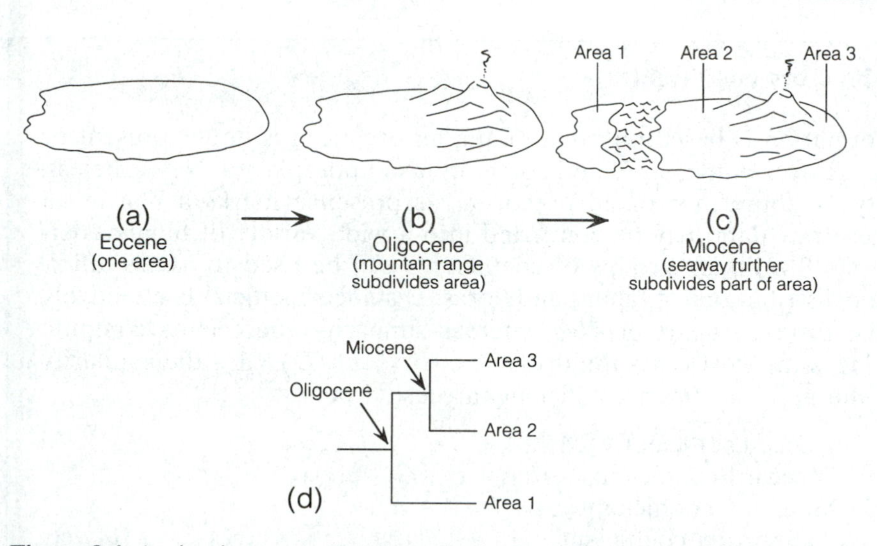

Figure 8.4 A vicariance model; through the Eocene, Oligocene and Miocene area (a) is first partitioned by a mountain belt (b) and further split by a seaway (c). Three areas (1, 2 and 3) are generated, reflected in the area cladogram. (Adapted from Grande, 1990.)

migration of taxa from the diverse biotas of the tropics. Alternatively some taxa may have originated at polar latitudes and migrated towards the equator during less harsh climatic conditions. A third model assumes that the modern distribution of polar taxa are merely the relicts of a past more cosmopolitan distribution.

8.4.2 Vicariance biogeography

Vicariance biogeography assumes a mobilist model based on plate tectonic theory (Fig. 8.4). These models are based on the studies of Leon Croizat; vicariance biogeography promotes the role of plate tectonics in the development of provinciality. Organisms develop *in situ* without significant dispersal taking place. The ranges of the taxa are modified by the break-up and dispersal of both continental and oceanic crust, rather than by the dispersal of the organisms themselves. Disjunct ranges and distributions are thus created by the fragmentation of once continuous provinces. Continents or island archipelagos may be split by sea-floor spreading; the divergent parts of these complexes carrying a range of biotas will gradually disperse the original province, with relative isolation generated by expanses of water. Alternatively intracontinental provinces may be split and fragmented by, for example, rising mountain ranges or developing rift valleys.

Box 8.2 Distance and similarity coefficients

Most biogeographic information is based on binary data; an organism is either present or absent, and so rare taxa are over-emphasized and common taxa underplayed. Nevertheless, irrespective of the rarity or abundance of an organism, its presence marks a flag in an overall taxon range. These raw data may be converted into a wide variety of binary coefficients for further analysis. Two main groups of coefficients can be used to assess affinities and palaeogeographic distributions of faunas and floras. Distance coefficients effectively measure, statistically, the distance apart of sites whereas similarity coefficients compute the closeness of assemblages. In most cases the distance coefficient (D) = 1 – the similarity coefficient (S). Four of the more common coefficients are listed below:

$$\text{Dice coefficient} = 2A/(2A + B + C)$$
$$\text{Jaccard coefficient} = A/(A + B + C)$$
$$\text{Simple Matching coefficient} = A + D/(A + B + C + D)$$
$$\text{Simpson coefficient} = A/(A + E)$$

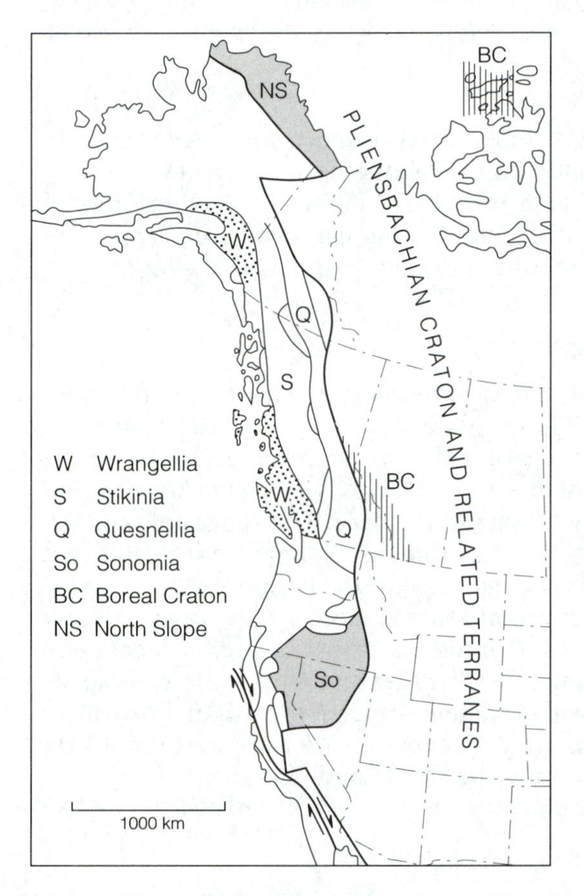

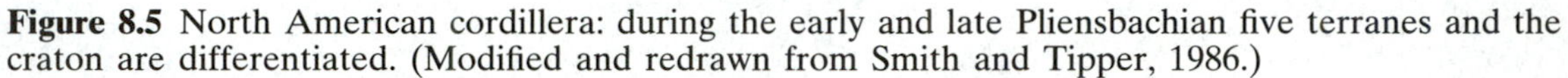
Figure 8.5 North American cordillera: during the early and late Pliensbachian five terranes and the craton are differentiated. (Modified and redrawn from Smith and Tipper, 1986.)

A is the number of taxa common to any two samples, *B* is the number in the first sample, *C* is the number in the second sample, *D* is the number of taxa absent from both samples and *E* is the smaller value of *B* or *C*.

These analyses will produce symmetrical matrices of distance or similarity coefficients. However two types of matrix can be generated. Intersite comparisons or *Q*-mode analysis provides a matrix of distances or similarities between localities or sites based on the distribution of the variates or taxa; *R*-mode analysis will generate a matrix of distances or similarities between taxa occurring across localities or sites which now act as the variates. The most efficient and informative analyses illustrate a cross plot of the *Q*- and *R*-mode clusters against the original, ordered data matrix (see also Chapter 7).

The North American Western Cordillera (see below) consists of a collage of terranes (Fig. 8.5) each with distinctive provincial signatures. Jurassic suspect terranes developed seaward of the North American craton carry ammonite faunas related to a northern circumpolar Boreal province and a southern Tethyan province. The changing provincial signatures of each terrane during the early Jurassic can monitor the movement of these microplates prior to docking (Smith and Tipper, 1986; Figs 8.6–8.7 and Table 8.1).

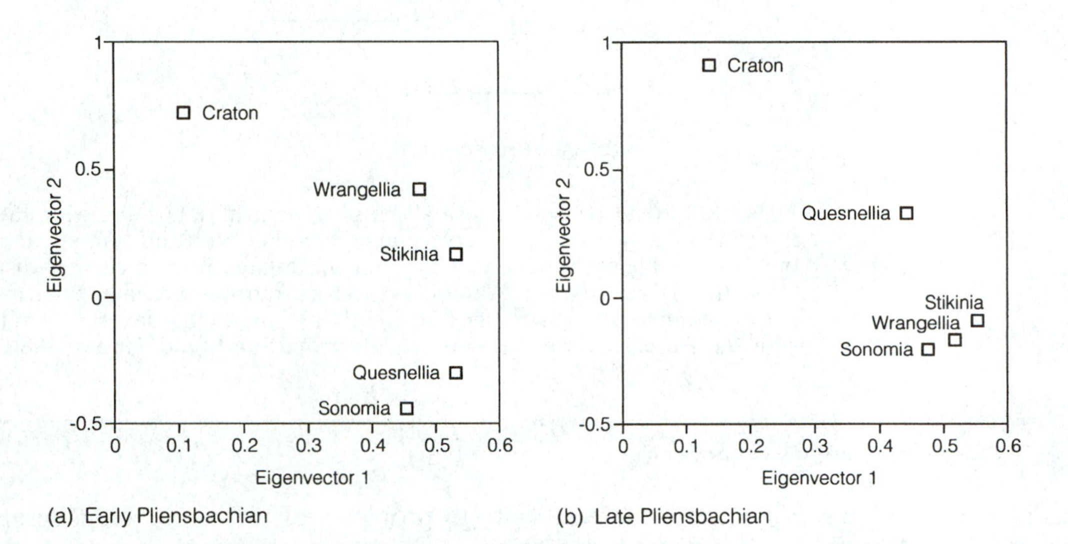

Figure 8.6 Ordination of small data set; the Q-mode cluster of terranes reflects the changing relative configurations of the terranes and Tethyan (Sonomia) and Boreal cratons between the early (a) and late (b) Pliensbachian. (Based on data from Smith and Tipper, 1986.)

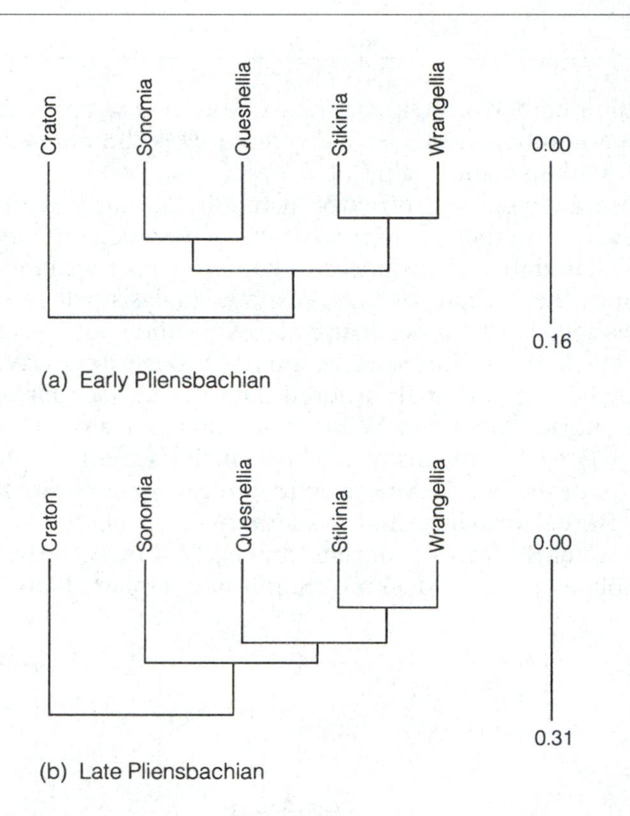

Figure 8.7 Dendrograms of a small biogeographical data set: during the early Pliensbachian (a) the displaced terranes together with the Sonomian part of the North American margin (with Tethyan faunas) form a cluster, distanced from the Boreal North American Craton; during the late Pliensbachian (b) the configuration has changed slightly as Quesnellia develops similarities with the Boreal Craton. (Based on data from Smith and Tipper, 1986.)

Box 8.3 Cladistics and biogeography

Cladistic methods have been applied specifically to problems of vicariance biogeography (Platnick and Nelson, 1978), although the technique works equally well as a method of organizing data sets without any such assumptions. Pattern or transformed cladism largely ignores the phylogenetic history of the taxa themselves but concentrates on their geographical distribution.

Taxonomic cladistics assumes any pair of organisms, at some stage in the past, shared a common ancestor (see Chapter 9). New taxa are recognized on the basis of derived features or apomorphies while these taxa also have shared features in common with their ancestors. When translated into biogeographical terms, and assuming a vicariance model, a now fragmented province shared a common origin with other provinces. Taxa that have appeared since fragmentation are analogous to 'autapomorphic' characters, now defining that particular province. Shared taxa, however, can be viewed as 'synapomorphies' and used to group together similar provinces that presumably were derived from a common province.

Table 8.1 Data matrices for a small biogeographical data set: the presence (1) and absence (0) of a range of ammonite taxa across five of the tectonic units annotated in Figure 8.5 for (a) the lower and (b) the upper Pliensbachian

(a) lower Pliensbachian	*Wrangellia*	*Stikinia*	*Quesnellia*	*Craton*	*Sonomia*
Amaltheus	0	0	0	0	0
Pleuroceras	0	0	0	0	0
Pseudoamaltheus	0	0	0	0	0
Becheiceras	1	0	0	0	0
Arieticeras	0	0	0	0	0
Fontanelliceras	0	0	0	0	0
Fuciniceras	0	0	0	0	0
Leptaleoceras	0	0	0	0	0
Lioceratoides	0	0	0	0	0
Protogrammoceras	0	0	0	0	0
Aveyroniceras	1	1	0	0	1
Prodactylioceras	0	0	0	0	1
Reynesoceras	0	0	0	0	0
Reynesocoeloceras	1	1	0	1	0
Acanthopleuroceras	1	1	1	0	1
Apoderoceras	1	0	0	0	0
Calliphylloceras	1	0	0	0	0
Coeloceras	1	1	0	0	0
Cymbites	0	0	0	0	0
Dayiceras	0	1	0	0	0
Dubariceras	1	1	1	0	1
Fanninoceras	0	0	0	0	0
Gemmellaroceras	1	1	0	0	0
Hyperderoceras	0	0	0	0	1
Gen nov	0	1	0	0	1
Metaderoceras	1	1	1	0	1
Phricodoceras	1	0	0	1	0
Phylloceras	0	1	0	0	0
Tropidoceras	1	1	0	1	0
Uptonia	1	1	0	1	0

Table 8.1b

(b) upper Pliensbachian	*Wrangellia*	*Stikinia*	*Quesnellia*	*Craton*	*Sonomia*
Amaltheus	1	1	1	1	0
Pleuroceras	0	0	0	1	0
Pseudoamaltheus	0	0	0	1	0
Becheiceras	1	1	0	0	0
Arieticeras	1	1	1	0	1
Fontanelliceras	1	0	0	0	0
Fuciniceras	1	1	1	0	0
Leptaleoceras	1	1	0	0	0
Lioceratoides	1	1	0	0	1
Protogrammoceras	1	1	1	1	1
Aveyroniceras	1	1	0	0	1
Prodactylioceras	1	1	0	0	1
Reynesoceras	1	0	1	0	0
Reynesocoeloceras	0	0	0	0	0
Acanthopleuroceras	0	0	0	0	0
Apoderoceras	0	0	0	0	0

Table 8.1 Continued

Calliphylloceras	0	0	0	0	0
Coeloceras	0	0	0	0	0
Cymbites	1	0	0	0	0
Dayiceras	0	0	0	0	0
Dubariceras	0	0	0	0	0
Fanninoceras	1	1	1	0	1
Gemmellaroceras	0	0	0	0	0
Hyperderoceras	0	0	0	0	0
Gen nov	0	0	0	0	0
Metaderoceras	0	0	0	0	0
Phricodoceras	0	0	0	0	0
Phylloceras	1	0	0	0	0
Tropidoceras	0	0	0	0	0
Uptonia	0	0	0	0	0

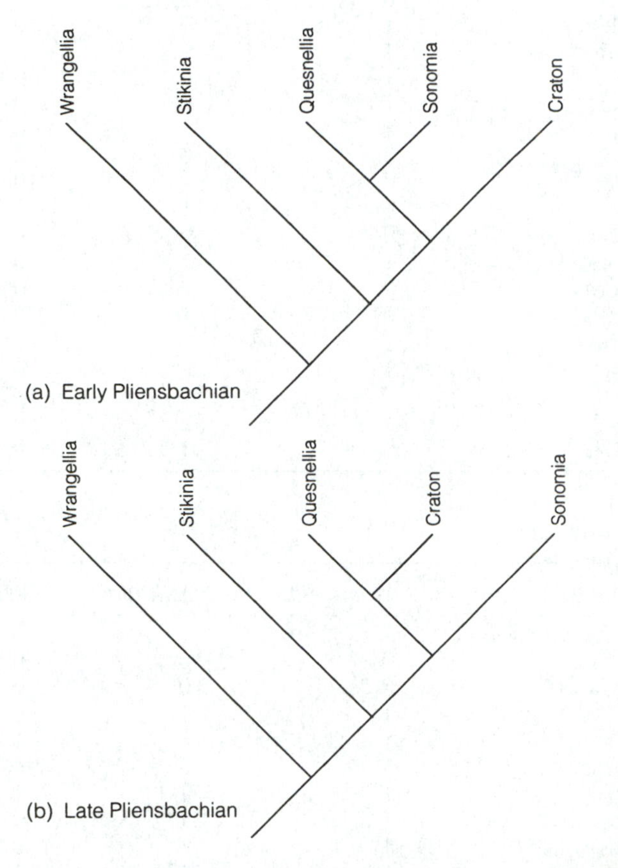

Figure 8.8 Cladistic analysis of a small data set: patterns similar to those of the dendrograms suggest a northward movement of Quesnellia during the early (a) and late (b) Pliensbachian. (Based on data from Smith and Tipper, 1986.)

8.5 RECOGNITION OF PAST BIOGEOGRAPHIC PROVINCES

A number of studies suggest that modern provinces based on species distributions are nevertheless comparable with fossil analogues based on genera (Jablonski *et al.*, 1985). Modern provinces are defined mainly on the distribution of living species of animals and plants. Ancient provinces have been recognized by a variety of qualitative and quantitative techniques, many noted above, but usually on the basis of the distribution of genera and higher taxonomic categories. The fossil record at the species level, however, is notoriously patchy and some authors have cast doubt on the direct equivalence of modern and ancient provinces. The numerical analysis of five modern provinces (the Arctic, Aleutian, Oregonian, Californian and Surian provinces) forming a latitudinal chain along the west coast of North America was based on the distribution of bivalves and gastropods (Campbell and Valentine, 1977). Both the Jaccard and Simpson coefficients were used to assess mutual similarities between the provinces. These provinces were recognized at the generic level and to a lesser extent at the familial level. The procedure was repeated for a longitudinal series of provinces developed within the temperate zones (the Oregonian, Californian, Virginian and Celtic provinces), with broadly similar results.

Box 8.4 Faunal provinces through time

Provincialism has varied throughout geological time (Fig. 8.9). Palaeobiogeographical signals have yet to be detected in the majority of Precambrian biotas. The Ediacara fauna of soft-bodied metazoan organisms has been used to test models for late Precambrian biogeography during the break-up of the supercontinent of Rodinia (Donovan, 1987). A number of assemblages within the earliest Cambrian Small Shelly or Tommotian fauna may be biogeographically controlled. Nevertheless the first clear signals have emerged from trilobite-dominated Cambrian faunas. Early Palaeozoic provinces have been recognized on the basis of benthic organisms including the brachiopods and trilobites together with microfossil groups such as the conodonts. In general terms the intense biogeographical signals of the Cambrian and Ordovician faunas are less clear during the Silurian following the amalgamation of many microcontinents with Laurentia during the late Silurian and Devonian.

During the Devonian the Old and New World realms relate to Europe and America, respectively, while the Malvinokaffric realm covered parts of South America, South Africa, China and Australia (Boucot, 1975). Late Palaeozoic provinces have been recognized with reference to rugose and tabulate corals, articulate brachiopods, bryozoans, ammonoids and crinoids (Bambach, 1990), when the Tethyan and Boreal realms were already differentiated. Diverse floral provinces related to climate are also recognized (Ziegler, 1990). Mesozoic provincial studies recognize the Tethyan (warm, circumtropical) and Boreal (cool, circumpolar) realms with a variety of fossils including the ammonites and belemnites. Cenozoic provinces developed following the breakdown of the Boreal and Tethyan systems in the late Cretaceous.

During the Palaeogene the broad outline of the modern continents and oceans had been established and by the Neogene the main continents had distinctive mammal faunas although the balance of individual faunas was modified by migrations such as the Great American Biotic Interchange. Modern island biotas have also developed with their own distinctive ecosystems (see, for example, Taylor *et al.*, 1979); some have peculiar large birds and rodents and small ungulates.

8.6 PALAEOCLIMATOLOGY

Both the distribution and diversity of organisms are at least partly controlled by climatic gradients. In particular many environments in the tropics such as biological reefs and rainforest ecosystems support very high-diversity communities. Moreover the geographic ranges of many plants and animals have a direct or indirect relationship to climatic and latitudinal gradients.

In general, transequatorial provinces are more extensive than those of the temperate latitudes; polar regions, where there is little overall variation in temperature, are possibly the least extensive. Biogeography and climatic gradients are related to patterns of changing biodiversity. In broad terms, low latitudes support high-diversity faunas; biodiversity decreases away from the tropics towards the poles. Nevertheless, superimposed on this simple working scenario is the control of environment on biodiversity. Studies on modern bivalve, bryozoan, coral and foraminifera faunas show marked increases in diversity towards the equator, and because many cooler-water species breed later in life, individuals may be larger than their tropical counterparts. High-diversity tropical biotas suggest there may be a tighter packing of ecological niches by stenotopic *K*-strategists; in contrast, low-diversity temperate biotas may be dominated by *r*-selected, eurytopic, trophic generalists (Hallam, 1994). In addition, in the tropics, there is a greater stability of resources such as light and nutrients throughout the year, in contrast to the greater seasonality at higher latitudes.

Latitudinal biodiversity gradients have probably existed throughout much of the Phanerozoic and have been demonstrated in a variety of fossil groups. A possible explanation involves the diversity pump (Valentine, 1984) where some high-latitude biotas are wiped out during episodes of cooling while others migrate towards lower latitudes, adapt and diversify; diversity thus increases at the low-latitude end of biotic ranges, usually in the tropics. Neogene biotic distributions, with diverse, relatively young tropical faunas, may be a result of such migrations of marine invertebrates driven by cooler conditions during the Palaeogene. Additionally the tropics provide more stable environments when compared with the seasonality of the temperate and polar

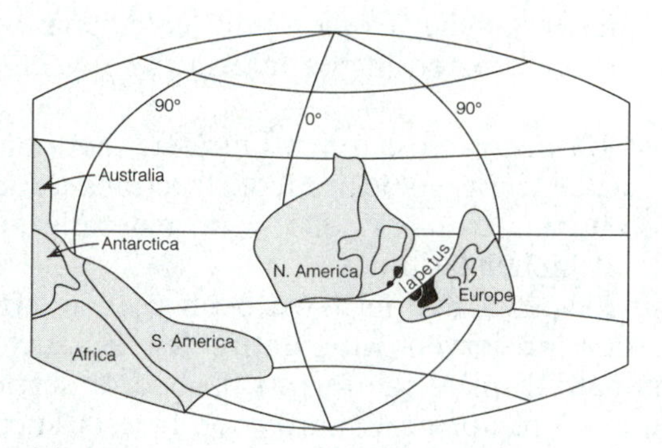

(a) 540 Ma Cambrian

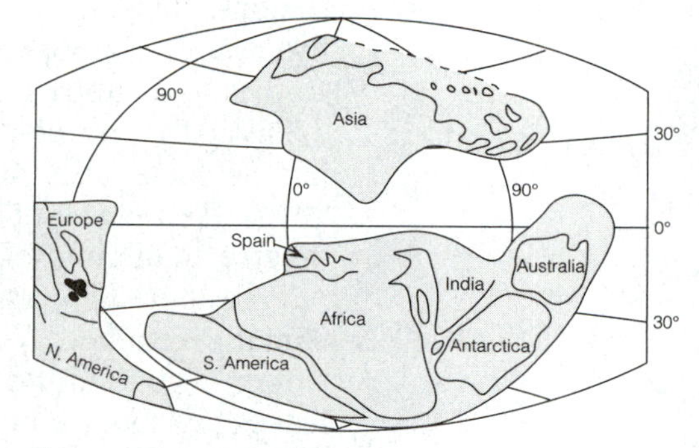

(b) 380 ± 30 Ma Devonian

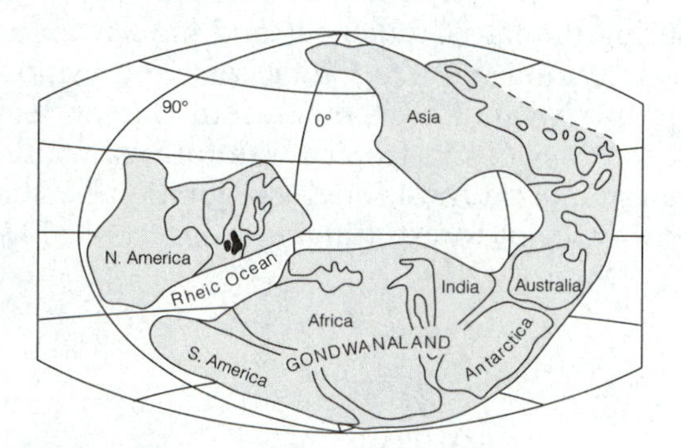

(c) 340 ± 30 Ma Carboniferous

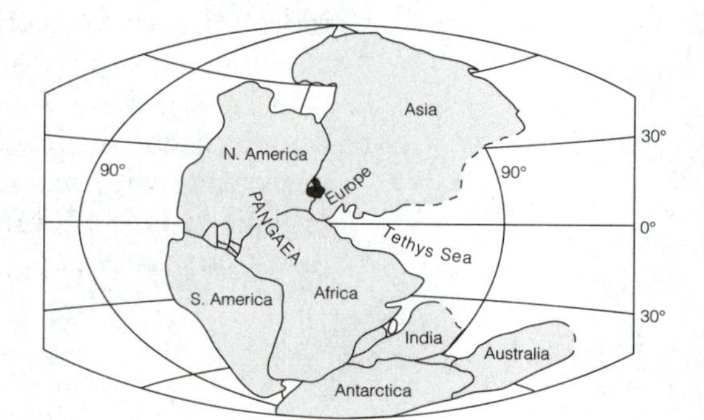

(d) 245 Ma Permian/Triassic

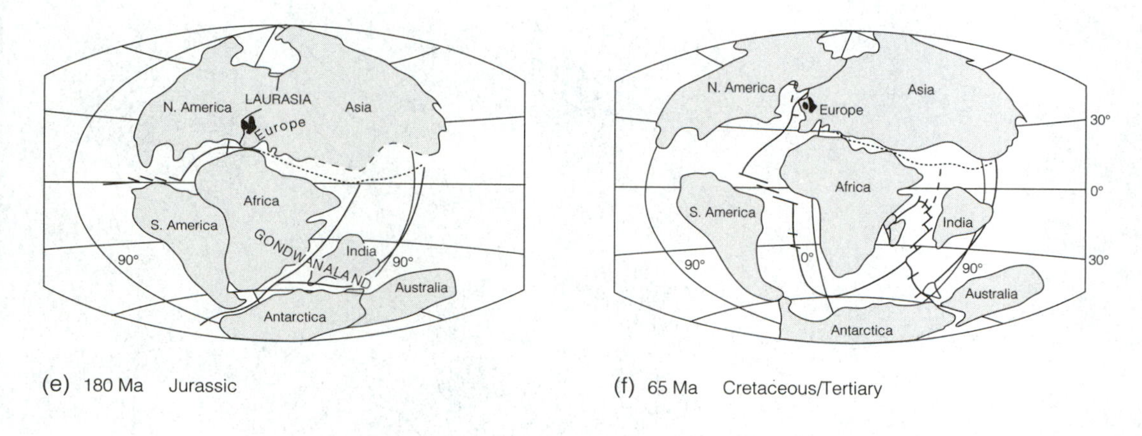

(e) 180 Ma Jurassic

(f) 65 Ma Cretaceous/Tertiary

— Constructive plate boundary - - - Destructive plate boundary

Figure 8.9 Collage of selected provinces through time: the early Palaeozoic was characterized by low and high-altitude provinces separated by the Iapetus Ocean, during the late Palaeozoic a Rheic Ocean separated Old and New World faunas, the Mesozoic was characterized by Boreal (high-latitude) and Tethyan (low-latitude) faunas as the world later approached its present configuration during the Cenozoic. (Modified from various sources.)

regions. During warm phases in the tropics and cold phases at the poles new opportunities are created at the interface between the tropics and subtropics.

Diversity gradients established across early Jurassic ammonoid faunas have also aided palaeogeographical reconstructions (Smith, 1990); Tethyan faunas, occupying tropical belts were generally more diverse than counterparts at higher latitudes.

Tree leaf shape, size and type of margin have been used to signal climatic regimes. Vegetation preserved with entire leaves, such as woody dicotyledons, dominate tropical rainforests. Lobed or serrated leaves are more common in more northerly hardwood forests, such as alder, beech, birch, elm and oak. Drip tips are commonly developed in the evergreen vegetation of humid environments. These leaf morphologies evolved during the late Cretaceous and have been used as palaeolatitude indicators during the late Cretaceous and Cenozoic (Fig. 8.10). Leaf size is also related to temperature: larger leaves usually occur in the understoreys of the tropical rainforests, with size decreasing with decreasing temperature. Modern vegetation has been organized into 12 different types of forest, ranging from tropical rainforests to tundras.

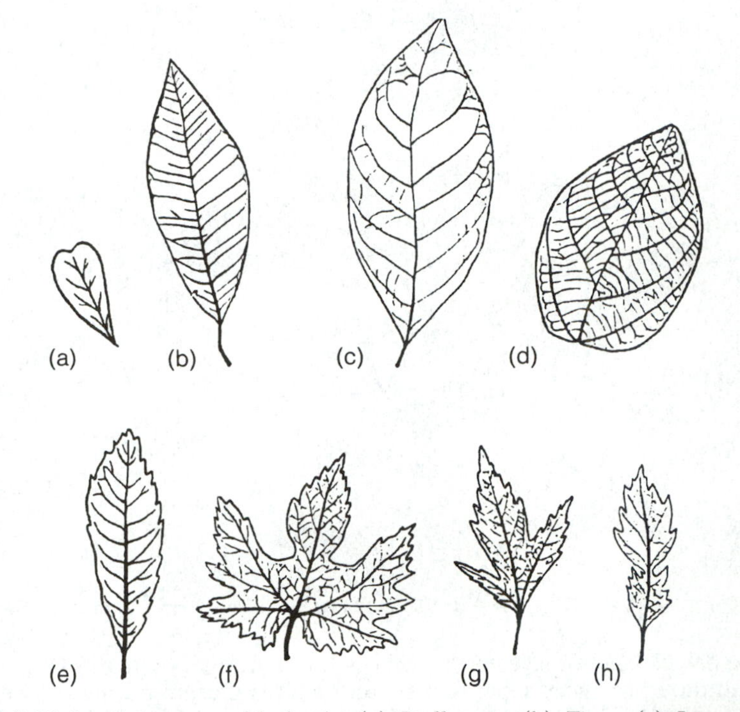

Figure 8.10 Leaf shapes and latitude: (a) *Dalbergia*, (b) *Ficus*, (c) *Laurus* and (d) *Apeiobopsis* from the Eocene of the Isle of Wight have entire margins typical tropical regions whereas (e) *Myrica*, (f) *Vitis*, (g) *Acer* and (h) *Quercus* from the Miocene of the South of France have dentate margins characteristic of temperate zones. (Redrawn from Fritel, 1903.)

8.7 PLATE MOVEMENTS

Despite the disbelief voiced over early palaeontology-based models for continental drift, modern plate tectonic and palaeogeographical reconstructions now rely heavily on biological data to both suggest and test plate distributions. Terrane models recognize that orogenic zones consist of a collage of tectonic units with unique geological histories mutually separated by tectonic structures such as faults. Virtually all orogens can now be interpreted in this way. Analyses of ancient biogeographical patterns and signatures have been and continue to be of fundamental importance in the generation of plate tectonic and terrane models for the Phanerozoic. Fossils have also been critical in distinguishing exotic blocks of crust, their origins and their tracks through time. Seaward of cratonic miogeoclines, oceanic terranes can move unconstrained and are often termed suspect. Faunal and floral data have helped fingerprint terranes, while their changing provincial signals can track their movements across latitudes. The North American Cordillera contains a number of suspect terranes, identified by their faunal signatures.

In the Early Palaeozoic orogen in Ireland, Harper and Parkes (1989) have described terranes located north of the Iapetus suture containing typically North American faunas, whereas terranes far to the south of the suture contain different faunas that developed marginal to the microcontinent of Avalonia. Some smaller terranes in central Ireland almost certainly evolved within the Iapetus Ocean itself, with their own distinctive, often endemic, faunas. All these terranes are bounded by faults; the fossil evidence suggests there has been considerable displacements between the blocks, since quite different faunas are now juxtaposed.

The majority of geographical barriers such as mountain ranges and ocean basins result directly from plate tectonic processes. Thus island chains associated with island arcs, spreading ridges or mantle hot spots may provide corridors for both terrestrial animals and plants while their relatively small, submerged shelves allow the passage of marine organisms (Fig. 8.12). Ocean basins, and especially subduction trenches, present formidable barriers for the majority of marine organisms. On land, mountain ranges resulting from continent–continent collision are efficient barriers to most terrestrial organisms. Valentine (1973) drew attention to a range of plate tectonic settings, including the evolution of spreading ridges, island arcs, subduction and transform zones, that may influence biological distributions. It is now recognized that in most cases provinciality develops through continental fragmentation and the controls of climatic or temperature gradients.

In general terms the highest provinciality is recorded during intervals of palaeocontinent dispersal where many provinces are associated with a variety of relatively small continents and oceans basins separated by latitudinal gradients (Valentine *et al.*, 1978); in contrast, fewer

Box 8.5 North American Cordillera

During the Mesozoic, high-diversity, low-latitude Tethyan faunas are generally quite distinct from the lower-diversity, high-latitude Boreal faunas and can be used to fingerprint high- and low-latitude terranes. The North American Cordillera is a mosaic of terranes (Fig. 8.11a), now plastered onto the west coast of the United States, but most probably originated at lower latitudes and were moved along the edge of the North American craton by strike-slip faulting. For example, in an east–west traverse across these terranes there is a progressive northward displacement of Tethyan type faunas of early Jurassic age; some of the more exotic, far-travelled terranes may have moved over 1300 kilometres. These American terranes hosting early Mesozoic Tethyan faunas may equally have originated in the Pacific west of the orogen as island complexes (Newton, 1987). Longitudinal rather than latitudinal terrane movement may have added these faunas to the Western Cordillera (Newton, 1988).

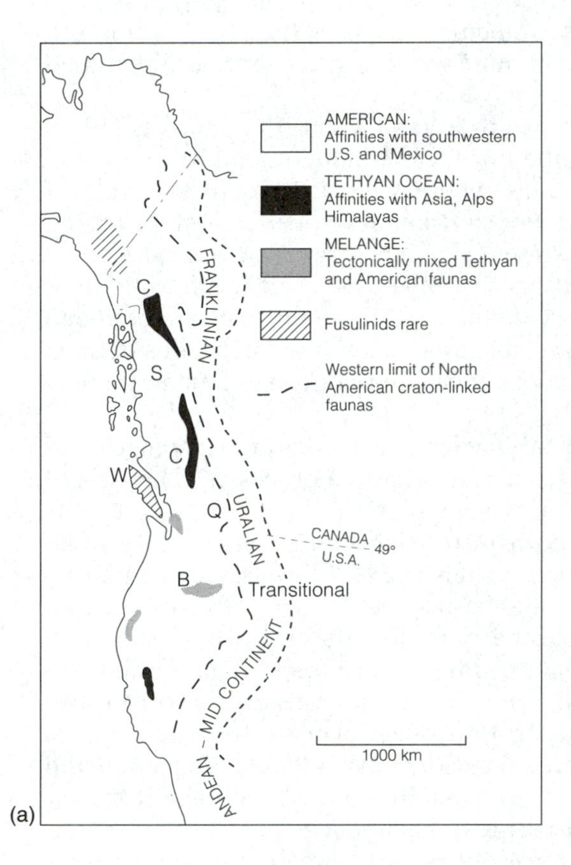

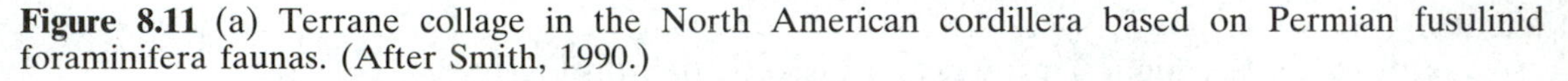
Figure 8.11 (a) Terrane collage in the North American cordillera based on Permian fusulinid foraminifera faunas. (After Smith, 1990.)

Another way of assessing the origins and tracks of these terranes involves plotting the boundaries between the provinces on the craton through time (Fig. 8.11b) (Hallam, 1986).

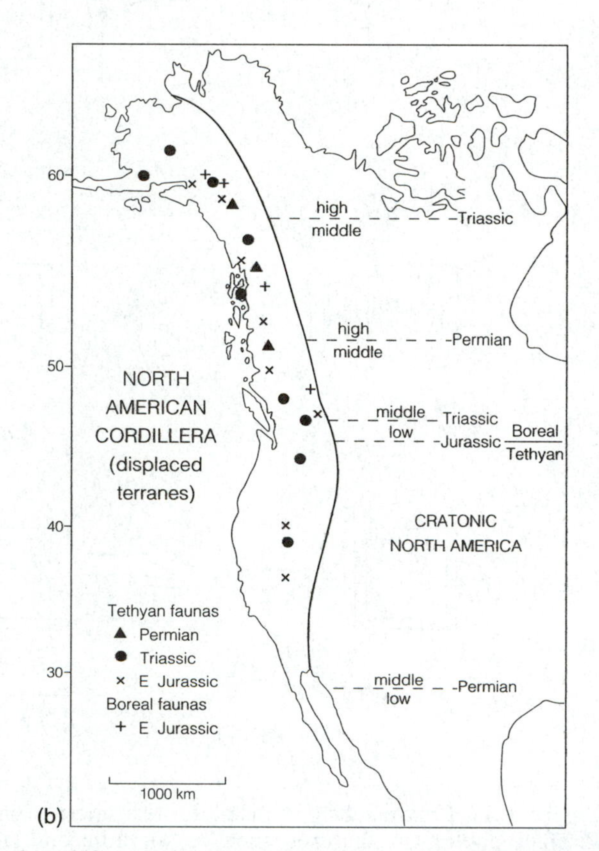

Figure 8.11 (b) Displaced faunas in terranes and provincial boundaries on the craton: postulated latitudinal boundaries on the craton during the Permian, Triassic and Jurassic are indicated: displaced faunas in marginal terranes suggest northward movement. (After Hallam, 1986.)

provinces are associated with periods of supercontinent assembly. The changing diversity of skeletal shelf benthos can be tracked through the Phanerozoic against changing continental configurations (Smith, 1990, fig. 8). The end Precambrian break-up of Rodinia generated, during the early Palaeozoic, a series of smaller continents including Avalonia, Baltica, Laurentia and Siberia together with the larger mass of Gondwana. The early–mid-Ordovician radiation of shelly benthos can be correlated with this interval of continental dispersion across a range of polar to equatorial latitudes following the late Precambrian break-up of Gondwana (Fig. 8.13). The massive drop in biodiversity at the Permian – Triassic boundary is coincident with the assembly of Pangaea while high diversities are associated with its breakup during the mid–late Mesozoic.

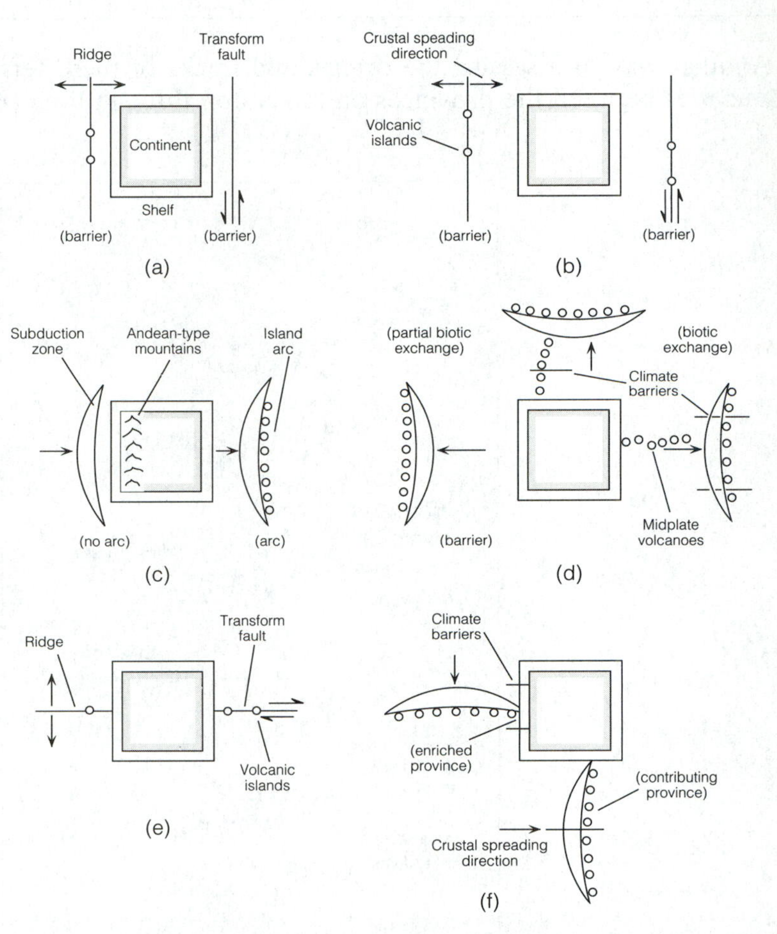

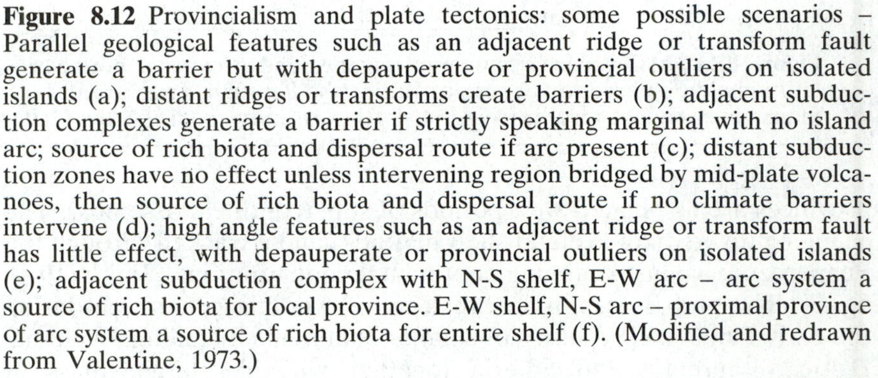

Figure 8.12 Provincialism and plate tectonics: some possible scenarios – Parallel geological features such as an adjacent ridge or transform fault generate a barrier but with depauperate or provincial outliers on isolated islands (a); distant ridges or transforms create barriers (b); adjacent subduction complexes generate a barrier if strictly speaking marginal with no island arc; source of rich biota and dispersal route if arc present (c); distant subduction zones have no effect unless intervening region bridged by mid-plate volcanoes, then source of rich biota and dispersal route if no climate barriers intervene (d); high angle features such as an adjacent ridge or transform fault has little effect, with depauperate or provincial outliers on isolated islands (e); adjacent subduction complex with N-S shelf, E-W arc – arc system a source of rich biota for local province. E-W shelf, N-S arc – proximal province of arc system a source of rich biota for entire shelf (f). (Modified and redrawn from Valentine, 1973.)

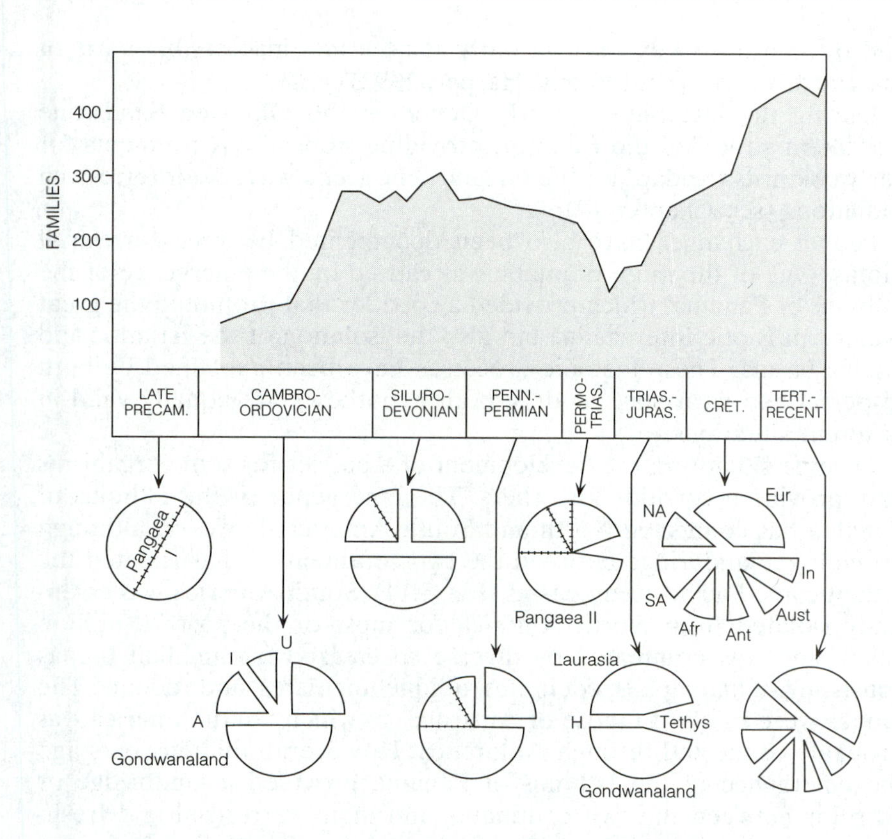

Figure 8.13 Changing family diversity of skeletal benthos through time in relation to plate configurations: high diversities are coincident with continental fragmentations, for example during the Ordovician, Devonian and the Cretaceous-Tertiary. Abbreviations: I – Iapetus Ocean, A – pre Appalachian-Variscan Ocean, U – pre-Uralian ocean, H – Hispanic corridor. (Redrawn from Smith, 1990.)

8.8 BIOGEOGRAPHY AND EVOLUTION

Both the present and past geographical configurations of the surface of the Earth have influenced evolutionary processes through time. The dynamic palaeogeography of the planet's surface in many instances may have helped promote evolutionary change. The early-mid-Ordovician radiation of skeletal organisms was coincident with intense magmatic and tectonic activity as Gondwana continued to break-up and island arcs and archipelagoes were generated in the world's oceans (Harper *et al.*, 1996). These were undoubtedly centres for speciation and possibly also macroevolutionary processes, setting the agenda for the radiation of the Palaeozoic marine biota. Unfortunately only a very small sample of the intra-oceanic islands of the Ordovician oceans still remain; these are generally found in mid-Palaeozoic mountain belts where the fossils are generally rare, metamorphosed, tectonized and often are now exposed in fairly remote and inaccessible parts of the Earth's crust (Bruton and Harper, 1992).

During the late Silurian–early Devonian the Old Red Sandstone Continent straddled the Equator, providing tropical environments for early colonists to adapt for life on land. The scene was set for terrestrial radiations (see Chapter 10).

Faunal exchanges have also been documented between terrestrial biotas. One of the more dramatic was caused by the emergence of the Isthmus of Panama, which provided a corridor that promoted the great American Biotic Interchange but also the isolation of the Atlantic and Pacific basins. These tectonic processes have been explained by both dispersal and vicariance models; both hypotheses are equally valid in different situations.

In some situations, the development of a barrier for some organisms may provide a corridor for others. The emergence of the Isthmus of Panama has connected North and South America; however, although providing a landbridge between the two continents, it has isolated the Atlantic and Pacific ocean basins (Fig. 8.14). South America was essentially isolated from North America for most of the past 70 million years, and was dominated by diverse specialized mammalian faunas consisting of marsupials, edentates, unique ungulates, and rodents. The faunas were similar to those of Australia, to which South America was probably connected through Antarctica. However, 3 million years ago the emergence of the Isthmus of Panama provided a landbridge or corridor between the two continents and many terrestrial and freshwater taxa were free to move across this isthmus. The Great American Biotic Interchange (GABI) allowed the North American fauna to invade the south and essentially wipe out many of the continent's distinctive, marsupial-dominated, mammalian populations (Webb, 1991). But some South American mammals were equally successful in the north and some such as the armadillo, opposum and porcupine still survive today in North America.

On the other hand the emergence of the isthmus also promoted changes in the marine faunas of the Caribbean basin. Contrary to the extinctions expected in these faunas, marked increases in the diversity of Caribbean molluscs have been documented (Jackson *et al.*, 1993). The emergence of the terrestrial landbridge and marine barrier initiated the upwelling of nutrients in the Caribbean area, with an increase in species diversity throughout the region. This radiation during the last few million years more than compensated for any extinctions in these marine faunas.

Figure 8.14 (opposite) Biogeographical change during the Great American Biotic Interchange (GABI): sloths, ant-eaters, armadillos, porcupines and opossums migrate north while jaguars, squirrels, sabre-tooths, elephants, deer, rabbits and wolves head south across the landbridge provided by the Isthmus of Panama, completed about 3 Ma. (Redrawn from Benton, 1990.)

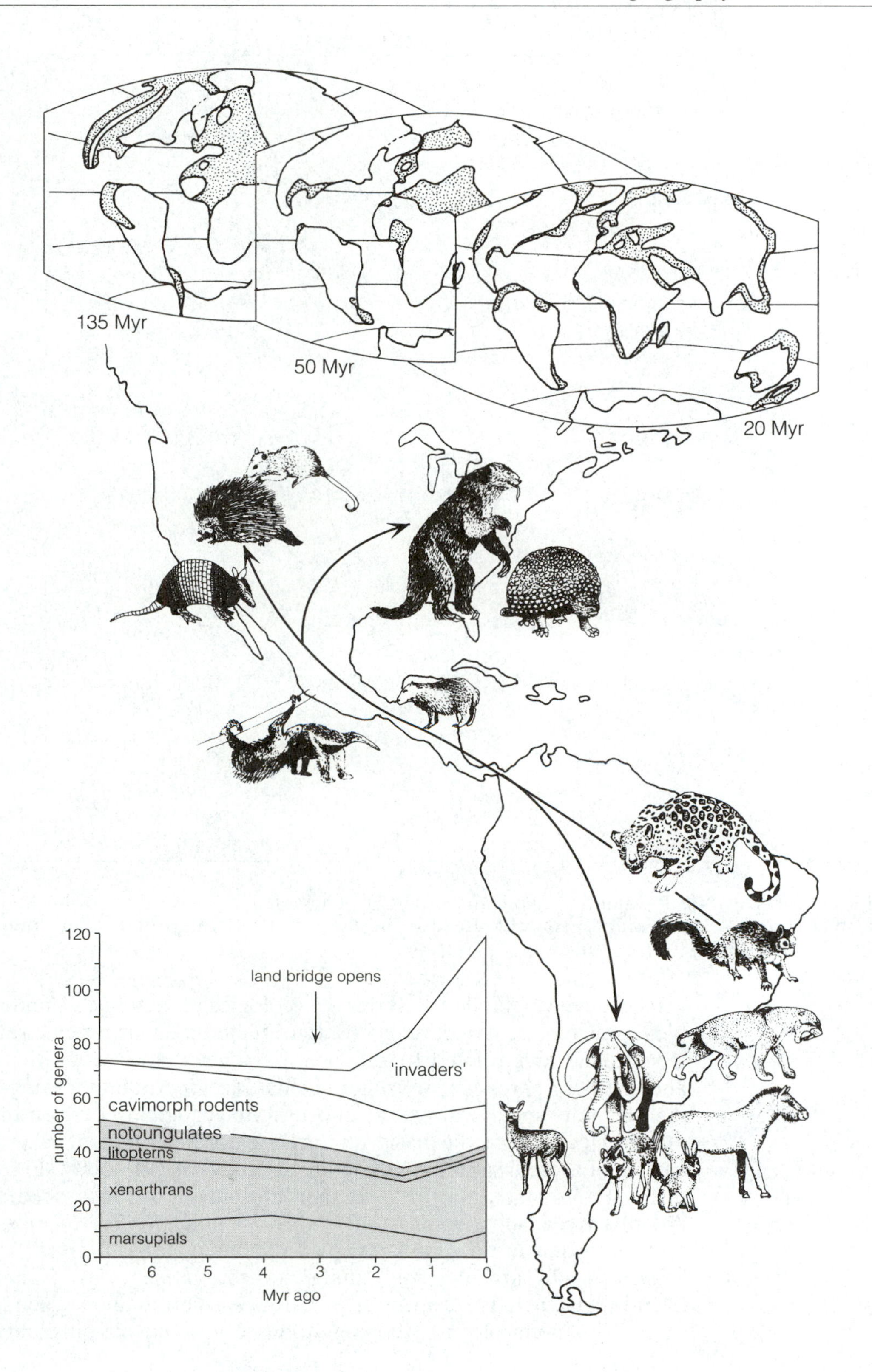
135 Myr
50 Myr
20 Myr
land bridge opens
'invaders'
cavimorph rodents
notoungulates
litopterns
xenarthrans
marsupials
number of genera
Myr ago
120
100
80
60
40
20
0
6
5
4
3
2
1
0

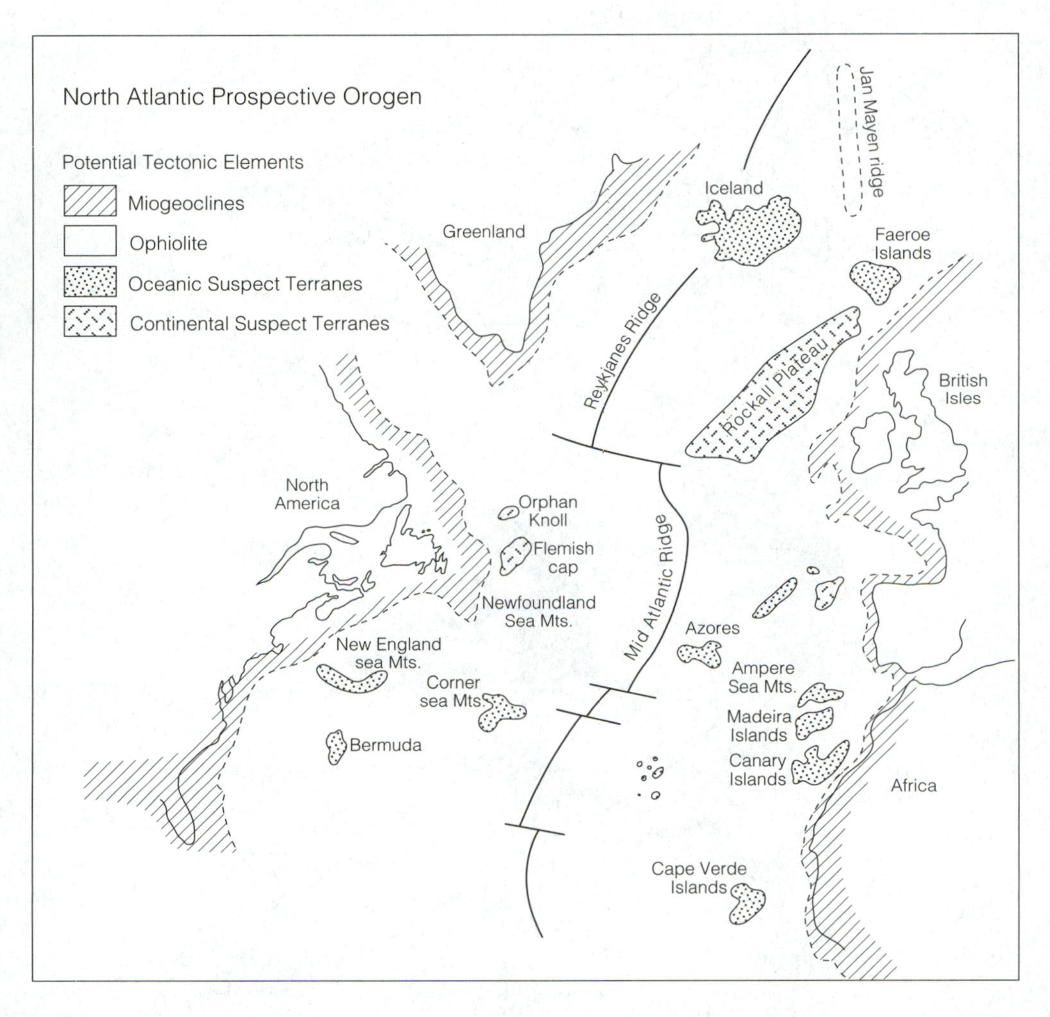

Figure 8.15 The present North Atlantic region: the ocean is punctuated by a variety of allochthonous continental blocks and volcanic islands, some destined to be the suspect terranes of a future mountain belt. (After Williams, 1984.)

Islands have fulfilled a series of biological functions. Island complexes may be generated in a variety of tectonic environments ranging from island arcs related to subduction, ridges associated with rifting and sea-floor spreading to island chains developed above a migrating mantle hot spot. Moreover allochthonous blocks of continental material may form the basis for larger, more permanent islands. A variety of these islands punctuate the Atlantic Ocean, today (Fig. 8.15) forming the materials for a further mosaic of terranes should the Atlantic Ocean close again in the future (Williams, 1984). Islands were also important in the geological past (Neuman, 1984).

Islands and archipelagos, similar to the present-day Galapagos Islands, have acted as centres of isolation, speciation and subsequently dispersal. Their roles as stepping stones for island-hopping migrants

Box 8.6 Changing diversity, immigration and extinction

Islands have provided researchers with self-contained laboratories for the study of the diversification and extinction of plants and animals; their study has formed the basis for MacArthur and Wilson's model of biotic equilibrium (Fig. 8.16). This model involves the dispersal, colonization and extinction of biotas in both natural and artificial island and island-type complexes. Initial intense immigration becomes balanced by an increasing extinction rate until an equilibrium point is reached where immigration more or less keeps pace with extinction. Clearly immigration rates are highest where the island is closest to a source of emigration, whereas extinction rates are greatest in the smallest areas. The model thus predicts a family of curves reflecting variations in the size of the island and its distance from a source of immigrants. There is, however, considerable debate regarding the applicability of island biogeography to other types of ecosystem, such as the tropical rainforests.

The island of Krakatau was shaken and partly dismantled by a series of powerful volcanic eruptions during late August 1883. Nevertheless the main remains of the complex, Rakata, was not lifeless for long. First, so-called aeolian plankton, wind-borne organisms, reached the island, followed by a whole variety of swimmers, animals on rafts, and those washed ashore by tidal waves. Many organisms arrived attached to others and of course many, such as birds, arrived by powered flight. Diversification and extinction was apparently at first fairly haphazard (Wilson, 1992); nevertheless the recolonization of the island was relatively sudden, if chaotic, as the animal and plant immigrants expanded to fill an ecological vacuum.

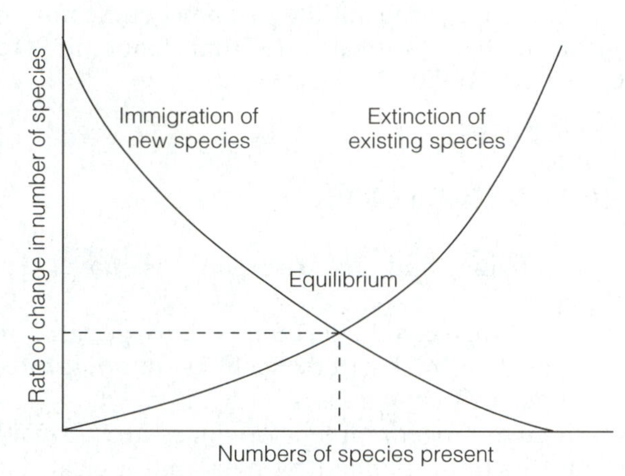

Figure 8.16 Equilibrium model for island biotas; the rate of immigration matches the rate of extinction and intersect at an equilibrium point. (Replotted from MacArthur and Wilson, 1963.)

has been demonstrated for a number groups of taxa. Moreover the endemism of island biotas can be enhanced and consolidated by the integration of island complexes (Rotondo *et al.*, 1981). Islands, however, have also provided models for the migration patterns, diversification and eventual extinction of biotas.

Whether of volcanic or continental origin, islands are mobile, associated with moving oceanic plates and many are relatively short-lived geological structures. Roles as 'Viking ships', carrying fossil organisms across latitudes, or as 'Noah's Arks', providing evolving lineages with cross-latitude transfers, have further complicated our understanding of palaeobiogeography of ancient oceanic systems. Nevertheless these structures were pivotal in the evolution and movement of many marine biotas through time, emphasizing the relationship between changing palaeogeography and biotic change.

8.9 BIOGEOGRAPHY AND EXTINCTION

A number of recent studies have explored the relationships between geographic range and extinction (Fig. 8.17). Longevity usually correlates with organisms having wide geographic spreads during so-called background intervals between major extinction events (Jablonski, 1986). Amongst the Triassic bivalve fauna, genera with endemic distribution were more prone to extinction at the end of the period than those with more cosmopolitan distributions (Hallam, 1981). Nevertheless during some focussed extinctions such as the end Cretaceous event both extinct and surviving genera had similar geographic spreads. But during the end Ordovician event many of the more cosmopolitan genera disappeared although some eurytopic, cross-provincial forms had a higher survival rate than endemic stenotopes (Owen and Robertson, 1995).

8.10 SUMMARY POINTS

- All living and fossil organisms have a defined, if variable, geographic range.
- Organisms can be arranged into geographic provinces mutually separated by barriers such as temperature gradients, mountain ranges or ocean basins.
- Modern biogeographic provinces are probably similar in composition to those defined by palaeontological data.
- Dispersal models invoke the migration and evolution of biotas from a site of origin.
- Vicariance models are based on the fragmentation of a province by geological processes.
- The distribution of fossil organisms can be analysed by a range of phenetic methods, including cluster analysis, seriation and also by cladistic analysis.
- Palaeobiogeography has provided basic data to suggest and test plate tectonic models.
- Changes in palaeogeography have promoted the interchange and

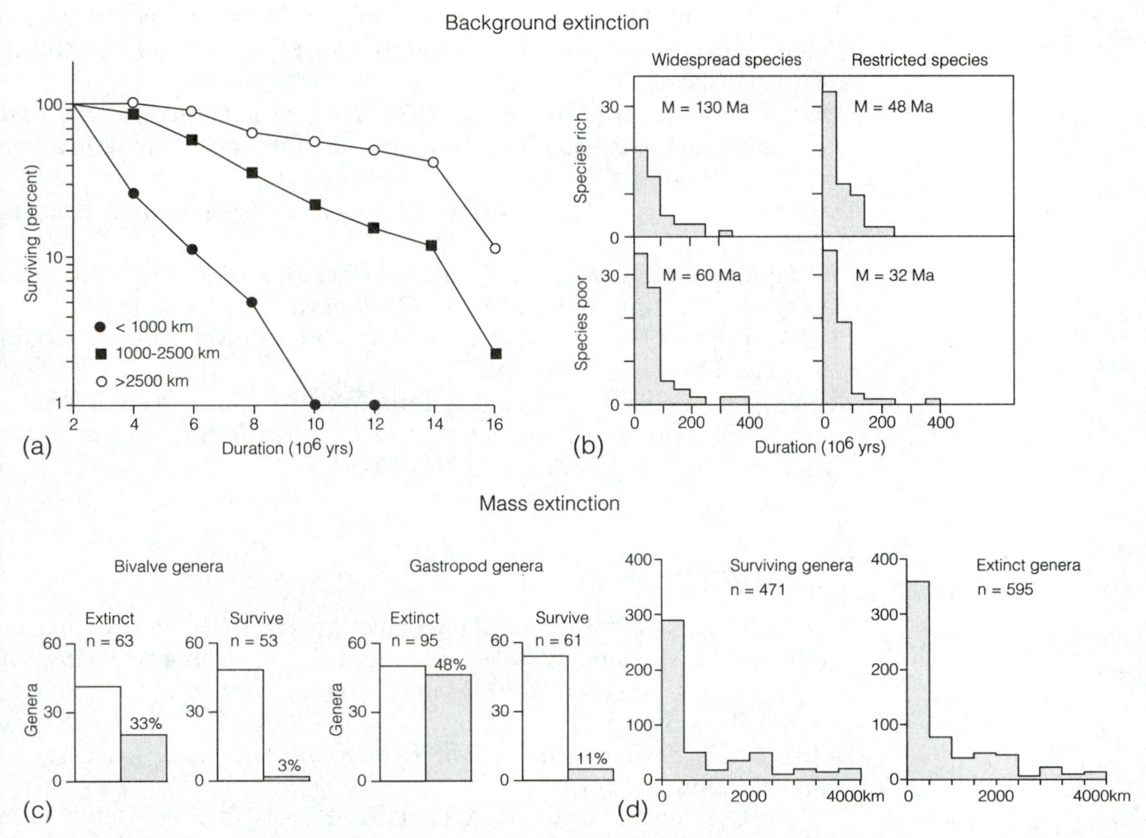

Figure 8.17 Biogeography and extinction: (a) During background intervals broad geographic range correlates with longevity, (b) During background intervals widespread genera are around longer than those with narrow distributions, (c) Differential extinctions of endemic (dark) and pandemic (light) mollusc genera during extinction events, (d) During the End Cretaceous event survivors and victims had comparable geographic distributions. (After Jablonski, 1986.)

migration of faunas and floras together with the radiation and extinction of taxa.

- Islands have fulfilled a variety of roles, providing loci for speciation and stepping stones for the migration of shallow-water marine biotas.
- The relationship between biogeography and extinction may help understand the nature of some extinction events.

8.11 FURTHER READING

Brenchley, P.J. (ed.). 1984. *Fossils and climate*. John Wiley & Sons, New York. 352 pp.

Briggs, J.C. 1974. *Marine zoography*. McGraw Hill, New York.

Briggs, J.C. 1987. Biogeography and plate tectonics. *Developments in Palaeontology and Stratigraphy* 10, 204 pp.

Cox, B.C. and Moore, P.D. 1993. *Biogeography an ecological and evolutionary approach*. 5th edition. Blackwell Scientific Publications, Oxford. 326 pp.

Gray, J. and Boucot, A.J. (eds) 1979. *Historical biogeography: plate tectonics and the changing environment.* Oregon State University Press, Corvallis, Oregon.

Hallam, A. (ed.) 1973. *Atlas of palaeobiogeography*. Elsevier, Amsterdam.

Middlemiss, F.A., Rawson, P.F. and Newall, G. 1971. Faunal provinces in space and time. *Special Issue Geological Journal* 4, 1–236.

Pianka, E.R. 1994. *Evolutionary ecology*. 5th edition. Harper Collins College Publishers. 486 pp.

Pielou, E.C. 1979. *Biogeography*. John Wiley & Sons, New York.

Vincent, P. 1990. *The biogeography of the British Isles. An introduction*. Routledge, London and New York.

8.12 REFERENCES

Bambach, R.K. 1990. Late Palaeozoic provinciality in the marine realm. *In:* McKerrow, W.S. and Scotese, C.R. (eds) *Palaeozoic Palaeogeography and Biogeography*. Geological Society Memoir 12, 307–323.

Boucot, A.J. 1975. *Evolution and Extinction Rate Controls*. Elsevier, Amsterdam, 427 pp.

Bruton, D.L. and Harper, D.A.T., 1992. Fossils in Fold Belts. *Terra Nova* 4, 179–183.

Campbell, K.A. and Bottjer, D. 1995a. Brachiopods and chemosymbiotic bivalves in Phanerozoic hydrothermal vents and cold-seep environments. *Geology* 23, 321–324.

Campbell, K.A. and Bottjer, D. 1995b. *Peregrinella*: an Early Cretaceous cold-seep-restricted brachiopod. *Paleobiology* 21, 461–478.

Campbell, C.A. and Valentine, J.W. 1977. Comparability of modern and ancient marine provinces. *Paleobiology* 3, 49–57.

Carlquist, S. 1974. *Island biology*. Columbia University Press, New York.

Crame, J.A. 1993. Bipolar molluscs and their evolutionary significance. *Journal of Biogeography* 20, 145–161.

Donovan, S.K. 1987. The fit of the continents in the Late Precambrian. *Nature* 327, 139–141.

Donovan, S.K. 1989. Taphonomic significance of the encrustation of the dead shell of Recent *Spirula spirula* (Linné) (Cephalopoda: Coleoidea) by *Lepas anatifera* Linné (Cirripedia: Thoracia). *Journal of Paleontology* 63, 698–702.

Donovan, S.K. 1992. Life on a log. *Rocks & Minerals* 67, 12–14.

Hallam, A. 1981. The end-Triassic extinction event. *Palaeogeography, Palaeoclimatology, Palaeoecology* 35, 1–44.

Hallam, A. 1986. Evidence of displaced terranes from Permian to Jurassic faunas around the Pacific margins. *Journal of the Geological Society of London* 143, 209–216.

Hallam, A. 1994. *An outline of Phanerozoic biogeography*. Oxford Biogeography Series. Oxford University Press, Oxford. 246 pp.

Hansen, T.A. 1980. Influence of larval dispersal and geographic distribution on species longevity in neogastropods. *Paleobiology* 6, 193–207.

Harper, D.A.T. and Parkes, M.A. 1989. Palaeontological constraints on the definition and development of Irish Caledonide terranes. *Journal of the Geological Society of London* 126, 723–725.

Harper, D.A.T., MacNiocaill, C. and Williams, S.H., 1996. The palaeogeography of early Ordovician Iapetus terranes: an integration of faunal and palaeomagnetic constraints. *Palaeogeography, Palaeoclimatology and Palaeoecology,* 121, 297–312.

Jablonski, D. 1986. Background mass extinctions: the alternation of macroevolutionary regimes. *Science* 231, 129–133.

Jablonski, D. and Lutz, R.A. 1983. Larval ecology of marine benthic invertebrates: paleobiological implications. *Biological Reviews* 58, 21–89.

Jablonski, D., Flessa, K.W. and Valentine, J.W. 1985. Biogeography and paleobiology. *Paleobiology* 11, 75-90.

Jackson, J.B.C., Jung, P., Coates, A.G. and Collins, L.S. 1993. Diversity and extinction of tropical American mollusks and emergence of the Isthmus of Panama. *Science* 260, 1624–1626.

Neuman, R.B. 1984. Geology and paleobiology of islands in the Ordovician Iapetus Ocean: review and implications. *Bulletin of the Geological Society of America* 95, 1188–1201.

Newton, C.R. 1987. Biogeographic complexity in Triassic bivalves in the Wallowa terrane, northwestern United States: oceanic islands, not continents, provide the best analogues. *Geology* 15, 1126–1129.

Newton, C.R. 1988. Significance of 'Tethyan' fossils in the American Cordillera. *Science* 242, 385–391.

Nopsca, F. Baron 1923. On the geological importance of the primitive reptilian fauna in the uppermost Cretaceous of Hungary; with a description of a new tortoise (*Kallokibotion*). *Quarterly Journal of the Geological Society of London* 79, 100–116.

Owen, A.W. and Robertson, D.B.R. 1995. Ecological changes during the end-Ordovician extinction. *Modern Geology* 20, 21–39.

Platnick, N.I. and Nelson, G. 1978. A method of analysis of historical biogeography. *Systematic Zoology* 27, 1–16.

Rotondo, G.M., Springer, V.G., Scott, G.A.J. and Schlanger, S.O. 1981. Plate movement and island integration-a possible mechanism in the formation of endemic biotas, with special reference to the Hawaiian islands. *Systematic Zoology* 30, 12–24.

Skelton, P.W. and Wright, V.P. 1987. A Caribbean rudist bivalve in Oman: Island-hopping across the Pacific in the Late Cretaceous. *Palaeontology* 30, 505–529.

Smith, P.L. 1990. Paleobiogeography and plate tectonics. *Geoscience Canada* 15, 261–279.

Smith, P.L. and Tipper, H.W. 1986. Plate tectonics and paleobiogeography: Early Jurassic (Pliensbachian) endemism and diversity. *Palaois* 1, 399–412.

Taylor, J.D., Braithwaite, C.J.R., Peake, J.F. and Arnold, E.N. 1979. Terrestrial faunas and habitats of Aldabra during the Pleistocene. *Philosophical Transactions of the Royal Society, London* B286, 47–66.

Valentine, J.W. 1973. *Evolutionary paleoecology of the marine biosphere*. Prentice-Hall, Englewood Cliffs, N.J. 511 pp.

Valentine, J.W. 1984. Neogene marine climate trends: implications for biogeography and evolution of the shallow-sea biota. *Geology* 12, 647–650.

Valentine, J.W., Foin, T.C. and Peart, D. 1978. A provincial model of Phanerozoic diversity. *Paleobiology* 4, 55–66.

Webb, S.D. 1991. Ecogeography and the Great American Interchange. *Paleobiology* 17, 266–280.

Wignall, P.B. and Simms, M.J. 1990. Pseudoplankton. *Palaeontology* 33, 359–378.

Williams, H. 1984. Miogeoclines and suspect terranes of the Caledonian – Appalachian orogen: tectonic patterns in the North Atlantic region. *Canadian Journal of Earth Sciences* 21, 887–901.

Williamson, D. 1992. *Larvae and evolution*. Chapman & Hall, London.

Wilson, E.O. 1992. *The diversity of life*. Allen Lane, the Penguin Press, London. 424 pp.

Wilson, J.T. 1966. Did the Atlantic close and reopen? *Nature* 211, 676–681.

Ziegler, A.M. 1990. Phytogeographic patterns and continental configurations during the Permian Period. *In:* McKerrow, W.S. and Scotese, C.R. (eds) *Palaeozoic Palaeogeography and Biogeography*. Geological Society Memoir 12, 363–379.

Evolutionary palaeoecology of the marine biosphere

This chapter describes the ecological changes that have occurred in the marine biosphere through Earth history and how the patterns may be analysed through palaeoecological studies. It discusses the ways in which the biosphere is filled following periodic evolutionary innovations and how diversification reflects the filling of new niches and the subdivision of established ones. Steady rates of species appearances tend to be balanced by steady extinction rates throughout long periods of time during which there are relatively modest changes to ecosystems. In contrast, mass extinctions are short-lived events when ecosystems are substantially restructured. The extent to which they change the course of evolution is still a subject of much debate.

9.1 THE EARLY HISTORY OF LIFE

The Earth was formed about 4600 Ma and in its initial stage of development it was a lifeless planet. The oldest rocks known are about 3800 my and therefore nothing is directly known about the first 800 my of Earth history, but it is likely that during this time there was a major episode of outgassing which led to the formation of the atmosphere and oceans. The early atmosphere probably lacked oxygen but contained nitrogen, water vapour and carbon dioxide, and possibly methane and ammonia. Other planets in the solar system probably had a similar atmosphere at an early stage, but unlike Venus which developed an atmospheric envelope of CO_2 and heated to about 500°C, and Mars which lost its atmosphere and froze, the Earth has maintained an equilibrium condition under which life could be sustained. The critical difference between the Earth and other planets is in the development of a biosphere which progressively modified the atmosphere by utilizing carbon dioxide during photosynthesis and so releasing oxygen. As a consequence life could evolve, protected from the effects of ultraviolet radiation, and in a relatively stable thermal regime.

The history of the biophere has involved the exploitation of mainly unfilled environments on the Earth's surface through major

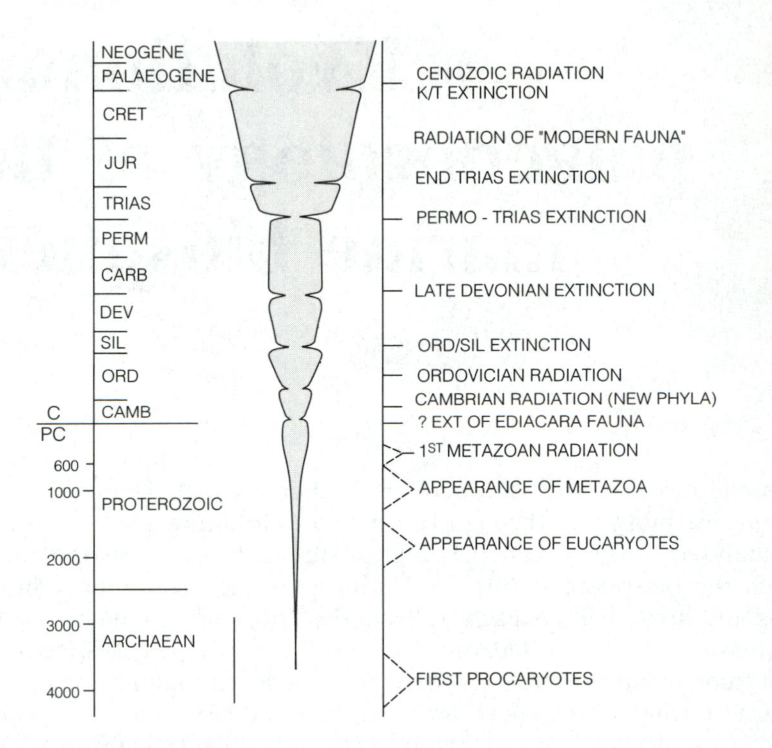

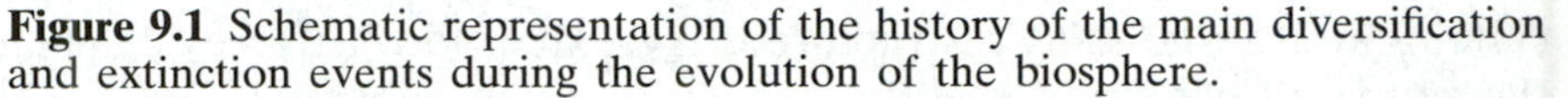

Figure 9.1 Schematic representation of the history of the main diversification and extinction events during the evolution of the biosphere.

evolutionary innovations, further filling those regions during periods of diversification and then the emptying of some spaces at times of mass extinction (Fig. 9.1). These patterns of innovation, diversification and extinction are considered below.

9.2 DIVERSIFICATION EVENTS IN EARTH HISTORY

9.2.1 The Origin of life and the earliest Prokaryota

Life originated in the hostile atmosphere of the early Earth. The oldest recognized fossils contributing to the early biosphere were **prokaryotic** organisms (cells without a nucleus or organelles and incapable of sexual reproduction) that were apparently anaerobic fermenters, living in an oxygen-free atmosphere and utilizing glucose that was converted to organic acids, ethanol, H_2 and CO_2. Both photoautotrophs and heterotrophs may have been present; the former, by using light as a source of energy, represent an important innovation in the use of CO_2 in carbon synthesis, but they used H_2, H_2S, or various organic substrates rather than H_20 in the reduction of CO_2 to produce cellular matter and consequently did not generate oxygen. The major prokaryotic innovation, that was to transform the atmosphere and hence the future biosphere, was the development of oxygen-producing photosynthesis. This event is recorded by the appearance of cyanobacteria (blue-green algae) that are preserved as tubular structures in rare circumstances but are more commonly seen as stromatolites which are laminated, dome-shaped accumulations of carbonate mud accreted by successive microbial mats, which may commonly have contained early cements producing a rigid structure. The earliest stromatolites date from about 3500 Ma and from then on they dominate the Precambrian fossil record because they have much the best preservation potential. Supporting evidence that photosynthetic activity started early in the Archaean comes from carbon stable isotopes. Photosynthetic organisms selectively take up more of the light isotope ^{12}C than ^{13}C, to give a $\delta^{13}C$ value of –27 ‰ and this low value has been detected in carbon from rocks as old as 3800 Ma. The evidence for an early Precambrian biota with groups other than cyanobacteria is slender. In the Archaean (>2500 Ma) there are a few records of bacteria other than cyanobacteria (e.g. in the Barberton Range in Swaziland, and localities in Western Australia) but the biological affinities or the biological origin of many of these has been questioned. It is not until the Proterozoic (2500–550 Ma) that diverse microbiotas are found preserved in cherts. As far as the evidence goes, cyanobacterial assemblages appear to have undergone little external morphological change throughout nearly 3000 my of the Precambrian, though judging by the disparity of gene sequences found in modern prokaryotes, it is likely that they had undergone substantial genetic evolution.

9.2.2 Appearance of the Eukaryota

The second great innovation amongst microbial biotas was the development of the **Eukaryota** (cells with a nucleus) in the Proterozoic between 2100 Ma and 1800 Ma, possibly associated with an increased level of atmospheric oxygen. The Eukaryota were capable of forming simple aggregates of cells and were generally capable of sexual reproduction, so promoting the spread of variation amongst the species. They are the forerunners of the multicellular higher plants and animals. The fossil record of the origins of the Eukaryota is unfortunately obscure because the differences in cell structure between prokaryotes and eukaryotes is not preserved during fossilization. The principal observable difference between the two groups is that eukaryotic cells tend to be larger. The most prevalent Precambrian eukaryotes are the acritarchs, which owe their preservation to their thick cell wall. They are typically 30–100 μm in diameter and may have smooth cell walls, be spinose or have more elaborate ornamentation. They are thought to be the planktonic encystment-stage of algae. They are recorded back to 1400 Ma and are common from 1000 Ma onwards until the appearance of the Vendian metazoans at about 650 Ma. Throughout the Mesoproterozoic and Neoproterozoic (~ 1600–550 Ma) biotas were characterized by both prokaryotic and eukaryotic species. However, a significant shift in the balance between the two groups appears to have occurred about the Mesoproterozoic/Neoproterozoic boundary (~1000 Ma), when the relative diversity and abundance of eukaryotes, particularly acritarchs, increased and the diversity of prokaryotes decreased (Knoll and Sergeev, 1995). The variety in the ornamentation of acritarchs and the appearance of various algal filaments, colonies and thalli indicate a substantial radiation amongst the eukaryotes at this time. The change in the prokaryote assemblages, on the other hand, is thought to be largely a reflection of changes in preservation potential. Most prokaryote biotas are related to a small range of peritidal facies and are preserved where these are affected by early cementation and silicification. There is evidence to suggest that in the Neoproterozoic there was a decrease in the proportion of facies which suffered early cementation, which reduced the chances of their preservation.

9.2.3 Appearance of the Metazoa

The organization of cells into complex multicellular animals and plants (Metazoa and Metaphytes) was to lead to the complex ecological systems that first developed within the oceans and then extended on to land. Morphological evolution through 3000 myr of the Precambrian appears to have been rather slow, but may have increased at the start of the Neoproterozoic. Then following the establishment of the more complex body plans of Metazoa, evolutionary rates, measured by changes in preserved morphology, appear to have been substantially higher and diversification more rapid. The fossil record is, however, not absolutely clear as to whether there was an immediate burst of metazoan or metaphyte evolution following their first appearance, or whether there was a substantial lag before diversification. The records of the earliest potential metazoan fossils are mainly in the form of trace fossils, which have been described from rocks as old as 2200 Ma (Crimes, 1994). The records of trace fossils from the period 2200 to 700 Ma are sparse and most have been reinterpreted as inorganic structures. However, an apparently backfilled burrow, referred to as *Bergaueria,* from rocks in the Mackenzie Mountains, Canada, dated as 800–1000 Ma, appears to be truly organic as do simple burrows (*Planolites*) from 700–959 Ma rocks in Namibia. Trace fossils are clearly very rare, or even absent, before 700 Ma, but the evidence as it stands suggests that coelomate metazoans, capable of burrowing, might have been present before 700 Ma (Crimes, 1994) and therefore there might have been cryptic metazoan evolution of more primitive forms from an even earlier time.

9.2.4 Broad patterns of metazoan diversification

The history of metazoan diversification and ecological evolution can be seen as successive waves of innovation followed by periods of stability which are terminated by mass extinctions. The phases of diversification in the Phanerozoic are thought to arise from a variety of stimuli or new opportunities. Biota may fill unexploited habitats following some major evolutionary innovation: the marine radiations in the late Precambrian and early Cambrian and the later invasions of terrestrial environments by plants and animals are examples of such events. Biota may exploit new niches within a habitat that is already colonized; the diversification of infaunal life styles in the Mesozoic is an example of this process. Biota may respond to major changes in the physical environment. For example, plate movements change the distribution of continents and hence the palaeobiogeography of the Earth, which may have consequences on global species diversity. Lastly, biotas may change in response to an evolutionary innovations amongst the fauna which change its trophic structure; the diversification of shell-crushing marine predators in the Mesozoic is thought to have triggered evolutionary changes amongst a variety of potential prey.

The largest bursts of innovation, signalled by the appearance of many new taxa, occurred when there were large numbers of nearly empty habitats to be exploited. Once most environments had been occupied, diversification was generally slower. The role of extinction was to empty niches and so provide opportunities for new developments. Background extinction allows species replacement within habitats, larger extinctions may affect communities throughout a region, while mass extinctions may destroy the entire structure of many ecosystems on a global scale. These patterns of ecological change are discussed in the following sections.

9.2.5 The Ediacara fauna

The first major radiation of Metazoa was that of the 'Ediacara fauna' which first colonized a wide range of marine environments in the Vendian, probably between 543 and 549 Ma. Elements of this non-skeletonized fauna were first described from Namibia but it received widespread attention when the rich and varied fauna was discovered in the Pound Quartzite at Ediacara in south Australia. The most common fossils resemble cnidarians and include a variety of medusoid-like forms; others forms have been regarded as early representatives of such later metazoans as annelids, arthropods and echinoderms. There are, however, fundamental differences of opinion as to the relationship of the Ediacara fauna to later taxonomic groups. One view is that the members of the fauna largely belong to known phyla and are part of the root stock or specialized offshoots from the root stock. Thus the fauna would consist of animals such as early annelids, cnidarians and echinoderms. A quite contrary view is that the biota belongs to an evolutionary clade that radiated and became extinct and so did not contribute to later evolutionary events. One exponent of this view is Seilacher (1992), who has emphasized the quilted nature of the fauna (Fig. 4.15), which is unlike nearly all later Metazoa; moreover, he suggested the animals belong in a kingdom of their own, the Vendobionta. This view was subsequently modified to suggest the Vendobionta were cnidarian-like organisms that belonged in a phylum, not a kingdom, of their own. The organisms were interpreted as being acellular, compartmentalized and filled with a 'plasmodial' fluid that gives them a rigidity similar to an air bed. Members of the fauna can be upright, reclining or 'sediment stickers' (Chapter 4, Fig. 4.13) and some were emersed within the sediment and apparently grew upwards as the sediment accumulated. Some of the Ediacara fossils are tens of centimetres across and many have a flat, often leaf-like shape. Such a morphology gives a large surface area suited to respiratory exchange of gases and uptake of nutrients. Between the radical interpretation of Seilacher and the most conservative views, other palaeontologists see the fauna as a mixture of clades, some that are extant and some that are extinct. Accordingly, the distinctive morphology of much of the fauna is interpreted as the diversification of less specialized clades that have not been preserved or are represented only by trace fossils (Runnegar, 1995). The contemporaneous ichnofauna includes a new variety of forms, that indicate the presence of a separate clade of coelomate, worm-like animals, that were capable of ploughing or burrowing through the sediment. At the end of the Vendian it appears that the Ediacara fauna largely disappeared, though some survivors are claimed to be present in the Burgess Shale fauna. Elements of the trace fossil fauna were also lost. Whether this is a major extinction that eliminated most or all of a major clade, or represents only the extinction of some more specialized elements of clades that further diversified in the Cambrian, is still a matter of controversy.

9.2.6 The early Cambrian evolutionary explosion

The appearance of the first animals with shells at the base of the Cambrian records the presence of a wide variety of new morphologies, acknowledged by the first records of about 10 new phyla, more than 50% of all recorded classes and many new orders. The earliest recorded skeletal fossils are anarbartids (elongated tubes with a distinctive trifoliate cross-section) and the teeth of protoconodonts (a group probably allied to the arrow worms). The following burst of metazoan evolution is marked by the appearance of a variety of small tubular and conical shells, many of which are of unknown affinities (Chapter 4, Fig. 4.14.), a variety of early molluscs, new types of acritarchs, trace fossils and archaeocyathids. The small shelly fauna diversified slowly through the Nemakit-Daldynian Stage then expanded rapidly in the Tommotian Stage and declined thereafter (Fig. 9.2). A second wave of innovation in the Atdabanian introduced varied inarticulate brachiopods, trilobites echinoderms and some new molluscs (Fig. 9.2). One interpretation of early metazoan evolution is that it developed as a succession of 'evolutionary faunas' each of which in turn diversified and then declined (Sepkoski, 1992). From this

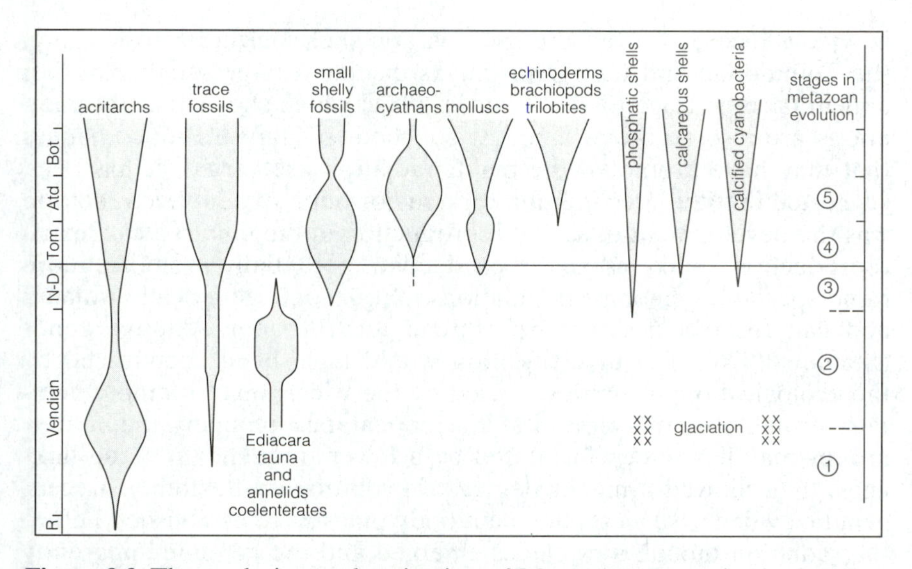

Figure 9.2 The evolution and extinction of Metazoan groups in the late Precambrian and early Cambrian. Abbreviation of stages: N-D, Nemakit-Daldynian; Tom., Tommotian; Atd., Atdabanian; Bot, Botomian; R., Riphaean. Evolutionary events: 1) Diversification of the Eukaryota, 2) diversification and acme of the Ediacara fauna, 3) appearance of first shelled fossils, 4) diversification of small shelly fauna and other groups, 5) appearance of new clades including part of the 'Cambrian' fauna. (Modified from Tucker, 1992.)

perspective the 'Ediacara fauna' is the first evolutionary fauna, the 'Tommotian fauna' with its small shelly fossils is the second evolutionary fauna and the 'Cambrian fauna' represented by the Atdabanian radiation is the third fauna. Two other faunas, the 'Paleozoic' and 'Modern' fauna, which were to become important later in the Phanerozoic, had many of their root stocks in the early Cambrian but did not diversify until much later (see section 9.2.7). Amongst these early evolutionary faunas the archaeocyathids, a group with a cup-shaped skeleton resembling that of a sponge, appear to have had a rather different evolutionary history in that, after a very rapid diversification in the Tommotian and Atdabanian, they abruptly declined and were virtually extinct before the upper Cambrian (Fig. 9.2).

Why, after such a long period of relatively slow morphological evolution amongst the Prokaryota and Eukaryota, did the metazoans diversify so explosively in a series of evolutionary bursts? Attempts to answer this question fall into three main categories, though they are not necessarily mutually exclusive. One view is that the radiations are largely a taphonomic artefact, a second is that they were determined by biotic changes and the third is that changes in the global environment were of most importance. It is unlikely that taphonomic factors arising from the development of a shelled biota with a much higher chance of preservation were the main influence on the record of the radiation. Of the three evolutionary faunas, the Ediacara fauna is widely recorded, even though it is non-skeletonized. Furthermore, the Tommotian and Cambrian faunas include a large number of first appearances of trace fossils (Crimes, 1994) which shows that the radiations are also recorded amongst soft-bodied faunas. Biotic changes that may have promoted the major radiation are varied. It has been suggested that the precondition for a major burst of adaptive radiation was the development of sexual reproduction in the eukaryotes. Sexual reproduction, unlike asexual reproduction, potentially promotes more rapid speciation amongst populations that become genetically isolated and can form clades that can spread into broader adaptive zones (Stanley, 1976). The diversification would have been encouraged by the ecological opportunities offered by the wide range of empty habitats available at this time, but the appearance of many major new groups may have been facilitated by a lower frequency of gene linkages, that allowed a higher degree of evolutionary flexibility. Recent genetic evidence suggests the main body plans were established before the radiation though new clades emerged and the genome apparently did not restrict the range of disparity amongst the fauna. It appears more likely that the post-Cambrian restriction of evolutionary novelty arose from the success by which the Cambrian fauna and its successors exploited most habitats and inhibited further major innovation. Specific triggers for the post-Vendian evolutionary bursts might have been innovations affecting the trophic structure in the marine world, particularly the appearance of herbivory and carnivory. Herbivores certainly appeared in the early Cambrian and may have contributed to adaptive diversity, but the commonly cited link between the late Precambrian decline of stromatolites and the expansion of herbivores that might have grazed on them is not as convincing as has been suggested; most stromatolites were probably calcified and resistant to grazing and stromatolites started to be reduced in numbers about 1500 my before the Cambrian but continued into the Cambrian with no marked decline at the time of radiation. The development of shells of different morphology and mineralogy in a variety of new clades might have been in response to the spread of carnivory and have been a cause of further diversification. A third view is that the Cambrian evolutionary 'explosion' was promoted by physical factors associated with unusually profound global environmental changes at this time. Atmospheric oxygen levels rising to a threshold which would sustain complex, active metazoan life in the oceans were certainly a pre-requisite for metazoan but there is no strong independent evidence that oxygen reached a threshold value at this time. There is, however, good geological evidence that the end of the late Precambrian Vendian glaciation marked a change in the world's climate from an extreme ice-house period to a greenhouse period in the Cambrian, with an associated rise in atmospheric levels of CO_2. Contemporaneously there was a change in the mineralogy of marine carbonates from dominantly aragonitic in the Precambrian to calcitic in the Cambrian which is thought to reflect the increased pCO_2 levels in the atmosphere (Tucker,

1992). Also, about the same time there were widespread movements of lithospheric plates, which resulted in continental rifting and the growth of new mid-ocean ridge systems and oceanic crust. Possibly associated with the new ridge systems and the melting of ice caps, sea-level rose and there was a global transgression which created widespread shallow cratonic seas where unusually large deposits of phosphorus were formed, particularly at the start of the Tommotian. The phosphorus is associated with cherts and black shales and is interpreted as indicating periods of intense upwelling of nutrients associated with new patterns of ocean circulation arising from contemporaneous changes in climate and ocean geography (Brasier, 1992).

In summary, the global sea-level rise may have offered a wide range of shallow marine habitats open for exploitation and the changing chemistry of the ocean may have facilitated the growth of protective shells. Changing global atmosphere may also have had an important role in the wider colonization of the marine world. Some of these physical changes may have determined the timing of the radiation, but the evolution of the eukaryotic cell was a pre-condition and the opportunities for colonization may be the key to the explosive radiation.

9.2.7 Diversification of the three great evolutionary faunas

As long ago as the middle of the last century that perceptive geologist John Phillips (1841) suggested that there were three great systems of organic life: Palaeozoic (old life), Mesozoic life (middle life) and Caenozoic life (new life) (see Chapter 1). In recent years this scheme has been modified using a vast compilation of data on the ranges of families that suggests that there were indeed three major Phanerozoic marine faunas, but these were a Cambrian fauna, a Palaeozoic fauna and a Modern fauna, each being characterized by families which had their greatest diversity in that period of time (Fig. 9.3) (Sepkoski, 1984, 1990). The Cambrian fauna (Fig. 9.4) is dominated by trilobites with associated hyolithids, eocrinoids, monoplacophorans, inarticulate brachiopods and a variety of crustaceans, which are only rarely preserved, but are recorded in Burgess Shale-type Lagerstätten. The fauna was dominated by sedentary or creeping epifaunal deposit-feeders, grazers or suspension-feeders. The Palaeozoic fauna (Fig. 9.4) introduced a rich variety of shelled epifaunal suspension-feeders, including articulate brachiopods, bryozoans, anthozoans, ostracodes and crinoids. The earliest ecologically complex reefs developed from associations between members of some of the new clades. In addition, the variety of trace fossils increased, mainly reflecting new infaunal and epifaunal deposit-feeders. The Modern fauna (Fig. 9.4) was characterized by bivalves, gastropods, echinoids, Malacostraca, Demospongia, osteichthian and chondrichthian fishes, reptiles and mammals. The constituent classes had a particularly wide variety of lifestyles, ranging from deep infaunal to epifaunal to nektonic and

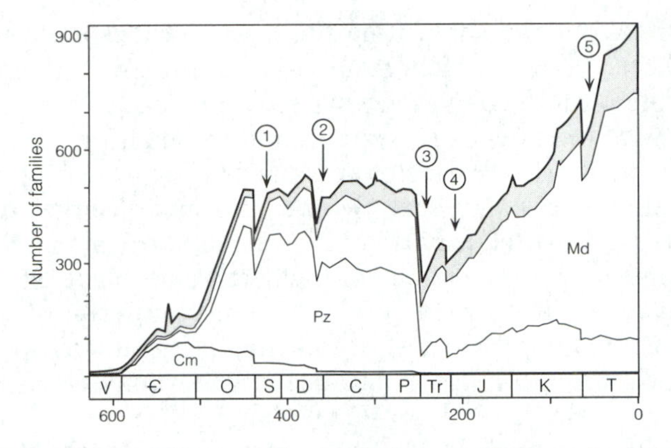

Figure 9.3 The three Phanerozoic evolutionary faunas, based on shelled marine families. The stippled area represents the diversity of families with rarely preserved members that lack well mineralized skeletons. Cm = Cambrian Evolutionary Fauna, Pz = Palaeozoic Evolutionary Fauna, Md = Modern Evolutionary Fauna. Numbers 1–5 are the five principal mass extinctions. (Modified from Seposki, 1990.)

included a full range of trophic activities. These ecological differences between the three evolutionary faunas indicate that there were periods of significant ecological restructuring.

One of the most challenging ideas in evolutionary ecology suggests that the successive rise and fall of these evolutionary faunas can be modelled to fit a relatively simple pattern (Sepkoski and Sheehan, 1983; Sepkoski, 1984, 1991). At first sight the three diversity curves (Figs 9.3 and 9.4) show little similarity, but Sepkoski has shown that the curves approximate closely to a mathematical model of diversification. The model is based on studies of island faunas that suggest that, in their early stage, homogeneous faunas will diversify exponentially but should later asymptotically approach an equilibrium plateau as the resource limits of the environment are approached (MacArthur and Wilson, 1967). The result is a simple sigmoidal curve, known to mathematicians as a logistic curve (Fig. 9.5a), similar to that seen for the early part of the Cambrian and Palaeozoic diversity curves in Figure 9.4. The analysis of the total Phanerozoic fauna is, however, more complex because the fauna is not homogeneous, but consists of three interacting faunas. Each fauna has its own rate of diversification (r), with $r_1>r_2>r_3$ and its own plateau of diversity (D), with $D_1<D_2<D_3$ (Figs 9.3 and 9.5). The faunas would be expected to interact with each other such that if one fauna diversified to bring the total diversity above the diversity plateau of another fauna, the latter would decline in compensation (see the decline of the Cambrian fauna as the Palaeozoic fauna climbs above its plateau, Fig. 9.3). The pattern of rapid expansion followed by a slower decline gives a characteristic asymmetric profile to the curve which can be modelled with coupled logistical equations of the form $\mathrm{d}D_i/\mathrm{d}t = r_i\, D_i\, (1-\Sigma D_j/\hat{D}_i)$ where D_i is the diversity of the ith evolutionary fauna at time t, r_i is its initial

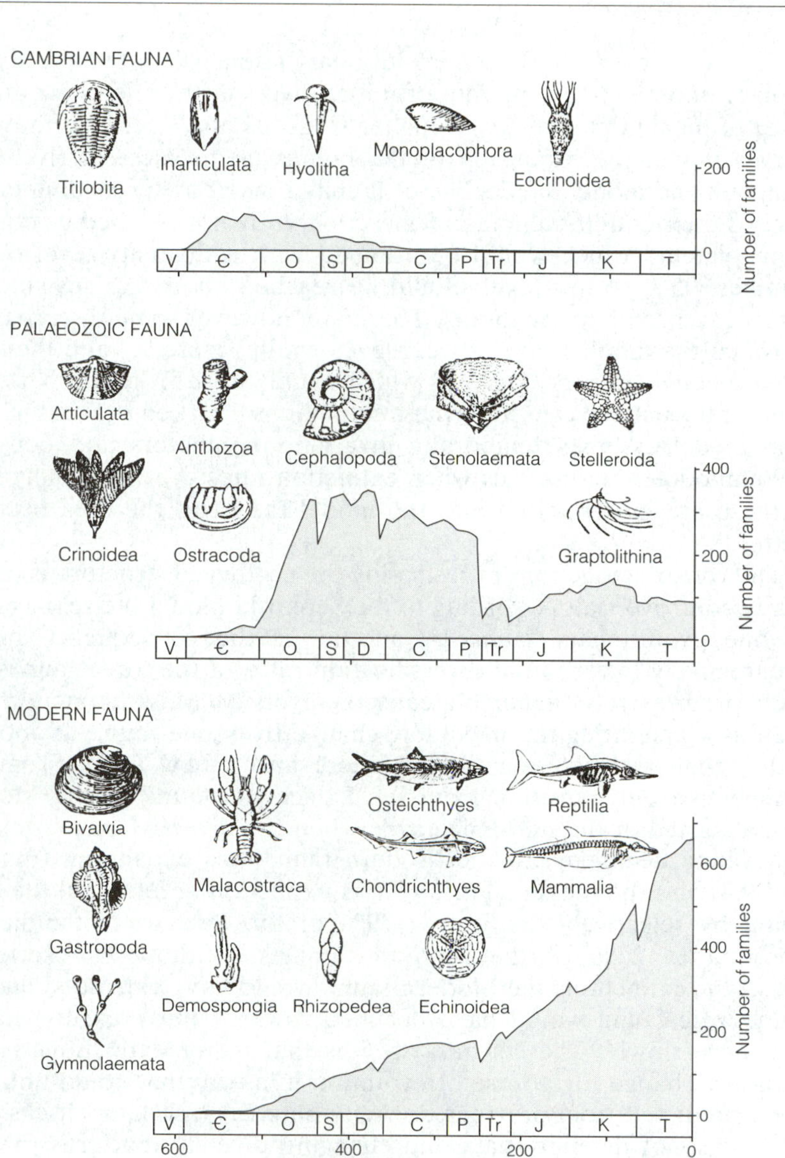

Figure 9.4 The three evolutionary faunas of the Phanerozoic, showing the characteristic members of the faunas: the Cambrian Fauna; The Palaeozoic Fauna; The Modern Fauna. (Modified from Sepkoski, 1990.)

diversification rate, $\hat{}D_i$ is its maximum or equilibrium diversity, and ΣD_j is the summed diversity at time t (Sepkoski, 1984).

$$\mathrm{d}D_i/\mathrm{d}t = \gamma i D_i\ (1 - \Sigma\ D_j/D_i) r_i$$

where D_i is the diversity of the ith evolutionary fauna at time t, r, is its initial diversification rate, D_i is its maximum or 'equilibrium' diversity, and ΣD_j is the summed diversity of all faunas at time t (Sepkoski, 1984).

This equation states that an evolutionary fauna will only diversify and replace another one if its rate of initial diversification is lower and its level of maximum diversity is higher (Figs 9.3 and 9.5). Note how the curves model the expansion of the faunas up to successively higher plateaux and model the decline of faunas 1 and 2 as succeeding faunas exceed their equilibria. Fauna 3 has apparently not reached its equilibrium plateau, which would be attained with a diversity level of 850 families; 95% of this level should be reached about 125 my into the future, according to the model. There are, however some discrepancies between the familial diversity curve shown in Figure 9.4 and the modelled diversity curves (Fig. 9.5) which mainly arise from the perturbations of the pattern caused by mass extinctions (marked 1–5 on Fig. 9.3). The model assumes temporally invariant parameters and does not accommodate the intervals when extinction rates were unusually high. If these are incorporated into the model the fit to the data becomes better.

The role of extinctions in reshaping the ecological structure is particularly controversial. According to the Sepkoski model the replacement of one evolutionary fauna by another is the consequence of the progressively lower initial diversification rates of the three faunas and their progressively higher plateaux of diversity. Mass extinctions are seen as accelerating the inevitable change from one fauna to another, rather than restructuring the biosphere into a new form. From this perspective the transition from the Palaeozoic fauna to the Modern fauna started in the mid-Palaeozoic when the diversity of the Palaeozoic fauna declined and the Modern fauna was expanding (Figs 9.3 and 9.4) and the massive Permo-Trias extinction accelerated the transition by selectively eradicating 79% of the Palaeozoic families, as opposed to 27% of the Modern families. Without the series of Mesozoic extinctions the Modern fauna would have expanded and the Palaeozoic fauna would have declined towards the present balance, but more slowly. The contrary view is that major extinctions fundamentally change the course of evolution in a way that could not have been predicted from previous diversity patterns (see later discussion).

Analysis of diversity data and changing diversity patterns through time is complex (see Box 9.1), so inevitably there have been criticisms of Sepkoski's model of Phanerozoic faunas. Criticism has been directed at the quality of the data, its analysis and to the theoretical basis of the model itself. Criticisms of the data have been particularly strong. The method by which organisms are classified into taxonomic units has a profound influence on the reconstruction of evolutionary patterns. It affects the record of appearances and extinction of taxa and assessments of taxonomic diversity. Traditionally organisms have been grouped into particular taxa on the basis of morphological and perceived phylogenetic relationship, but with no consistent methodology. Consequently traditional taxa are commonly not monophyletic, but include taxa from phylogenetically unrelated clades and even more commonly they are paraphyletic, being derived from a common

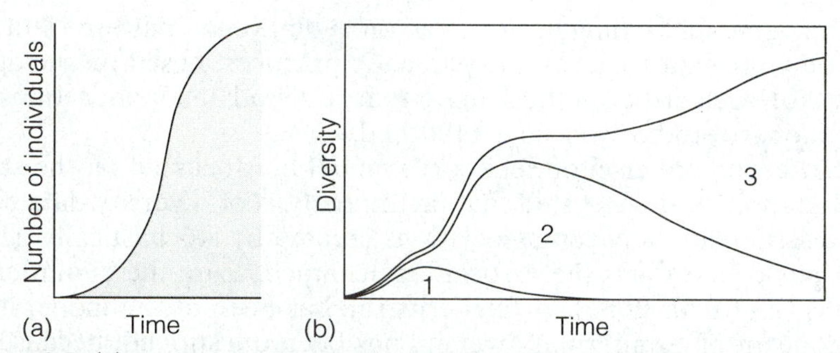

Figure 9.5 (a) A logistic curve of diversification. (b) A three phase kinetic model of the diversification of three faunas which in turn show decreasing rates of diversification but an increasing height to the plateau of diversity. The model should be compared with the diversity curves of three evolutionary faunas shown in Figure 9.6. (Modified from Sepkoski, 1984.)

ancestor but including some but not all of the descendants. The stratigraphic ranges of paraphyletic taxa are commonly truncated because their descendants are placed in another taxon. This is well illustrated by the early Palaeozoic echinoderms, which have conventionally been regarded as having substantially disappeared by the end of the Cambrian, but are seen to have flourished into the Ordovician if the Cambrian taxa are recognized as being paraphyletic and their descendants are included (Fig. 9.6). In a challenging analysis of systematic methods, Smith (1994) suggested that traditional taxonomic units are

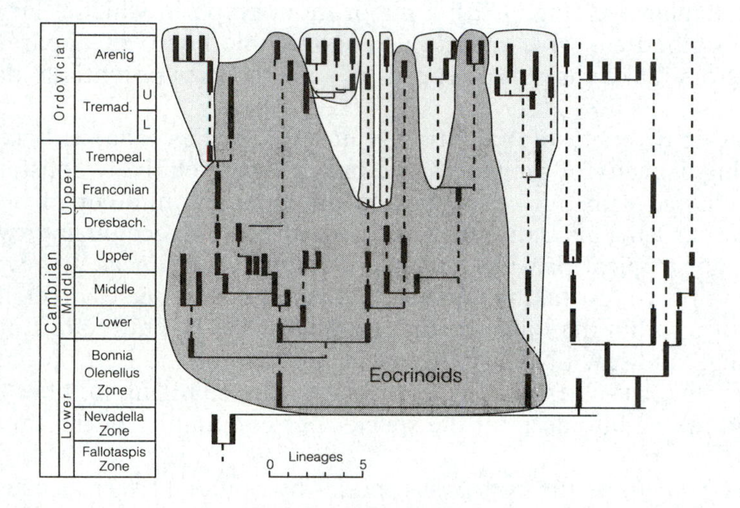

Figure 9.6 Phylogenetic tree of early Palaeozoic echinoderms. The eocrinoids, a member of Sepkoski's Cambrian fauna, appear to decline and then terminate in the early Ordovician (dense shading). However, the eocrinoids are a paraphyletic group and if all the derived clades are included (light shading) the early Palaeozoic crinoids are seen to maintain a high diversity into the Ordovician. (From Smith, 1994.)

likely to give misleading information on evolutionary patterns and that it is only through a change in systematic practices, based on the application of a cladistic methodology, that a sound taxonomic basis for evolutionary studies will be established.

The second criticism of Sepkoski's model has focussed on the statistical analysis. It is suggested that factor analysis of diversity data could not discriminate between associations created by stochastically generated phylogenies and the associations characterizing the evolutionary faunas (Hoffman, 1985). A final criticism has been of the model itself; the concept of equilibrium diversity has been questioned, because the extrapolation from models of island biogeography to create models of global diversity is considered invalid. However, a vigorous defence of 'evolutionary faunas' and diversity dependent models, which addresses the criticisms described above, has been made by Sepkoski (1991).

Box 9.1 Measuring changing diversity

Ecological associations are composed of species, therefore the best measure of changing diversity is one that is based on a census of the number of species present at any one period of time. Ostensibly this should be merely a matter of counting the number of fossil species recorded in different periods throughout the Phanerozoic. Unfortunately the recorded fossils represent only a part of the complete fossilized fauna, and the relationship between the recorded fauna and the total fauna is not constant but is variably biased by several factors. For example, the chance of collecting fossils will be influenced by the size of outcrop area and the volume of the rocks being sampled (Fig. 9.7a and b). The number of species described will also be affected by the number of palaeontologists who have worked on rocks of a particular age (Fig. 9.7d); a major monograph in which a large number of new species are described can give rise to a 'monographic burst' of apparent evolution. These and other factors make the simple census of fossil species potentially unreliable.

The empirical model of species diversity of Valentine *et al.* (1978) (Fig. 9.8a) is based on the premise that the carrying capacity of biogeographic provinces is relatively constant and that global diversity will change with time according to the changing number of biogeographic provinces (see Chapter 8). This initial premise is modified by recognizing two types of provinces: high diversity tropical provinces and low diversity provinces at lower latitudes. The model is further improved by taking into account increases in species diversity per community at particular times within the Phanerozoic (Bambach, 1977). Fairly arbitrary estimates of the diversity of high and low diversity provinces are assigned, based on data from the Recent, and changing diversity is estimated according to the changing number of the provinces of each type, and taking into account the species per community of the time-period being considered.

An alternative approach, referred to as the concensus model by Signor (1985), because it involved four of the protagonists in the debate over diversity patterns, is based on the observed similarity between estimated diversity change with time in four independent data sets: namely, trace fossil diversity, numbers of decribed species, within-habitat species richness and numbers of genera and numbers of families (Sepkoski *et al.*, 1981) (Fig. 9.8b).

The similarity in the pattern in the four sources of data is persuasive, but it could merely reflect a common source of error, arising from the kind of sampling biases described above (Signor, 1985).

A third estimate of changes in relative diversity that is commonly used is that of Sepkoski (1979) based on family diversity (Fig. 9.9a; see also section 9.2.7). The advantage of using family data as a surrogate for species data is that the record of families has been determined from a very large data base, so that for a particular segment of geological time it is likely to be relatively complete and the data are therefore less affected by the sampling biases that affect species data. The disadvantage of family data as a reflection of species diversity is that families vary greatly in the number of species that they contain and it is arguable whether one accurately reflects the other. An even more severe criticism is that many families are paraphyletic (they are incomplete clades that do not include all branches and all descendants) and therefore many of the 'family' extinctions are pseudo-extinctions, whereby one family appears to become extinct because its descendants are placed in a successor family that has been given a different name.

None of the estimates described above directly uses species data to obtain a diversity curve of species through time. In an attempt to obtain a more direct estimate of changing species diversity, Signor (1985) calibrated the distribution data for the described number of species against the estimated diversity of the Cenozoic and adjusted the data for the effects of sampling intensity on the distribution (Fig. 9.9b).

9.2.8 Diversity independent models of Phanerozoic diversity

In a robust criticism of diversity dependent models, Hoffman (1985) suggested that 'rates of origination and extinction of taxa may depend primarily on a whole complex of physical and ecological factors rather than on general laws'. Accordingly the history of the biosphere should be seen as predominantly influenced by perturbations of the physical environment and adjustments of biological interactions, with little influence being exerted from existing diversity. Changes in the global physical environment may include climatic change, atmospheric change, eustatic shifts in sea-level and changes in ocean circulation. Biotic changes include the extent to which species divide up the resources of ecosystems (niche partitioning) and different organizations of the feeding chain (changes in trophic structure). One view of diversity changes in the biosphere driven by physical and biotic factors is that of Valentine (1989) who suggests that the increasing expansion and complexity of ecosystems, as reflected in the increase in species numbers, is largely determined by the increased number of biogeographic provinces and diversification within palaeocommunities.

9.2.9 The Origins of diversity

Diversification can occur when evolutionary innovations allow biota to invade empty ecospace. For example, many of the major innovations

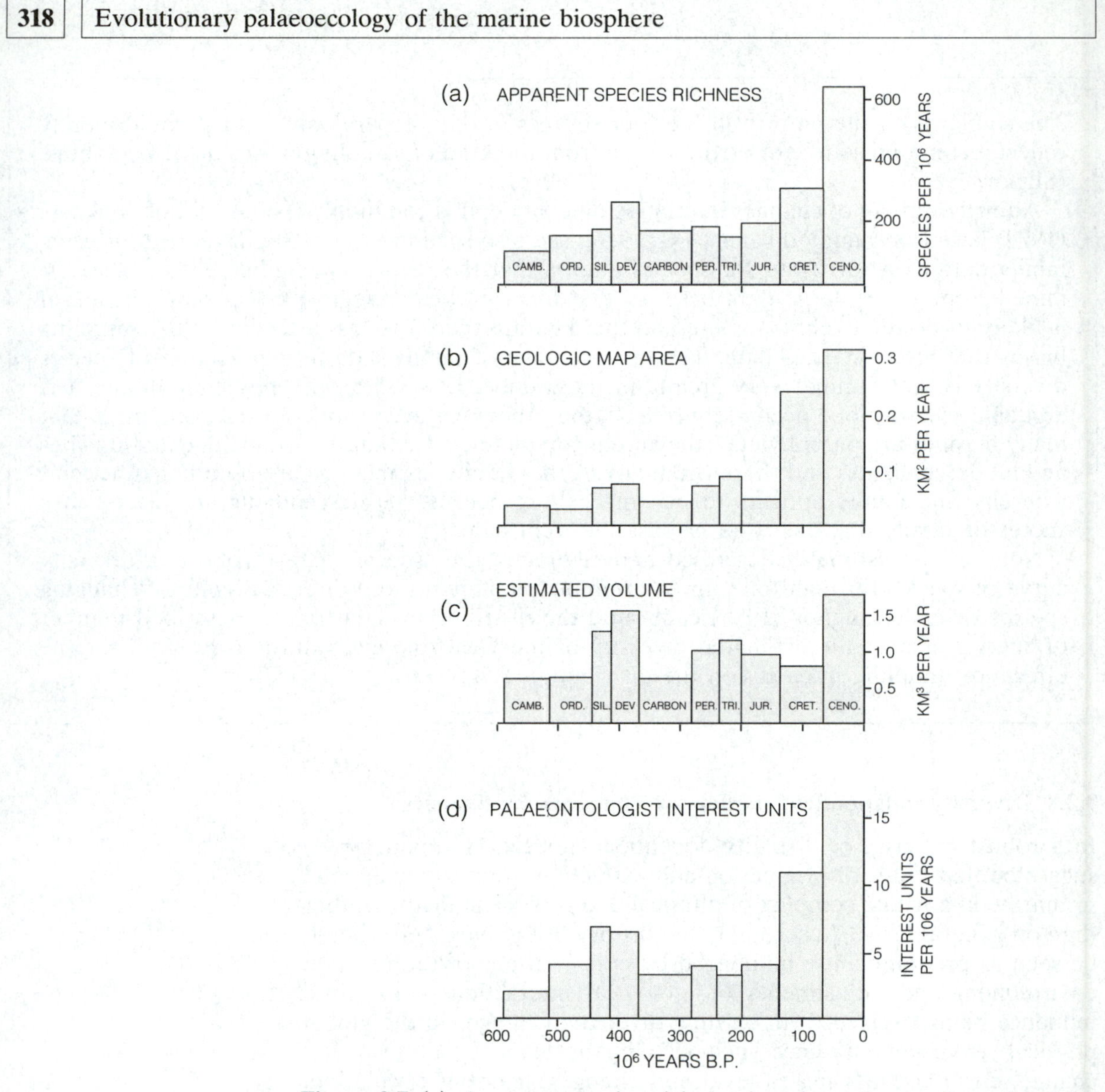

Figure 9.7 (a) Apparent species richness based on the number of species recorded relative to the time span of each system. (b) The area of outcrop of each system. (c) The estimated volume of rock available for inspection. (d) The intensity of palaeontological study of each system, based on the declared interests of palaeontologists in *The Directory of Palaeontologists of the World*. (Modified from Signor, 1985.)

amongst marine faunas appeared during the early Cambrian radiation, when adaptations to occupy unexploited marine habitats rapidly produced a wide range of morphological variation (disparity) which is registered taxonomically by the first records of many phyla and classes. At the same time there was increased interaction between organisms, and longer and more complex food chains started to develop, with

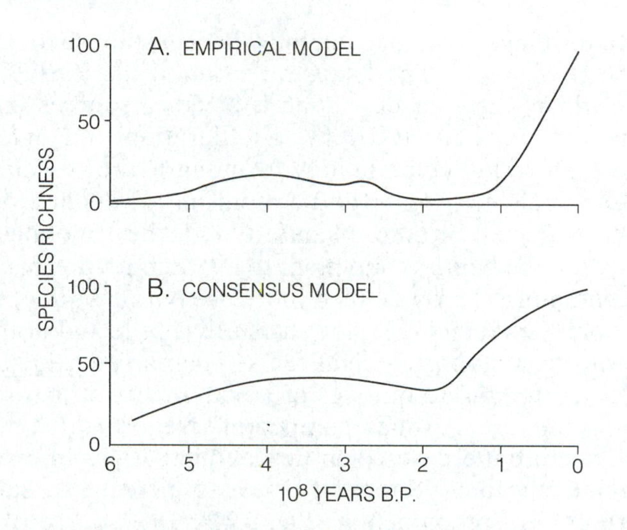

Figure 9.8 Two models of species richness through the Phanerozoic. These models represent changes in standing species-richness relative to the Recent. A. Empirical model (Valentine, 1970). B. consensus Model. (Modified from Signor, 1985.)

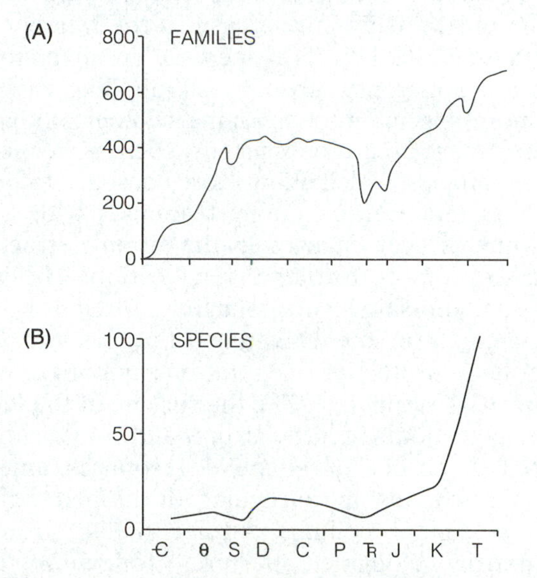

Figure 9.9 Diversity of skeletonized marine animals. A. Family diversity through time, commonly used as a proxy for variation in relative species diversity. B. Estimated percentage species variation. (Modified from Signor, 1985.)

carnivores at the top of the trophic structure, as has been shown from studies of the Burgess Shale fauna (Conway Morris, 1986). Later Phanerozoic innovation and diversification took place within the major clades and the fossil record suggests that the general increase in diversity later in the Phanerozoic proceeded by adaptations within a range of body plans, that had already been established (Valentine, 1969; Erwin *et al.*, 1987). The major radiation in the Early Ordovician established many new clades, that utilized resources in a more specific manner, particularly by the proliferation and diversification of suspension-feeding communities. Amongst the sessile benthos, brachiopods, corals, bryozoans and crinoids all diversified rapidly, the graptolites diversified in the plankton and the trilobites radiated amongst the mobile benthos and plankton. Communities became more numerous and more clearly defined in their environmental preferences and their more varied and complex structure is reflected in the appearance of coral-stromatoporoid-algal reefs. The early Ordovician radiation established the broad outlines of a community structure that was to last until the end-Permian extinction. Nevertheless, within this ecological structure there were substantial fluctuations in overall diversity with relatively long periods of diversity increase or stability and short, sharp periods of extinction (Fig. 9.21). One source of increased diversity develops at the habitat level, within local environments on the shelf (α diversity), and this approximately parallels changes in the total diversity of the group. The implication is that within-habitat diversification can be a major contributor to overall diversification.

The overall increase in the number of species through the Phanerozoic appears to be mirrored by an increase in the number of species within communities (Fig. 9.10). The increase in diversity within a community does not apparently arise by subdividing existing niches and so packing more species into the same ecological space, but by adaptive innovations that create new niches, such as deeper burrows, or new predatory strategies, which allow more ecospace to be exploited (Bambach, 1985). Diversification amongst animals with a particular mode of feeding may trigger changes in the trophic structure within communities and so promote further diversification. The widespread development of shell crushing (durophagous) predatory behaviour amongst marine animals in the Devonian, Jurassic and Cretaceous apparently generated protective adaptations amongst their prey (Signor and Brett, 1984; Vermeij, 1977); the success of the large marine reptiles, such as placodonts, ichthyosaurs and mosasaurs appears to have promoted a variety of adaptive responses amongst their molluscan prey. Gastropods in particular developed antipredatory adaptations such as tighter coiling, sturdier shells, strong external sculpture and narrow elongate apertures following the mainly Cretaceous diversification of shell-crushing animals such as teleost fish, sharks, rays, decapod crustaceans and shell-boring gastropods. Contemporaneously, an increased intensity of grazing apparently promoted a trend towards infaunal life styles amongst marine benthos.

There is some evidence to suggest that major evolutionary innovation preferentially appeared in inshore environments and spread out across shelf regions. This trend is not pronounced amongst Cambrian faunas except for trace fossils which show meandering and spiral traces appearing in shallow marine areas in the late Precambrian and early

Median Number of Species	High Stress Environments	Variable Nearshore Environments	Open Marine Environments
CENOZOIC	8.5	39	61.5
MESOZOIC	7.5	17	25
UPPER PALEOZOIC	8	16	30
MIDDLE PALEOZOIC	9.5	19.5	30.5
LOWER PALEOZOIC	7	12.5	19

Figure 9.10 Changes in the species diversity in communities with time. Communities in variable nearshore and open marine environments show stepped increases in diversity between the Lower and Mid-Palaeozoic and between the Mesozoic and Cenozoic but nearshore high-stress communities remain nearly constant. (From Bambach, 1977.)

Cambrian, but spreading in the Ordovician into deep-water environments where they remained a characteristic part of the ichnofauna thereafter (Crimes, 1994). The trend can be seen amongst the Palaeozoic fauna which arose inshore in the early Ordovician and spread outwards to partly displace the Cambrian fauna (Sepkoski and Sheehan, 1983). Amongst later faunas higher taxa appear to preferentially originate in nearshore habitats and in certain latitudinal belts, particularly the tropics, though it has been suggested that this pattern is an artefact of sampling.

9.3 EXTINCTION

9.3.1 Patterns of extinction

Background rates of origination and extinction apparently tend to keep global species diversity at a relatively constant level. However, there are times when extinction rates are raised to varying degrees and these, particularly the episodes of mass extinction, play an important role in restructuring ecosystems.

The average life-span of a species has been estimated to be between about 2 and 11 my, depending on the group being considered, the geologic period being investigated and factors such as the geographic range of the species. Species become extinct when their total interbreeding population has been eliminated. This is particularly likely if the breeding population is small and localized, whereas widespread species with large populations are hard to eliminate. Severe climatic events, habitat disruption, introduction of new predators or parasites and disease are known to eradicate local populations (Ford, 1982) and are probably causes of extinction of entire species that have limited geographic ranges. Other proposed causes of extinction include competition of two species for a similar niche, changes in ocean current systems and both positive and negative changes in nutrient supply. However, it is unclear which, if any, are the most prevalent causes of extinction. Knowledge of the extinction of recent species is distorted by the prevailing influence of mankind. Clearly many species have become extinct by habitat disruption, as in the tropical rainforests, and by human predation, as in the case of the moas of New Zealand, but these causes of extinction may not be typical of other periods of geological time. Low levels of background extinction characterize much of the geologic record, but there is a spectrum of peaks of extinction (Table 1, Jablonski, 1991a) which reaches an estimated level of 96% of all species in the end-Permian extinction. Some of the larger extinctions have been characterized as **mass extinctions,** which are recognized as being distinct on the basis of their pattern, duration, breadth and magnitude. To qualify as a mass extinction the event should be confined to a short interval of geological time, it should affect a wide variety of clades occupying a wide spectrum of habitats and it should eradicate a high proportion of species. Precise values for these parameters have never been set, but an analysis of magnitude as measured by family-level extinction suggested that there were five episodes of extinction that stood above all others and could be characterized as mass extinctions (Fig. 9.11). All five mass extinctions satisfy the criteria of short duration and large breadth, as indeed do several others which have been referred to as mass extinctions in the literature. A periodicity of extinction peaks with a period length of 26.2 ± 1 my has been identified in the Mesozoic and Cenozoic, but not in the Palaeozoic, based on a statistical analysis of familial and generic extinctions (Raup and Sepkoski, 1986). The biological reality of this pattern has been questioned because it has been shown that for some groups as little as 25% of the extinctions are 'real', the remaining 75% being composed of extinctions of paraphyletic groups (pseudoextinctions) and other spurious data (Patterson and Smith, 1989).

9.3.2 Analysing patterns of extinction

An accurate determination of duration and pattern, magnitude and breadth of extinctions is an important part of understanding the nature and cause of a major extinction. The duration of all major extinctions is less than 10 my, most are less than 5 my and the K/T extinction could have lasted as little as a few thousand years. Estimates of the duration of a particular extinction commonly vary because the pattern of extinction is not clear. The 'pattern' of an extinction refers to whether it occurs at a single horizon implying that it was sudden, or at several horizons and was stepped, or was gradual (Fig. 9.12). The

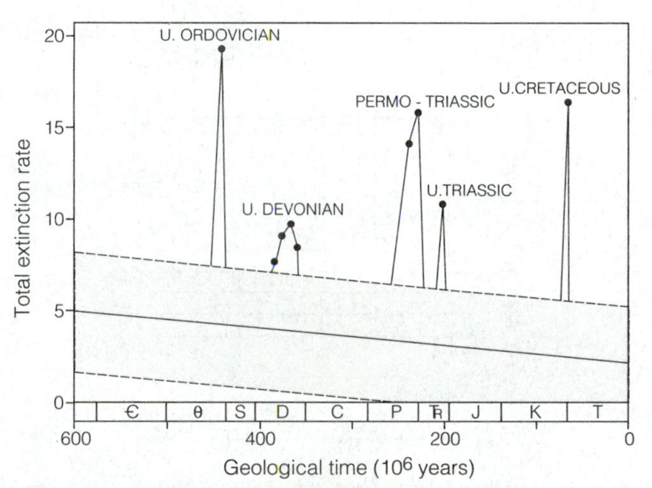

Figure 9.11 Total extinction rate (extinctions per million years) through time of families of marine invertebrates and vertebrates. Four statistically significant peaks of extinction are present and a late Devonian peak is strong but not statistically significant. Background extinction is shown by the regression line which is fitted to points having extinction rates less than eight families per 10^6 years. The dashed lines define the 95% confidence level for the regression. (From Raup and Sepkoski, 1982.)

perception of pattern is strongly influenced by the preservation of species within a sequence and the adequacy of the sampling. For example, an extinction can appear to have been sudden in a sequence where there is a hiatus in sedimentation which truncates the ranges of some species (Fig. 9.13c). Conversely, because the recorded range of a species in any one section is likely to be less than the full stratigraphic range of that species, it is common for apparent extinctions to precede a sudden extinction event, making it appear gradual (**the Signor-Lipps effect**) (Fig. 9.13b). The Signor-Lipps effect may be particularly enhanced where an extinction horizon occurs at the top of a regressive sequence in which species diversity decreases upwards for environmental reasons. Extinctions that appear to have occurred in a stepwise fashion can also be the result of the way samples are distributed in the rock record. Where fossils are concentrated at sporadic horizons within an otherwise barren sequence, each horizon may contain the last record of several species in that section. As a consequence the disappearances will appear to be stepped (see the 125 m level in Fig. 9.13b).

Studies of extinction patterns made on a single section or confined to a few sections within one region suffer from the sampling problems outlined above and may also give a local rather than global picture of extinction. To counter the sampling problems the extinction interval should have been sampled on a global scale, but this requires high-resolution stratigraphy and exceptionally accurate correlation to do so. Conventional biostratigraphy commonly proves inadequate in

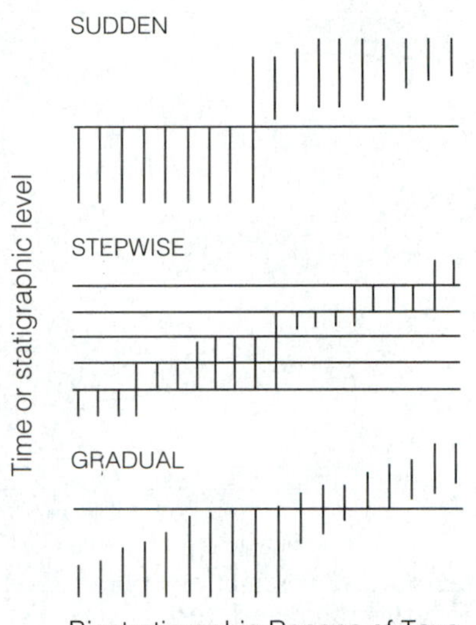

Figure 9.12 Three different patterns of extinction.

correlating between different facies and widely dispersed sections so the most promising approach to this correlation problem is the use of prominent peaks or troughs in the stable isotope record that reflect changes in ocean state that are commonly global and synchronous.

Estimates of the magnitude of species extinctions are hazardous because of the problems of adequate sampling. The same criticism applies to some extent to genera, but less so to families. But families are not necessarily monophyletic groups, but are commonly paraphyletic or polyphyletic, and their disappearance may have little relevance to true extinction. A quite different source of pseudo-extinctions arises from taxa that disappear from the geological record at an extinction horizon, but appear again at a substantially higher horizon. These taxa that appear to arise from the grave are known as **Lazarus taxa.** When they are not identified their ancestors make a false contribution

Figure 9.13 (opposite) Experiments to show how different patterns of extinction may develop. (A) The experiments are based on the stratigraphic ranges for 21 ammonite lineages at Zumaya, Spain. The vertical scale is metres below the Cretaceous-Tertiary boundary. Fossil occurrences are marked by horizontal ticks. The histogram on the right shows the changing number of lineages with time. (B) All the fossil occurrences above the 100m level are eliminated to simulate a sudden extinction. Note that diversity declines gradually towards the extinction horizon although the modelled extinction was sudden. The last recorded occurrence of several species that are known to have ranged up to the 100m level are below the 100m level and so appear to have become extinct earlier (the Signor-Lipps Effect). Note also a spurious extinction step has appeared at the 125m level. (C) A preservational gap is imposed on the data. The effect is to produce an apparent extinction at the 125m level. Lineages 16, 18, 19, 21 disappeared at the hiatus, but reappeared as Lazarus taxa after the pseudo-extinction event. (From Raup, 1989.)

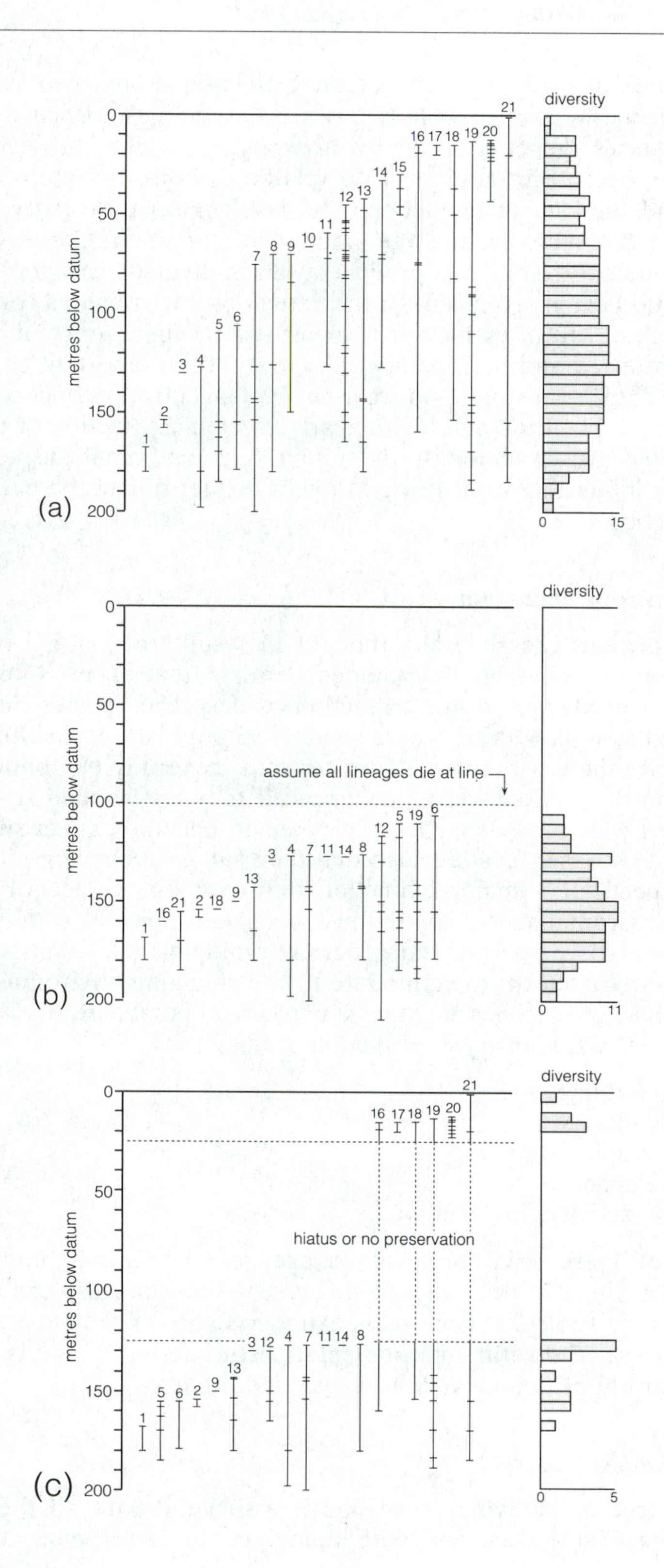
diversity
metres below datum
(a)
(b)
(c)
assume all lineages die at line
hiatus or no preservation

to estimates of the magnitude of an extinction. Some estimates of faunal change at extinction horizons are based solely on changes in overall species, generic or family diversity. This can, however, be misleading because total diversity is related to both the rate of origination and the rate of extinction. The level of taxic diversity is the amount of origination minus the amount of extinction. Consequently a fall in origination rate can produce a fall in diversity that looks like an extinction event, even though the extinction rate remains constant.

The breadth of an extinction is measured by the variety of clades that are affected and their ecological variety. It is important to know whether both terrestrial and marine biotas suffer extinction and whether both plankton and benthos are affected. Extinction of particular trophic groups amongst the fauna (e.g. epifaunal suspension-feeders or infaunal deposit-feeders) might further reflect the nature of the extinction.

9.3.3 Causes of extinction

Mass extinctions are generally thought to result from global perturbations of the physical environment, rather than from biological causes. To produce a major extinction, widespread species must be eliminated as well as those with a small local population. To eliminate such species the environmental stresses must generally be abnormally severe and their effect will be enhanced if they are applied suddenly and over a wide area. If many species are to become extinct simultaneously the stresses must be such that they cut across ecological lines. These aspects of a major extinction influence our choice of likely causes. Environmental changes must be geographically widespread, environmentally pervasive and generally rapid so that faunas do not migrate away from or accommodate to the changing environment.

The physical changes to the environment that are commonly proposed as causes of mass extinction are:

1. sea-level changes
2. climatic changes
3. oceanographic changes
4. bolide impact
5. volcanic activity.

Some of these environmental changes are commonly intimately linked (e.g. climatic, oceanographic and sea-level changes) and as can be seen from Table 9.1, each mass extinction could have been associated with one of several environmental perturbations. This has led to a wide variety of opinions about cause and effect.

9.3.3.1 Sea-level changes

Some degree of bathymetric change is associated with all the mass extinctions (Table 9.1), and with many of the lesser ones as well (e.g. the Pliensbachian/Tithonian, the Cenomanian/Turonian, and the Oligocene/Miocene extinctions). Change in sea-level can affect the biotic world at different ecological levels; it can change the number of biogeographic provinces by opening or closing seaways and so changing global palaeogeography, it can modify habitat diversity and hence the variety of communities by changing the depth of epicontinental seas, and it can change the amount of habitable area and so affect species abundance. A close relationship between habitable area and species richness has been shown to exist on many scales, varying from decapod species on coral heads to birds and amphibians plus reptiles on islands. The relationship can be modelled by a species/area curve (Fig. 9.14a) fitted by the equation $S = kA^{z}$ where S is the number of taxa, A is the area, z is the slope of the curve, which varies according to the taxa being considered, and k is the intercept on the y axis which is thought to be approximately related to environmental factors such as climate. The equation does not uniquely model a species to area relationship, but instead could model a species to habitat-diversity relationship. This alternative raises the question in conservation planning as to whether a single large conservation area preserves more species than several small, heterogeneous ones. Habitat heterogeneity is a factor that intuitively seems likely to be correlated with diversity, so it may be better to view the model when applied to large areas, as a species/area/habitat-heterogeneity relationship. The main criticisms of the model as applied to geological extinctions are:

1. that a cause and effect relationship has never been satisfactorily demonstrated;
2. no species/area relationship emerged from an analysis of changing bivalve diversity during the end Eocene/Oligocene regression in the Gulf Coast region (USA) (Hansen, 1987) and there is no relationship between the many substantial sea-level changes in the Quaternary and species extinction (Valentine and Jablonski, 1991); and
3. extinctions do not necessarily correspond with maximum regressions but may correspond more closely with the early phase of a transgression and its associated anoxia (see oceanographic change, section 9.3.3.3.). A close correlation between ammonoid diversity and sea-level change was identified by House (1989) and interpreted as indicating a species/area effect, but because sea-level is linked with other environmental changes this specific link is debatable.

9.3.3.2 Climatic change

All mass extinction events coincide with some degree of climatic change (Table 9.1), as do some of the smaller ones, such as the Eocene/Oligocene and Miocene events. Change in temperature is generally believed to be the most potent climatic force causing extinction, though other aspects of climate, such as aridity and rainfall, have

Table 9.1

		LATE ORDOVICIAN	*LATE DEVONIAN*	*PERMO-TRIAS*	*LATE TRIAS*
TIMING OF EVENTS		2 phases ~ 1 My apart	extinctions over 3my. main extinctions over 2my with peak near end	3 or more events within 3 my. Extinctions concentrated in 2my with peak at the end	2 Late Trias events over 15my+. A late Carnian phase and a major late Norian (Trias-Iur boundary) event
	FAMILIES	12%	14–22%	52%	12%
MAGNITUDE	GENERA	61%	55%	84%	47%
	SPECIES	85%	82%	96%	76%
BREADTH (major clades)		Most clades: graptolites, trilobites, brachiopods, bryozoans corals, ostracods, conodonts	Most clades: tabulate and rugose corals, stromatoporoids, conodonts, forams, brachiopods, fishes	All clades: total ext. tabulate and rugose corals, goniatites, trilobites, blastoids, 3 crinoid groups, orthoids, productids	Many clades: ammonoids, brachiopods, conodonts, bivalves, bryozoans, gastropods, vertebrates, plants in Carnian and Norian
BREADTH (ecological groups)		Most groups: phytoplankton, zooplankton, sessile benthos mobile benthos	Most groups: most affected- tropical faunas, reefs, shallow benthos marine, more than few fishes	(All ecological groups) particularly zooplankton, sessile susp. feeders, high-level carnivores, forms with planktotrophic larvae	Major changes in reptilian ecology. Communities with ammonoids and bivalves modified by major extinction
CAUSES CITED		Lowered sea-level climatic cooling anoxia oceanographic change eutrophication bolide impact	Sea-level change, climatic cooling, anoxia bolide impacts	High/low temperatures sea-level fall, bolides, anoxia, trace element poisoning, cosmic radiation, volcanism	Sea-level changes, anoxia, climatic change, bolides
CAUSES CURRENTLY FAVOURED		Combined cooling oceanographic change, eutrophication, anoxia	Linked climate, sea-level, anoxia, linked bolide impact, climate, sea-level, anoxia.	Multiple linked and unlinked causes. Sea-level-volcanism-climate or anoxia	Anoxia (end Norian) increased aridity (late Trias)
STABLE ISOTOPE RECORD	$\delta^{13}C$	+~6‰ at 1st extinction –6‰ at 2nd extinction	+ 3.9‰ at Lr and U. Keilwasser events	Late Permian +3.5‰ shift –4‰ at P–T boundary	–2‰ at boundary
	$\delta^{18}O$	+3.5‰ at 1st extinction –3.5‰ at 2nd extinction		–0.5 to 4.5‰ at boundary	large negative shift

Table 9.1

		CENOMANIAN-TURONIAN	*CRETACEOUS-TERTIARY*	*EOCENE-OLIGOCENE*
TIMING OF EVENTS		Moderate stepwise extinctions over 1–3my, culminating in a major event over about 0.5my	Nearly instantaneous or progressive over 10s to 100s Ky, to a sharp peak	Stepped over several My
MAGNITUDE	FAMILIES	7%	11%	
	GENERA	26%	47%	15%
	SPECIES	53%	76%	35%
BREADTH (major clades)		Many clades: ammonites, nannofossils, dinoflagellates, ostracods, bivalves including rudists	Most clades: ammonites, dinosaurs extinct, planktic forams, brachiopods, coccoliths	Many clades: planktic-benthic forams nannoplankton, ostracodes, bivalves, gastropods, echinoids, plants, mammals
BREADTH (ecological groups)		Plankton and nekton sessile epifaunal benthos	Phyto- and nannoplankton and feeders on plant matter. Low latitude plants. Land living tetrapods	Equatorial migration of planktonic and nektonic faunas and plants
CAUSES CITED		Anoxia, bolides, sea-level change	Bolide impact, climatic change, sea-level fall, ocean chemistry	Cooling, eutrophication bolide impacts, volcanism
CAUSES CURRENTLY FAVOURED		Anoxia or linked causes: bolide impacts-climatic change-ocean chemistry	Bolide impact, or climatic change and associated changes	Cooling, comet showers
STABLE ISOTOPE RECORD	$\delta^{13}C$	<+3‰ at main extinction interval	<–3‰ at extinction	
	$\delta^{18}O$	fluctuating, <2‰ + and- recorded	–0.5‰ followed by +3‰	+2‰ near boundary

Table 9.1 Continued

	LATE ORDOVICIAN	*LATE DEVONIAN*	*PERMO-TRIAS*	*LATE TRIAS*
SEA-LEVEL CHANGE (timing and magnitude)	50–100m fall mainly postdating 1st extinction 50–100m rise at 2nd extinction	Fluctuations through Frasnian with a Late Frasnian regression and Frasnian/Fammenian transgression	Prolonged sea-level fall during early phases of extinction. Sharp rise at main extinction phase	Late Trias regression followed by a rapid late Norian transgression
CLIMATE	Cooling of ocean waters, including tropics by several °C at 1st extinction. Similar warming at 2nd extinction	Climatic conditions uncertain. Possible rapid cooling	Pronounced warming trend following Perm.-Carb glaciation Equatorial aridity.	Increasingly arid in late Trias, but with wetter phases in Carnian. Moister climates in early Jurassic
ANOXIA	Widespread anoxia coincides with 2nd extinction	Widespread anoxia coincides with the Keilwasser extinctions	Widespread anoxia in low and high latitudes at main extinction event	Regional anoxia at end Trias extinction event
BOLIDE IMPACT	Iridium anomally reported at extinction level; not confirmed on re-analysis	Several weak iridium anomalies microtektites. At least 3 large Devonian impact craters	Iridium spikes in boundary sections in China, but not confirmed. Spikes in Carnic Alps questionably bolide	Manicougan impact crater (Quebec), dated as late Triassi
VOLCANISM	Subdued	Subdued	Extensive basaltic volcanism in Siberian Traps in late Permian	Subdued

Table 9.1 Continued

	CENOMANIAN-TURONIAN	*CRETACEOUS-TERTIARY*	*EOCENE-OLIGOCENE*
SEA-LEVEL CHANGE (timing and magnitude)	Sea-level fall below event, sharp rise at main extinction event	Fall at or near K-T boundary	Fluctuations
CLIMATE	mixed evidence of climatic fluctuation	Increase in seasonality near K-T boundary. Shift in floral belts. Possibly a short cool event at boundary	Pronounced cooling
ANOXIA	Widespread at main event	Only localized anoxia	
BOLIDE IMPACT	Modest iridium spikes in late Cenomanian-early Turonian. Some microtektites	Boundary clay layer with marked iridium spike, shocked quartz, microtektites	Microtektite layers, but only one associated with an extinction horizon
VOLCANISM	Subdued	Extensive basaltic volcanism in Deccan Traps, India, at about the K-T boundary	African Rift Valley lavas

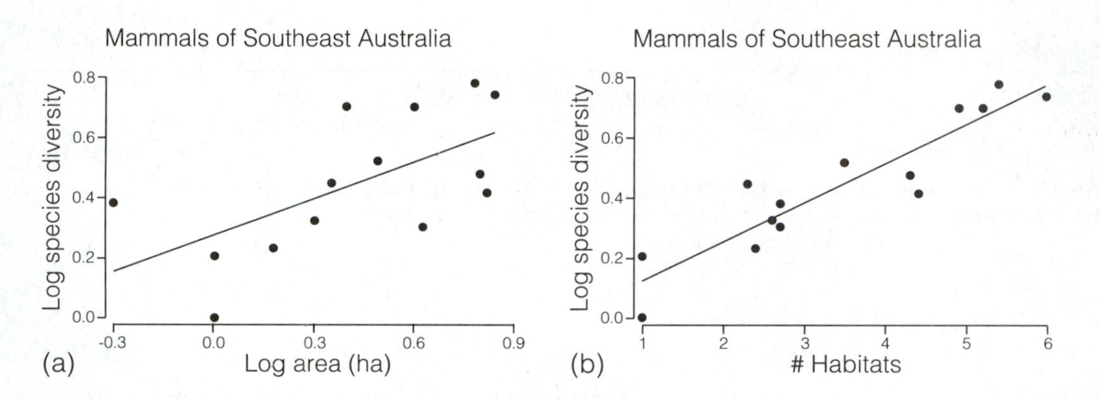

Figure 9.14 (a) Species-area curve for small mammals in parts of Australia. (b) A species-habitat curve for the same small mammal data produces a better fit. (From Rosenzweig, 1995.)

been proposed in specific instances. Evidence for climatic cooling in the geological record includes widespread extinction amongst tropical faunas, including the disappearance of reefs, the migration of warm-adapted taxa towards the equator, contemporaneous continental glaciation and $\delta^{18}O$ isotope records. All species have temperature limits to their viability, consequently temperature is clearly an important control on the geographical distribution of species. Sudden changes of temperature will put animals under stress and instances where local populations are eradicated by a severe winter are common. On the other hand, most marine animals have a tolerance of short-term temperature changes greater than those normally experienced in the sea, and the elimination of whole species by a short-term shift in temperature is probably rare (Clarke, 1993). Longer-term climatic shifts are generally accommodated by migration of the biota in parallel with the movement of climatic belts. However, where a barrier blocks migration, extinction may result. An example of this is the extinction of bivalves trapped in the Gulf of Mexico on the western Atlantic by the barrier of Florida, as climatic belts migrated equatorwards during Plio-Pleistocene global cooling (Stanley, 1986).

When there has been long-term climatic change, such as that which caused the cooling of Southern Ocean waters during the Cenozoic, organisms commonly accommodate to the changing temperature. On the other hand there is evidence that transitions between one climatic state and another can be so rapid that organisms are unlikely to adapt. If global temperature change was both rapid and substantial enough to eliminate entire climatic belts then extinction would potentially be large and ecologically pervasive. This would be particularly so if tropical regions cooled substantially because they have a very diverse and highly adapted fauna that is particularly susceptible to physical perturbation.

There has been widespread criticism of the claim that climate is the 'common cause' of mass extinctions. Critics point to the relatively weak climatic changes associated with some extinctions and the fact that some climatic changes appear to post-date the extinctions they are supposed to have caused. Furthermore, there was no mass extinction associated with the Quaternary glaciation. On the other hand the the close correlation in time between the onset of glaciation and extinction in the end-Ordovician, the end-Oligocene and mid-Miocene extinctions does strongly suggest there might be a causal relationship in these instances. Increase in temperature is not generally seen as a likely cause of major extinction, but there is evidence that a brief episode of global warming at the end of the Palaeocene created warm deep water, associated with which there was a large extinction of benthic foraminifera (Kennett and Stott, 1991).

9.3.3.3 Oceanographic change

Interest in the possibility that oceanographic changes could play an important role in mass extinctions has been generated in recent years by the recognition that significant stable isotope excursions of $\delta^{13}C$ are associated with all the mass extinctions (Table 9.1). Peaks and troughs of $\delta^{13}C$ values indicate that there have been changes in production and cycling of organic carbon in the oceans (see Chapter 2, Box 2.3). These changes are likely to be related to variations in the production, sedimentation and storage of organic carbon resulting from changes in the structure and circulation of the oceans initiated by temperature changes, sea-level changes and changes in the palaeogeography of ocean basins.

A strong case has been made for a link between rising sea-level, anoxia and mass extinction (Hallam, 1986). Anoxic sediments are found at, or close to, the extinction level in all mass extinctions and several lesser ones, such as the Pliensbachian/Tithonian and Cenomanian/Turonian events. The anoxic sediments, which are generally dark grey to black shales, are associated with transgressive sequences formed in interior cratonic basins, shelves and more rarely deeper basins. $\delta^{13}C$ values from the anoxic part of the sequence generally show a positive excursion, reflecting either high marine productivity and/or increased sedimentation of organic matter. Low levels of dissolved oxygen in sea-water greatly reduce the diversity of animals that animals can live in the environment (see Chapter 6). Thus localized anoxia would cause mortality amongst local populations, but it would probably require anoxia throughout an entire basin to cause a regional extinction of species. Whether anoxia has ever been developed on a global scale, sufficient to cause a mass extinction is debatable. It is, of course, not an extinction mechanism that would affect terrestrial as well as marine ecosystems.

Changes in oceanic structure and circulation, initiated by temperature changes have also been proposed for some extinctions. The evidence for the proposed link comes from positive excursions in $\delta^{13}C$. The isotope shift implies that organic carbon is being removed from shallow marine waters either by being incorporated in an increased amount of living matter or because organic matter is sinking into deep waters or being sedimented. Where a carbon isotope excursion is not associated with widespread organic-rich black shales, an increase in productivity is favoured. High levels of organic productivity are associated with high nutrient (eutrophic) ecosystems which characteristically have a high abundance but low diversity, so that widespread eutrophication of the oceans could potentially result in high levels of extinction. An increase in nutrient cycling would be expected when there was a change from the stratified oceans typical of greenhouse periods to the vigorous bottom circulation and upwelling characteristic of ice-house climates. Evidence of high nutrient levels and high abundance of a relatively few planktonic species would help to support a eutrophication model of extinction.

9.3.3.4 Bolide impact

The discovery of a horizon with high levels of iridium coincident with the end-Cretaceous extinction transformed thinking about the role that extra-terrestrial bodies might have on evolution. The significance of iridium is that it is one of several platinum group elements (including Ru, Rh, Pd, Os and Pt) that are found in extremely low amounts in the Earth's crust, but in higher amounts in extra-terrestrial material. The implication of the iridium 'spike' is that the iridium was sedimented following a bolide impact (bolides are asteroids, comets and other extra-terrestrial bodies). Further evidence supporting a bolide impact came from the presence of shock-metamorphosed quartz and microtektites. Shock-metamorphosed quartz is identified by its two or more sets of intersecting shock lamellae; it is associated with most bolide impact sites and is also recorded from nuclear testing sites. It can also be associated with massive volcanic eruptions but the wide distribution and multilamellar stucture found in the K/T boundary layer have not been demonstrated amongst volcanic deposits. Microtektites are glassy millimetre-sized droplets created by melting on impact, and can be distributed continent-wide in the vapour cloud. Their mineralogy relates them to the impact site and potentially allows that to be identified. Commonly associated with microtektites are microspherules, about 1 mm in diameter, which are hollow or composed of replacement minerals such as sanidine, clays and pyrite. Some of these objects are probably impact-related but others may be diagenetically altered biogenic remains.

The commonest bolides that impact the Earth are comets and asteroids. The size of an impact crater depends particularly on what type of body is impacting and its velocity. The well-studied Montagnais crater in offshore Nova Scotia is 45 km in diameter and is estimated to have formed by impact from a 3.4 km cometary nucleus. Studies of cratering over a longer period suggest that 45 km craters were created on average about every 10 my. Although it has been suggested that the spectrum of extinction peaks from small to large might be related to a range of impact intensities, this is not supported by the substantial Montagnais event at 50.8 Ma, which did not apparently result in a peak of extinction. If larger impacts are required for major extinction they will be rare events, on the order of several tens of millions of years (Jansa, 1993).

Criticisms of bolide impact as a cause of mass extinctions are mainly of two kinds. One set of criticisms is directed at the evidence used to identify a bolide impact. It is claimed that no piece of evidence is unique to a bolide. Iridium spikes can have several origins; they may arise at condensed horizons in a sedimentary sequence, iridium can be concentrated organically, as for example in stromatolites, and it may occur at raised levels in volcanic deposits. Shocked metamorphic quartz and some microspherules could also come from volcanic activity. Nevertheless, the K/T iridium layer is particularly persuasive as a candidate for a bolide impact (though not to everyone!) because it not only has all the diagnostic features, but is remarkably widespread in both marine and terrestrial environments.

The second kind of criticism is directed towards the ability of an extra-terrestrial object to cause mass extinctions in general, or to have been the cause of a specific mass extinction. The arguments in these cases hinge mainly on the specific nature of a mass extinction and are discussed further below (section 9.3.4.5).

9.3.3.5 Volcanism

The role of volcanism in mass extinction has commonly been promoted as an alternative interpretation to the data presented for bolide impact, though as volcanologists have assembled information about the incidence of major volcanism through geologic time and its possible effects, a more general case is emerging.

Large eruptions which might affect the global environment are of two types; they may be mainly acidic, extremely explosive eruptions, commonly associated with caldera formation, or they may be more subdued outpourings of basaltic material, commonly in the form of plateau basalts.

Large explosive volcanic eruptions loft fine volcanic ash and sulphurous gases into the stratosphere where the gases form minute droplets, called aerosols. The particles and aerosols absorb and scatter incoming solar radiation and so decrease the heat energy reaching the Earth. Studies of historic eruptions suggest that there is a correlation between these and short-term climatic cooling (Rampino,1991). More dramatically, the vast Toba eruption in Sumatra at 75,000 BP, which ejected about 50 km^3 of volcanic material, coincides with the onset of the last glaciation. However, whether there is a cause and effect relationship is debatable.

Possibly more significant as a cause of major extinctions are the enormous eruptions of plateau basalts that have occurred intermittently in the past. During major lava eruptions convective plumes rising above fire fountains, carry large amounts of sulphur-rich gases into the stratosphere, and generate sulphuric acid aerosols, which if carried around the globe by stratospheric winds could have severe climatic and biological effects. The eruptions of vast volumes of basaltic magmas are approximately coeval with both the end-Permian and K/T extinctions. The late-Cretaceous flood basalts of the Deccan Traps of India, with an estimated volume of more than two million cubic kilometres, have been associated with the K/T iridium layer, shocked quartz, microspherules and the mass extinction. Criticisms of this association are that the features of the iridium layer are more tenuously associated with volcanism than with bolide impact, and there are uncertainties about the exact timing of the volcanism relative to the extinction horizon. The evidence for an association between major extrusions of late Permian lavas and the mass extinction rests mainly on their synchroneity. A statistical correlation between extinction events and major episodes of flood basalt volcanism, with a periodicity of about 30 my, tantalizingly suggests that there could be a causal relationship (Stothers and Rampino, 1990), but the detailed relationship has not been explored.

9.3.3.6 Discussion

Analyses of the causes of extinction have varied, from those that propose a single cause for all extinctions, such as bolide impacts, to those that identify a single cause for a particular extinction, to those that identify multiple causes, such as sea-level change, climate and ocean state which might be interrelated (see Joachimski and Buggisch, 1993, for the late Devonian), or unrelated and just coincident in their timing. Commonly it has not been possible to isolate one cause of extinction from two or more contenders. However, when it is possible to determine the precise timing of the physical changes relative to the biological extinctions it is possible to exclude as 'causes' those physical changes that post-date the extinctions.

9.3.4 The causes and ecology of the major extinctions

Seven of the major extinctions are discussed in the following section and a summary of data for each one is presented in Table 9.1.

9.3.4.1 Late Ordovician extinction

In terms of magnitude this was a mass extinction, with the loss of an estimated 12% of families, 61% of genera and 85% of species. Nevertheless very few major clades disappeared and although most communities lost some of their constituent members, there appears to have been no major restructuring of the marine biosphere; most Silurian communities look much like those of the Ordovician in comparable environments.

Extinctions occurred abruptly twice separated by about 0.5–1 my (Fig. 9.15). Extinction affected most or all clades, it nearly eradicated the graptolites and was severe amongst other groups including the trilobites, brachiopods (50%+ genera) (Harper and Rong Jia-yu, 1995) and conodonts (Table 9.1). The effects of the extinction were ecologically widespread but were particularly severe amongst the zooplankton (graptolites, conodonts and pelagic trilobites) (Brenchley, 1989). Similar high levels of extinction may have affected the phytoplankton but for preservational reasons the record is less clear. Benthic communities lost members, particularly those in deeper water, but most were not wholly destroyed. Extinction was severe amongst the benthic faunas of tropical regions and appears to have been greater than at high latitudes. The first phase of extinction correlates closely with a fall in sea-water temperature, recorded in a positive $\delta^{18}O$ isotope excursion and the onset of marine regression both reflecting a major growth-phase of the Gondwanan ice-caps. A synchronous $\delta^{13}C$ shift of up to +7‰ implies a major change in carbon cycling in the oceans, reflecting either greatly increased productivity or the loss of organic matter from surface waters which was stored either in deep marine waters or as organic-rich muds on the sea-floor (Brenchley *et al.,* 1995). It is suggested that a major cause of the first phase of extinction was the abrupt fall in the temperature of marine waters which extended into the vulnerable tropical habitats. In addition, the isotopic evidence suggests that there might have been eutrophication of the surface ocean waters caused by upwelling of cold water generated in polar regions, resulting in an increase in productivity, but a decrease in diversity.The second phase of extinction is associated with an abrupt rise in sea-level, a rise in sea-water temperatures, a negative $\delta^{13}C$ shift of 5‰ and widespread anoxia. Some clades that had just survived the first extinction were eradicated, the *Hirnantia* fauna, which radiated after the first extinction, largely disappeared, corals suffered heavy extinction and reefs temporarily disappeared. Factors that favoured survival during the first phase of extinction were clearly no guarantee against extinction when the environmental changes were different (Owen and Robertson, 1995). Identifying one single cause for this phase of extinction is difficult. Rising temperature might have affected the *Hirnantia* fauna, which was possibly a cool water fauna, but is unlikely to have been harmful to reefs. Anoxia is widespread in temperate regions, but less so amongst the tropical carbonate environments where the corals and reefs existed. A fall in productivity and nutrient levels implied by the carbon isotope record is unlikely to have been associated with a decline in reefs that favour low nutrient levels. It appears likely that change in temperature and oceanic state and particularly widespread anoxia, were the major causes of the second phase of extinction (Brenchley *et al.*, 1995), but the temporary demise of reefs might have been related to the deep flooding of shelves and platforms at the start of the Silurian.

The environmental changes that in sum caused the mass extinction appear to be linked to the climatic changes that caused the growth

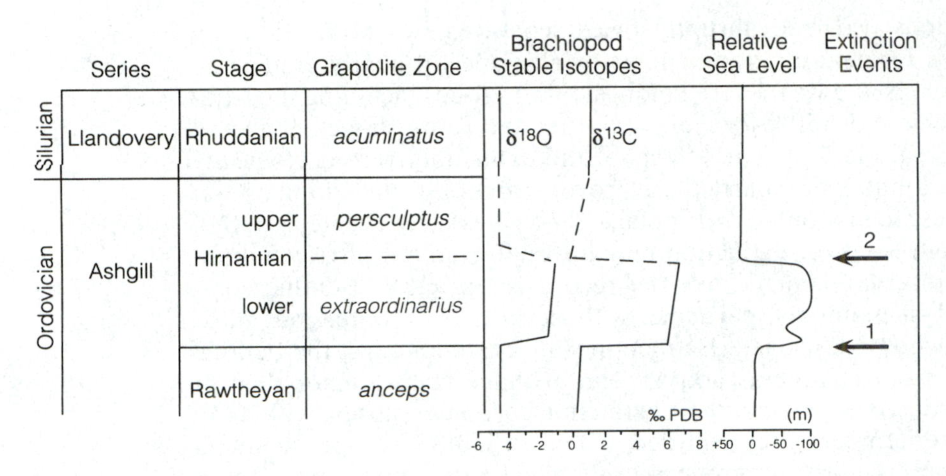

Figure 9.15 The late Ordovician extinction occurred in two phases; the first coincided with a fall in sea-level and positive carbon and oxygen excursions, the second phase coincided with a rise in sea-level and a negative shift in carbon and oxygen.

and demise of the Gondwanan icecaps. Thus changes in sea-level, temperature levels in ocean waters, ocean circulation, nutrient fluxes and carbon cycling may all have the same ultimate cause and have contributed in varying degrees to the extinction. However, it is the fast rate at which the icecaps grew and then disappeared which appears to be exceptional and sets this glaciation apart from others, and it might be the exceptionally fast rate of environmental change that was a crucial factor in promoting widespread extinction.

9.3.4.2 Late Devonian extinction

Estimates of the duration of this extinction in the late Frasnian have varied from 15 my to 0.5 my, but a value of about 3 my is now generally accepted. The extinction appears to be stepped with a pronounced final peak occupying a short interval in the Upper Keilwasser Horizon (Fig. 9.16). Nearly all clades suffered major extinction and it is estimated that between 14% and 22% of families disappeared, while 55% of genera and 82% of species were lost. The extinction affected both the plankton, including foraminifera, conodonts and ammonoids and the sessile benthos, such as corals stromatoporoids, brachiopods and bryozoans (Table 9.1). Tropical faunas were more severely affected than those in high latitudes and coral-stromatoporoid reefs largely disappeared from the geological record for a long period. A particularly striking aspect of this extinction is that ecosystems continued to flourish throughout the earlier extinction events because origination of species remained high. However, diversity took a sudden fall in the final event because both the extinction rates were relatively high and origination levels were low (McGhee, 1989).

There are three prominent hypotheses concerning the cause of extinction. One identifies global cooling as a major cause based on the pattern of biotic changes, particularly the extinction amongst atrypoid brachiopods and the demise of reefs in high latitudes (Copper, 1986). It has been suggested that the climatic deterioration might be associated with a short period of glaciation recorded on Gondwanaland, but the synchroneity of the two events is not well constrained. A second hypothesis associates the extinction with anoxia which is reflected in several dark shale horizons in the upper Frasnian. The peak of extinction falls in the Upper Keilwasser Horizon (Fig. 9.16), which is the thickest of the anoxic shales. The two main Frasnian anoxic horizons correlate with strongly positive $\delta^{13}C$ excursions (Fig. 9.16), which are interpreted as reflecting the storage of carbon on the sea-floor. According to the anoxic ocean model, anoxia was initiated during transgressive phases in the Frasnian as warm saline waters accumulated on tropical to sub-tropical shelves. Oxidation of organic matter in the descending saline waters reduced the amount of dissolved oxygen to low levels, so producing anoxic intermediate and bottom waters where the residual organic carbon could accumulate. By sedimenting carbon and removing it from the oxidative surface waters, the amount of CO_2 that was returned to the atmosphere decreased and the falling CO_2 levels produced a negative greenhouse effect. Thus the anoxic events would have been coupled with global cooling, giving a two-pronged attack on the global fauna (Joachimski and Buggisch, 1993). There are, however, problems with anoxia being the major factor, because although carbon-rich shales showing a positive $\delta^{13}C$ shift are widespread in Europe, they appear to be absent in many places elsewhere. A third hypothesis is based on the presence of horizons with raised levels of iridium and some with microtektites: it is suggested that there were several extra-terrestrial impacts close to the Frasnian-Famennian boundary interval, which caused rapid drops of temperature on a global scale. However, the raised levels of iridium are not high – much less than that at the K/T boundary – and the most convincing microtektite layers post-date the main extinction horizons (McGhee, 1996). Clearly the impact hypothesis, like the other two, is far from proven.

9.3.4.3 End-Permian extinction

This is the largest of the mass extinctions in the Earth's history and eliminated a large part of the preceding Palaeozoic fauna. The duration of the extinction is a matter of dispute. Sections through the Permo-Triassic boundary are not common and many appear incomplete because of a global late Permian regression. Diversity decreased from the Capitanian into the basal Trias, a period of several million years (Fig. 9.17), but extinction was apparently concentrated in two main phases, the second, at the Permo-Triassic boundary, being the larger. Some 52% of all families, 84% of all genera and 96% of all species

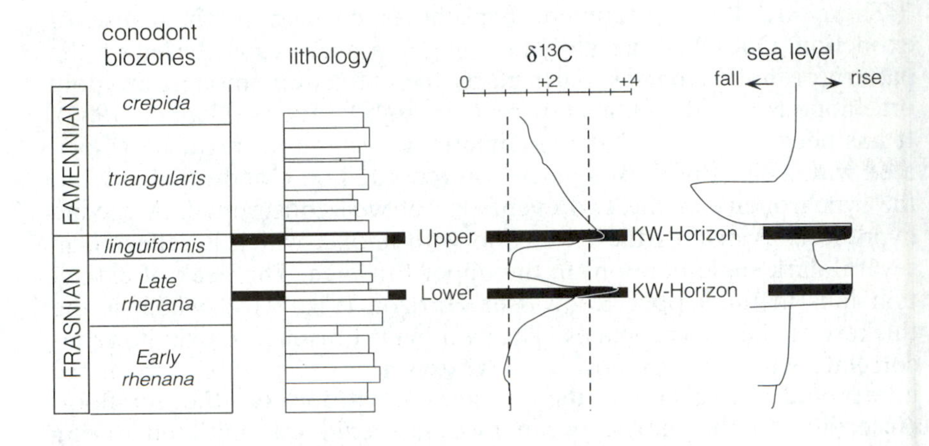

Figure 9.16 Two of the main phases of the late Devonian extinction coincided with black shale intervals (the Lower and Upper Keilwasser horizons, (KW), both of which coincide with a rise in sea-level and show a positive $\delta^{13}C$ excursion. (From Joachimski and Buggisch, 1993.)

are estimated to have been lost in the extinction (Table 9.1), though this might prove to be an over-estimation, because an increasing number of Lazarus taxa are being discovered in the Trias. The intensity of the extinction differed considerably between groups; several major groups disappeared completely, e.g. tabulate and rugose corals, conulariids, eurypterids, leperditiid ostracodes, goniatitic ammonoids, strophomenides, orthides, productides, blastoids, three groups of crinoids together with trilobites, whilst some groups such as nautiloids, sponges, non-fusulinid foraminifera, conodonts and some gastropods were hardly affected. The Palaeozoic fauna was particularly hard hit and lost 79% of families, as opposed to 27% of families amongst the Modern fauna. Terrestrial faunas appear to have suffered high levels of extinction, but plants seem less affected. There is, however, an intriguing spike in the abundance of spores of the fungus *Timpanicysta* at the P/T boundary, which might reflect widespread decay and collapse of terrestrial ecosystems. Amongst marine faunas all parts of ecosystems suffered to some degree, but zooplankton and epifaunal suspension-feeders and taxa with planktotrophic larvae were particularly affected (Erwin, 1993, 1994).

There were many changes in the physical environment near the end of the Permian and consequently there has been a multitude of suggestions as to the cause of the end-Permian extinction. Provinciality was reduced as a result of the assemblage of the supercontinent of Pangea. The regression near the end of the Permian was probably the largest in the Phanerozoic and it exposed a greater area of platform and shelf than at any other time and reduced the habitable area and habitat diversity of shelf seas. Contemporaneously large volumes of evaporites were deposited, there was an overall global warming, but possibly a cold phase at the P/T boundary; there were vast outpourings of plateau basalts, there was widespread marine anoxia and oxygen, carbon and sulphur isotopes indicate important changes in the world ocean (Fig.9.17). Extra-terrestrial impacts have also been suggested, but the evidence for these is slender. Different authorities have selected one or more of these factors as of critical importance. Erwin (1994) rejects some of the proposed mechanisms as the single cause of mass extinction and prefers to see the extinction as being multi-causal. Reduced provinciality is rejected because the lowest degree of provinciality does not coincide with the P/T extinction. Volcanism is unlikely to be the major cause because the volumes appear insufficient to cause the necessary environmental destruction, and also the peak of the volcanism post-dates the extinction. Climatic cooling is rejected because Permo-Carboniferous glaciation had apparently ended before the extinction and there is no evidence of a cold period in the $\delta^{18}O$ record. Anoxia is rejected as a single cause because it was regional, not global, and post-dates at least some of the extinctions. Instead Erwin suggested a three-phase extinction. The first phase was initiated by habitat loss as the regression drained the continental interiors and narrowed the shelves. The second phase resulted from the increased continentality that ensued and induced climatic instability. This, coupled with volcanic activity and increasing levels of CO_2, produced environmental degradation and ecological collapse. The third phase of extinction then occurred with the onset of the early Triassic marine trangression that promoted anoxia in marine environments and eradicated many coastal terrestrial environments. The potential role of anoxia has been emphasized by evidence that it was particularly widespread and affected regions in both low and high latitudes, and indications that the extinction was concentrated in the transgressive interval (Wignall and Twitchett, 1996). However, the robustness of this and other models still requires testing against a more precisely timed schedule of extinctions and environmental change.

9.3.4.4 Late Triassic mass extinction

This is one of the more poorly documented mass extinctions, principally because there are so few informative marine boundary sections. The extinction is generally seen as occurring at the Trias/Jurassic boundary, but a strong case has been made that it occurred in two or possibly three phases, spaced over more than 20 my, with one phase affecting faunas in the late Scythian, a second affecting reef faunas and terrestrial vertebrate faunas in the Late Carnian, and a third and largest one at the end of the Norian, i.e. at the Trias/Jurassic boundary (Fig. 9.18). Although the Carnian extinctions may have been somewhat dispersed through time, the end-Norian event appears abrupt. Most clades were affected by the extinction (Table 9.1), though the evidence for phytoplankton extinction is not well documented. Some important elements of the Palaeozoic fauna that had survived the end-

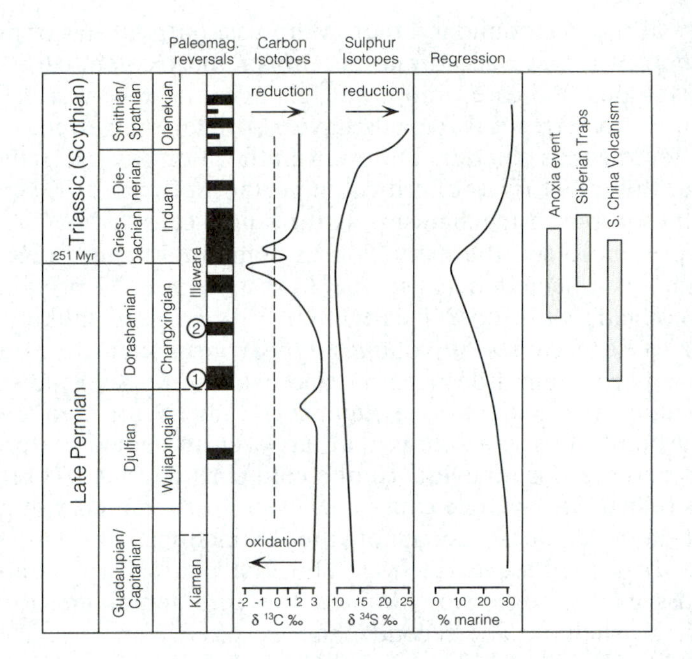

Figure 9.17 The Permo-Triassic extinction related to a range of environmental changes. Note particularly the coincidence with a period of lowered sea-level, a negative $\delta^{13}C$ shift, anoxia in the oceans and extensive extrusion of plateau basalts. (Modified from Erwin, 1994.)

Permian extinction, such as conodonts, murchisonacean gastropods and all but one genus of spiriferides, disappeared at this level and some palaeontologists see the Triassic extinction as being almost as important as the P/T extinction in clearing away the remnants of the Palaeozoic fauna. Amongst the rest of the fauna, ammonites were severely reduced, possibly to a single genus, gastropods were hard hit and about half of the genera and most species of bivalves were lost and reefs composed of large metazoans largely disappeared. Many vertebrate families became extinct, opening the way for the dominance of the dinosaurs in the Mesozoic and there was a modest extinction amongst plants.

It is important to note that the extinction affected both marine and terrestrial faunas and resulted in major losses amongst the zooplankton (ammonites and conodonts) and suspension feeding benthos. The causes of extinction that have commonly been invoked are sea-level change, climatic change, anoxia and bolide impact. The case for a bolide impact that might have affected both marine and terrestrial faunas is that the Manicouagan impact crater in Quebec, about 100 km in diameter, is one of the largest recorded in the Phanerozoic and is of approximately the right age. However, no contemporaneous iridium spike or other evidence of impact fall-out has been recorded close to the Trias/Jurassic boundary. Sea- level fell to particularly low levels in

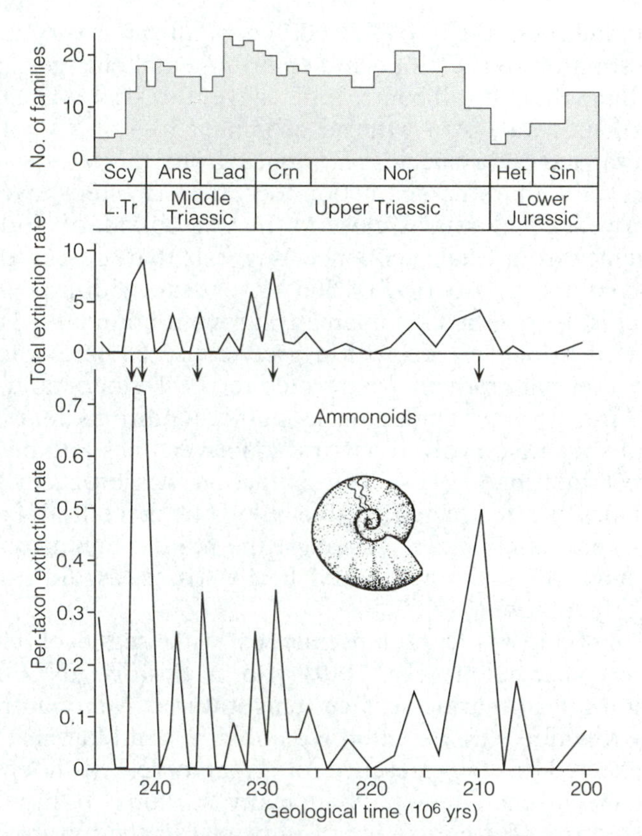

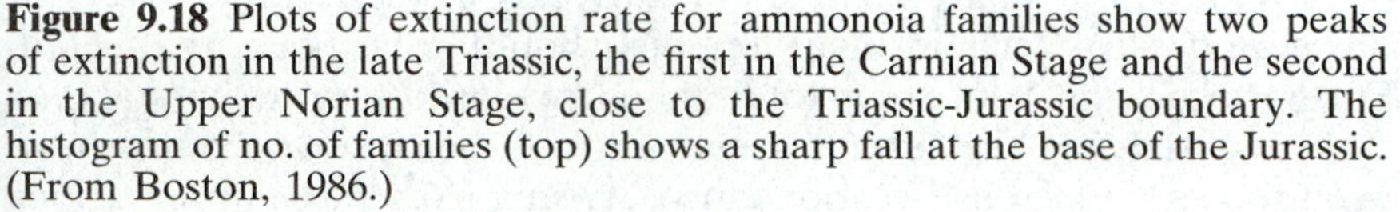

Figure 9.18 Plots of extinction rate for ammonoia families show two peaks of extinction in the late Triassic, the first in the Carnian Stage and the second in the Upper Norian Stage, close to the Triassic-Jurassic boundary. The histogram of no. of families (top) shows a sharp fall at the base of the Jurassic. (From Boston, 1986.)

the late Trias, large areas of continents were exposed and climates were generally continental, as reflected in extensive red beds. At the start of the Jurassic, sea-level rose rapidly, climates became more equable and there was a spread of anoxia during the transgression. One interpretation of the extinction is that marine habitat diversity was greatly reduced during the regression and then again during the transgression as anoxia spread. The contemporaneous loss of terrestrial faunas is ascribed to the loss of coastal environments during the transgression (Hallam, 1990) or to climatic change, particularly increasing aridity (Benton, 1990).

9.3.4.5 The Cenomanian/Turonian extinction

Sections in which this extinction level are recorded are widely distributed throughout the world and have been studied in particular detail in the Chalk of northwest Europe and the Western Interior Basins of the USA. The probability that small-scale cyclicity in the Chalk reflects a Milankovitch periodicity of ~20,000 years allows a very high stratigraphic resolution to the biotic and environmental changes and makes it a rewarding, though still controversial, event to study. The extinction was one of moderate size with an estimated loss of 7% of families, 26% genera and 53% species. Evidence from the Western Interior Basins suggests that the extinction occurred in steps over several millions of years, with a peak close to the C/T boundary (Elder, 1991), while evidence from Chalk sequences suggests that most of the extinction was confined to a period of 500 Ky or less, with a peak level of extinction just below the Cenomanian/Turonian boundary. The extinction followed a long period with a greenhouse climate when oceans had a low thermal gradient from poles to the Equator and from the surface to deep-water. Under these stable conditions many lineages would probably have evolved narrow adaptive zones and been extinction-prone (Kauffman, 1995). The extinction was broad in that most marine clades were affected (Table 9.1), but terrestrial faunas and floras survived mainly intact. Amongst the marine faunas both phytoplankton and zooplankton suffered heavy losses, as did both sessile and mobile benthos.

The extinction was contemporaneous with sea-level changes, a positive $\delta^{13}C$ excursion (Fig. 9.19) and a positive $\delta^{18}O$ excursion suggesting a fall in marine surface temperatures. Many authors associate the extinction with an extensive anoxic event (Jarvis *et al.*, 1988) that is reflected in the presence of organic-rich sediments at the Cenomanian/Turonian boundary in many sections. It has, however, been pointed out that anoxia is not universal in the marine waters of the period, so species in many regions should have been unaffected. Furthermore, the $\delta^{13}C$ excursion implies increased productivity and/or carbon burial, but the former seems unlikely because coccoliths have been shown to decrease in abundance. Arising from these observations an alternative extinction mechanism might be starvation resulting from the increased sedimentation of organic carbon and associated burial of nutrients (Paul and Mitchell, 1994). An alternative view of the extinction comes from very detailed studies of sequences in the Western Interior of the USA made by Kauffman and his associates (Kauffman, 1995). Here the extinction appears to have started in the early Mid-Cenomanian with the loss of major parts of the tropical reef ecosystems, but to have been particularly concentrated within about 500,000 years of the Cenomanian-Turonian boundary. Within this interval the extinction appears to have progressed in a series of short-lived episodes, most of which were associated with major fluctuations in trace elements, including raised iridium levels, stable isotope excursions and changes in the amount of organic carbon. It is difficult to isolate one of these environmental perturbations as the single cause of extinction and Kauffman (1995) prefered to interpret the extinction in terms of linked effects. Each phase he suggests was triggered by a bolide impact, which disturbed the oceans and caused advection of trace elements from the previously stable bottom waters. The result

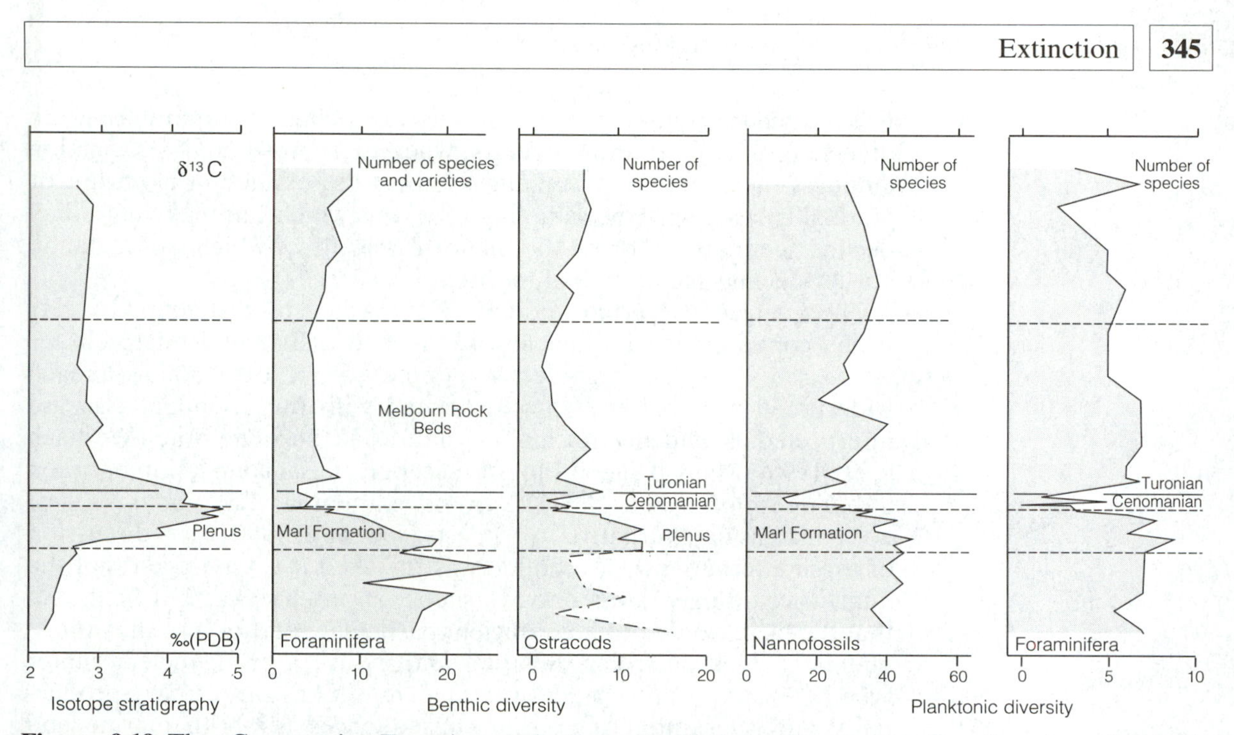

Figure 9.19 The Cenomanian-Turonian extinction recorded in changes in the diversity of benthic foraminifera and ostracodes and planktonic foraminifera and nannofossils. The decline in diversity coincides with a positive $\delta^{13}C$ excursion, but data suggests that the benthic diversity started to decline before the planktonic diversity. (Modified from Jarvis *et al.*, 1988.)

appears to have been higher productivity, which increased the export of carbon from the surface of the oceans and so expanded the oxygen minimum layer, causing widespread anoxia. Associated climatic fluctuations and climatic changes increased the environmental stress on the various ecosystems. In summary, the key factors that lead to this being mass extinction appear to have been the stable and narrowly adapted state of the preceding fauna, the rapidity of environmental change during each phase of extinction, the cumulative effects of interrelated environmental stresses and the cumulative effect of successive phases of extinction, which initially affected tropical ecosystems and successively spread to affect sub-tropical and temperate faunas.

9.3.4.6 The end-Cretaceous extinction (the K/T event)

This is the most studied and most controversial of all the mass extinctions. The controversy centres around the question as to whether the extinction was instantaneous, and therefore consistent with a catastrophe created by a bolide impact (see section 9.3.3.4), or was gradational or stepped up towards a peak of extinction, consistent with progressive environmental deterioration. The pattern and duration of an extinction are critical to the identification of the mechanism of extinction. Several groups, such as the ammonites and the dinosaurs were undergoing a long-term decline, but both the ammonites (Ward and Macleod, 1988) and the dinosaurs (Sheehan *et al.*, 1991) disappeared abruptly at the

K/T boundary. Other groups such as the pelagic foraminifera were diverse until close to the boundary, when there was a major extinction amongst the group. Determining whether the extinction is abrupt or gradual is beset with the sampling problems which can make a gradual decline seem abrupt or by the Signor Lipps effect which can make an abrupt decline seem gradual (section 9.3.3.2).

The Cretaceous/Tertiary boundary is marked by the boundary clay which contains high iridium levels, shocked quartz and microtektites which are all widely regarded as good evidence of a bolide impact; widespread charcoal and soot associated with the boundary clay are interpreted as evidence for major wildfires at the same time (Wolbach *et al.*, 1990). Many palaeontologists accept a causal connection between a bolide impact and the K/T mass extinction, but some suggest volcanism as an alternative hypothesis (see Smit, 1990, for a discussion of these alternatives). If a bolide was involved it is envisaged that the impact would generate a massive shock, an intense wave of heat, and that the blast would project sufficient particulate matter into the atmosphere to block out solar radiation and induce a brief nuclear-winter type of event. There would be immediate effects on primary productivity and substantial ecological consequences for both marine and terrestrial ecosystems. A negative excursion of $\delta^{13}C$ is thought to reflect the greatly reduced productivity in the oceans, referred to by Hsü and McKenzie (1985) as a 'Strangelove ocean', while the negative $\delta^{18}O$ values suggest that ocean temperatures may have risen shortly after the impact as a consequence of raised CO_2 levels in the atmosphere arising from the mass mortality, decay of dead organisms and the destruction of vegetation by wildfires (Smit, 1990). This 'catastrophic' model accommodates a large body of data, but there are some discrepant observations which favour more gradual ecological change across the K/T boundary. Although late Cretaceous nannoplankton decline sharply in diversity at the boundary, Keller *et al.* (1993) and Keller and Perch-Nielsen (1995) have presented extensive data, some of which suggest a progressive decline heralding the more abrupt extinction event and other data that show survival of smaller, more cosmopolitan species into the Tertiary (Fig. 9.20). Furthermore, Perch-Nielsen *et al.* (1982) have used the contrast in $\delta^{13}C$ values across the K/T boundary to show that some of the 'Cretaceous' nannofossils found in lowermost Tertiary sediments, which had been regarded as being reworked, have the low values of Tertiary shell material and therefore survived the extinction event. Other data from high latitudes indicate that the plankton extinction at the K/T boundary was subdued at most in these regions (Keller *et al.*, 1993). It may be that the search for a single cause implies conflict between the data where none exists. There is substantial evidence of a gradual late Cretaceous decline in diversity amongst such groups as the dinosaurs, the ammonites and probably the nannoplankton. Superimposed on this decline there appears to have been a bolide impact which was associated with a catastrophic extinction event that terminated some clades, greatly

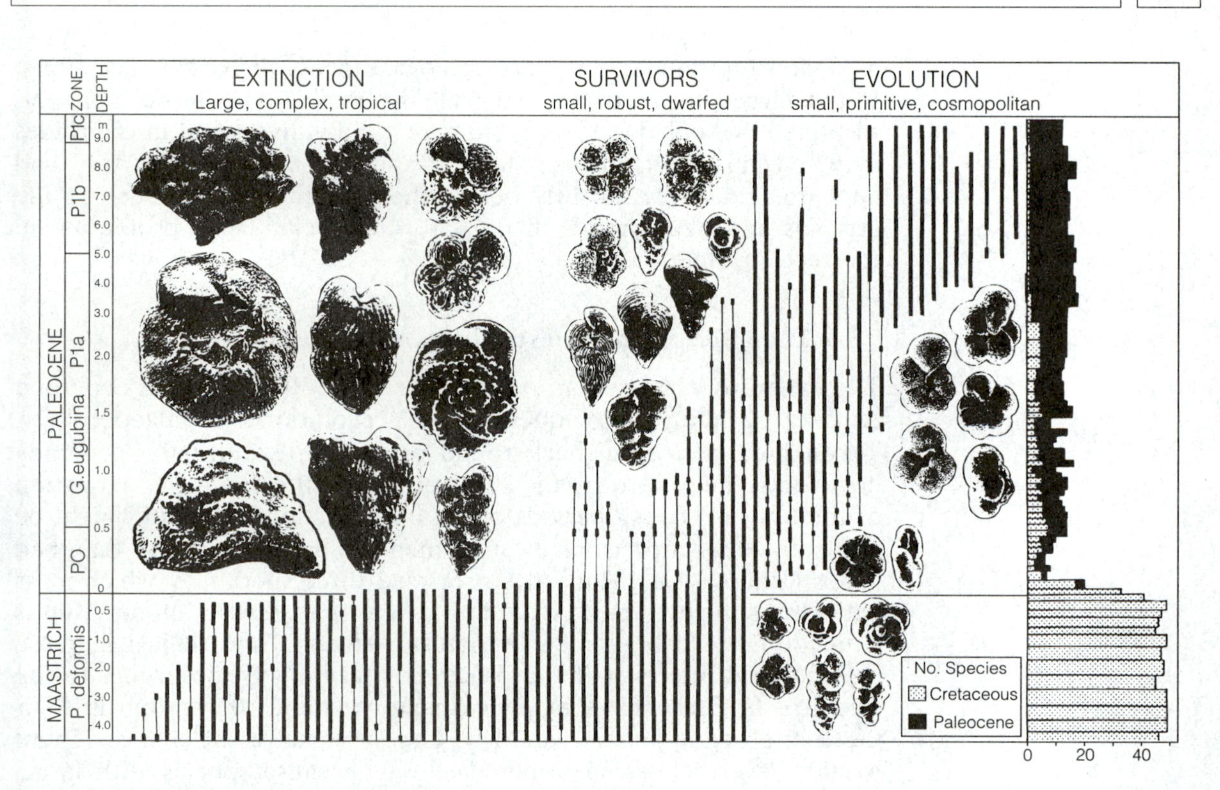

Figure 9.20 Stratigraphic ranges of planktonic foraminifera across the K/T boundary at El Kef. Tropical, large, complex, deep and intermediate water dwellers disappear at, or before the K-T boundary, but smaller, cosmopolitan surface-dwellers survive. New Tertiary taxa are small, simple and unornamented. (Modified from Keller and Perch-Nielson, 1995.)

reduced others, but had relatively little affect on the biota of high-latitude regions.

9.3.5 Recovery from mass extinctions

9.3.5.1 Patterns of survival and radiation

Most major extinctions culminate in a **'survival interval'** when there was a trough in taxonomic diversity. Within this interval there may be an initial period (a **lag phase**) when there are only rare appearances, diversity is very low and the dominant taxa (**disaster species**) appear to be those that can flourish in the stressful conditions that were associated with the preceding extinction. An example of a disaster species is the inarticulate brachiopod *Discinisca sp.* which dominated the lag phase of the Cenomanian/Turonian extinction. This species is normally rare or absent in marine waters, but attained dominance in stressed post-extinction nearshore environments. In the later part of a survival interval (the **rebound phase**) some new species start to appear, and others reappear as Lazarus taxa, having survived the extinction in refugia. In the Cenomanian/Turonian event many of these are opportunists and some are ecological generalists (Harries, 1993). In the succeeding '**recovery interval**', diversification is more rapid and sustained. New clades appear and may rapidly diversify, but old clades may re-establish themselves too. In some instances clades that had appeared relatively recently before the extinction event, such as the atrypids and pentamerids in the late Ordovician, may proliferate in the recovery interval.

9.3.6 The effect of major extinctions on evolution

The extent to which major extinctions redirect the course of evolution is one of the major questions of evolutionary palaeoecology. Throughout periods of background extinction, evolution is generally believed to be guided by the survival of the fittest and the extinction of the unfit over long periods of natural selection, though it might be that some extinction is a matter of chance – bad luck, rather than bad genes (Raup, 1991). Raup has proposed three modes by which mass extinctions might occur. According to the 'Fair Game' mode, a mass extinction would be guided by natural selection and so just increase the rate and intensity of the process. Alternatively, according to the 'Field of Bullets' mode, major extinctions would be random in their destruction and survival would be largely a matter of chance. There would, nevertheless, be some selectivity because species with large, widespread populations would have a greater chance of surviving the bullets. Under the third 'Wanton Extinction' mode, there is selectivity, but species do not survive because they are well adapted to 'normal' conditions but because they withstand the abnormal conditions during the extinction event. Extinction under the 'Field of Bullets' or the 'Wanton extinction' modes could potentially alter the course of evolution by leaving a residual fauna that consists of chance remnants of pre-existing communities, so creating opportunities for new adaptations and new community structures.

According to the model of Sepkoski (1984), mass extinctions accelerate inherent evolutionary trends established amongst evolutionary faunas as they diversify to an equilibrium plateau or are declining in accordance with the diversification of the succeeding fauna. Accordingly, the mass extinctions at the end of the Ordovician and in the late Devonian could be seen as temporarily reduced diversity below the equilibrium level. Likewise the end-Permian extinction could be interpreted as temporarily accelerating the disappearance of the Palaeozoic fauna and halting the radiation of the Modern fauna and resetting the radiation to a lower level of diversity. Some support for this interpretation comes from the patterns of diversity change in the Palaeozoic and Modern faunas. The Palaeozoic fauna had been declining in diversity from the late Devonian onwards while the Modern fauna diversified. This trend is accelerated amongst the Palaeozoic fauna when an estimated 79% of species became extinct, as opposed to 27% of the Modern fauna. It has been commonly observed that a mass extinction commonly abruptly terminates a clade that was already in decline, e.g. ammonites and dinosaurs at the K/T boundary, and this could be viewed as a natural consequence of an environmental 'catastrophe' acting on an extinction-prone clade. If this is so, the extinctions mainly enhanced trends that were already established and were possibly inherent to the evolutionary fauna to which they belong.

The degree of ecological disruption associated with major extinctions appears to vary considerably between different events so the degree of evolutionary disruption might vary accordingly. Some extinctions apparently modified the ecological structure of the biosphere for only a short period, before the general community structure was re-established, though commonly with different taxa and less commonly with new clades, e.g. the end-Ordovician and Cenomanian/Turonian extinctions. More substantial ecological disruption, as in the late Devonian and K/T extinctions, saw the disappearance of important clades which were not directly replaced by ecologically equivalent clades in the recovery period. The largest extinction, that close to the Permo-Triassic boundary, led to the reduction and disappearance of a large part of the long-established Palaeozoic fauna and destroyed large parts of the Palaeozoic benthic sessile suspension-feeding communities, which were never to be replaced by communities of a similar nature. The extent to which the end-Permian extinction merely accentuated trends already inherent in the preceding history of Palaeozoic faunas is still a matter of debate. However, detailed studies of Permian gastropods by Erwin (1993) showed that although the group as a whole did not suffer severe extinction, the disappearance of some clades, e.g. the Omphalotrochidae and the platycerids, and the expansion of new groups of more modern aspect, radically altered the overall nature of the gastropod fauna. Furthermore, these changes were not foreshadowed by a diversity decline preceding the extinction, nor by increasing diversity amongst those that were to be a constituent part of the modern fauna. This evidence suggests that for these groups extinction redirected the course of evolution, even though in other groups it may have accelerated existing trends towards extinction.

9.3.7 Radiation, stasis and extinction

There is a growing perception that communities have undergone stages of diversification, stasis and extinction. The scales of change are very variable from the major radiations of the early Cambrian and early Ordovician to the minor radiations that follow minor extinction events. Extinction events too appear to range in scale through a wide spectrum from the five or so mass extinctions to the numerous perturbations that may affect ecosystems globally in a minor way, or to extinction events that are confined to a few groups. It is the periods of ecological stasis, when communities undergo little change in diversity, that are progressively being recognized and again these can be characterized in terms of differences of degree. They vary from the overall stability in the general nature of communities of the Palaeozoic fauna throughout the interval from Ordovician to the Permo-Triassic boundary, to the much shorter periods when there was virtually no change over a few million years in the generic or species composition of communities between minor extinction events (Brett and Baird, 1995). The implication is that ecological evolution is not a steady process, but passes through periods when there is little change, punctuated by episodes of extinction of varying severity which are followed by phases of ecological recovery. Communities in times of stasis are seen by some to have an integrated structure, with interdependence between species. In contrast to this condition the communities of the late Quaternary and Holocene are unstable associations of species that respond individually to environmental change and shift their habitat to different degrees. The extent to which communities are integrated units or loose associations adjusted to similar environmental controls is still a matter of debate. However, periods of stasis followed by episodes of rapid change do appear to form a common pattern amongst marine benthic faunas generally and this is mirrored in the historical pattern of reefs formed by colonial organisms (Fig. 9.21). However, extinctions do not necessarily affect all organisms equally. Recent studies of reefs have shown that there is one history of reefs with colonial organisms that respond to one set of environmental pressures and a different history for reefs largely constructed of microbial carbonates that respond to different kinds of environmental change. Although the pattern of change amongst marine faunas shown in Figure 9.21 may well be partly orchestrated by major perturbations of the global environment, it also appears likely that some elements of the marine biota danced to a different tune.

9.4 SUMMARY POINTS

- The evolution of prokaryotes which could synthesize carbon from CO_2 releasing oxygen, generated an atmosphere that was to retain the Earth in a climatic equilibrium necessary for the sustained evolution of life.
- Prokaryotes show little detectable morphological evolution for more than 1500 my, though evidence from DNA suggests substantial evolution.
- The evolution of the eukaryotic cell allowed sexual reproduction and a potentially faster rate of evolutionary change.
- There appears to have been a significant radiation amongst the eukaryotes close to the Mesoproterozoic-Neoproterzoic boundary (~1000 Ma) when a variety of large acritarchs appeared. There was apparently a complementary decline in prokaryote diversity but this is believed to be the result of unfavourable conditions for preservation.

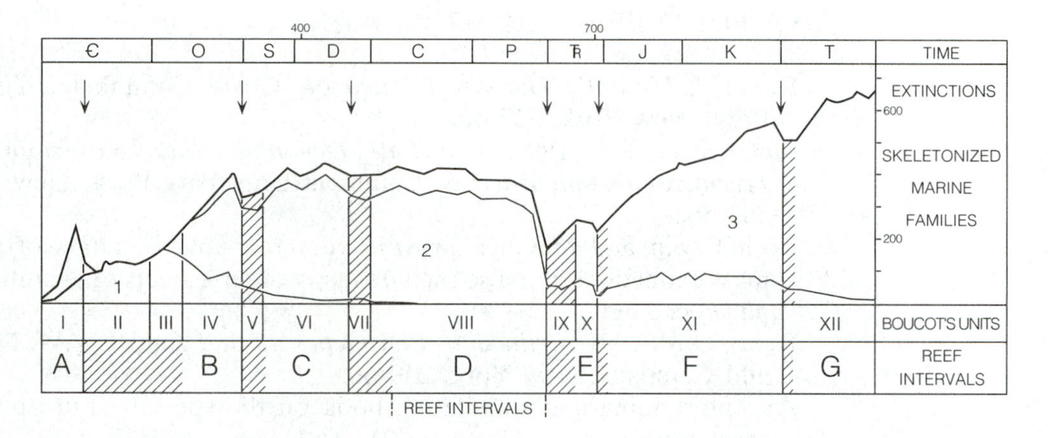

Figure 9.21 A model showing 12 intervals of community evolution (Ecologic-Evolutionary Units (EEUs) of Boucot, 1983, shown with Roman numerals). They relate to three periods when the three evolutionary faunas radiated (EEU I, III and IV and EEUs, XI and XII), followed by a time of stasis (EEUs II, VI and VIII). Any of these phases could be punctuated by major extinctions (indicated by arrows), after which there was an interval characterized by a recovery fauna (EEUs V, VII and IX). Reef faunas reflect the general state of contemporaneous benthic faunas and are shown with letters A–G. (Modified from Sheehan 1991.)

- The first metazoan radiation is recorded by the Ediacara fauna (~650–600 Ma), an association of enigmatic organisms. There is disagreement as to whether they are part of the root stock of typical Cambrian phyla or are clades that radiated and mainly became extinct before the Cambrian.
- Following the major Cambrian radiation, the Phanerozoic history of the biosphere records further periods of radiation, periods of stability (stasis) and episodes of major extinction.
- Three evolutionary faunas are recognized by Sepkoski, each of which is believed to diversify to a plateau and then decline in diversity as the succeeding fauna diversifies. The diversity patterns of the faunas have been described according to a kinetic model, modified to take into account the episodes of extinction.
- Alternative models of Phanerozoic diversity change are diversity independent.
- Major extinctions appear to result from severe physical perturbations that are global, cut across ecological boundaries and are generally rapid in their effects. Change in sea-level, climate, anoxia and the chemistry of the oceans, commonly accompany major extinctions and evidence for bolide impacts accompany some.
- It is generally difficult to identify a single cause for all mass extinctions. Some appear to have a single cause, most result from the cumulative effects of several causes which may be genetically linked.
- Mass extinctions are followed by periods of recovery and radiation, each with a characteristic fauna.

9.5 FURTHER READING

Erwin, D.H. 1993. *The Great Paleozoic Crisis*. Columbia University Press, New York. 327 pp.

McGhee, G.R.J. 1996. *The Late Devonian Mass Extinction: The Frasnian/Famennian Crisis*. Columbia University Press, New York. 303 pp.

Both Erwin and McGhee provide very readable accounts of major mass extinctions, but also include many other aspects of evolutionary palaeoecology.

Raup, D.M. 1991. *Extinction: Bad Genes or Bad Luck*? W.W. Norton and Company, New York. 210 pp.

An entertaining and stimulating book on the special nature of mass extinctions.

Skelton, P. (ed.) 1993. *Evolution: a biological and palaeontological approach*. Addison-Wesley Publishing Co., Wokingham. 1064 pp.

A very useful sourcebook for information on most aspects of evolution.

Smith, A.B. 1994. *Systematics and the fossil record*. Blackwell Scientific Publications, Oxford. 223 pp.

A challenging book at quite an advanced level analysing the effects of systematics and sampling on perceptions of evolutionary patterns.

Valentine, J.W. 1973. *Evolutionary palaeoecology of the marine biosphere*. Prentice Hall Inc, Eaglewood Cliffs. 511 pp.

This book was published more than 20 years ago, but it is so good that it still well worth reading.

9.6 REFERENCES

Bambach, R.K. 1977. Species richness in marine benthic habitats through the Phanerozoic. *Paleobiology* 3, 152–167.

Bambach, R.K. 1985. Classes and adaptive variety: the ecology of diversification in marine faunas through the Phanerozoic. *In:* Valentine, J.W. (ed.) *Phanerozoic Diversity Patterns.* Princeton University Press, Princeton, 191–253.

Benton, M.J. 1990. End-Triassic. *In:* Briggs, D.E.G. and Crowther, P.R.(eds) *Palaeobiology: a Synthesis*, Blackwell Scientific Publications, Oxford, 194–198.

Boucot, A.J. 1983. Does evolution take place in an ecological vacuum? *Journal of Paleontology* 57, 1–30.

Brasier, M.D. 1992. Nutrient-enriched waters and the early skeletal fossil record. *Journal of the Geological Society, London* 137, 621–630.

Brenchley, P.J. 1989. The late Ordovician extinction. *In:* Donovan, S.K. (ed.) *Mass Extinctions: Processes and Evidence.* Columbia University Press, New York, 104–132.

Brenchley, P.J., Carden, G.A.F. and Marshall, J.D. 1995. Environmental changes associated with the 'first strike' of the late Ordovician mass extinction. *Modern Geology* 20, 69–82.

Brett, C.E. and Baird, G.C. 1995. Coordinated stasis and evolutionary ecology of Silurian to Middle Devonian faunas in the Appalachian Basin. *In:* Erwin, D.H. and Anstey, R.L. (eds), *New Approaches to Speciation in the Fossil Record*. Columbia University Press, New York, 285–314.

Clarke, A. 1993. Temperature and extinction in the sea: a physiologist view. *Paleobiology* 19, 499–518.

Conway Morris, S. 1986. The community structure of the Middle Cambrian Phyllopod bed (Burgess Shale). *Palaeontology* 29, 423–467.

Copper, P. 1986. Frasnian-Famennian mass extinction and cold water oceans. *Geology* 14, 835–839.

Crimes, T.P. 1994. The period of early evolutionary failure and dawn of evolutionary success: the record of biotic changes across the Precambrian-Cambrian boundary. *In:* Donovan, S.K. (eds) *The Palaeobiology of Trace Fossils*. J. Wiley and Sons, Chichester, 105–133.

Elder, W.P. 1991. Molluscan extinction patterns across the Cenomanian-Turonian Stage boundary in the western interior of the United States. *Paleobiology* 15, 299–320.

Erwin, D.H. 1993. *The Great Paleozoic Crisis*. Columbia University Press, New York, 327 pp.

Erwin, D.H. 1994. The Permo-Triassic extinction. *Nature* 367, 231–236.

Erwin, D.H., Valentine, J.W. and Sepkoski, J.J.Jr. 1987. A comparative study of diversification events: the early Palaeozoic versus the Mesozoic. *Evolution* 41, 1177–1186.

Ford, M.J. 1982. *The Changing Climate; Responses of the Natural Flora and fauna*. George Allen and Unwin, London, 190 pp.

Hallam, A. 1986 The Pliensbachian and Tithonian extinction events. *Nature* 319, 765–768.

Hallam, A. 1990. The end-Triassic mass extinction event. *In:* Sharpton, V.L. and Ward, P.D. (eds), *Global catastrophes in Earth history; An interdisciplinary conference on impacts, volcanism and mass mortality*. Geological Society of America Special Paper, 247, Geological Society of America, 577–583.

Hansen, T.A. 1987. Extinction of Late Eocene to Oligocene molluscs: Relationship to shelf area, temperature changes and impact events. *Palaios* 2, 69–75.

Harper, D.A.T. and Rong, Jia-yu. 1995 Patterns of change in the brachiopod faunas through the Ordovician-Silurian interface. *Modern Geology* 20, 83–100.

Harries, P. 1993 Diversification after the Cenomanian-Turonian extinction: *Cretaceous Research* 14, 563–583.

Hoffman, A. 1985. Biotic diversification in the Phanerozoic: diversity independence. *Palaeontology* 28, 387–392.

House, M.R. 1989. Ammonoid extinction events. *Philosophical Transactions of the Royal Society of London* B 325, 307–326.

Hsü, K.J. and McKenzie, J.A. 1985. A 'strangelove' ocean in the earliest Tertiary. *In:* Sundquist, E.T. and Broecker, W.S. (eds), *Natural Variations; Archaean to Present.* American Geophysical Union Geophysical Monograph, **32,** 487–492.

Jablonski, D. 1989. The biology of mass extinction: a paleontological view. *Philosophical Transactions of the Royal Socity of London,* B 325, 357–368.

Jablonski, D. 1991. Extinctions: a paleontological perspective. *Science* **253**, 754–757.

Jansa, L.F. 1993. Cometary impacts into ocean: their recognition and the threshold constraint for biological extinction. *Palaeogeography, Palaeoclimatology, Palaeoecology* 104, 271–286.

Jarvis, I., Carson, G.A., Cooper, M.K.E., Hart, M.B., Leary, P.N., Tocher, P.A., Horne, D. and Rosenfeld, A. 1988. Microfossil assemblages and the Cenomanian-Turonian (late Cretaceous) oceanic oxygen event. *Cretaceous Research* 9, 3–103.

Joachimski, M.M. and Buggisch, W. 1993 Anoxic events in the late Frasnian – Causes of the Frasnian-Famennian faunal crisis. *Geology* 21, 675–678.

Kauffman, E.G. 1995. Global change leading to biodiversity crisis in a greenhouse world: the Cenomanian-Turonian (Cretaceous) mass extinction. *In:* Stanley, S.M., Kennett, J.P. and Knoll, A.H. (eds), *Effects of Past Global Change on Life.* National Academy Press, Washington, 47–71.

Keller, G., Barrera, E., Schmitz, B. and Mattson, E. 1993. Gradual mass extinction, species survivorship and long-term environmental changes across the Cretaceous-Tertiary boundary in high latitudes. *Geological Society of America Bulletin* 105, 979–997.

Keller, G. and Perch-Nielsen, K.v.S. 1995. Cretaceous-Tertiary (K/T) mass extinction: Effect of global change on calcareous microplankton. *In:* Stanley, S.M., Kennett, J.P. and Knoll, A. H (eds), *Effects of past global change on life.* National Academy Press, Washington, 72–93.

Kennett, J.P. and Stott, L.D. 1991. Abrupt deep-sea warming, palaeoceanographic changes and benthic extinctions at the end of the Palaeocene. *Nature* 353, 225–229.

Knoll, A.H. and Sergeev, V.N. 1995. Taphonomic and evolutionary changes across the Mesoproterozoic-Neoproterozoic transition. *Neues Jahrbuch für Paläontologie* 195, 289–302.

MacArthur, R.H. and Wilson, E.O. 1967. *The Theory of Island Biogeography*. Princeton University Press, Princeton, 203 pp.

McGhee, G.R.J. 1989. The Frasnian-Famennian extinction event. *In:* Donovan, S.K. (ed), *Mass extinctions: Processes and evidence.* Columbia University Press, New York, 133–151.

McGhee, G.R.J. 1996. *The Late Devonian Mass Extinction: The Frasnian/Famennian Crisis*. Columbia University Press, New York, 303 pp.

Owen, A.W. and Robertson, D.B.R. 1995. Ecological changes during the end-Ordovician extinction. *Modern Geology* 20, 21–39.

Patterson, C. and Smith, A.B. 1989. Periodicity in extinction: the role of systematics. *Ecology* 70, 802–811.

Paul, C.R.C. and Mitchell, S.F. 1994. Is famine a common factor in marine mass extinctions? *Geology* 22, 679–682.

Perch-Nielsen, K., McKenzie, K.J. and He, Q. 1982. Biostratigraphic and isotopic stratigraphy and the catastrophic extinction of calcareous nannoplankton at the Cretaceous/Tertiary boundary *Geological Society of America Special Paper* 190, 353–371.

Phillips, J. 1841. *Figures and descriptions of Palaeozoic fossils of Cornwall, Devon and West Somerset, Observed in the Course of the Ordnance Survey of that District*. Longman, London,

Rampino, M.R. 1991. Volcanism, climatic change, and the geologic record. *Sedimentation in Volcanic Settings.* Society of Economic Paleontologists and Mineralogists, Tulsa, 45, 9–18.

Raup, D.M. 1989. The case for extra-terrestrial causes of extinction. *Philosophical Transactions of the Royal Society of London* B 325, 421–435.

Raup, D.M. 1991. *Extinction: Bad Genes or Bad Luck*? W.W. Norton and Company, New York, 210 pp.

Raup, D.M. and Sepkoski, J.J.Jr. 1986. Periodic extinction of families and genera. *Science* 231, 833–836.

Runnegar, B. 1995. Vendobionta or Metazoa? Developments in understanding the Ediacara 'fauna'. *Neues Jahrbuch für Geologie und Paläontologie* 195, 303–318.

Seilacher, A. 1992. Vendobionta and Psammocorallia: lost constructions of Precambrian evolution. *The Journal of the Geological Society* 149, 607–613.

Sepkoski, J.J.Jr. 1979. A kinetic model of Phanerozoic taxonomic diversity: II. Early Palaeozoic families and multiple equilibria *Paleobiology*, 5, 222–251.

Sepkoski, J.J.Jr. 1984. A kinetic model of Phanerozoic taxonomic diversity. 111. Post-Paleozoic families and mass extinctions. *Paleobiology* 10, 246–267.

Sepkoski, J.J.Jr. 1990. Evolutionary Faunas. *In:* Briggs, D.E.G. and Crowther, P.R. (eds), *Palaeobiology: a Synthesis.* Blackwell Scientific Publications, Oxford, 37–41.

Sepkoski, J.J.Jr. 1991. Diversity in Phanerozoic oceans: a partisan review. *In:* Dudley, E.C. (ed.), *The unity of evolutionary biology.* Proceedings of ICSEB IV, Discorides Press, Portland, Oregon, 210–236.

Sepkoski, J.J.Jr. 1992. Proterozoic-Early Cambrian diversification of metazoans and metaphytes. *In:* Schopf, J.W. and Klein, C. (eds), *The Proterozoic Bioshere: A Multidisciplinary Study.* Cambridge University Press, Cambridge,

Sepkoski, J.J.Jr., Bambach, R.J., Raup, D.M. and Valentine, J.W. 1981, Phanerozoic marine diversity and the fossil record. *Nature* 293, 435–437.

Sepkoski, J.J.Jr. and Sheehan, P.M. 1983. Diversification, faunal change, and community replacement during the Ordovician radiations. *In:* Tevesz, M.J.S. and McCall, P.J. (eds), *Biotic Interactions in Recent and Fossil Benthic Communities.* Plenum, New York, 673–717.

Sheehan, P.M. 1991. Patterns of synecology during the Phanerozoic. *In:* Dudley, E.C. (ed.), *The Unity of Evolutionary Biology.* Proceedings of ICSEB IV. Discorides Press, Portland, 103–118.

Sheehan, P.M., Fastovsky, D.E., Hoffman, R.G., and Gabriel, D.L. 1991. Sudden extinction of the dinosaurs: latest Cretaceous, Upper Great Plains, U.S.A. *Science* 254, 835–839.

Signor, P.W. 1985. Real and apparent trends in species richness through time. *In:* Valentine, J.W. (ed.), *Phanerozoic diversity patterns.* Princeton University Press, Princeton, 129–150.

Signor, P.W. 1990. Patterns of diversification. *In:* Briggs, D.E.G. and Crowther, P.R. (eds), *Palaeobiology- a Synthesis.* Blackwell Scientific Publications, Oxford, 130–135.

Signor, P.W. and Brett, C.E. 1984. The mid-Paleozoic precursor to the Mesozoic marine revolution. *Paleobiology* 10, 229–245.

Smit, J. 1990. Meteorite impact, extinctions and the Cretaceous-Tertiary boundary. *Geologie Mijnbouw* 69, 187–204.

Smith, A.B. 1994. *Systematics and the fossil record.* Blackwell Scientific Publications, Oxford, 223 pp.

Smith, A.B. and Patterson, C. 1988. The influence of taxonomic methods on the perceptions of patterns of evolution. *Evolutionary Biology* 23, 127–216.

Stanley, S.M. 1976. Ideas on the timing of metazoan diversification. *Paleobiology* 3, 209–219.

Stanley, S.M. 1986. Anatomy of a regional mass extinction: Plio-Pleistocene decimation of the Western Atlantic bivalve fauna. *Palaios* 1, 17–36.

Stothers, R.B. and Rampino, M.R. 1990. Periodicity in flood basalts, mass extinctions, and impacts. A statistical view and model. *Geological Society of America Special Paper* 247, 9–18.

Tucker, M.E. 1992. The Precambrian-Cambrian boundary: seawater chemistry, ocean circulation and nutrient supply in metazoan evolution, extinction and biomineralisation. *Journal of the Geological Society* 149, 655–668.

Valentine, J.W., 1969. Patterns of taxonomic and ecological structure of the shelf benthos during Phanerozoic time. *Palaeontology* **12**, 684–709.

Valentine, J.W. 1970. How many marine invertebrate fossil species? A new approximation. *Journal of Paleontology* 44, 410–415.

Valentine, J.W. 1989. Phanerozoic marine faunas and the stability of the Earth system. *Palaeogeography, Palaeoclimatology, Palaeoecology (Global and Planetary Change Section)* 75, 137–155.

Valentine, J.W. and Jablonski, D. 1991. Biotic effects of sea level change: the Pleistocene test. *Journal of Geophysical Research* 96, No. B4, 6873–6878.

Valentine, J.W., Foin, T.C. and Peart, D. 1978. A provincial model of Phanerozoic marine diversity. *Paleobiology* 4, 55–66.
Vermeij, G.J. 1977. The Mesozoic marine revolution; evidence from snails, predators and grazers. *Paleobiology* 3, 245–258.
Ward, P.D. and MacLeod, K. 1988. Macrofossil extinction patterns at Bay of Biscay Cretaceous-Tertiary boundary sections. *Contributions to Lunar and Planetary Institute* 673, 206–207.
Wignall, P.B. and Twitchett, R.J. 1996. Oceanic anoxia and the end Permian mass extinction. *Science* **272**, 1155–1158.
Wolbach, W.S., Gilmour, I. and Anders, E. 1990. Major wildfires at the Cretaceous/Tertiary boundary. *Geological Society of America Special Paper* 247, 391–400.

10 Fossil terrestrial ecosystems

Terrestrial ecosystems have been populated by a succession of spectacular and diverse animals and plants throughout most of the Phanerozoic. The shape and patterns of change are similar to those of the marine biosphere (see Chapter 9), though not always synchronous. Terrestrial biotas have expanded, filled and exploited ecological space following the appearance of adaptive innovations and new guilds of animals and plants; this process is illustrated during the various colonizations of the land and conquests of the air. Moreover new adaptations have also allowed land biotas to occupy new niches in areas already colonized. Land biotas have responded through time to changes in both the environment and to biological factors such as competition and predation. Terrestrial environments provide a large ecospace for potential life; successive rapid adaptive radiations and the subsequent fine-tuning of adaptations have helped fill the space. Nevertheless extinction events have, as in the marine ecosystem, periodically emptied many niches, providing new evolutionary possibilities for subsequent radiations.

10.1 INTRODUCTION

The changing Phanerozoic landscape is ideally suited for studies of long-term evolutionary palaeoecology. The green *Cooksonia* meadows of the Silurian and much of the Devonian with small arthropods gave way to scenes of clumsy amphibians working the gymnosperm forests of the late Devonian and early Carboniferous, culminating in the lush multi-storied vegetation of the later Carboniferous patrolled by arthropods like the large, mobile *Arthropleura*. Mesozoic vegetation with herbaceous pteridophytes and woody gymnosperms had coniferous canopies. Dinosaurs dominated the Mesozoic landscapes while later in the era, by the mid-Cretaceous, angiosperms added colour to the vegetation and established coevolutionary relationships with insect pollinators. The diverse mammal faunas of the Cenozoic, and the spread of grasslands and grazers, led eventually to australopithecines roaming the savannas of East Africa, forming the basis for the track of human evolution. The evolution of the terrestrial ecosystem presents interactions between the biotas themselves and environmental change within an evolutionary framework.

10.2 INITIAL ADAPTATIONS AND THE EARLY TERRESTRIAL RECORD

The early record of terrestrial life is generally poor since many organisms lacked mineralized tissues. The first appearance of many groups in the fossil record probably, significantly, post-dates their actual origins. The initial land invasions were controlled by two main factors: first, the types and adaptations of organisms preparing to invade the land and secondly, the suitability of the ambient atmosphere and available terrestrial environments. Microbes almost certainly laid the foundation for subsequent waves of terrestrial invaders, preparing organic seed beds across an otherwise barren latest Precambrian and earliest Palaeozoic landscape. Pioneer terrestrial micro-organisms, surviving and reproducing in drops of interstitial water, were probably initially anaerobic. By the Ordovician, soils with the burrows of possible millipedes occurred together with the spores of primitive land plants, although macrofossils of both millipedes and land plants are not reported until the early and mid-Silurian, respectively.

Terrestrial organisms have, nevertheless, remained physiologically aquatic, shielded from the atmosphere by a variety of protective systems. This fact is the basis of the concept of 'hypersea' where terrestrial organisms retain their former aquatic environments within their bodies (McMenamin and McMenamin, 1994), rather like the concept of a space suit filled with fluid instead of air. Some groups returned or may have remained in the water, pursuing life strategies in freshwater environments (Gray, 1988).

The evolving atmosphere provided its own constraints on evolving terrestrial life forms. Not only is oxygen required for the metabolism of aerobic organisms, the partial pressure of oxygen controls the ozone shield, screening out harmful ultra-violet radiation. An ozone screen, adequate for the maintenance and survival of terrestrial life, may have been established by the early Proterozoic (Kasting, 1993); such a screen could have developed at 10^{-2} of the present oxygen levels on the planet. But apart from organisms adapted to life at high altitudes, modern animals and plants no longer require protection from ultra-violet radiation.

The partial pressure of atmospheric carbon dioxide is a major control on climate, its build-up promoting the greenhouse effect. Although the late Precambrian glaciations suggest the prevalence of ice-house conditions, there were dramatic climatic and oceanic changes near the Precambrian–Cambrian boundary coincident with first the appearance of the soft-bodied Ediacara fauna, and secondly, at the base of the Cambrian, the radiation of skeletal organisms (see sections 9.2.5 and 9.2.6). The lack of bioturbation by late Precambrian moneran communities permitted the burial of large amounts of organic carbon, tying up atmospheric CO_2. Warmer climates, however, with higher levels of atmospheric oxygen probably characterized conditions at the Precambrian–Cambrian boundary interval. Marine epibenthic animals with protective and supportive skeletons were ready for terrestrialization as were some types of algae.

Following the end Proterozoic break-up of Rodinia (see Chapter 8), many of the continents were on the move during the early Palaeozoic. One area of the crust, at least, provided adequate habitats for evolving land biotas. The Caledonian mountain-building event, climaxing during the late Silurian and early Devonian, sutured the Baltica and microcontinents of Avalonia onto Laurentia, with the destruction of the Iapetus Ocean. The resulting supercontinent straddled the equator and a range of nonmarine environments developed in response to the rising Caledonian Mountains. Many of the key early terrestrial faunas and floras are found along or adjacent to the track of the Caledonian–Appalachian orogeny, coincident with the construction of the Old Red Sandstone or Laurussian continent.

Terrestrial assemblages, including both animals and plants, are now known from a variety of mid-Palaeozoic localities. Mid-Devonian biotas from Gaspé, Québec (Pragian), Rhynie, NE Scotland (Pragian), Alken-an-der-Mosel, Germany (Emsian) and Gilboa, New York State (Givetian) include a variety of invertebrates together with lycopsid, rhyniopsid and progymnospermopsid plants. One of the oldest biotas, from Ludford Lane in the Anglo-Welsh area, includes a Prídolí fauna dominated by arachnids, myriapods and eurypterids together with a flora of rhyniophytoids together with *Cooksonia* and *Nematothallus* washed into inter- and subtidal lag deposits (Jeram *et al.*, 1990). This allochthonous land-derived biota with predatory arthropods, however, indicates that complex and sophisticated terrestrial ecosystems were well established by the late Silurian; initial land-based ecosystems were clearly post-dated by their current fossil record.

10.2.1 Plants

During their early history (Fig. 10.1) land plants developed a variety of adaptations linked to the demands of a new life style. Three main life and physiological strategies were pursued to ensure survival. First, ephemeral and opportunistic life styles have evolved to avoid drought, second, the rehydration of cytoplasm provided an extreme tolerance to desiccation, and third an internally hydrated microenvironment within the plant helped maintain a 'space suit' effect (Edwards and Burgess, 1990). In addition structural modifications to allow 1. the intake of water and nutrients, 2. gaseous exchange, 3. mechanical support and anchorage, together with 4. innovative reproductive strategies, were developed in some of the earliest land plants, the rhyniophytoids. In the more typical tracheophytes, however, xylem permits the transport of water through a vascular system, phloem allows the move-ment of manufactured nutrients, waxy cuticles reduce evaporation and stomata promote gaseous exchange and transport (Edwards and Burgess, 1990). The plant cell walls are strengthened by cellulose and lignin.

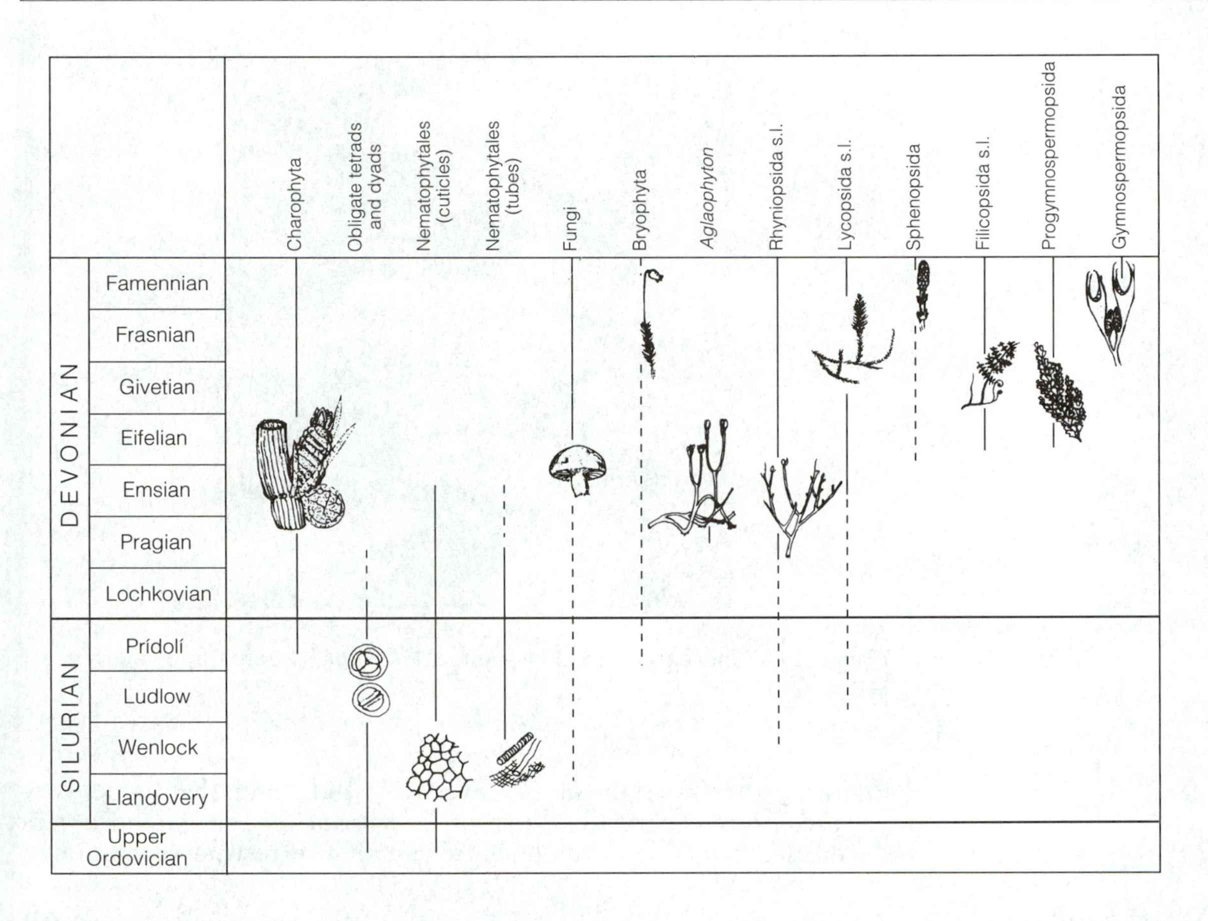

Figure 10.1 The fossil record of major plant invasions of the land from the Ordovician-Devonian. (Replotted from Edwards and Burgess, 1990.)

The earliest land plants may have survived without lignified fluid-conducting cells, but as these plants grew larger, support and water-conduction became essential. Reproductive strategies had to be drastically modified. Land-plant reproductive systems were designed to function in water-deficient environments; spores evolved that were light with a resistant waxy coating or cuticle, resistant to breakage and abrasion, ready for dispersal by wind and animals.

Recognition of land plants in the fossil record should be straight-forward. These plants have conducting tracheids, stomata, spore tetrads and spores with trilete markings. Nevertheless, considerable controversy has surrounded the identification of the earliest land plants, possibly since the spores of a number of marine plant groups have morphologically similar spores to those of the true land plants. Moreover the disparate parts of these early plants never occur together. Early Palaeozoic land surfaces were almost certainly coated with a green scum of cyano- and eubacteria. Mid- and late-Ordovician land floras probably had thalloid-type plants with a cuticle and spores

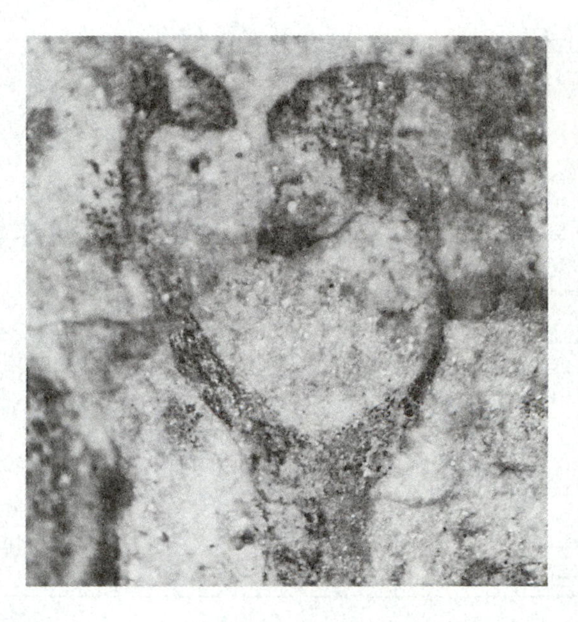

Figure 10.2 The early vascular plant *Cooksonia*. (Photograph courtesy of Dianne Edwards.)

following the life style of the liverwort. But from the Llandovery onwards, the *Salicornia*-like turf of *Cooksonia* (Fig. 10.2) and others set the scene for the development of a green terrestrial vegetation.

10.2.2 Animals

Two well-known groups of invertebrates have successfully colonized land: the arthropods and the gastropod molluscs (Fig. 10.3), in addition to rotifers, and various worms. Both groups developed body-support systems and a variety of organs to cope with air-breathing respiration. Equally significant, and perhaps more spectacular, was the colonization of land by the vertebrates; the skeletons of *Acanthostega* and *Ichthyostega* have formed the basis for a number of functional studies, while early tetrapod footprints reveal something of the behaviour of these early amphibians.

10.2.2.1 Arthropods

Centipedes, millipedes and a specialized group of myriapods, the arthropleurids, together with the trigonotarbids, scorpions and possibly the eurypterids, had appeared on land by the Devonian. Some groups returned to the water to pursue aquatic life modes. Some living insects, for example, are basically terrestrial but with secondary adaptations for aquatic life styles. They all have a dendritic network of vessels capable of supplying every part of the body with oxygen. The first

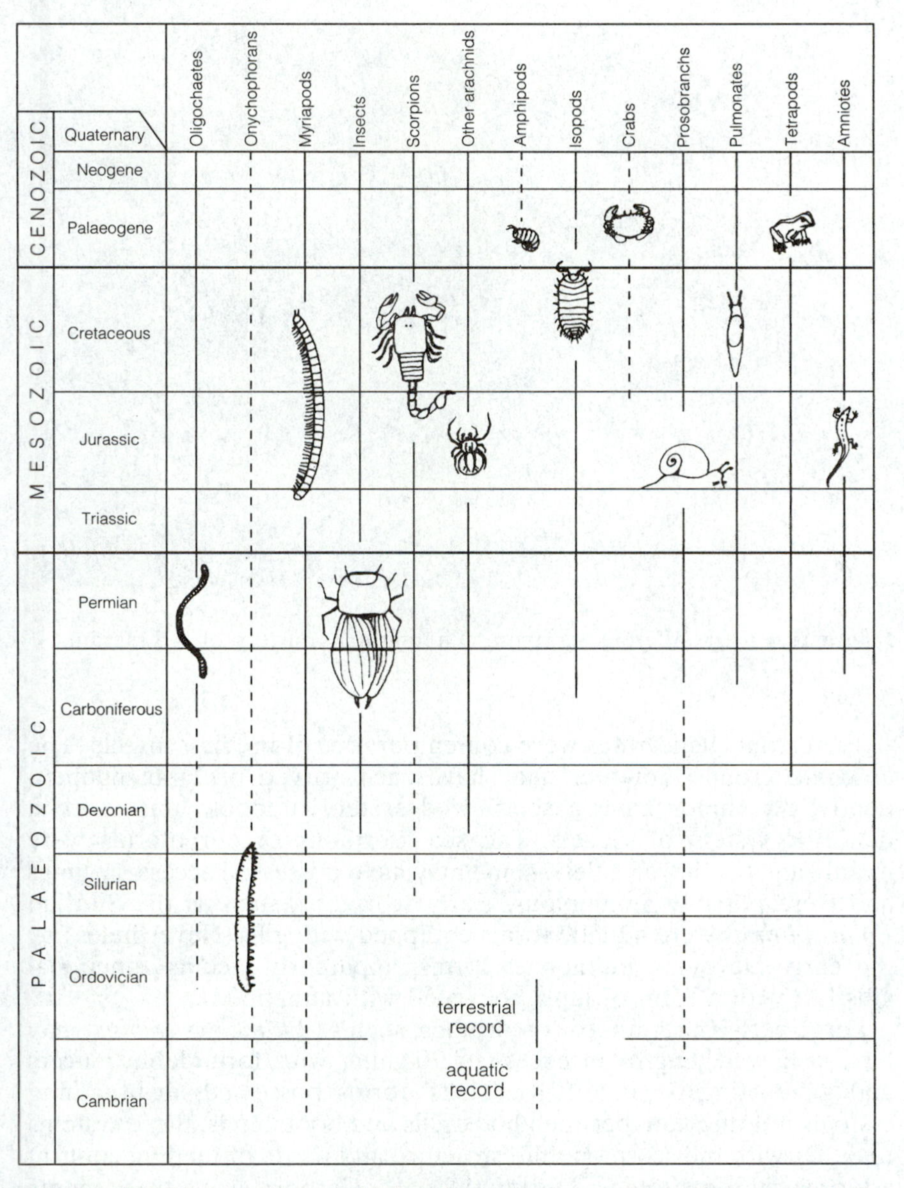

Figure 10.3 The fossil record of major animal invasions of the land during the Phanerozoic. (Replotted from Selden and Edwards, 1989.)

terrestrial insects were probably derived from land-based onychophorans, a group dating back to the Cambrian. Although insects were present in late Silurian assemblages, diverse faunas were not conspicuous until the Carboniferous. The first insects were wingless forms. However, by the mid-Carboniferous, winged taxa had evolved, initially with stiff, unfolded wings; folded wings appeared later in the period during a spectacular diversification of the flying insects.

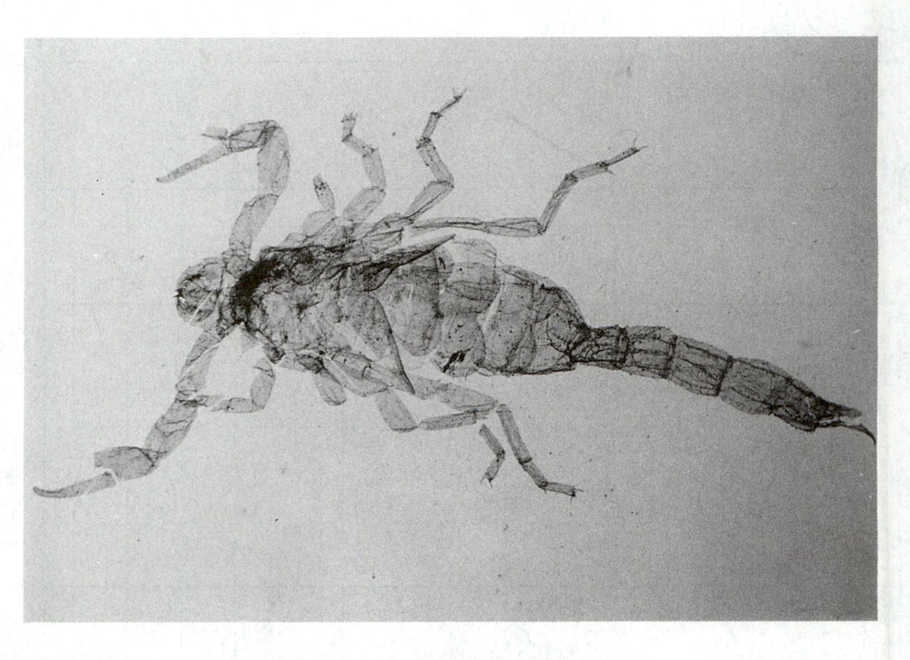

Figure 10.4 Carboniferous scorpion. (Photograph courtesy of A.J. Jeram.)

Terrestrial chelicerates were contemporaries of the first insects. The arachnid groups, however, may have each arrived on land independently. Arachnids have a specialized system of book lungs with a dendritic system of vessels. The sea scorpions or eurypterids were mainly aquatic; nevertheless, some may have possessed accessory lungs and were possibly amphibious. Early scorpions, such as the Silurian *Palaeophonus* were aquatic forms equipped with gills. Nevertheless by the early Devonian pulmonate forms, apparently lacking functional gills but with a form of lung, coexisted with aquatic taxa.

Large early Carboniferous scorpions, such as *Pulmonoscorpius*, may have achieved lengths in excess of 700 mm, with formidable pincers and giant stingers (Fig. 10.4). These forms possessed air-breathing systems intermediate between book gills and book lungs; these systems together with the scorpion limb structure and mode of feeding confirm a terrestrial life strategy. During the early Carboniferous these giants were the dominant predators in terrestrial ecosystems; the rise of tetrapod predators during the period may have forced the scorpions to adapt to more inconspicuous, nocturnal habits, lurking in burrows or thick vegetation during daytime (Jeram, 1990).

10.2.2.2 Molluscs

Land snails, including both the prosobranchs and the pulmonates or lung-bearing snails together with the slugs, were first represented in late Carboniferous faunas. Litter and shade provided by the lush vegetation of the Carboniferous was probably essential for the development of these mollusc biotas. These groups have a form of lung, developed from a modified part of the mantle cavity, rich in blood vessels.

Some of the oldest reported terrestrial snail faunas, dominated by the prosobranch *Dawsonella*, occur in upper Carboniferous fresh-water limestones and within fossil tree stumps at Joggins, Nova Scotia. It was not until the late Jurassic and Cretaceous that land snails were again apparently conspicuous members of the terrestrial ecosystem, when the prosobranchs and pulmonates formed the basis for the large variety of modern land snails.

10.2.2.3 Vertebrates

The first amphibians appeared during the late Devonian. Although their bones are rare, they are widespread, recorded from Australia, Greenland, Scotland and Russia. Two fascinating amphibious tetrapods are known from the late Devonian (Famennian) of Greenland. A third, *Ichthyostegopsis*, is known only from skulls. The Greenland taxa, *Ichthyostega* and *Acanthostega* are both known from relatively complete skeletons. Both are ichthyostegids but have significant anatomical differences. *Ichthyostega* is about a metre long with strong shoulders, a robust rib cage supported by four limbs, the two hind limbs appearing more like flippers than legs (Fig. 10.5); the tetrapod also had a flat, broad, fish-like head and a small fin was developed on the tail and seven digits on each hand. There are many similarities between *Ichthyostega* and the lobe-finned fishes; it is considered to be the most primitive of all the ichthyostegids. The smaller *Acanthostega* had relatively weak arms and wrists together with eight digits on each hand; moreover this tetrapod possessed gills, a fish-like acoustic system and a strong, flexible tail. The majority of these structures suggest an aquatic life-mode; the limbs initially evolved to aid locomotion over reefs and other obstacles on the sea-floor. The initial phases of terrestrialization of the lobe-finned fishes, then, appears to have taken place underwater in some of the most remarkable examples of pre-adaptation, with limbs developing from the pectoral and pelvic fins of a rhipidistian ancestor.

Air breathing is achieved in the bony fishes and tetrapods by a series of inflatable air sacs with moist linings. These lungs were already present in the osteichtheyan fishes.

These early tetrapods had, in fact, six, seven or eight digits, rather than the more normal five. Why eventually five digits were eventually selected is not known; however, environmental constraints rather than strict genetic programming may have been an important factor.

In addition to body fossils the early tetrapod record is also illustrated by trace fossils. Recently a spectacular trace fossil assemblage has been documented from upper Devonian strata on Valentia island, Co. Kerry, western Ireland. The Kerry footprints, comprising the

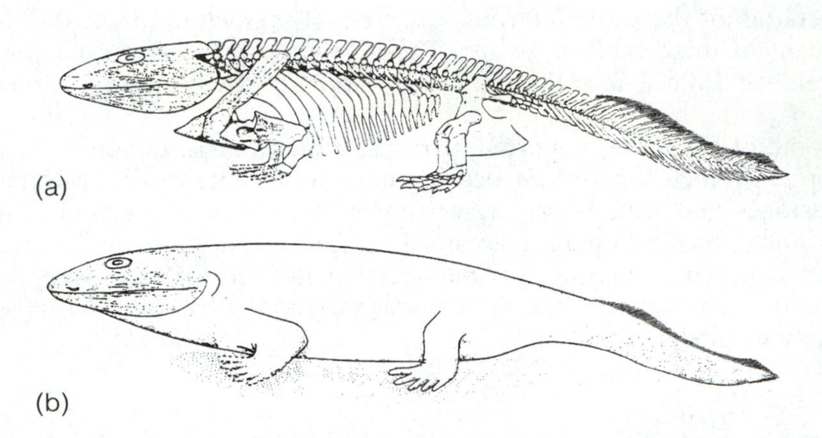

Figure 10.5 The first amphibians: skeletal and actual reconstructions of *Ichthyostega*. (After Milner, 1990.)

sigmoidal trackway of a quadruped in red nonmarine sediments (Fig. 10.6), provide certain proof that land had been colonized by the vertebrates during the late Devonian (Stössel, 1995).

10.3 CONQUEST OF THE AIR

The airways presented further opportunities for expansion. Although few animals spend their entire lives on the wing, the air contains vast volumes of ecological space for exploitation. Many organisms can

Figure 10.6 Late Devonian trackway from Valentia Island, Co Kerry. (D.A.T. Harper, original.)

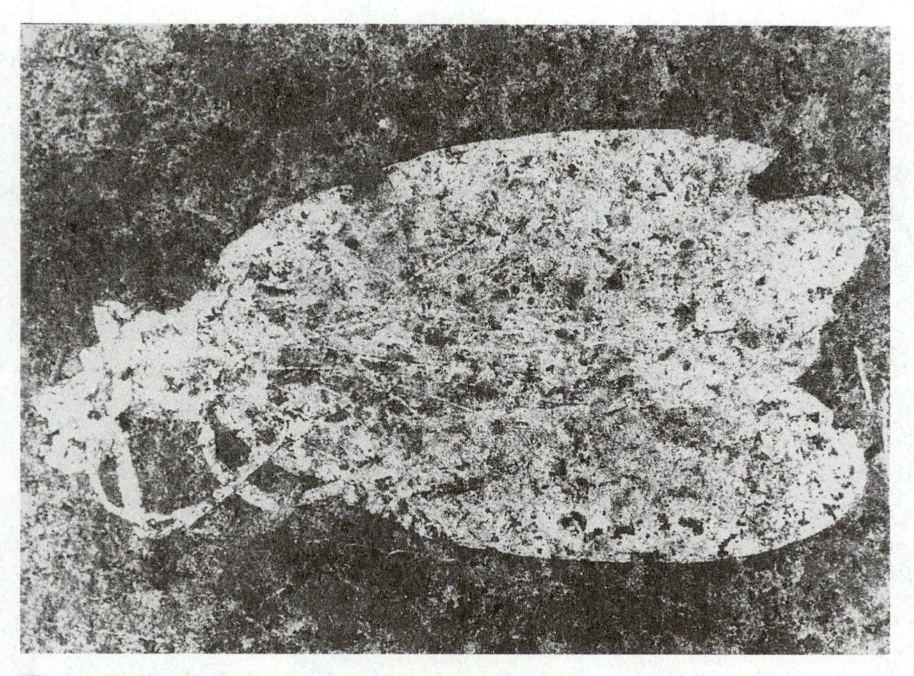

Figure 10.7 Mid-Carboniferous (Namurian) winged insect from the upper part of Clare Shales, County Clare. (Photograph courtesy of E. Jarzembowski.)

sustain passive flight by parachuting and gliding. Two main groups, however, the arthropods and the vertebrates, have developed powered flight. Arthropod flight is restricted to the insects and probably arose during the Devonian, although the first active fliers, such as the Palaeoptera, without folded wings and the Neoptera, with folded wings, are not definitively recorded until the mid-late Carboniferous (Fig. 10.7).

Three groups of vertebrates have developed flight independently. The reptilian pterosaurs were important members of the planet's air fleet during the late Triassic–late Cretaceous. Wing membranes, consisting of skin stretched over stiffened fibres, extended from the outer digit on the hand along the forelimb. Some of the most spectacular pterosaurs, such as *Quetzalcoatlus*, were large with wingspans approaching 15 m, probably adapted for soaring. The first birds, such as *Archaeopteryx*, had feathered wings, fused clavicles, features considered to be typically avian (Fig. 10.8). Nevertheless *Archaeopteryx* lacked an ossified sternum and acrocoracoid brace; some authors have argued that the pectoral muscles could not be firmly attached to the chest region of the animal, whereas without an acrocoracoid brace a strong upstroke necessary for sustained flight is not possible. Nevertheless, bats also lack both an ossified keeled sternum and the acrocoracoid brace. A number of other characters, such as teeth and the hind limbs, are more typical of its dinosaur ancestry. Extended wings with feathers need not have been initially adapted for flight.

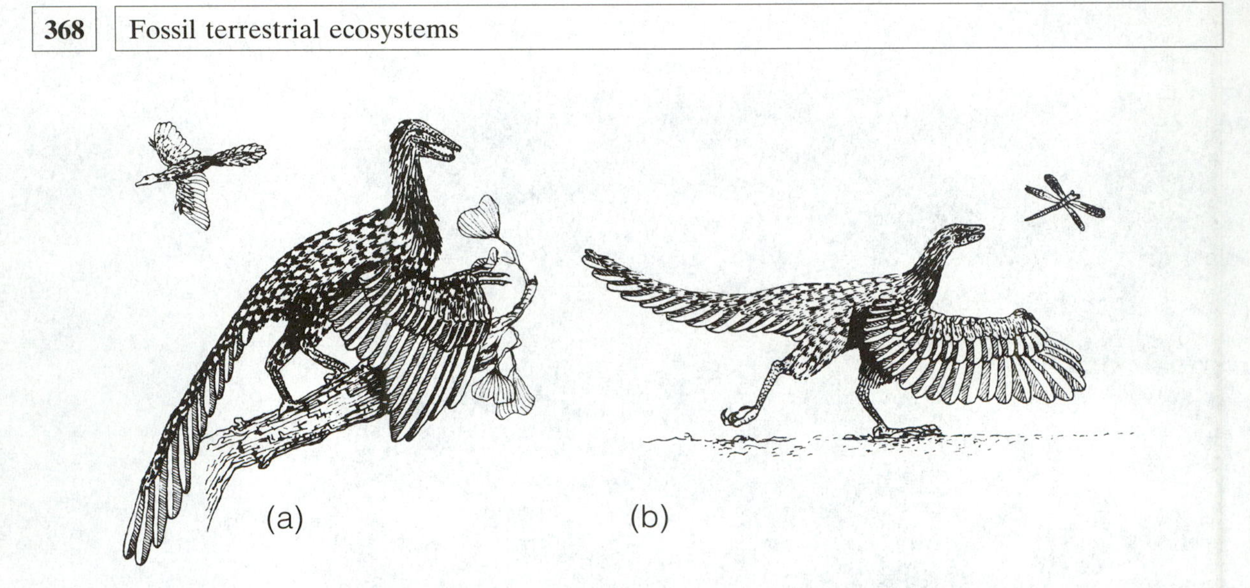

Figure 10.8 Two models for the origin of flight in *Archaeopteryx*: a. Arboreal model with *Archaeopteryx* as a tree dweller hopping then flying from branch to branch, b. *Archaeopteryx* as a ground dweller chasing and leaping up after insects. (Redrawn from Benton, 1990.)

Rather the wings may have been developed as insect traps which acquired a quite different primary function.

Finally the bats first appeared during the Eocene; in contrast to birds and pterosaurs, the wing membrane of bats is also attached to the hind limbs and thus ground locomotion is restricted. The bats probably evolved from a gliding ancestor with nocturnal, arboreal habits and an insectivorous diet.

10.4 TRENDS

Since the mid-1800s the marine faunas of the Phanerozoic have been organized, successively, into the familiar Palaeozoic, Mesozoic and Cenozoic eras. This concept of progressive change in Phanerozoic faunas as a whole was extravagantly enhanced by Sepkoski's evolutionary faunas; but the Cambrian, Palaeozoic and Modern faunas are defined, exclusively, in terms of marine taxa (see Chapter 9). Can similar patterns be detected in the Phanerozoic development of terrestrial faunas and floras?

Terrestrial floras have been grouped into a number of successive evolutionary phases (Fig. 10.9) (Niklas *et al.*, 1983). The mid-Ordovician–early Silurian interval was probably dominated by a variety of post-algal, pre-tracheophyte land plants, prior to the arrival of the first true vascular colonists. The Silurian to mid-Devonian flora (Flora A) was dominated by the early vascular plants. Flora B appeared in the late Devonian and diversified during the Carboniferous, Permian and Triassic; this flora was dominated by pteridophytes. Flora C included the gymnosperms. The final group (Flora D) is dominated by the

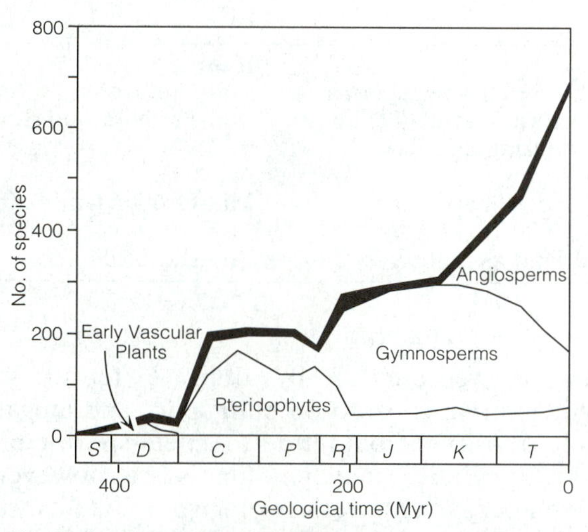

Figure 10.9 Evolutionary terrestrial plant floras: analysis of the distribution of vascular land plant diversity through the Phanerozoic reveals four main groups, each with shared structural and/or reproductive grades. The analysis is based on the distributions of approximately 18,000 fossil plant species; the thick solid line at the top of the graph represents non-vascular plants and *incertae sedis*. (Replotted from Niklas *et al.*, 1983.)

angiosperms or flowering plants, appearing first during the Cretaceous and steadily diversifying during the Cenozoic. Obviously these evolutionary floras are not coincident with either Phillips' or Sepkoski's divisions of the Phanerozoic marine biota (see Chapter 9), but allowing for the temporal displacement, the patterns are not too dissimilar.

This discordance between the appearance and diversification of successive marine faunas and land-based plant floras has been recognized for some time. The Phanerozoic tracheophyte radiations have formed the basis for definition of a series of so-called planetary floras, the Palaeophytic, Mesophytic and Cenophytic (Table 10.1); these divisions are analogous to but not coeval with the Palaeozoic, Mesozoic and Cenozoic divisions of the evolving marine realm. Traditionally the Palaeophytic ranges from the early Devonian to the early Permian and its floras are dominated by lower vascular plants. The Mesophytic continues from the late Permian to the early Cretaceous with predominantly gymnosperm plants. Finally the Cenophytic interval from the late Cretaceous to the present is characterized by angiosperms. Major advances in the understanding of early terrestrial plants, however, have forced changes in this classification (Gray, 1993). The Palaeophytic has been extended downwards to the base of the middle Ordovician but ranges upwards to its traditional summit in the middle Permian. Three subdivisions have been established, the Eoembryophytic (mid-Ordovician to pre-latest early Silurian), Eotracheophytic (latest early Silurian–early Gedinnian) and Eutracheophytic (late Gedinnian–mid-Permian). The Eoembryophytic interval lacks tracheophyte

Table 10.1 Classification of land floras

Evolutionary Floras	Characteristics	Range
CENOPHYTIC	Seed-bearing, flowering angiosperms	Mid-Cretaceous – Recent
MESOPHYTIC	Dominated by gymnosperms including advanced cone-bearing and seed-producing forms	Mid-Permian – early Cretaceous
PALAEOPHYTIC	Dominated by spore-bearing land vascular plants	Mid-Ordovician – early Permian
EPEIROPHYTIC	Algal-microbial land assemblages	Pre-Mid-Ordovician

megafossils but is represented by obligate tetrad spores. The Eotracheophytic interval is based on both spore and megafossil material, marking the transition from tetrad to trilete-sporing plants during the late Llandovery. Entire, matching plants are, however, unknown until the late Wenlock, when a variety of rhyniophytoids were present. Finally the Eutracheophytic interval established the tracheophyte land strategy, leading to the diverse pteridosperm vegetation of the later Palaeozoic.

Terrestrial tetrapod faunas can also be organized into broad evolutionary phases (Fig. 10.10) (Benton, 1985). The late Palaeozoic fauna (Fauna I) ranged from the latest Devonian to the earliest Jurassic and was dominated by labyrinthodont amphibians, mammal-like reptiles and anaspids. The Mesozoic fauna (Fauna II) was characterized by early diapsids, dinosaurs and pterosaurs and ranged from the early Triassic to the late Cretaceous. The Cenozoic fauna (Fauna III) contains much more familiar animals which dominate the modern landscape: frogs, salamanders, lizards, snakes, turtles, crocodiles, birds and mammals all diversified during the Tertiary although the fauna as a whole ranges from the mid-Triassic to the present day.

10.5 TERRESTRIAL ECOSYSTEMS THROUGH TIME

This general commentary on changing terrestrial communities through time is complemented with a series of specific case histories. In contrast to marine faunas, terrestrial assemblages are much fewer in the fossil record. These assemblages, however, have been studied in considerable detail by a wide range of experts to provide a series of data spikes illustrating the evolution and expansion of terrestrial ecosystems.

Land plants probably appeared prior to terrestrial animals, although there is no direct evidence. The subsequent evolution of land vegetation involved the development of dispersal strategies for spores and seeds, the effective pollination of plants, and within the forests, competition in both expanding canopies and undergrowth for light and space. Arthropod herbivory was established in the late Palaeozoic; evidence of chewing, leaf mining and sucking includes chewed leaves, galls and leaf mines from the Carboniferous onwards. Tetrapod herbivory

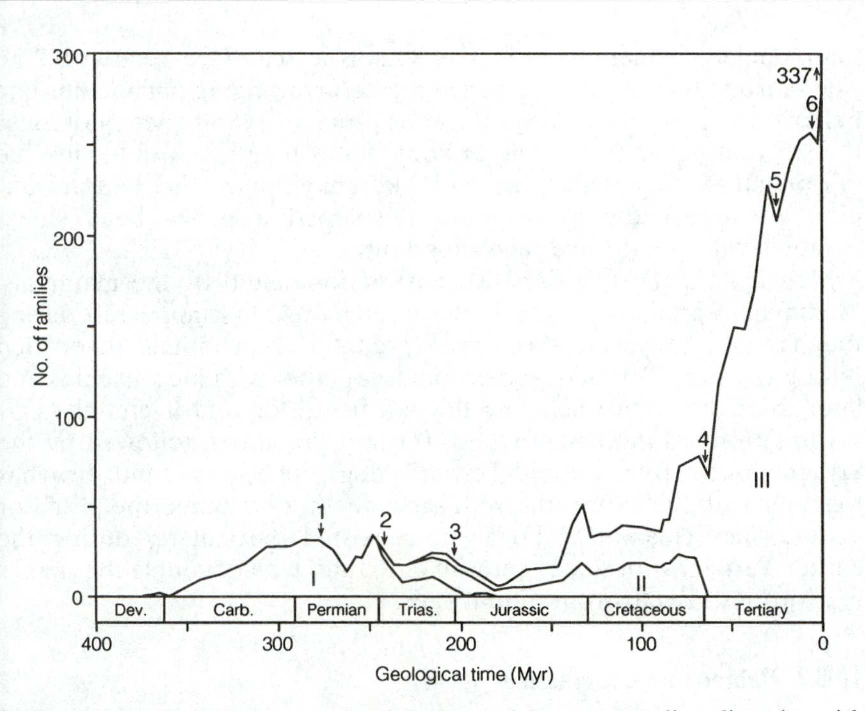

Figure 10.10 Evolutionary terrestrial tetrapod faunas: standing diversity with time for families of terrestrial tetrapods. Six apparent mass extinctions are sequentially numbered: 1. early Permian, 2. late Permian–early Triassic, 3. late Triassic, 4. late Cretaceous, 5. early Oligocene and 6. late Miocene. (Replotted from Benton, 1985.)

advanced with changes in browse height, browse line and food-processing structures. Predator systems evolved later with initially the scorpions filling these niches on land. Tetrapod amphibians probably displaced the scorpions during the Carboniferous to establish themselves as the first terrestrial predatorial vertebrates.

Predation generally caps the food chain. Terrestrial predation by vertebrates has been described in terms of a succession of changing relationships of megadynasties (Massare and Brett, 1990). This is another method of viewing the evolution of terrestrial communities, through changes at the top of the food pyramid (Fig. 10.11). Megadynasty I (Carboniferous–early Permian) was characterized by primitive amphibians and reptiles, most notably the sail-lizard *Dimetrodon*, and high predator/prey ratios. Megadynasty II (early Permian–mid-Triassic) was much more varied with possible endothermic, mammal-like therapsids and lower predator/prey ratios.

Megadynasty III (late Triassic–Cretaceous) included the age of the dinosaurs, with relatively low predator/prey ratios. Amongst the bipeds the agile slender coelurosaurs contrasted in size with the later giant carnosaurs such as *Allosaurus* and *Tyrannosaurus*. Obvious prey were the giant herbivorous sauropods. Nevertheless, an effective arms race had already begun on land, comparable but even more spectacular than similar developments in the Mesozoic seas (see section 9.2.9). The sauropods evolved large size, a muscular, swinging tail and herding behavioural patterns, all advantageous protective strategies. Moreover the stegosaurs had a covering of bony plates together with a club-like tail punctuated by spines; the tank-like ankylosaurs also had armour plating whereas the ceratopsians developed a heavy head shield together with one to five menacing horns.

Megadynasty IV (Tertiary–Recent) is dominated by the mammals. Mammalian predators, such as the insectivores, first appeared during the Triassic; however, the larger predatorial mammals diversified during the early Tertiary. Arctocyonids together with mesonychids and later the Creodonta (including the cat, bear, dog and hyena-like carnivores) formed the first wave of Tertiary predators, followed by the Hyaenodonts. By the mid-Tertiary dogs, bear-dogs and hyaenas together with the cats, some with sabre teeth, dominated the predator guilds. These mammals, however, coexisted particularly during the earlier Tertiary with some raptorial birds, and with of course the snakes and various insectivorous amphibians.

10.5.1 Palaeozoic ecosystems

Complex terrestrial ecosystems appeared more than 200 my later than the first varied metazoan marine communities. Nevertheless during the mid- and late Palaeozoic, terrestrial environments were colonized by vascular plants and air-breathing tetrapod vertebrates while terrestrial arthropods radiated during the transitions from green meadows to lush forests during the Devonian. Within the space of 200 million years the origin, assembly and modernization of sophisticated ecosystems had occurred on land. Complex physiologies were developed in animals and plants to cope with the acquisition and internal transport of water and nutrients. A simple food chain was established where arthropod and tetrapod detritivores and herbivores linked primary food sources with carnivores.

Ordovician communities have been described from palaeosols such as the Dunn Point Formation of Nova Scotia where spore tetrads and burrows imply some form of plant cover coexisted with an arthropod fauna. These communities were probably dominated by a variety of microbes and microbial mats together with plants at or near the vascular grade and bryophytes, lichens and fungi. These primitive land plants had a limited genetic diversity, some tolerance to desiccation and a short vegetative cycle. These community types probably had a near global distribution.

Although many Silurian land plants have been retrieved from marine facies, many palaeosols contain plant biotas. The first land-based vascular plants, like *Cooksonia*, were seedless, coating the landscape with a green turf but lacking an extensive root system; anchorage and the intake of water was achieved through a rhizome. By the early–mid-Devonian many floras had cosmopolitan distributions. Heterospory

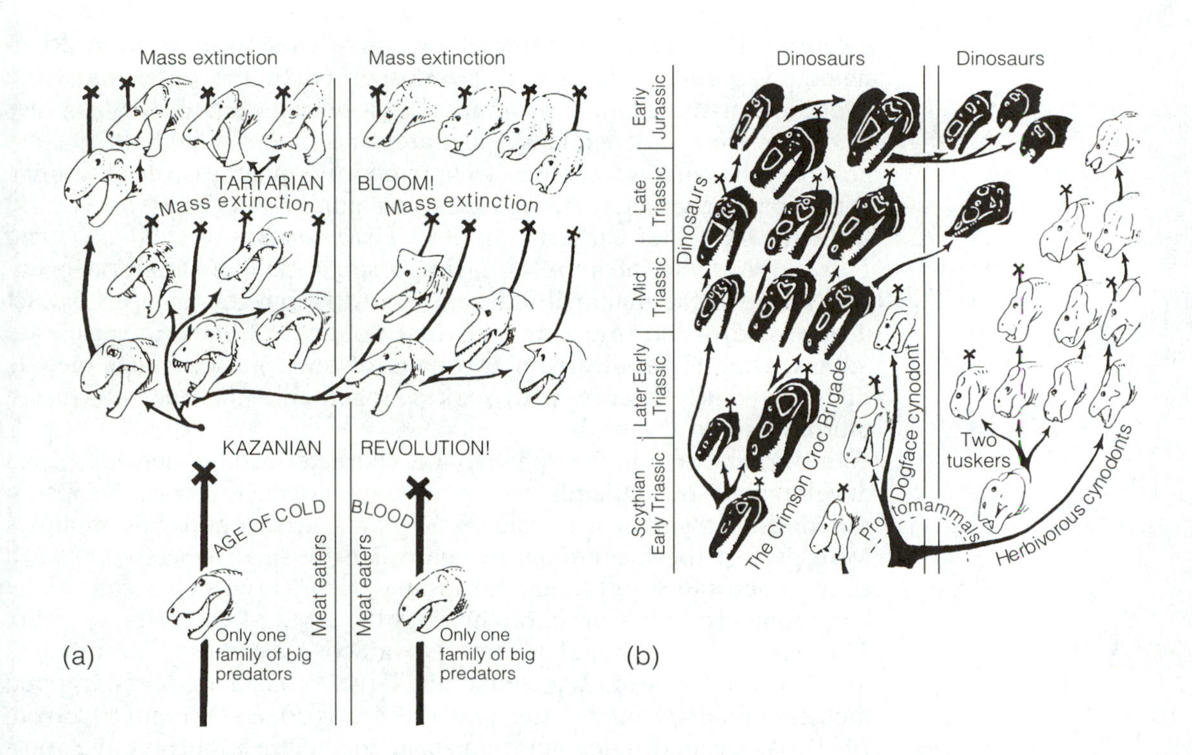

Figure 10.11 Early predator–prey dynasties: A. Early Permian ecosystem dominated by slowly-evolving ectothermic predators; appearance of probable-endothermic theraspids triggered a mid-Permian (Kazanian) revolution with the appearance of five families of predators and four of herbivores; the Tatarian Bloom was sparked by the radiation of surviving protomammals. B. In Triassic ecosystems the mammal-like reptiles were replaced by archosaurs in predator guilds (left); early archosaurs including the so-called crimson crocodiles (erythrosuchians) were replaced by dinosaurs during the late Triassic early Jurassic, the dinosaurs also invaded herbivore guilds at the expense of the mammal-like reptiles. Each head represents a single family. (Modified from Bakker, 1986.)

preceded the appearance of the progymnosperms, primitive gymnosperm-type plants with a fern-like reproductive strategy.

Subsequent plant evolution accelerated through the development of new architectures and structures. Many Devonian plants evolved a main trunk together with secondary vascular tissues. These features together with the development of roots permitted taller plants, firmly anchored into the soil. The style and shapes of leaves diversified.

Late Devonian and early Carboniferous vegetation had a much more modern aspect. The giant clubmosses, *Lepidodendron* and *Sigillaria*, grew to nearly 40 m in height, whereas horsetails like *Calamites* and tree ferns such as *Psaronius* reached nearly 15 m. Giant millipedes in excess of 2 m in length together with large dragonflies patrolled the forests; cockroaches, springtails and stoneflies together with centipedes, spiders and scorpions were content in the leafy refuse of the forest floor and probably also above in the canopy.

During the late Carboniferous, extensive coal swamps were dominated by seedless plants and gymnosperms. In the vertebrates the evolution of the cleidoic or amniote egg permitted the early reptiles to depart from a dependence on water and diversify into three main groups. The anapsid reptiles lack temporal openings; this primitive condition is characteristic of fishes and amphibians and was carried through into the earliest reptiles, *Hylonomus*, *Westlothiana* and *Paleothyris*. Synapsid reptiles having a single pair of temporal openings included the mammal-like reptiles. Although the group is extinct it gave rise to the true mammals that dominate Cenozoic vertebrate faunas. Diapsid reptiles with two pairs of temporal openings include the crocodiles, lizards and snakes together with the dinosaurs, pterosaurs and the birds.

In Euramerica, the Westphalian was characterized by the widespread development of wetlands and peat dominated by ferns, lycopsids, sphenopsids together with some seed plants. Arthropods and molluscs were part of these communities, patrolled by small insectivores such as the microsaurs and larger carnosaurs such as the pelycosaurs. The large temnospondyls probably marked the apex of the food pyramid. The classic cliff section at Joggins, Nova Scotia exposes a Westphalian forest whose tree trunks, including the large lycopsid *Sigillaria*, trapped the amphibian *Dendrerpeton* and the tetrapod *Hylonomus* (Carroll, 1994). Associated mudstones, adjacent to the trees, display the footprints of over ten vertebrate genera, together with the trails of giant *Arthropleura*.

Climatic changes during the Permian promoted a series of developments in the terrestrial flora. The giant clubmosses and horsetails of the hitherto marshy lowlands declined, while conifers, cycads, ginkgos and seed ferns migrated downhill from upland terrains to occupy the desiccating flood plains and swamps. The early Permian vertebrate fauna was dominated by carnivorous pelycosaurs or sail-lizards such as *Dimetrodon*.

The early Permian red bed sequences of the American Midcontinent have yielded a remarkable set of terrestrial assemblages, with a number of bizarre animals. The Geraldine Bonebed contains a well-preserved fauna of amphibians and reptiles together with the giant horsetail, *Calamites*, conifers and ferns and seed ferns; the climate was tropical, with seasonal rains providing ponds and rivers across an otherwise arid landscape. Insects and ostracodes together with freshwater sharks and lungfish lived within and above the ponds; the areas around these ponds, however, were inhabited by tetrapods. Large herbivores such as *Diadectes* lived together with larger predators such as *Euryops* and the sail-lizard *Dimetrodon*. Chunks of fossil charcoal suggest wildfires sporadically ripped through the vegetation around these ponds, periodically driving the animal life into the sanctuary of water.

One of the more bizarre inhabitants of these ponds and rivers, *Diplocaulus*, had a head shield like a boomerang (Cruikshanks and Skews, 1980). *Diplocaulus* may have skulked in the mud, making use

Box 10.1 Rhynie biota

The remarkable Rhynie biota was preserved in cherts deposited from silica-rich fluids that gushed as hot springs from an epithermal gold-bearing reservoir (Trewin, 1993). It is the oldest example of an entire terrestrial ecosystem preserved more or less in place; moreover, because of silicification, the Rhynie assemblage is one the first terrestrial flora where detailed plant structures can be studied. During the early Devonian the Rhynie area of Aberdeenshire was dominated by alluvial plains with occasional ponds developed against a backdrop of waning volcanism (Fig. 10.12). Silicification occurred in stages; nevertheless a variety of the animal and plant communities inhabiting both the flood plains and ponds together with the communities that featured in the heated sinter pools are preserved. The Rhynie biota provides a window, if slightly atypical, on this critical early phase of land colonization. A range of tracheophyte plants together with algae and fungi are preserved together with a variety of arthropods including some of the earliest arachnids, the trigonotarbid *Palaeocharinus*. The most common plants are *Rhynia*, *Aglaophyton*, *Horneophyton* and *Asteroxylon* that may have coexisted in marsh-like conditions, probably occupying moist patches, reaching heights of 200–300 mm. The branchiopod arthropod *Lepidocaris* signalled periods of flooding and the ponding of an otherwise dry landscape. The majority of the arthropods were, however, terrestrial. The arachnids *Palaeocharinus* and *Palaeocharinoides* probably fed on other arthropods lying in ambush within hollow spore cases and stems. The collembolid *Rhyniella* probably munched spores or micro-organisms, whereas both *Rhyniella* and the mite *Protacarus* were probably detritivores. The fauna, like other late Silurian–late Devonian assemblages, is dominated by predatory and detritivore arthropods; herbivory developed later. There is no doubt the environment was terrestrial punctuated by pools where aquatic organisms periodically thrived.

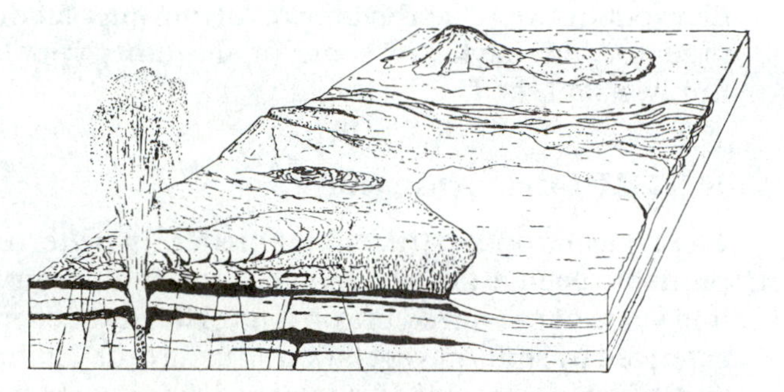

Figure 10.12 The Rhynie landscape during the early Devonian. (Redrawn from Trewin, 1993.)

of currents to take off silently from the riverbed to surprise shoals of fishes with its rather unpleasant face.

The mid-Permian radiation of predatory therapsids included the diverse bear-sized, dome-headed anteosaurs, trundling through the gymnosperm forests. The synapsid *Dimetrodon*, a cold-blooded mammal-like reptile, remained the major tetrapod predator; specialized dragon

Box 10.2 East Kirkton Lake

The East Kirkton assemblage is preserved in Dinantian (lower Carboniferous) carbonates, deposited in a variety of shallow-water facies adjacent to hot mineral springs. The assemblage is diverse and mixed; both aquatic and terrestrial communities are preserved at this unique site in the Lothian Region of Scotland (Clarkson *et al.*, 1993). Sediments were deposited in a shallow lake surrounded by a diverse vegetated landscape punctuated by volcanic cones (Fig. 10.13). The temperature of the lake probably fluctuated during cycles of hot-spring activity. Although the lake appears to have been uninhabitable for most of the time, the surrounding lush gymnosperm and pteridosperm forests were homes to a variety of terrestrial animals; the animals may have been washed into the lake or perhaps driven there during periodic forest fires. Giant scorpions headed the invertebrate food chain, above harvestmen, millipedes and mites. The vertebrate fauna is particularly significant since it contains some of the oldest known amphibians and reptiliomorphs. Temnospondyls such as the 30 cm long *Balanerpeton* probably resembled the large terrestrial salamander *Dicamptodon*; it lived around the margins of the lake where it could attack relatively large prey. Snake-like amphibians, the aïstopods, are represented by *Ophiderpeton* which probably hung out around the lake, feeding on arthropods and small vertebrates. The most common tetrapods were the anthracosaurs and included *Eldeceeon*, *Silvanerpeton* and *Westlothiana*. All three were predators; *Silvanerpeton* was probably aquatic whereas the other two occupied terrestrial habitats. The lake had a rich fish life when at its deepest; sharks together with acanthodian, actinopterygian and sarcopterygian fishes occupied various levels within the lake waters.

flies and spiders preyed on other insects. Late Permian sabre-toothed dicynodonts were herbivores, evolving this feeding strategy after the predatory life styles of some of the other reptilian groups had been firmly established.

10.5.2 Mesozoic ecosystems

The typical trophic structure of modern-type terrestrial ecosystems had probably been established by the Permian during the first 150–200 million years of organized life on land. The next 200 million years was characterized by shifts in vegetation structure, developments in the functions and life strategies of the tetrapod herbivores together with the fine tuning of interactions between animals and plants. Two major extinctions within the era had a profound effect on terrestrial ecosystems: the end-Triassic and the end-Cretaceous events exercised a significant control on the evolutionary direction of vertebrate land communities.

Triassic biotas evolved against a background of dry and highly seasonal climates. Gymnospermous seed plants increased in diversity together with scale-leaved conifers and thick-cuticled seed ferns. The more water-efficient diapsid reptiles increased in abundance. Herbivores with foraging reaches of 1–2 m dominated, while a huge variety of insects occupied a variety of niches in and on the floors of

Figure 10.13 The East Kirkton Lake: *Westlothiana*, the stem-amniote tetrapod relaxes on a rock adjacent to a scorpion pincer about 600 mm long; in the background the dense forest is dominated by arborescent gymnosperms and pteridosperms. (Redrawn from Clarkson *et al.*, 1993.)

the developing forests. Cynodonts and rhynchosaurs were relatively low browsers. The first dinosaurs swept across Pangaea during the late Triassic. *Eoraptor*, a small, agile, bipedal carnivore and the larger *Herrerasaurus*, lived together with rhynchosaurs and synapsids in Argentina accounting for about a third of the carnivores, in relatively stable communities. The prosauropod dinosaurs, such as *Plateosaurus*, were facultative bipeds commonly reaching 10 m in length; more significantly these bipeds had a foraging reach of up to 4 m. Nevertheless this strategy contrasted with the low browsing strategy adopted by the ornithischian dinosaurs. The development of a variety of food-processing modes, for example, pulping, slicing, crushing and grinding, characterized herbivore evolution. Herbivory, perhaps due its complexity, has usually arisen after carnivory. Dinosaur carnivores thus probably predated the herbivores.

During the late Triassic, browsing heights approaching 4 m aided the feeding patterns of herbivores inhabiting much more open ground.

Box 10.3 Nýřany Lake

One of the most spectacular fossil lake deposits, dominated by amphibians, occurs in the upper Carboniferous of Czechoslovakia (Milner, 1980). This lacustrine ecosystem, preserved as coalified mudstone and shale, was characterized by anoxic bottom conditions. The inhabitants of the Nýřany Lake are spread across three main ecological associations (Fig. 10.14): an open-water and lacustrine association, dominated by fishes together with anthracosauroid and loxommatid amphibians; the shallow-water and swamp-lake association with amphibians, small fishes, land plants and other plant debris; and finally the terrestrial-marginal association with microsaur amphibians and primitive reptiles. The trophic structure has been reconstructed in some detail for these palaeocommunities; in the open-water environments fishes, such as the spiny acanthodians, fed on plankton but were themselves attacked by the amphibians, presumably at the top of the food chain. In terrestrial environments plant material was consumed by a variety of invertebrates, including insects, millipedes, spiders, snails and worms; these provided food and nutrients for a range of amphibians, themselves prey for larger animals, mainly the larger amphibians and reptiles.

One of the most interesting Nýřany amphibians, *Branchiosaurus*, was about 75 mm in length with a tadpole-like body. *Branchiosaurus* has figured in studies of heterochrony as a classic example of paedomorphosis, the transfer of juvenile characters of the parent being passed on to the adult stages of the daughter populations.

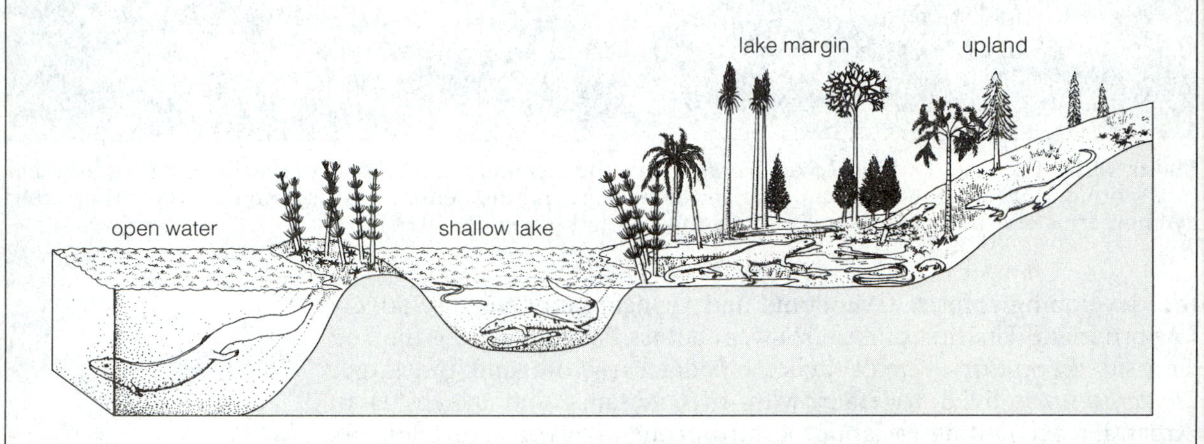

Figure 10.14 The Nýřany Lake: late Carboniferous amphibian-dominated communities. Four main habitats are indicated, from left to right: open water (*Baphetes*), shallow water *Ophiderpeton*, *Sauropleura*, *Microbrachis*, *Scincosaurus*, lake margin (*Gephyrostegus*, *Amphibamus*, *Phlegethontia*, *Ricnodon*), possible uplands (*Scincosaurus*). (Redrawn from Benton, 1990.)

Laurentian floras were characterized by large conifers, cycladaleans, bennettitaleans, ferns and spenopsids. Diffuse canopies arose from the primitive, dense woodlands supplemented by a variety of low shrubs. Gondwanan vegetation lost the characteristic *Glossopteris* flora and was dominated by seed ferns. Late Triassic and early Jurassic faunas and floras from North Carolina and Virginia (Olsen *et al.*, 1978) together with Nova Scotia (Olsen *et al.*, 1987) are associated with lacustrine environments; these biotas are dominated by fishes and aquatic reptiles together with insects, crustaceans and a flora dominated by conifers, cycads, ferns and lycopods. Carnivorous theropods of various sizes capped the food chains of the early Jurassic ecosystems. Complex and sophisticated interactions characterized these lake communities where many of the animals and plants had a more modern aspect.

During the Jurassic, Pangaea began to break up and disperse. By the end of the period the surface of the planet was punctuated by a variety of continents, microcontinents and ocean basins. Herbaceous pteridophytes and woody gymnosperms continued; however, conifers were the most diverse of the trees, although ginkgos were still important. Sawflies, leaf and plant hoppers, shield and plant bugs populated the canopies and undergrowth. While on the ground the early Jurassic vertebrate communities included cynodonts, prosauropods and small ornithischians. But the arrival of the giant, high-browsing sauropods dramatically altered the herbivore–vegetation balance. These dinosaurs had huge calorific requirements, probably offset by relatively low metabolic rates. Vegetation had to develop rapid regrowth strategies or alternatively toxic foliage to survive this massive disruption.

The late Jurassic sustained more arid climates, across Laurasia, probably due to a change in monsoonal circulation. Giant sauropods like *Apatosaurus* and *Diplodocus* coexisted with the giant carnosaurs such as *Allosaurus* and *Megalosaurus* and the armoured dinosaurs such as *Stegosaurus*. The first definitive bird, *Archaeopteryx*, with strong similarities to the therapod dinosaurs *Compsagnathus* and *Deinonychus*, added a further new dimension to terrestrial communities.

The earliest Cretaceous floras lacked angiosperms, having similarities with the older Jurassic plant communities, although some advanced gymnosperms had flowers. Although the palynomorphs of flowering plants are common in low-latitude assemblages, typical under-storey tree and shrub habits were not well developed until the Cenomanian. Angiosperm diversity then exploded. Late Cretaceous floras were characterized by 50–80% flowering plants, possibly at the expense of the cycadophytes and ferns. Nevertheless, the diversity of conifers, and possibly much of the Mesozoic treeline, remained largely unchanged.

Amongst the tetrapods the stegosaurs declined in importance as there was a general decrease in the number of high browsers. This life strategy was probably assumed by bipeds such as the hadrosaurid ornithopods that radiated during the mid-Cretaceous. Ceratopsians such as *Triceratops* were armed to the teeth. These groups almost certainly foraged in herds, promoting vast destruction of vegetation; the disturbed undergrowth may have favoured the spread of weedy plants such as the angiosperms as more substantial vegetation was extensively eaten.

The early Cretaceous sediments of the English Wealden in south-east England and on the Isle of Wight have yielded some of most

Box 10.4 Otter assemblage

The Otter Sandstone is exposed between Budleigh Salterton and Sidmouth on the south Devon coast. The sediments were deposited during the mid-Triassic in damp tracts bordering river channels and floodplains populated by a variety of animals and plants (Fig. 10.15a). The diverse fauna is dominant by the pig-like herbivorous rhynchosaurs, rather clumsy quadrupeds with scissor-like jaws (Spencer and Isaac, 1983). These lived together with large fish-eating amphibians, small lizards and carnivorous forerunners of the dinosaurs, the thecodontians and large scorpions. Several food chains have been established for this and a number of other mid-Triassic ecosystems (Benton *et al.*, 1994). A flora of coniferalean and equisetalean plants was munched by the rhynchosaurs and small reptiles. Fishes including the large *Dipteronotus* provided food for temnospondyl amphibians and the smaller reptiles together with the rhynchosaurs; these in turn were devoured by archosaurs such as *Bromsgroveia*. The composition of these mid-Triassic faunas is variable (Fig. 10.15b); however, the Devon fauna is dominated by *Rhynchosaurus* itself.

Figure 10.15 (a) Reconstruction of the mid-Triassic scene at Otter, south Devon; in the middle ground two *Rhynchosaurus* are shadowed by a pair of rauisuchians behind; *Dipteronotus* fishes are monitored by a temnospondyl amphibian while in the foreground a large scorpion approaches a pair of procolophonids on the rocks.

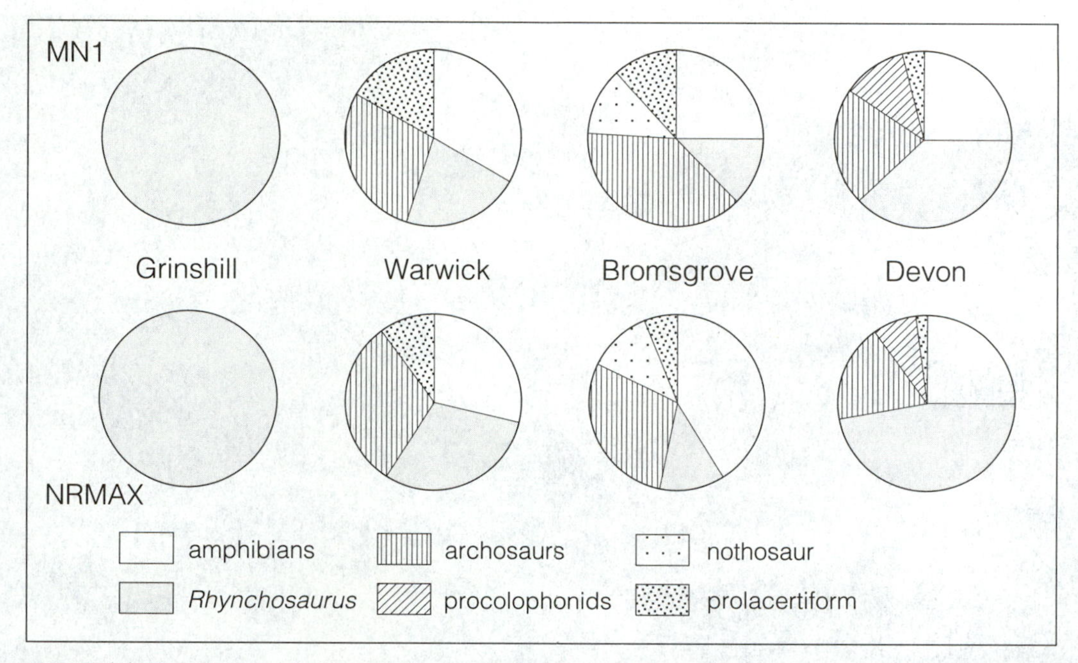

Figure 10.15 (b) Pie charts of the relative abundance of taxa from the main mid Triassic localities in England. (Redrawn and replotted from Benton *et al.*, 1994.)

BOX 10.5 Cotswold assemblage

Although middle Jurassic strata in England yielded the first named dinosaur, *Megalosaurus*, reported by Buckland in 1824, data are generally sparse on mid-Jurassic terrestrial communities. Intensive investigation of Bathonian lacustrine sediments from Gloucestershire has provided a unique view of mid-Jurassic life around an ancient Cotswold pond (Metcalfe *et al.*, 1992). Amphibians, crocodiles, fresh-water fishes and turtles swam in the warm waters of the lake (Fig. 10.16a); but the remains of dinosaurs, mammal-like reptiles together with mammals themselves, washed into the lake, were preserved in the lake sediments. The vertebrates occur together with lycopod spores, their plants provided a lush marginal vegetation; marine shell debris, probably overwash deposits, are dominated by oysters whereas autochthonous fresh-water gastropods and ostracodes probably lived in and around the lake. The fauna of the Cotswold pond, however, has a more global significance. The primitive representatives of the amphibians, crocodilians, turtles, dinosaurs, pterosaurs and mammal-like reptiles were very much in decline by the mid-Jurassic; more modern-type amphibians, such as the frogs and salamanders, the lizards together with the ankylosaur, sauropod, stegosaur and large theropod dinosaurs and also the pterodactyls and birds were now appearing on the scene. The Cotswold fauna contains early examples of frogs and salamanders, lizards and pterosaurs together with an abundance of mammal-like reptiles such as *Stereognathus*, similar to forms described from the Stonesfield Slate in the mid-19th century.

Figure 10.16 (a) Reconstruction of the scene in and around the Cotswold Lake, Gloucestershire about 165 Mya.

complete dioramas of early Cretaceous landscapes, replete with crocodiles, turtles, mammals and fishes together with plants and insects and of course, dinosaurs. The giant, heavy sauropods and the armour-plated stegosaurs, common during the late Jurassic, were less important during the early Cretaceous. The Wealden region was covered by a lush vegetation of ferns and trees together with an undergrowth of cycads; the region was swampy, punctuated by lagoons and rivers (Fig. 10.18). The large ornithopods, for example *Iguanodon,* reaching 10 m in length and equipped with an efficient chewing technique, dominated the herbivore fauna; these techniques were a significant advance on sauropod food-processing which involved the digestion of more or less complete vegetation. The agile and smaller *Hypsilophodon* probably lived within trees, its long tail providing balance for more acrobatic movements. Diverse carnivores, such as the bizarre clawed *Baryonyx*, had a varied diet of fishes together with small reptiles and mammals.

Late Cretaceous localities in North America and Mongolia have diverse dinosaur communities dominated by the duck-billed dinosaurs, the hadrosaurs; these large herbivorous bipeds blasted air through their nasal passages for communication, while their distinctive crests aided the recognition of species. The heavily armoured quadrupedal ceratop-

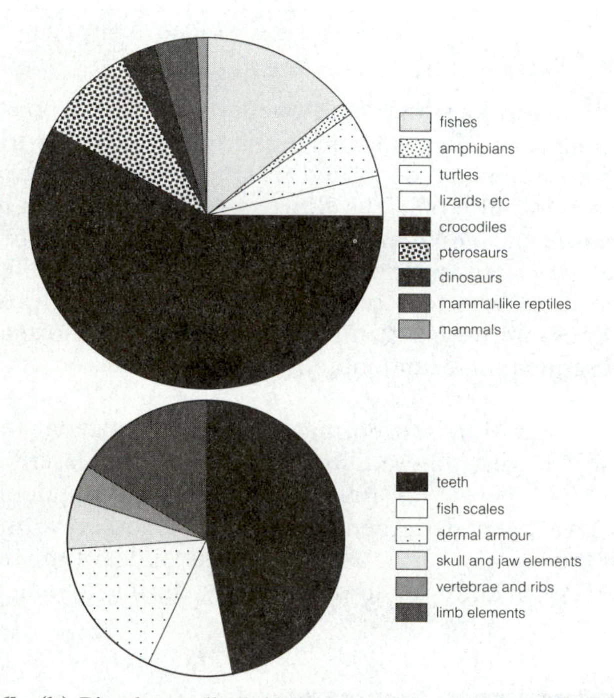

Figure 10.16b (b) Pie charts displaying the relative numbers of animals in the fauna based on a sample size of 15,000 microvertebrate specimens (top) and the main types of fossil material identified (bottom). (Redrawn from Metcalfe *et al.*, 1993.)

sians, equipped with clubs, horns and shields, rumbled through the undergrowth like tanks. The ankylosaurs were coated with a bony armour and a formidable club-like tail. *Tyrannosaurus* completed the picture; often cited as vicious predator, recent studies suggest it functioned as a scavenger. However, by the end of the Cretaceous the dinosaurs were extinct (Box 10.7).

10.5.3 Cenozoic ecosystems

Floras from the earliest Tertiary are known mainly from pollen. During this period flowering plants, insects and birds diversified while the mammals radiated from small insectivores to the larger carnivores and herbivores. Of the three groups of extant mammals, the monotremes, the marsupials and the placentals, it was the last group that diversified to dominate. Although the initial mammal radiation occurred during the first 5 million years of the Tertiary, the angiosperm radiation lasted about 10 my. The birds and the mammals were the major terrestrial survivors following the end-Cretaceous event. In some of the earliest Tertiary mammal faunas, for example the Bug Creek faunal-facies, a high diversity of marsupials and multituberculates together with a primate have been recorded from mainly microvertebrate data (Archibold, 1981).

Box 10.6 Morrison assemblage

The Morrison Formation consists of sandstones and mudstones which were deposited over vast areas of Colorado and Wyoming, extending into neighbouring states. The formation contains a wide range of nonmarine sediments, deposited in lakes, rivers and swamps; it preserves a part of a late Jurassic terrestrial biota. The Morrison fauna was dominated by giant sauropods such as *Camarasaurus* and *Diplodocus* together with *Stegosaurus* and the carnivorous *Allosaurus* and was associated with a flora dominated by conifer trees. Dragonflies and pterosaurs flew overhead, while crocodiles and turtles probably took advantage of areas of standing water. Dinosaurs were first found in the Morrison Formation, by chance, during 1877. These and subsequent finds undoubtedly changed 19th-century perceptions of the geological past.

Apart from the bones of dinosaurs the Morrison Formation has a spectacular terrestrial ichnofauna (Fig. 10.17). The world's largest dinosaur trackway has been described from the Morrison Formation of Colorado, where over 100 individual trackways along nearly 4 km of a single bedding plane have been described. Sauropods together with bipedal ornithopods and theropods skirted the margins of the Morrison Lake; most of the dinosaurs seemed to have been heading WNW, mainly in herds, possibly during a long-distance migration.

Figure 10.17 The dinosaur freeway. (Photograph courtesy of M.G. Lockley.)

Figure 10.18 An early Cretaceous Wealden landscape: *Iguanodon* munches horsetails in the background, while in the middle distance a patrolling *Megalosaurus* has yet to spot a small group of *Hypsilophodon* feeding in the foreground. (Redrawn from McKerrow, 1978.)

Changes in the Tertiary mammal faunas can also be linked to the development of mammal guilds; groups of often unrelated animals pursued a range of identical life strategies evolving a series of similar, convergent adaptations. Burrowing and arboreal guilds, for example, became a feature of the Cenozoic mammal fauna as terrestrial life continued to expand into new habitats.

During the Palaeocene and Eocene a variety of new insects including the butterflies and moths provided additional pollination for the angiosperms while guilds of insectivores with specialized teeth and rodents developed and ungulates, such as the horse, evolved larger body size. The Oil Shale of Grube Messel, near Frankfurt has preserved some of the most complete mid-Eocene vertebrate fossils and certainly the best-preserved mammals (Franzen, 1990). The mode of preservation has permitted study of the hair, skin, internal organs and even the stomach contents of a range of early mammals. The

Box 10.7 Hell Creek assemblages

The development of the Hell Creek Formation and equivalent units in Alberta, Montana and Wyoming spans the Cretaceous–Tertiary boundary and contains rich dinosaur and mammal faunas, providing much material for debate on the mode and timing of the extinction of the dinosaurs. The lower parts of these units contain a diverse dinosaur fauna with *Albertosaurus, Tyrannosaurus, Dromaeosaurus, Velociraptor, Saurornitholestes, Paronychodon, Thescelosaurus, Edmontosaurus, Pachycephalosaurus, Ankylosaurus* and *Triceratops*. In the upper part, ungulate mammals appear. Detailed analysis of the Hell Creek biotas suggested that dinosaur extinction was gradual, mainly demonstrated by the decreasing abundance of dinosaur teeth per metric ton of sediment through the interval (Sloan *et al.*, 1986). There is also a reduction in taxonomic diversity; 30 dinosaur genera were present in the region 8 million years prior to the end of the period, only 12 remained just beneath the boundary; however, between seven and 11 apparently survived into the Palaeocene. An alluvial channel above the local base of the Tertiary contains unreworked teeth of mammals and Palaeocene pollen, together with the teeth of seven species of dinosaurs. Through the higher parts of these units the mammal fauna diversifies, with species of *Protungulatum, Mimatuá, Baioconodon* and *Oxyprimus* dominating. Consequently a range of factors may have contributed to the decline of the dinosaurs. A cooling of the climate together with a drop in sea-level may have induced a greater seasonality and a decline in terrestrial vegetation; moreover, invading herbivorous mammals arriving from Asia may have provided the dinosaurs with some competition (Sloan *et al.*, 1986).

Both the data and interpretation have been hotly contested. Critics have pointed out that the teeth are in Palaeocene channels incising the Cretaceous sequence; the dinosaur remains are probably outwash deposits, eroded from the underlying top Cretaceous strata (Smit and Van Der Kaars, 1984).

outlines of the soft parts of bats and frogs, the latter showing details of blood vessels, are also preserved. The shale contains a diverse fossil vegetation of beech, laurel, oak, palms and vines together with rarer conifers and water lilies; frogs, toads and salamanders occurred together with crocodiles, tortoises and snakes. Life in and around a series of ponds in the humid tropics is signalled by this varied biota. Mammals are, however, relatively rare. Small herbivorous ungulates such as the artiodactyles (relatives of modern cattle and deer) and perissodactyles (early horses and tapirs) were washed into the lake where they were rapidly entombed in anoxic muds. Primitive insectivores occur together with creodonts, rodents and lemur-like primates. The tiny but remarkable omnivore *Leptictidium* spurted around on its hind legs balanced by a long tail rather like a kangaroo. A few Asian and South American taxa occur together with European forms, indicating that the Messel lake may have been part of a migration route during the Eocene.

During the cooler Oligocene a number of phases of extinction cleared away the early Tertiary mammal groups (Fig. 10.20). The

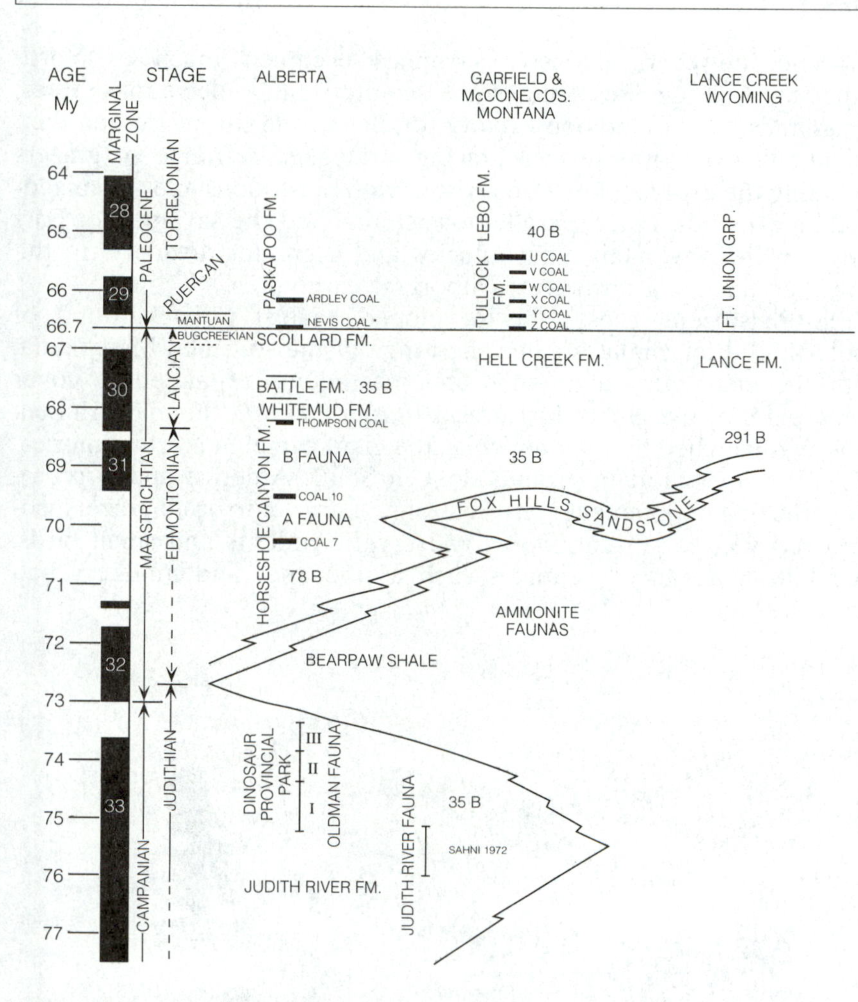

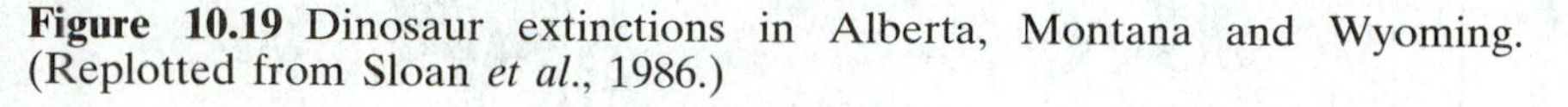
Figure 10.19 Dinosaur extinctions in Alberta, Montana and Wyoming. (Replotted from Sloan *et al.*, 1986.)

tropics were more restricted. Woodlands of both coniferous and deciduous trees spread across temperate latitudes where pig-like animals the size of sheep browsed and grazed these wooded areas and adjacent more open grasslands. Cat, dog and hyena-like carnivores roamed the plains. Hyracoids were common in Africa whereas the notoungulates occupied areas of South America.

The warmer climate of the Miocene has been associated with tectonic events along the western Cordillera, the Andes, the Alps and the Himalayas. Moreover these events restricted and partially closed the Tethys Ocean. Land bridges promoted the intercontinental migrations of terrestrial faunas and generated the upwelling of nutrient-laden waters in temperate latitudes. A wide variety of animals, including aardvarks, horses, hyraxes, pigs and rodents were on the move. Upwelling supplies of nutrients may have encouraged the radiation of marine mammals such as the seals and whales. By the late

Miocene, however, extensive savannas occupied much of North America with grasses and some conifers and deciduous trees. Grasslands offered the opportunity for changes in life mode and diet, essentially from browsing to grazing strategies. A range of grazers including the evolving horses coexisted with mastodons and the mixed-feeding camelids. Later, parallel forms inhabited the savannas of East Africa with convergent morphologies and a greater diversity. In the skies, large soaring birds were important carnivores.

Plio-Pleistocene ecosystems developed against a background of cooler and drier climates when the range of the so-called Megafauna, with the elephants, mammoths and mastodons, expanded to cover many parts of the globe. In Eurasia *Elasmotherium*, the giant rhinoceros, was matched by *Titanotylopus*, the giant camel in North America, *Megatherium*, the giant ground sloth in South America and *Diprotodon*, the giant marsupial, in Australia. Large cursorial grazers co-occurred with larger carnivores and cryptic rodents and small birds. In addition advanced primates such as the apes and monkeys had evolved (Fig. 10.21).

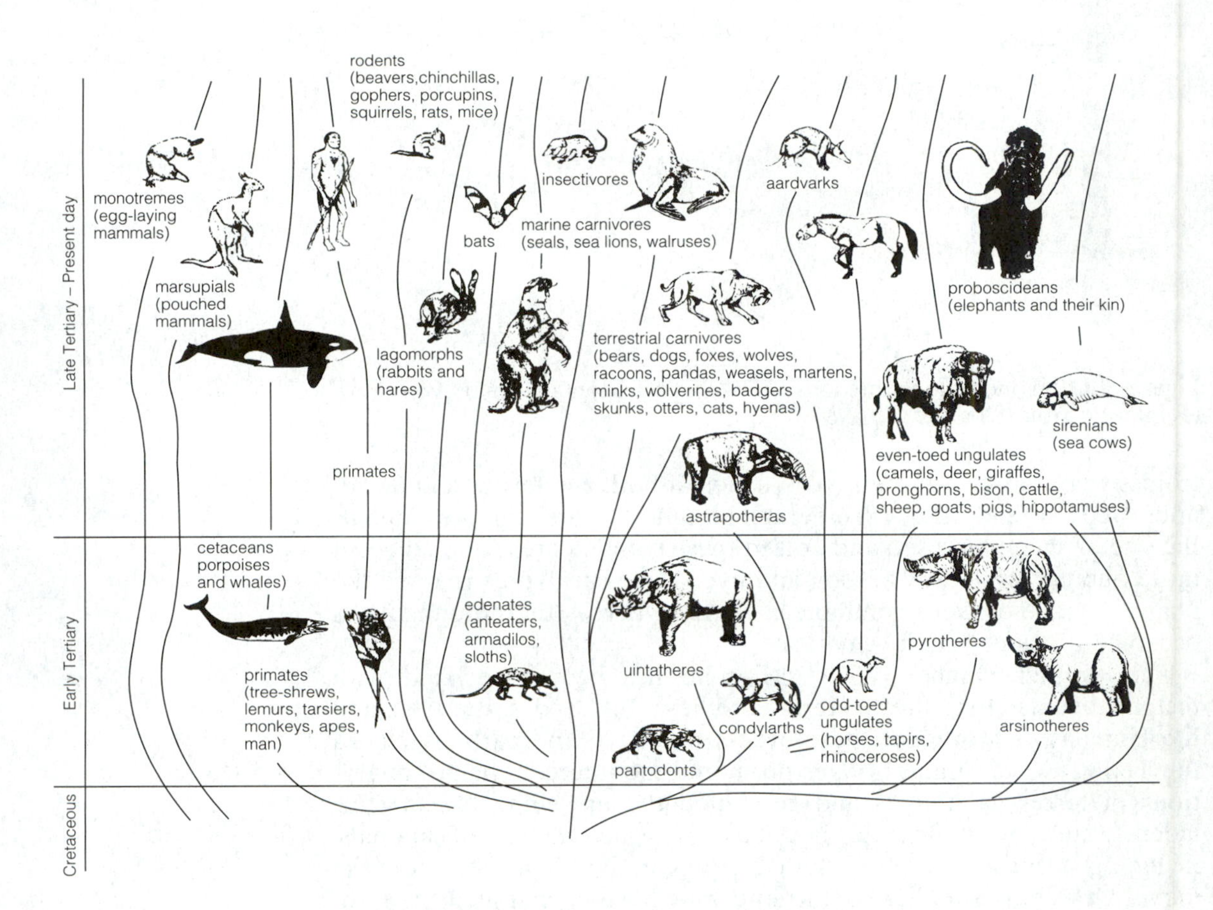

Figure 10.20 Key lineages in the Tertiary mammal ecosystem. (Redrawn from Nisbet, 1991.)

The early hominids appear to form two distinct lineages based on the genera *Australopithecus* and *Homo*. One of the oldest known hominids, *Australopithecus afarensis*, or Lucy, was excavated in 1974 from sediments deposited over 3 Ma in the Hadar region of Ethiopia; the same species was probably responsible for the footprints recorded from the 3.75 Ma ash at Laetoli (see Chapter 5). Twenty years later a similar skull, tagged as 'Lucy's cousin', was recovered from the same region in sediments about 3.9 Ma. Although the Leakey family and its associates argued that both *Australopithecus* and *Homo* were present in the original finds from Hadar, the Johanson group have maintained that the Ethiopian material is monospecific, representing the earliest member of the hominid lineage. The australopithecines had flattish faces, variably developed teeth, jaw bones and cranial crests together with relatively small brain cases. Between 3–1 Ma up to four species of *Australopithecus* may have coexisted with early *Homo* species such as *H. habilis* in East Africa. The australopithecines pursued diets of tough plant material while *H. habilis*, with larger brain capacities, was already developing a tool-based technology. Late Pliocene climatic change and tectonic events across the East African rift system may have encouraged faunal diversity in the hominids, creating a variety of niches; similar diversification is apparent in the African antelope fauna (Vrba, 1980).

Many lines of evidence, such as the footprints at Laetoli, signalling bipedalism, stone tools at Olduvai (2 Ma) and fire at Zhoukoudian (0.5 Ma), together with the construction of figurines and cave paintings (35,000 BP) define the early cultural track of the hominids. By 10,000 BP humans had a near global distribution; agricultural and technological developments were becoming more important elements of hominid evolution than morphological changes in the lineage itself.

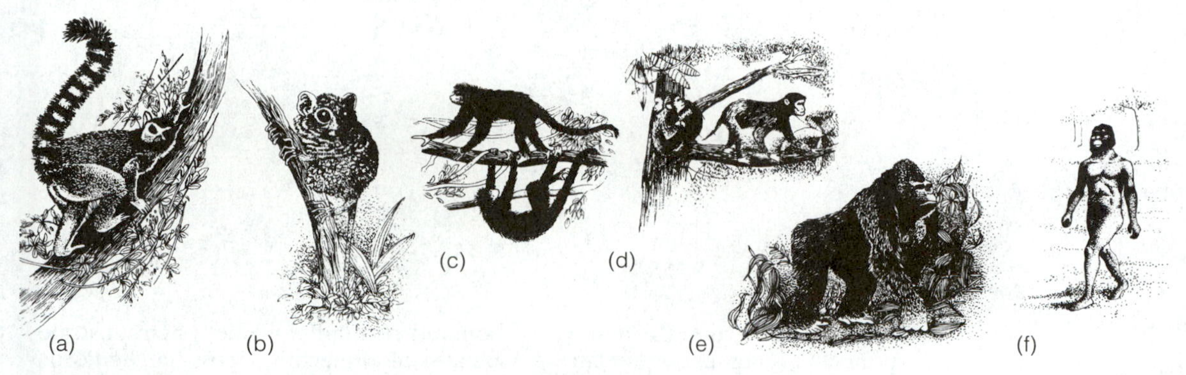

Figure 10.21 Primate evolution: (a) ring-tailed lemur, (b) spectral tarsier, (c) spider monkey, (d) rhesus monkey, (e) gorilla and (f) early human, *Australopithecus*. (Redrawn from Benton, 1990.)

10.6 MASS EXTINCTIONS

Three major events affected the terrestrial ecosystem. The effects of the end-Permian, end-Triassic and end-Cretaceous extinctions were locally devastating; the older late Devonian event apparently had less effect. Apart from the obvious loss of life each event did empty, ecologically, niches and dramatically reset evolutionary agenda.

The Permian extinctions amongst the terrestrial biotas appeared to have been extended yet significant (Erwin, 1990). Terrestrial vertebrates disappeared during a series of extinctions which affected both amphibians and reptiles. The end-Permian event mainly affected the theraspid mammal-like reptiles. The Permian floras of the tropics were dominated by the broad-leaved pteridophytes, cordaites and pecopterid ferns while the *Glossopteris* flora of pteridosperms characterized high latitudes associated with Gondwana. Between the early Permian and mid-Triassic plant diversity decreased by 50% when the typical Palaeophytic flora transitted through a mixed flora to typical Mesophytic floras with conifers, cycads and gingkoes together with new groups of pteridophytes and pteridosperms. This change occurred, globally, over an interval of about 20 million years. Drier climates during the Triassic encouraged the conifers to colonize lowland areas hitherto occupied by the pteridosperms and pteridophytes. This restructuring of vegetation together with drier conditions may have influenced vertebrate communities through the boundary interval.

Triassic events were also extended. At least two phases have been reported (Benton, 1990). Traditional analyses suggested the dinosaurs

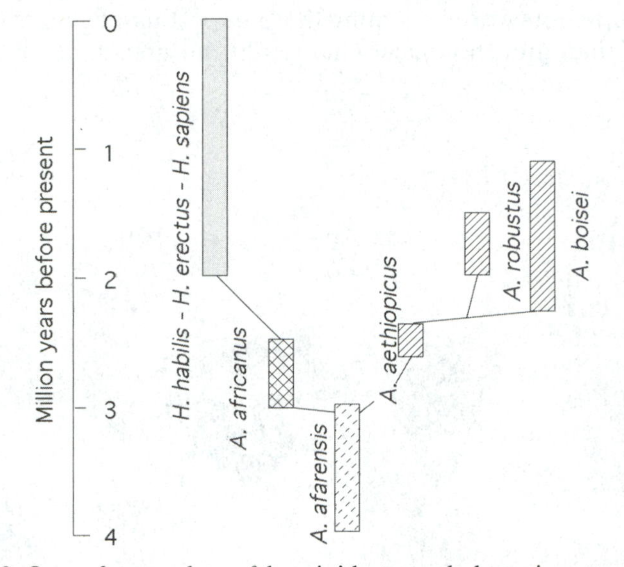

Figure 10.22 One of a number of hominid stratophylogenies: stem australopithecine (diagonal ornament), gracile australopithecines (cross-hatched ornament), robust australopithecines (diagonal-lined ornament), *Homo* lineage (blank ornament). (Plotted from various sources.)

Box 10.8 East African Savanna

The East African savannas, during the last few million years, have provided the environments and habitats for developing hominid life (Fig. 10.22). *Homo habilis*, one of the more advanced human species, was first described from Olduvai associated with a stone tools, the Oldowan industry. The handyman, *Homo habilis*, may yet comprise two different species; nevertheless, the remains of this toolmaker are associated with the bones of buffalo, elephant, hippo, pig and zebra. Whether a predator or a scavenger, *Homo habilis* clearly relied on these animals, employing a hunter-gathering strategy from bases near lakes and rivers. Moreover the handyman may have developed speech capabilities to complement facial expressions and gestures. On the basis of these attributes *Homo habilis* was perhaps more clever and certainly more skilful than the contemporary australopithecines, evolving a relatively sophisticated industrial culture together with more advanced communication techniques. Subsequently, *H. erectus*, including Java and Peking man, appearing about 1 Ma had a widespread distribution and formed the basis of the Palaeolithic culture.

in some way out-competed contemporary mammal-like reptiles, rhynchosaurs and thecodontians. No long-term decline in these last groups can be clearly correlated with a radiation in the dinosaurs. In fact dinosaurs diversified during the early Norian (latest Triassic) following the disappearance of the rhynchosaurs and the sharp decline of the thecodontians and mammal-like reptiles. The change in vertebrate faunas was matched by changes in plant communities. For example the Gondwanan *Dicroidium* flora was apparently replaced by conifers and bennettitaleans.

The end-Cretaceous event was shorter and since it involved the dinosaurs, apparently more dramatic. There is a very sudden change in floras at the boundary which is marked by an abundance of fern spores. This fern spike suggests a rapid, opportunistic recolonization of the land by ferns following the devastation of the late Cretaceous vegetation. The late Cretaceous flora may have been destroyed by wildfires ignited by either asteroid impact or volcanic eruption. Apart from the appearance, initially, of a fern spike, vegetation was displaced relative to climate zones rather than removed, suggesting an increase in seasonality in early Tertiary climates.

Most authorities agree that the dinosaurs disappeared during the terminal Cretaceous extinction event. Irrespective of the two competing interpretations of the Hell Creek fauna, dinosaurs were in a state of serious decline during the latest Cretaceous, although their final eradication may have been fairly sudden. Only 12 species of dinosaurs in fact reached the boundary (Halstead, 1990). Mammal distributions across the boundary have been described from teeth. In general terms mammal faunas expanded into vacant ecospace during a restructuring of vertebrate terrestrial communities. The multituberculates, with chisel-like incisors, continued through the boundary interval with little change whereas herbivores such as the condylarthrans diversified during the early Tertiary; this radiation was tracked by a later Tertiary radiation of the insectivore proteutherians together with radiations in the carnivores and primates.

10.7 THE MODERN ECOSYSTEM

Many ecologists accept that we are living during the sixth major extinction event. Prehistoric civilizations were responsible for initiating the large-scale destruction of habitats and the widespread introduction of exotic animals. More recently, for example, the extinction of large mammals and flightless birds in Africa, Australia, North America, Madagascar and New Zealand can be significantly correlated with the arrival of humans in these areas. Longer-term climatic changes may eventually be associated with industrialization driving an increase in the abundance of greenhouse gases and the destruction of the ozone screen. The main immediate threat, however, involves the destruction of forest and vegetation, removing with it abundant and diverse ecosystems. While this is a global problem some areas are in more danger than others and some are the loci of endemic taxa facing imminent extinction: western California, western Ecuador, central Chile, western Amazonia, Atlantic Brazil, the Ivory Coast, Tanzania, Cape Province, Madagascar, western Ghats, Sri Lanka, eastern Himalayas, Peninsular Malaysia, northern Borneo, the Philippines, southwestern Australia and New Caledonia are currently at most risk (Wilson, 1992). In the high diversity reef ecosystems, coral bleaching is now commonplace on tropical structures, generated by a range of stress factors including fluctuating temperature, chemical pollution and the influx of freshwater, all initiated by human interference.

The Red Data Books, compiled by the International Union for Conservation of Nature and Natural Resources (IUCN), have cited habitat destruction (73%), introduction of new species (68%), chemical pollutants (38%), species hybridization (38%) and over-harvesting (15%) as the main factors for species decline (Wilson, 1992). Even in Ireland, where industrialization is minimal, the vertebrate fauna has suffered through the following agencies: agricultural intensification and abandonment, forrestation, peat extraction, destruction of woodland, land reclamation and coastal development, fishing intensification, changes in water quality, chemical and oil pollution, together with hunting and fishing (Whilde, 1993).

Some estimates suggest up to 100 acres of tropical rainforest are destroyed each minute, while industrial waste is continuously pumped into the atmosphere. There is little doubt that changes in agricultural and industrial processes are generating acid rain, changes in the chemical composition of the planet's atmosphere and oceans, fluctuations in the planet's temperature together with a depletion of the ozone layer. This evolving ecological Armageddon may yet, without major changes in human attitudes, be an actual test of many causes of extinction.

10.8 SUMMARY POINTS

- The planet contains a huge range of terrestrial environments and habitats; relatively few have a good potential for fossil preservation, although there are exceptions.
- Plants (vascular), invertebrates (arthropods, molluscs, worms) and vertebrates (amphibians) developed a range adaptations, probably before and during the Silurian and Devonian, for life on land.
- Flight was developed diachronously, in a variety of modes, across various groups, for example the insects (including the dragonflies, butterflies, moths and wasps), reptiles (including the birds and pterosaurs) and the mammals (bats) from the mid-Carboniferous onwards.
- The changing distributions of vascular land plant and tetrapod vertebrate families through time display similar, though not synchronous, patterns to those of the marine invertebrates.
- Palaeozoic terrestrial biotas were established by the Silurian and developed to include vascular land plants, a range of invertebrates including various arthropods, molluscs and worms together with amphibians.
- During the Mesozoic terrestrial ecosystems were dominated by evolving communities of dinosaurs and pterosaurs together with gymnosperm forests; flowering plants appeared during the early-mid-Cretaceous and developed relationships with insect pollinators.
- The early Cenozoic (Paloeogene) landscape was dominated by the sudden radiation of mammal faunas, mainly groups of placentals forming a variety of guilds; vegetation dominated by angiosperms recovered more slowly following the end-Cretaceous extinction event.
- During the late Cenozoic (Neogene) the widespread development of grasslands co-occurred with the evolution of grazing rather than browsing strategies; horses and mastodons together with camellids roamed the grasslands and similar forms occupied the savannas, patrolled by cat and dog-like carnivores.
- The Phanerozoic terrestrial ecosystem was periodically devastated and reset by a series of extinction events; the end-Permian, late Triassic and end-Cretaceous events were major, the early Permian, early Oligocene and late Miocene more minor.
- Humans evolved during the last 4 million years to dominate modern terrestrial ecosystems; developing agriculture, industry, science and technology have significantly altered the planet's biosphere and will affect its future.

10.9 FURTHER READING

Behrensmeyer, A.K., Damuth, J.D., DiMichele, W.A., Potts, R., Sues, H.-D. and Wing, S.L. 1992. *Terrestrial ecosystems through time. Evolutionary paleoecology of terrestrial plants and animals*. University of Chicago Press, Chicago and London. 568 pp.

Benton, M.J. 1990. Vertebrate palaeontology. Chapman & Hall, London. 377 pp.

Carroll, R.J. 1987. Vertebrate palaeontology and evolution. W.H. Freeman, New York.

Gordon, M.S. and Olson, E.C. 1995. *Invasions of the land. The transitions of organisms from aquatic to terrestrial life.* Columbia University Press, New York. 312 pp.

Leakey, R. and Lewin, R. 1992. *Origins reconsidered*. Abacus, London. 375 pp.

10.10 REFERENCES

Archibald, J.D. 1981. The earliest known Palaeocene mammal fauna and its implications for the Cretaceous-Tertiary transition. *Nature* 291, 650–652.

Benton, M.J. 1985. Mass extinction among non-marine tetrapods. *Nature* 316, 811–814.

Benton, M.J. 1990. End-Triassic. *In*: Briggs, D.E.G. and Crowther, P.R. (eds), *Palaeobiology – a synthesis*. Blackwell Scientific Publications, Oxford, 194–198.

Benton, M.J., Warrington, G., Newell, A.J. and Spencer, P.S. 1994. A review of British Middle Triassic terapod assemblages. *In:* Fraser, N.C. and Sues, H.-D. (eds) *In the shadow of the dinosaurs*. Cambridge University Press, Cambridge, 131–160.

Carroll, R.L. 1994. Evaluation of the geological age and environmental factors in changing aspects of the terrestrial vertebrate fauna during the Carboniferous. *Transactions of the Royal Society of Edinburgh: Earth Sciences* 84, 427–431.

Clarkson, E.N.K., Milner, A.R. and Coates, M.I. 1993. Palaeoecology of the Viséan of East Kirkton, West Lothian, Scotland. *Transactions of the Royal Society of Edinburgh: Earth Science* 84, 417–425.

Cruikshank, A.R.I. and Skews, B.W. 1980. The functional significance of nectridian tubular horns (Amphibia: Lepospondyli). *Proceedings of the Royal Society of London* B 209, 513–537.

Edwards, D. and Burgess, N.D. 1990. Plants. *In*: Briggs, D.E.G. and Crowther, P.R. (eds) *Palaeobiology – a synthesis.* Blackwell Scientific Publications, Oxford, 60–64.

Erwin, D.H. 1990. End-Permian. *In*: Briggs, D.E.G. and Crowther, P.R. (eds) *Palaeobiology – a synthesis*. Blackwell Scientific Publications, Oxford, 187–194.

Franzen, J.L. 1990. Grube Messel. *In*: Briggs, D.E.G. and Crowther, P.R. (eds) *Palaeobiology – a synthesis*. Blackwell Scientific Publications, Oxford, 289–294.

Gray, J. 1988. Evolution of the freshwater ecosystem: the fossil record. *Palaeogeography, Palaeoclimatology, Palaeoecology* 62, 1–214.

Gray, J. 1993. Major Palaeozoic land plant evolutionary bio-events. *Palaeogeography, Palaeoclimatology, Palaeoecology* 104, 153–169.

Halstead, L.B. 1990. Cretaceous – Tertiary (Terrestrial). *In*: Briggs, D.E.G. and Crowther, P.R. (eds) *Palaeobiology – a synthesis*. Blackwell Scientific Publications, Oxford, 203–206.

Jeram, A.J. 1990. When scorpions ruled the world. *New Scientist*, June, 52–55.

Jeram, A.J., Selden, P.A. and Edwards, D. 1990. Land animals in the Silurian: arachnids and myriapods from Shropshire, England. *Science* 250, 658–661.

Kasting, J.F. 1993. Earth's early atmosphere. *Science* 259, 920–926.

McMenamin, M.A.S. and McMenamin, D.L.S. 1994. *Hypersea*. Columbia University Press, New York.

Massare, J.A. and Brett, C.E. 1990. Terrestrial [Predation]. *In*: Briggs, D.E.G. and Crowther, P.R. (eds), *Palaeobiology – a synthesis*. Blackwell Scientific Publications, Oxford, 373–376.

Metcalfe, S.J., Vaughan, R.F., Benton, M.J., Cole, J., Simms, M.J. and Dartnall, D.L. 1992. A new Bathonian (Middle Jurassic) microvertebrate site, within the Chipping Norton Limestone Formation at Hornsleasow Quarry, Gloucestershire. *Proceedings of the Geologists' Association* 103, 321–342.

Milner, A.R. 1980. The tetrapod assemblage from Nýřany, Czechoslovakia. *In:* Panchen, A.L. (ed.) *The terrestrial environment and the origin of land vertebrates* Systematics Association Special Volume 15, Academic Press, London, pp. 439–96.

Niklas, K.J., Tiffney, B.H. and Knoll, A.H. 1983. Patterns in vascular land plant diversification. *Nature* 303, 614–616.

Olsen, P.E., Shubin, N.H. and Anders, M.H. 1987. New early Jurassic tetrapod assemblages constrain Triassic-Jurassic terapod extinction event. *Science* 237, 1025–1029.

Olsen, P.E., Remington, C.L., Cornet, B. and Thomson, K.S. 1978. Cyclic change in Late Triassic lacustrine communities. *Science* 201, 729–733.

Selden, P.A. 1990. Invertebrates. *In*: Briggs, D.E.G. and Crowther, P.R. (eds), *Palaeobiology – a synthesis*. Blackwell Scientific Publications, Oxford, 64–68.

Sloan, R.E., Rigby, J.K., Van Valen, L.M., Gabriel, D. 1986. Gradual dinosaur extinction and simultaneous ungulate radiation in the Hell Creek Formation. *Science* 232, 629–633.

Smit, J. and Van Der Kaars, S. 1984. Terminal Cretaceous extinctions in the Hell Creek area, Montana: Compatible with catastrophic extinction. *Science* 223, 1177–1179.

Spencer, P.S. and Isaac, K.P. 1983. Triassic vertebrates from the Otter Sandstone Formation of Devon, England. *Proceedings of the Geologists' Association* 94, 267–269.

Stössel, I. 1995. The discovery of a new Devonian trackway in SW Ireland. *Journal of the Geological Society of London* 152, 407–413.

Trewin, N.H. 1993. Depositional environment and preservation of biota in the Lower Devonian hot-springs of Rhynie, Aberdeenshire, Scotland. *Transactions of the Royal Society of Edinburgh: Earth Sciences* 84, 433–442.

Whilde, A. 1993. *Threatened mammals, birds, amphibians and fish in Ireland*. Irish Red Data Book 2: Vertebrates. HMSO, Belfast. 224 pp.

Wilson, E.O. 1992. *The diversity of life.* Allen Lane, Penguin Press, London. 424 pp.

Vrba, E.S. 1984. Patterns in the fossil record and evolutionary processes. *In* M.-W. Ho and Saunders, P.T. (eds) *Beyond Neo-Darwinism: an introduction to the new evolutionary paradigm.* 115–142. Academic Press, London.

Index